Communications
in Computer and Information Science 974

Commenced Publication in 2007
Founding and Former Series Editors:
Phoebe Chen, Alfredo Cuzzocrea, Xiaoyong Du, Orhun Kara, Ting Liu,
Dominik Ślęzak, and Xiaokang Yang

More information about this series at http://www.springer.com/series/7899

Yury Evtushenko · Milojica Jaćimović
Michael Khachay · Yury Kochetov
Vlasta Malkova · Mikhail Posypkin (Eds.)

Optimization and Applications

9th International Conference, OPTIMA 2018
Petrovac, Montenegro, October 1–5, 2018
Revised Selected Papers

 Springer

Editors
Yury Evtushenko 🆔
FRC CSC RAS
Dorodnitsyn Computing Centre
Moscow, Russia

Milojica Jaćimović
University of Montenegro
Podgorica, Montenegro

Michael Khachay 🆔
Krasovsky Institute of Mathematics
and Mechanics
Yekaterinburg, Russia

Yury Kochetov 🆔
Sobolev Institute of Mathematics
Siberian Branch of the Russian Academy
of Sciences
Novosibirsk, Russia

Vlasta Malkova
FRC CSC RAS
Dorodnitsyn Computing Centre
Moscow, Russia

Mikhail Posypkin 🆔
FRC CSC RAS
Dorodnitsyn Computing Centre
Moscow, Russia

ISSN 1865-0929   ISSN 1865-0937 (electronic)
Communications in Computer and Information Science
ISBN 978-3-030-10933-2   ISBN 978-3-030-10934-9 (eBook)
https://doi.org/10.1007/978-3-030-10934-9

Library of Congress Control Number: 2018966328

Preface

This volume contains the refereed proceedings of the 9th International Conference on Optimization and Applications (OPTIMA 2018)[1]. Organized annually since 2009, the conference attracted a significant number of researchers, academics, and specialists in many fields of optimization, operations research, optimal control, game theory, and their numerous applications in practical problems of operations research, data analysis, and software development.

The broad scope of OPTIMA made it an event where researchers involved in different domains of optimization theory and numerical methods, investigating continuous and discrete extremal problems, designing heuristics and algorithms with theoretical bounds, developing optimization software and applying optimization techniques to highly relevant practical problems, can meet together and discuss their approaches and results. We strongly believe that this may lead to the cross-fertilization of ideas between researchers elaborating on modern optimization theory and methods and employing them for valuable practical problems.

The conference was held during October 1–5, 2018, in Petrovac, Montenegro, at the picturesque Budvanian Riviera on the azure Adriatic coast. By tradition, the main organizers of the conference were the Montenegrin Academy of Sciences and Arts, the Dorodnicyn Computing Centre FRC CSC RAS, and the University of Evora. This year, the key topics of OPTIMA were grouped into five tracks:

(1) Mathematical Programming
(2) Combinatorial and Discrete Optimization
(3) Optimal Control
(4) Optimization in Economy, Finance, and Social Sciences
(5) Applications

The Program Committee (PC) and the reviewers of the conference included 132 well-known experts in continuous and discrete optimization, optimal control and game theory, data analysis, mathematical economy, and related areas from leading institutions of 25 countries including Argentina, Australia, Austria, Azerbaijan, Belgium, Bulgaria, Canada, Czech Republic, France, Germany, Greece, Italy, Kazakhstan, Lithuania, Montenegro, Poland, Portugal, Russia, Serbia, Sweden, Ukraine, The Netherlands, UK, and USA. This year we received 168 submissions mostly from Russia but also from Australia, Azerbaijan, Belgium, China, Kazakhstan, Montenegro, Nigeria, Poland, Portugal, Romania, Serbia, Ukraine, and the USA.

Out of 103 full papers considered for review (65 abstracts and short communications were excluded due to formal reasons) only 36 papers were selected by the PC for publication. Thus, the acceptance rate for this volume was about 35%. Each submission was reviewed by at least three PC members or invited reviewers, experts in their fields,

---

[1] http://www.agora.guru.ru/display.php?conf=optima-2018.

in order to supply detailed and helpful comments. To encourage young researchers, the Program Committee recommended to include 41 papers in the supplementary volume after their short presentation and discussion at the conference and subsequent revision.

The conference featured three invited lectures as well as several plenary and keynote talks. The invited lectures were:

- Prof. Panos M. Pardalos (University of Florida, USA), "Data uncertainty and Robust Machine Learning Optimization Models"
- Prof. Suresh P. Sethi. (University of Texas, USA), "Feedback Stackelberg Games for Dynamic Supply Chains with Cost Learning"
- Prof. Radu Ioan Bot. (University of Vienna, Austria), "A General Double-Proximal Gradient Algorithm for d.c. Programming"

The plenary talks were presented by;

- Prof. Yuri Nesterov (Université Catholique de Louvain, Belgium) on "Soft Clustering by Convex Electoral Models"
- Prof. Nenad Mladenovic (Serbian Academy of Sciences and Arts, Serbia) on "Quality Functions for Detecting Communities on Complex Networks with VNS"
- Prof. Alexey Tret'yakov (Siedlce University, Poland) on 'New perspective on some classical Results in optimization'
- Prof. Mikhail Posypkin (Dorodnicyn Computing Centre FRC CSC RAS, Russia) on 'Piecewise-linear bounding functions in univariate global optimization'
- Prof. Oleg Burdakov (Linköping University, Sweden) on "Spectrum-Based Limited Memory Algorithms"
- Prof. Vladimir Jaćimović (University of Montenegro, Montenegro) on "Distributed Consensus on Homogeneous Spaces"
- Prof. Adil Erzin (Sobolev Institute of Mathematics, Russia) on "Approximation Algorithm for the Network Capacitated Facility Location Problem"

We would like to thank all the authors for submitting their papers and the members of the Program Committee for their efforts in providing exhaustive reviews. We would also like to express special gratitude to all the invited lectures and plenary speakers.

October 2018

Yury Evtushenko
Milojica Jaćimović
Michael Khachay
Yury Kochetov
Vlasta Malkova
Mikhail Posypkin

Organization

Program Committee Chairs

Milojica Jaćimović	Montenegrin Academy of Sciences and Arts, Montenegro
Yury G. Evtushenko	Dorodnicyn Computing Centre FRC CSC RAS, Russia
Maksat Kalimoldayev	Institute of Information and Computational Technologies, Kazakhstan

Program Committee

Samir Adly	University of Limoges, France
Kamil Aida-zade	Institute of Control Systems of ANAS, Azerbaijan
Alexander Afanasiev	Institute for Information Transmission Problems RAS, Russia
Anatoly Antipin	Dorodnicyn Computing Centre FRC CSC RAS, Russia
Baski Balasundaram	Oklahoma State University, USA
Radu Ioan Bot	University of Vienna, Austria
Oleg Burdakov	Linköping University, Sweden
Sergiy Butenko	Texas A&M University, USA
Vladimir Bushenkov	University of Evora, Portugal
Igor Bykadorov	Sobolev Institute of Mathematics, Russia
Olga Druzhinina	FRC CSC RAS, Russia
Adil Erzin	Novosibirsk State University, Russia
Francisco Facchinei	University of Rome La Sapienza, Italy
Alexander Gasnikov	Moscow Institute of Physics and Technology, Russia
Alexander Gornov	Institute of System Dynamics and Control Theory, Russia
Ivan Ivanov	Sofia University, Bulgaria
Vladimir Jaćimović	University of Montenegro, Montenegro
Igor Kaporin	Dorodnicyn Computing Centre, FRC CSC RAS, Russia
Alexander Kelmanov	Sobolev Institute of Mathematics, Russia
Michael Khachay	Krasovsky Institute of Mathematics and Mechanics, Russia
Oleg Khamisov	Melentiev Energy Systems Institute, Russia
Vera Kovacevic-Vujcic	University of Belgrade, Serbia
Yury Kochetov	Sobolev Institute of Mathematics, Russia
Pavlo Krokhmal	University of Arizona, USA
Dmitri Kvasov	University of Calabria, Italy
Alexander Lazarev	Trapeznikov Institute of Control Sciences, Russia
Alexander Lotov	Dorodnicyn Computing Centre, FRC CSC RAS, Russia
Nevena Mijajlović	University of Montenegro, Montenegro
Nenad Mladenovic	Mathematical Institute, Serbian Academy of Sciences and Arts, Serbia
Angelia Nedich	University of Illinois at Urbana Champaign, USA

Yuri Nesterov	CORE Université Catholique de Louvain, Belgium
Nicholas Olenev	Dorodnicyn Computing Centre FRC CSC RAS, Russia
Panos Pardalos	University of Florida, USA
Alexander Pesterev	Trapeznikov Institute of Control Sciences, Russia
Boris Polyak	Trapeznikov Institute of Control Sciences, Russia
Yury Popkov	Institute for Systems Analysis, FRC CSC RAS, Russia
Igor Pospelov	Dorodnicyn Computing Centre FRC CSC RAS, Russia
Mikhail Posypkin	Dorodnicyn Computing Centre FRC CSC RAS, Russia
Oleg Prokopyev	University of Pittsburgh, USA
Yaroslav Sergeyev	University of Calabria, Italy
Angelo Sifaleras	University of Macedonia, Greece
Alexander Shananin	Moscow Institute of Physics and Technology, Russia
Mathias Staudigl	Maastricht University, The Netherlands
Michel Thera	University of Limoges, France
Tatiana Tchemisova	University of Aveiro, Portugal
Anna Tatarczak	Maria Curie-Skłodowska University, Poland
Alexey Tret'yakov	Siedlce University, Poland
Stan Uryasev	University of Florida, USA
Frank Werner	Otto von Guericke University, Germany
Adrian Will	National Technological University, Argentina
Vitaly Zhadan	Dorodnicyn Computing Centre FRC CSC RAS, Russia
Anatoly Zhigljavsky	Cardiff University, UK
Julius Žilinskas	Vilnius University, Lithuania
Jakov Zinder	University of Technology, Australia
Tatiana Zolotova	Financial University under the Government of the Russian Federation, Russia
Vladimir Zubov	Dorodnicyn Computing Centre, FRC CSC RAS, Russia
Anna Zhykina	Omsk State Technical University, Russia

Additional Reviewers

Aleksander Adelshin	Ivet Galabova	Alexander Kazakov
Dmitry Arkhipov	Alexander Gnusarev	Sergey Khamidullin
Sergey Astrakov	Jacek Gondzio	Vladimir Khandeev
Artem Baklanov	Vasily Gorbachukvasy	Mikhail Khrustalev
Maksim Barketau	Jorge Gotay	Andrey Kibzun
Vladimir Berikov	Mikhail Gusev	Konstantin Kobylkin
Radu Ioan Bot	Sergey Hrushev	Stepan Kochemazov
Sébastien Bubeck	Antti Hyvärinen	Artyom Kondakov
Ilya Chernyh	Alexey Istomin	Alexander Kononov
Gennadii Demidenko	Evgeny Ivanko	Polina Kononova
Maciej Drozdowski	Igor Izmest'ev	Julia Kovalenko
Tatiana Filippova	Victor Adrian Jimenez	Mikhail Kovalyov
Alexander Fominyh	Julia Kallrath	Konstantin Kudryavtsev

Sergey Kumkov	Yuri Ogorodnikov	Vadim Shmyrev
Tatyana Levanova	Hao Pan	Evandro de Souza
Bertrand M. T. Lin	Victor Pan	Petro Stetsyuk
Natalia Loukachevitch	Anna Panasenko	Vitaly Strusevich
Yajun Lu	Roman Plotnikov	Maxim Sviridenko
Nicolas Majorel Padilla	Alexander Plyasunov	Ivan Takhonov
Valeriy Marakulin	Leonid Popov	Ilia Tarasov
Oxana Matvijchuk	Artem Pyatkin	Alexander Tarasyev
Andrey Melnikov	Hosseinali Salemi	Valentin Tkhai
Zhuqi Miao	Yaroslav Sergeyev	Oxana Tsidulko
Simone Naldi	Dmitry Serkov	Elias Tsigaridas
Christoph Neumann	Vladimir Servakh	Anton Ushakov
Olga Nevzorova	Alexander Sesekin	Hamidreza Validi
Ekaterina Neznakhina	Sergey Sevastyanov	Igor Vasilyev
Ioanis Nikolaidis	Jiří Sgall	Yinyu Ye
Vladimir Norkin	Natalia Shamray	Vitaly Zhadan
Evgeni Nurminski	Vladimir Shenmaier	Yakov Zinder

Organizing Committee Chairs

Milojica Jaćimović	Montenegrin Academy of Sciences and Arts, Montenegro
Maksat Kalimoldayev	Institute of Information and Computational Technologies, Kazakhstan
Mikhail Posypkin	Dorodnicyn Computing Centre FRC CSC RAS, Russia

Organizing Committee

Natalia Burova	Dorodnicyn Computing Centre FRC CSC RAS, Russia
Alexander Golikov	Dorodnicyn Computing Centre FRC CSC RAS, Russia
Alexander Gornov	Institute of System Dynamics and Control Theory SB RAS, Russia
Vesna Dragović	Montenegrin Academy of Sciences and Arts, Montenegro
Vladimir Jaćimović	University of Montenegro, Montenegro
Mikhail Khachay	Krasovsky Institute of Mathematics and Mechanics, Russia
Alexander Kelmanov	Sobolev Institute of Mathematics, Russia
Yury Kochetov	Sobolev Institute of Mathematics, Russia
Vlasta Malkova	Dorodnicyn Computing Centre FRC CSC RAS, Russia
Nevena Mijajlović	University of Montenegro, Montenegro
Oleg Obradovic	University of Montenegro, Montenegro
Nicholas Olenev	Dorodnicyn Computing Centre FRC CSC RAS, Russia
Tatiana Tchemisova	University of Aveiro, Portugal
Yulia Trusova	Dorodnicyn Computing Centre FRC CSC RAS, Russia
Svetlana Vladimirova	Dorodnicyn Computing Centre FRC CSC RAS, Russia
Victor Zakharov	FRC Computer Science and Control RAS, Russia
Ivetta Zonn	Dorodnicyn Computing Centre FRC CSC RAS, Russia
Vladimir Zubov	Dorodnicyn Computing Centre FRC CSC RAS, Russia

Invited Talks

A General Double-Proximal Gradient Algorithm for D.C. Programming

Radu Ioan Bot

University of Vienna, Austria
radu.bot@univie.ac.at

Abstract. The possibilities of exploiting the special structure of d.c. programs, which consist of optimizing the difference of convex functions, are currently more or less limited to variants of the DCA methods. These assume that either the convex or the concave part, or both, are evaluated by one of their subgradients.

In this talk we propose an algorithm which allows the evaluation of both the concave and the convex part by their proximal points. Additionally, we allow a smooth part, which is evaluated via its gradient.

For this algorithm we show that every cluster point is a solution of the optimization problem. Furthermore, we show the connection to the Toland dual problem and prove a descent property for the objective function values of a primal-dual formulation of the problem. Convergence of the iterates is shown if this objective function satisfies the Kurdyka - Lojasiewicz property. In the last part, we apply the algorithm to an image processing model.

Keywords: d.c. programming · Numeric optimization

Data Uncertainty and Robust Machine Learning Optimization Models

Panos M. Pardalos

University of Florida, USA
pardalos@ufl.edu

Abstract. This talk presents robust chance-constrained support vector machines (SVM) with second-order moment information and obtains equivalent semidefinite programming (SDP) and second-order cone programming (SOCP) reformulations. Three types of estimation errors for mean and covariance matrix are considered and the corresponding formulations and techniques to handle these types of errors are presented. A method to solve robust chance-constrained SVM with large scale data is proposed based on a stochastic gradient descent method.

Keywords: Semidefinite optimization · Support vector machines
Second-order cone programming

Feedback Stackelberg Games for Dynamic Supply Chains with Cost Learning

Suresh P. Sethi

The University of Texas at Dallas, USA
sethi@utdallas.edu

Abstract. We consider a decentralized two-period supply chain in which a manufacturer produces a product with benefits of cost learning, and sells it through a retailer facing a price-dependent demand. The manufacturer's second-period production cost declines linearly in the first-period production, but with a random learning rate. The manufacturer may or may not have the inventory carryover option. We formulate the resulting problems as two-period Stackelberg games and obtain their feedback equilibrium solutions explicitly. We then examine the impact of mean learning rate and learning rate variability on the pricing strategies of the channel members, on the manufacturer's production decisions, and on the retailer's procurement decisions. We show that as the mean learning rate or the learning rate variability increases, the traditional double marginalization problem becomes more severe, leading to greater efficiency loss in the channel. We obtain revenue sharing contracts that can coordinate the dynamic supply chain. In particular, when the manufacturer may hold inventory, we identify two major drivers for inventory carryover: market growth and learning rate variability. Finally, we demonstrate the robustness of our results by examining a model in which cost learning takes place continuously.

Keywords: Two-period supply chains · Stackelberg games

Contents

Optimal Control

Applications

Mathematical Programming

New Perspective on Slack Variables Applications to Singular Optimization Problems

Yuri Evtushenko[1,4] , Vlasta Malkova[1], and Alexey Tret'yakov[1,2,3(✉)]

[1] Dorodnicyn Computing Centre, FRC CSC RAS,
Vavilov st. 40, 119333 Moscow, Russia
`evt@ccas.ru`
[2] Faculty of Sciences, Siedlce University, 08-110 Siedlce, Poland
`tret@ap.siedlce.pl`
[3] System Research Institute, Polish Academy of Sciences,
Newelska 6, 01-447 Warsaw, Poland
[4] Moscow Institute of Physics and Technology (State University),
9 Institutskiy per., Dolgoprudny, Moscow Region 141701, Russia

Abstract. This paper is devoted to a new approach for solving nonlinear programming (NLP) problems for which the Kuhn-Tucker optimality conditions system of equations is singular. It happens when the strict complementarity condition (SCC), a constrained qualification (CQ), and a second-order sufficient condition (SOSC) for optimality is not necessarily satisfied at a solution. Our approach is based on the construction of p-regularity and on reformulating the inequality constraints as equality. Namely, by introducing the slack variables, we get the equality constrained problem, for which the Lagrange optimality system is singular at the solution of the NLP problem in the case of the violation of the CQs, SCC and/or SOSC. To overcome the difficulty of singularity, we propose the p-factor method for solving the Lagrange system. The method has a superlinear rate of convergence under a mild assumption. We show that our assumption is always satisfied under a standard second-order sufficient optimality condition.

Keywords: Degeneracy · Nonlinear programming · p-factor method Superlinear convergence · 2-regularity

1 Introduction

The paper is devoted to the nonlinear programming (NLP) problem with inequality constraints:

$$\text{minimize} \quad f(x) \qquad \text{(NLP)}$$
$$\text{subject to} \quad g_1(x) \le 0, \ldots, g_m(x) \le 0, \tag{1}$$

where f and g_j are smooth functions from R^m to R.

© Springer Nature Switzerland AG 2019
Y. Evtushenko et al. (Eds.): OPTIMA 2018, CCIS 974, pp. 3–20, 2019.
https://doi.org/10.1007/978-3-030-10934-9_1

The Lagrangian for problem (1) is defined as

$$\mathcal{L}(x, \lambda) = f(x) + \sum_{j=1}^{m} \lambda_j g_j(x),$$

where $\lambda = (\lambda_1, \ldots, \lambda_m)$ is vector of Lagrange multipliers. The Kuhn-Tucker (KT) conditions [1] are satisfied at x^* with some $\lambda^* \in \mathbb{R}^m$ if

$$\mathcal{L}'_x(x^*, \lambda^*) = f'(x^*) + \sum_{j=1}^{m} \lambda_j^* g_j'(x^*) = 0, \tag{2}$$
$$\lambda_j^* g_j(x^*) = 0, \qquad \lambda_j^* \geq 0, \qquad g_j(x^*) \leq 0, \qquad \text{for all } j = 1, \ldots, m.$$

The point x^* at which relations (2) are satisfied is called a *stationary point* or a *KT point*. The *set of active constraints* at x^* is defined as

$$A(x^*) = \{j = 1, \ldots, m \mid g_j(x^*) = 0\}.$$

There are various methods for solving the NLP problem based on the KT system. However, these approaches usually require some constraint qualification (CQ) and a second-order sufficient condition (SOSC). Namely, the SOSC in [2] states that there exist a Lagrange multiplier vector λ^* satisfying (2) and a scalar $\nu > 0$ such that

$$\omega^\top \mathcal{L}''_{xx}(x^*, \lambda^*)\omega \geq \nu \|\omega\|^2, \quad \text{for all } \omega \text{ such that} \tag{3}$$
$$g_j'(x^*)\omega \leq 0, \qquad\qquad\qquad \text{for all } j \in A(x^*).$$

The SOSC in [3] is given in the following form:

$$\omega^\top \mathcal{L}''_{xx}(x^*, \lambda^*)\omega \geq \nu \|\omega\|^2, \quad \text{for all } \omega \text{ such that}$$
$$g_j'(x^*)\omega \leq 0, \qquad\qquad\qquad \text{for all } j \in A_0(x^*), \tag{4}$$
$$g_j'(x^*)\omega = 0, \qquad\qquad\qquad \text{for all } j \in A_+(x^*),$$

where for any λ^* satisfying (2), $A_+(x^*, \lambda^*) = \{i \in A(x^*) \mid \lambda_i^* > 0\}$, $A_+(x^*) = \bigcup_{\lambda^*} A_+(x^*, \lambda^*)$; and $A_0(x^*) = A(x^*) \backslash A_+(x^*)$.

In the paper, we refer to (3) as to the SOSC, but the same arguments are also true for SOSC (4). We also introduce the following modified condition: assume that there exist a Lagrange multiplier vector λ^* satisfying (2) and a scalar $\nu > 0$ such that

$$\omega^\top \mathcal{L}''_{xx}(x^*, \lambda^*)\omega \geq \nu \|\omega\|^2, \quad \text{for all } \omega \text{ such that} \tag{5}$$
$$g_j'(x^*)\omega = 0, \qquad\qquad\qquad \text{for all } j \in A(x^*).$$

Observe that this modified condition (5) is weaker that (3). At the same time, it is sufficient for the convergence properties of the method presented in the paper. Let us note that (5) is *not* a second-order sufficient condition for optimality.

In this paper, we propose a new approach to solving the problem (1). Our approach is based on the construction of a p-regularity introduced earlier in [4–7], and on reformulating the inequality constraints as equalities. The last technique is aimed to apply results known for problems with equality constraints and is used, for example, by Bertsekas in [8] for regular inequality constrained problems. Namely, by introducing slack variables $s_1, \ldots, s_m$, we get the equality constrained problem:

$$\operatorname*{minimize}_{(x,s)} \quad f(x)$$
$$\text{subject to} \quad g_1(x) + s_1^2 = 0, \ldots, g_m(x) + s_m^2 = 0. \tag{6}$$

As is mentioned in [8], this approach is straightforward and is useful in some contexts. If x^* is a local minimum for (1), then (x^*, s^*) is a local minimum for (6), where $s^* = ((-g_1(x^*))^{1/2}, \ldots, (-g_m(x^*))^{1/2})$. Moreover, in [2] it notes that this approach is quite attractive because of its universality and simplicity, however, it requires the strong assumption (SOSC and SCC). In the paper, we describe an approach that is based on this technique and does not require a strong restrictive assumption.

A slight modification on the following lemma was derived in [9].

Lemma 1. *Let x^* be a feasible point for problem (1). Then x^* is a local minimizer to (1) if and only if (x^*, s^*) is a local minimizer to problem (6), where $s^* = (s_1^*, \ldots, s_m^*)$,*

$$s_j^* = (-g_j(x^*))^{1/2}, \qquad j = 1, \ldots, m. \tag{7}$$

The necessary optimality conditions for the equality constrained problem (6) can be stated as the following Lagrange system of $(n+2m)$ equations in $(n+2m)$ unknowns (x, s, λ):

$$F(x, s, \lambda) = (L'(x, s, \lambda))^\top = (L_x'(x, s, \lambda), L_s'(x, s, \lambda), L_\lambda'(x, s, \lambda))^\top = 0_{n+2m}, \tag{8}$$

where

$$L(x, s, \lambda) = f(x) + \sum_{j=1}^m \lambda_j (g_j(x) + s_j^2)$$

and $\lambda \in \mathbb{R}^m$ is a Lagrange multiplier. System (8), referred to as the *Lagrange system*, can be solved by Newton's method or by any *Newton-like method* described, for example, in [8, Sect. 4.4.2]. We will refer to the Newton method applied to some form of the Lagrange system as the *Lagrange-Newton method*. If the KT conditions, the strict complementarity condition, and the SOSC hold for (1) at a regular local minimum x^*, then the matrix $L''(x, s, \lambda)$ is invertible in a neighborhood of (x^*, s^*, λ^*) and the Lagrange-Newton method quadratically converges to (x^*, s^*, λ^*). However, if the strict complementarity does not hold and/or the active constraint gradients are linearly dependent at x^*, then the Lagrange system (8) is singular at (x^*, s^*, λ^*); i.e. the matrix $F'(x^*, s^*, \lambda^*) = L''(x^*, s^*, \lambda^*)$ is not invertible, and the Lagrange-Newton method loses its convergence properties. To overcome the difficulty of singularity, we propose the p-factor method for solving the Lagrange system (8). The method has a superlinear rate of convergence to the solution of the NLP problem. We only make a mild assumption that the Lagrange system (8) satisfies the *p-regularity condition* at (x^*, s^*, λ^*) *along some element $h \neq 0$.*

Using the squared slack variables itself would have the drawbacks in the non-regular case. For example, as mentioned above, Lagrange system (8) is singular at (x^*, s^*, λ^*) if problem (1) is nonregular. Moreover, because the nonregular system (8) might have many solutions in a neighborhood of the point (x^*, s^*, λ^*), there is no guarantee of getting the correct sign of the Lagrange multiplier λ^*. However, the slack variable technique combined with the p-factor approach avoids this drawback and leads to the new method with a superlinear rate of convergence derived in the paper. We should mention that our consideration is local and we assume that an initial approximation (x^0, s^0, λ^0) is given in a small neighborhood of the point (x^*, s^*, λ^*), where λ^* is a Lagrange multiplier satisfying (2). Our approach can be viewing as a modification of system (8), $\Phi(x, s, \lambda) = 0$, which has a unique solution in the neighborhood of (x^*, s^*, λ^*). Therefore, even though the nonregularity of optimality conditions (8) corresponding the equality constrained problem might have many solutions, the modified system has a unique solution, which identifies the solution of NLP problem (1). Moreover, the uniqueness of the solution of the system $\Phi(x, s, \lambda) = 0$ provides the correct sign of the Lagrange multiplier.

To compare our approach with others, we would like to note that our assumption is weaker than the SOSC assumed in [2,3]. Namely, we prove that if the SOSC (3) holds, then the Lagrange system (8) satisfies the 2-regularity condition along some element $h \neq 0$.

The organization of the paper is as follows. In Sect. 2, we recall some definitions of the p-regularity theory [7] for finite dimensional spaces and describe the p-factor method for solving singular nonlinear equations. We give some auxiliary results in Sect. 3. In Sect. 4, we describe a new approach to solving NLP (1). Then, in Sect. 5, we prove that the second-order sufficient condition (3) is a sufficient condition for the Lagrange system (8) to be 2-regular at (x^*, s^*, λ^*) along some vector h.

Notation 1. Let p be a natural number and let $B \colon \mathbb{R}^n \times \mathbb{R}^n \times \cdots \times \mathbb{R}^n$ (with p copies of $\mathbb{R}^n$) $\to \mathbb{R}^q$ be a continuous symmetric p-multilinear mapping. The p-form associated to B is the map $B[\cdot]^p \colon \mathbb{R}^n \to \mathbb{R}^q$ defined by $B[x]^p = B(x, x, \ldots, x)$, for $x \in \mathbb{R}^n$. If a function $f \colon \mathbb{R}^n \to \mathbb{R}$ is a differentiable function, the row vector of its first-order partial derivatives at a point $x \in \mathbb{R}^n$ will be denoted by $f'(x) \colon \mathbb{R}^n \to \mathbb{R}^n$. If a mapping $F \colon \mathbb{R}^n \to \mathbb{R}^q$ is a class C^p, we let $F^{(p)}(x)$ be the pth derivative of F at the point x (a symmetric multilinear map of p copies of $\mathbb{R}^n$ to $\mathbb{R}^q$) and the associated p-form (also called the *p-order mapping*) is $F^{(p)}(x)[h]^p = F^{(p)}(x)(h, h, \ldots, h)$. Then $F^{(i)}(x)[h]^i$ is a vector-valued object and $F^{(i)}(x)[h]^{i-1}$ is a matrix-valued object. Further, $\operatorname{Ker} \Lambda = \{x \in \mathbb{R}^n \mid \Lambda x = 0\}$ denotes the null-space (kernel) of a given linear operator $\Lambda \colon \mathbb{R}^n \to \mathbb{R}^q$, and $\operatorname{Im} \Lambda = \{y \in \mathbb{R}^q \mid y = \Lambda x \text{ for some } x \in \mathbb{R}^n\}$ is its image space. In addition, we use the notation $\operatorname{Ker}^i F^{(i)}(x) = \{h \in \mathbb{R}^n \mid F^{(i)}(x)[h]^i = 0\}$.

We denote the scalar product of two vectors x and y in the same space by $\langle x, y \rangle$. We define the distance of a vector $x^* \in \mathbb{R}^n$ to a set $X \subset \mathbb{R}^n$ by

$$\operatorname{dist}(x^*, X) = \inf_{x \in X} \|x - x^*\|,$$

where $\| \cdot \|$ denotes the Euclidean norm.

2 The p-Factor Method for Solving Singular Nonlinear Equations: Elements of the p-Regularity Theory

In this section, we recall some definitions of the p-regularity theory [7] for finite dimensional spaces and describe the p-factor method for solving singular nonlinear equations.

Throughout this section, we consider the nonlinear equation:

$$F(x) = 0, \tag{9}$$

where $F \colon \mathbb{R}^n \to \mathbb{R}^q$ is a sufficiently smooth nonlinear mapping. The mapping F can be represented as the q-vector of functions $F_i(x) \colon \mathbb{R}^n \to \mathbb{R}$, $i = 1, \ldots, q$; i.e., $F(x) = (F_1(x), \ldots, F_q(x))^\top$. Let $x^* \in \mathbb{R}^n$ be a solution to (9).

The mapping F is called *regular* at x^* if

$$\operatorname{Im} F'(x^*) = \mathbb{R}^q, \tag{10}$$

or, in other words,

$$\operatorname{rank} F'(x^*) = q.$$

The mapping F is called *nonregular (irregular, degenerate)* if the regularity condition (10) is not satisfied.

First, we recall the definition of the p-regular mapping and of the p-factor operator. Denote the image of the first derivative of F evaluated at x^* by $Y_1 = \operatorname{Im} F'(x^*)$. Let P_2 be a $q \times q$ matrix of the orthoprojector onto subspace $(Y_1)^\perp$ in $\mathbb{R}^q$, and

$$Y_j = \operatorname{Im}\left(P_j F^{(j)}(x^*)[h]^{j-1} \right), \qquad j = 2, \ldots, p, \qquad h \in \mathbb{R}^n,$$

where P_j is the orthoprojector onto subspace $(Y_1 \oplus \cdots \oplus Y_{j-1})$ in $\mathbb{R}^q$. We assume that there exists a number p such that

$$\mathbb{R}^n = Y_1 \oplus \cdots \oplus Y_p. \tag{11}$$

The number p is chosen as the minimum number for which (11) holds. We divide the consideration into two subsections: we consider the general case of any value of p in Sect. 2.1 and the specific case of $p = 2$ in Sect. 2.2.

2.1 General Case

The p-factor operator plays the central role in the p-regularity theory. We give the following definition of the p-factor operator.

Definition 1. *The linear operator $\Psi_p(h) \colon \mathbb{R}^n \to \mathbb{R}^q$, defined by*

$$\begin{aligned}
\Psi_p(h) = F'(x^*) &+ \frac{1}{2!} P_2 F''(x^*)[h] + \frac{1}{3!} P_3 F'''(x^*)[h]^2 + \ldots \\
&+ \frac{1}{p!} P_p F^{(p)}(x^*)[h]^{p-1}, \qquad h \in \mathbb{R}^n,
\end{aligned} \tag{12}$$

is called the p-factor operator.

Remark 1. There exist various algorithms for constructing matrices $P_2, \ldots, P_p$ at any point x in some sufficient small neighborhood of the point x^*. One of the algorithms will be described in Sect. 4. Moreover, the multipliers $\frac{1}{j!}$, $j = 2, \ldots, p$, in (12) can be omitted.

Now we are ready to introduce another important definition in the p-regularity theory.

Definition 2. *The mapping $F \colon \mathbb{R}^n \to \mathbb{R}^q$ is called p-regular at x^* along an element $h \in \mathbb{R}^n$, $h \neq 0$, if*

$$\operatorname{Im} \Psi_p(h) = \mathbb{R}^q.$$

Definition 3. *Let*

$$H_p = \{h \in \mathbb{R}^n \mid F'(x^*)h = 0, \ P_j F^{(j)}[h]^j = 0, \ j = 2, \ldots, p\}. \tag{13}$$

The mapping $F \colon \mathbb{R}^n \to \mathbb{R}^q$ is called p-regular at x^* if $H_p = \{0\}$ or F is p-regular at x^* along every $h \in H_p \setminus \{0\}$.

Because in the rest of the paper F is the mapping from the space $\mathbb{R}^n$ into the same space $\mathbb{R}^n$, we state an interesting property of such a mapping in the following lemma.

Lemma 2. *If $F \colon \mathbb{R}^n \to \mathbb{R}^n$, that is $q = n$, and F is p-regular at x^*, then $H_p = \{0\}$.*

Proof. Assume on the contrary that there exists $h \in H_p$, $h \neq 0$, such that F is p-regular at x^* along the vector h and consider the linear operator $\Psi_p(h) \colon \mathbb{R}^n \to \mathbb{R}^n$ defined in (12). By Definition 2, $\Psi_p(h)$ is regular. Hence, because $q = n$, there is no z such that $\Psi_p(h)[z] = 0$. On the other hand, let $z = h$. By the definition of the set H_p, we get that $\Psi_p(h)[z] = 0$, that is a contradiction.

The following property of the p-regular mapping $F \colon \mathbb{R}^n \to \mathbb{R}^n$ is used in Sect. 3.1 for a choice of the parameter $\mu(x)$.

Lemma 3. *Let $F \colon \mathbb{R}^n \to \mathbb{R}^n$, $F \in C^{p+1}(\mathbb{R}^n)$, $F(x^*) = 0$ and F be p-regular at x^*. Then there exists a neighborhood U of the point x^* in $\mathbb{R}^n$ and $C > 0$ such that*

$$\|F(x)\| \geq C \|x - x^*\|^p \qquad \forall x \in U. \tag{14}$$

Proof. The proof of (14) is by contradiction. Assume that there exists a sequence $\{c_k\} \to 0$ such that for any k there exists $x^k \in B_{1/k}(x^*)$ such that

$$\|F(x^k)\| < c_k \|x^k - x^*\|^p. \tag{15}$$

Then the sequence $\{(x^k - x^*)/\|x^k - x^*\|\}$ approaches some unit vector $\bar{h}$ as $k \to \infty$ (we denote by $\{(x^k - x^*)/\|x^k - x^*\|\}$ any convergent subsequence of this sequence without change of the notation). By the Taylor expansion

$$
\begin{aligned}
F(x^k) &= F(x^*) + F'(x^*)(x^k - x^*) + \cdots + \frac{1}{p!}F^{(p)}(x^*)(x^k - x^*)^p \\
&\quad + o(\|x^k - x^*\|^{p+1}) \\
&= \left(F'(x) + \cdots + \frac{1}{p!}F^{(p)}(x^*)(x^k - x^*)^{p-1} \right)(x^k - x^*) \\
&\quad + o(\|x^k - x^*\|^{p+1}).
\end{aligned}
\tag{16}
$$

Dividing the last equality by $\|x^k - x^*\|$ and taking limits, we get by (15) that $\bar{h} \in \operatorname{Ker} F'(x^*)$. Then, multiplying both parts of (16) by the matrix P_2, dividing by $\|x^k - x^*\|^2$, and taking limits, we get that $h \in \operatorname{Ker}^2 P_2 F''(x^*)$. Continuing the process, we get that $h \in H_p$ that is a contradiction. Hence, we proved (14).

Let $F \colon \mathbb{R}^n \to \mathbb{R}^n$. Introduce mapping $\Phi \colon \mathbb{R}^n \to \mathbb{R}^n$ as

$$
\begin{aligned}
\Phi(x) &= F(x) + \frac{1}{2!}P_2 F'(x)[h] + \frac{1}{3!}P_3 F''(x)[h]^2 + \ldots \\
&\quad + \frac{1}{p!}P_p F^{(p-1)}(x)[h]^{p-1}, \qquad h \in \mathbb{R}^n,
\end{aligned}
\tag{17}
$$

where h is any vector in $\mathbb{R}^n$ such that F is p-regular at x^* along h, that is

$$
\begin{aligned}
\operatorname{Im} \Big(F'(x^*) &+ \frac{1}{2!}P_2 F''(x^*)[h] + \frac{1}{3!}P_3 F'''(x^*)[h]^2 + \ldots \\
&+ \frac{1}{p!}P_p F^{(p)}(x^*)[h]^{p-1} \Big) = \mathbb{R}^n.
\end{aligned}
$$

Assume that a point x^0 is given in a sufficient small neighborhood of the point x^*. Then the following iterative method is called the *p-factor method*:

$$
x^{k+1} = x^k - [\Phi'(x^k)]^{-1}[\Phi(x^k)], \qquad k = 0, 1, \ldots .
\tag{18}
$$

The p-factor method for solving nonlinear equations was introduced in [10] in a slightly different form. The following theorem states the convergence properties of the p-factor method.

Theorem 1. *Let x^* be a solution of (9). Let $F \in C^{p+1}(\mathbb{R}^n)$ and let F be a p-regular at the point x^* along some vector $h \in \mathbb{R}^n$, $h \neq 0$. Then there is a neighborhood $U(x^*)$ of x^* in $\mathbb{R}^n$ such that for any $x^0 \in U(x^*)$, the sequence $\{x^k\}$ generated by the p-factor method (18) converges to x^* and*

$$
\|x^{k+1} - x^k\| \leq C\|x^k - x^*\|^2, \qquad k = 0, 1, \ldots ,
$$

where $C > 0$ is some constant.

Proof. Because P_j is the orthoprojector onto subspace $(Y_1 \oplus \cdots \oplus Y_{j-1})^\perp$, $j = 1, \ldots, p$, then $\Phi(x^*) = 0$. Moreover, because $\Phi'(x^*) = \Psi_p(h)$ and $\operatorname{Im} \Psi_p(h) = \mathbb{R}^n$, the matrix $\Phi'(x^*)$ is invertible.

Thus the p-factor method given in (18) is an application of Newton's method to system $\Phi(x) = 0$ in a sufficiently small neighborhood of x^*. Then the statement of the theorem follows from the properties of Newton's method [8, Proposition 1.4.1].

In the following subsection, we give a specific form of the p-factor operator (12) and of the p-factor method for the case of $p = 2$. We need the specific form for our consideration in the following sections of the paper.

2.2   Case of $p = 2$

The following definition of a specific form if Definition 1 for the of $p = 2$.

Definition 4. *The linear operator $\Psi_2(h) \colon \mathbb{R}^n \to \mathbb{R}^n$,*

$$\Psi_2(h) = F'(x^*) + P_2 F''(x^*)h, \qquad h \in \mathbb{R}^n,$$

is said to be the 2-factor operator, where P_2 is a matrix of the orthoprojector onto $(\operatorname{Im} F'(x^*))^\perp$, which is an orthogonal complementary subspace to the image of the first derivative of F evaluated at x^*.

The next definition is a specific form of Definition 2. We use this definition in the proof of Theorem 3.

Definition 5. *The mapping $F \colon \mathbb{R}^n \to \mathbb{R}^n$ is called 2-regular at x^* along an element $h \in \mathbb{R}^n$ if*

$$\operatorname{Im} \Psi_2(h) = \mathbb{R}^n. \tag{19}$$

In the case of $p = 2$, the 2-factor Newton's method is the following iterative method:

$$x^{k+1} = x^k - \left[F'(x^k) + P_2 F''(x^k)[h] \right]^{-1} \left[F(x^k) + P_2 F'(x^k)[h] \right], \quad k = 0, 1, \ldots. \tag{20}$$

In a slightly different form the 2-factor Newton's method was first introduced in [11]. Different realizations of the 2-factor Newton's method are considered in [12].

3   Auxiliary Results

3.1   Technique for Estimating the Linearly Independent Rows of the Jacobian

In this section, we assume that $F \colon \mathbb{R}^n \to \mathbb{R}^n$ is in C^2. The goal of this section is to describe a technique for estimating the set of indices $\{i_1, \ldots, i_r\} \subset \{1, \ldots, n\}$ such that the vectors $F'_{i_1}, \ldots, F'_{i_r}$ are linearly independent. At the same time, the rank of the Jacobian $F'(x^*)$ is estimated. The technique was presented in [13], but we give it here for the completeness of our consideration. We assume that the point x^* is not known and base the estimation on the information about $F(x)$ at some point x in a sufficiently small neighborhood of x^*. We use this technique in Sect. 4 to construct the mapping Φ introduced in (17). We also refer to this technique in Sect. 3.2. The technique is based on the following lemma derived in [14].

Lemma 4. *Assume that $F \colon \mathbb{R}^n \to \mathbb{R}^n$ is in C^2 and that*

$$\operatorname{rank} F'(x^*) = r \leq n. \tag{21}$$

Assume also that there exists a small neighborhood U' of the point x^* and a function $\bar{\mu} \colon U' \to \mathbb{R}$ such that

$$\bar{\mu}(x^*) = 0$$

and

$$c\|x - x^*\| \leq \bar{\mu}(x) \leq C\|x - x^*\|, \qquad \forall x \in U',$$

where c, $C > 0$, $c \leq C$, are some constants. Then for any $\varepsilon \in (0, 1)$, there exists a neighborhood $U \subset U'$ of x^* such that the following hold with $\mu(x) = \bar{\mu}(x)^{1-\varepsilon}$:

1. *For any index $i \in \{1, \ldots, n\}$, the equality $F'_i(x^*) = 0$ holds if and only if*

$$\|F'_i(x)\| < \mu(x) \qquad \forall x \in U.$$

2. *For any index set $\{i_1, \ldots, i_q\} \subset \{1, \ldots, n\}(2 \leq q \leq r)$, the vectors $F'_{i_1}(x^*), \ldots, F'_{i_q}(x^*)$ are linearly independent if and only if the separation condition holds:*

$$\operatorname{dist}\left(F'_{i_k}(x), L^q_{i_k}(x)\right) \geq \mu(x) \qquad \forall k = 1, \ldots, q, \quad \forall x \in U,$$

where $\operatorname{dist}\left(F'_{i_k}(x), L^q_{i_k}(x)\right)$ is the distance between the vector $F'_{i_k}(x)$ and the subspace $L^q_{i_k}(x)$ given by

$$L^q_{i_k}(x) = \operatorname{span}\left\{F'_{i_1}(x), \ldots, F'_{i_{k-1}}(x), F'_{i_{k+1}}(x), \ldots, F'_{i_q}(x)\right\}, \quad q \geq 2, \quad x \in \mathbb{R}^n.$$

3. *For any set of indices $\{i_1, \ldots, i_k\} \subset \{1, \ldots, n\}(1 \leq k < r)$ such that the vectors $F'_{i_1}(x^*), \ldots, F'_{i_k}(x^*)$ are linearly independent and for any index $i_{k+1} \in \{1, \ldots, n\} \backslash \{i_1, \ldots, i_k\}$, the vectors $F'_{i_{k+1}}(x^*), F'_{i_1}(x^*), \ldots, F'_{i_k}(x^*)$ are linearly independent if and only if*

$$\operatorname{dist}\left(F'_{i_{k+1}}(x), \operatorname{span}\left\{F'_{i_1}(x), \ldots, F'_{i_k}(x)\right\}\right) < \mu(x) \qquad \forall x \in U.$$

Now we recall briefly the technique for choosing a set of indices $\{i_1, \ldots, i_r\} \subset \{1, \ldots, n\}$ such that the vectors $F'_{i_1}(x^*), \ldots, F'_{i_r}(x^*)$ are linearly independent, where r is defined in (21).

Technique for Estimating the Linearly Independent Rows of the Jacobian $F'(x^*)$. Assume that a point x^0 is given in a sufficiently small neighborhood of x^*. Assume also that there is a function $\mu(x)$ satisfying assumption of Lemma 4.

1. We choose the first index $i_1 \in \{1, \ldots, n\}$ such that

$$\|F'_{i_1}(x^0)\| \geq \mu(x^0).$$

2. Assume that the index set $\{i_1, \ldots, i_k\}\ (1 \leq k)$ is chosen such that the vectors $F'_{i_1}(x^*), \ldots, F'_{i_k}(x^*)$ are linearly independent. To choose the next index $i_{k+1} \in \{1, \ldots, n\} \setminus \{i_1, \ldots, i_k\}$, we verify the following conditions:

$$\|F'_{i_{k+1}}(x^0)\| \geq \mu(x^0)$$

and

$$\text{dist}\left(F'_{i_{k+1}}(x^0), \text{span}\left\{F'_{i_1}(x^0), \ldots, F'_{i_k}(x^0)\right\}\right) \geq \mu(x^0).$$

In general, the choice of the function $\mu(x)$ depends on the specific of the problem. Lemma 3 implies the choice of the function $\mu(x)$ in the case when F is p-regular at x^*. As follows from the lemma, if we define $\mu(x)$ as $\mu(x) = \|F(x)\|^{1/(p+1)}$ then the assumptions of Lemma 4 hold.

Remark 2. The technique described in this section also estimates the rank of the Jacobian $F'(x^*)$.

3.2 Transformation of the System of Equations

In this section, we recall another technique originally proposed in [13]. The technique is used in the following sections of the paper to simplify constructing an orthoprojector.

Consider the system of equations:

$$F(x) = (F_1(x), \ldots, F_n(x))^\top = 0. \tag{22}$$

Assume that the vectors $F'_1(x^*), \ldots, F'_r(x^*)$ are linearly independent. To determine these vectors, we use the technique described in Sect. 3.1.

We reduce the system (22) to the following one

$$G(x) = \begin{pmatrix} G_1(x) \\ \vdots \\ G_r(x) \\ G_{r+1}(x) \\ \vdots \\ G_n(x) \end{pmatrix} = 0, \tag{23}$$

where

$$G'_i(x^*) = 0, \qquad i = r+1, \ldots, n,$$

and the solution x^* of (22) remains a solution of (23).

Here is one of the possible realizations of such a transformation:

$$G(x) = B(x)F(x), \tag{24}$$

where $B(x)$ is an $n \times n$ matrix defined by

$$
B(x) = \begin{pmatrix}
1 & \cdots & 0 & 0\,0\ldots0 \\
\vdots & & \vdots & \vdots\,\vdots & \vdots \\
0 & \cdots & 1 & 0\,0\ldots0 \\
-\alpha_1^{r+1}(x) & \cdots & -\alpha_r^{r+1}(x) & 1\,0\ldots0 \\
-\alpha_1^{r+2}(x) & \cdots & -\alpha_r^{r+2}(x) & 0\,1\ldots0 \\
\vdots & & \vdots & \vdots\,\vdots & \vdots \\
-\alpha_1^{n}(x) & \cdots & -\alpha_r^{n}(x) & 0\,0\ldots1
\end{pmatrix},
\tag{25}
$$

and the coefficient row-vectors $\alpha^k(x) = (\alpha_1^k(x), \ldots, \alpha_r^k(x))$, $k = r+1, \ldots, n$, are determined as

$$
\left(\alpha^k(x)\right)^{\top} = (\Gamma(x))^{-1}\begin{pmatrix} \langle F_k'(x), F_1'(x)\rangle \\ \vdots \\ \langle F_k'(x), F_r'(x)\rangle \end{pmatrix}, \qquad k = r+1, \ldots, n,
$$

$$
\Gamma(x) = \left(\langle F_i'(x), F_j'(x)\rangle\right)_{i,j=1,\ldots,r}.
$$

Obviously, under such a transformation, the solution x^* to system (22) remains a solution to (23). Moreover, the orthoprojector P_2 onto $\left(\operatorname{Im} G'(x^*)^{\perp}\right)$ is a diagonal matrix $P_2 = \operatorname{diag}(p_i)_{i=1}^n$, where

$$
p_i = \begin{cases} 0, & j = 1, \ldots, r; \\ 1, & j = r+1, \ldots, n. \end{cases}
$$

We are using the definition of P_2 to simplify the construction of the orthoprojector in Sect. 4.

Finally, we briefly describe how the transformation described in this section can be used in the numerical realization of the method proposed in the paper as well as for other numerical methods.

Assume that an initial approximation x^0 is given in a sufficiently small neighborhood of the solution x^*. Then matrix $B(x^0)$ is given by (25), and mapping $\widetilde{G}(x)$ can be defined as

$$
\widetilde{G}(x) = B(x^0)F(x).
\tag{26}
$$

By construction, if x^* is a solution of $F(x) = 0$, then it is also a solution of $G(x) = 0$. Because in the small neighborhood of the point x^*,

$$
\|B(x^0) - B(x^*)\| \le \delta
$$

for some $\delta \ge 0$, the mapping $\widetilde{G}(x)$ defined by (26) can be used instead of mapping $G(x)$ introduced in (24).

4   The 2-Factor Newton's Method

In this section, we propose a new approach to solving the nonlinear programming (NLP) problem (1). First, we convert the problem (1) to a problem with only equality constraints by introducing additional variables s_j and replacing the inequalities $g_j(x) \leq 0$, $j = 1, \ldots, m$, by

$$g_j(x) + s_j^2 = 0, \qquad s_j \in \mathbb{R}, \qquad j = 1, \ldots, m.$$

The introducing of the additional variables transforms the problem (1) to the following one:

$$\begin{aligned} \underset{(x,s)}{\text{minimize}} \quad & f(x) \\ \text{subject to} \quad & c_j(x,s) = g_j(x) + s_j^2 = 0, \qquad j = 1, \ldots, m. \end{aligned} \tag{27}$$

The Lagrangian for problem (27) is defined as

$$\begin{aligned} L(x, s, \lambda) &= f(x) + \sum_{j=1}^{m} \lambda_j c_j(x, s) \\ &= f(x) + \sum_{j=1}^{m} \lambda_j g_j(x) + \sum_{j=1}^{m} \lambda_j s_j^2, \end{aligned}$$

where $\lambda = (\lambda_1, \ldots, \lambda_m)$.

Introduce the notation

$$\Lambda = \operatorname{diag}(\lambda_j)_{j=1}^{m}, \qquad S = \operatorname{diag}(s_j)_{j=1}^{m}, \qquad s^2 = (s_1^2, \ldots, s_m^2)^{\top},$$

$g(x) = (g_1(x), \ldots, g_m(x))^{\top}$, and $e = (1, 1, \ldots, 1)^{\top}$. Let x^* be a solution of (1), and let

$$A(x^*) = \{j = 1, \ldots, m \mid g_j(x^*) = 0\}$$

be the active index set at the point x^*. By applying the first-order necessary conditions for equality constraints to problem (27), we obtain a system of $(n + 2m)$ equations in the $(n + 2m)$ unknowns x, s, and λ:

$$F(x, s, \lambda) = (L'(x, s, \lambda))^{\top} = \begin{bmatrix} L_x'(x, s, \lambda)^{\top} \\ L_s'(x, s, \lambda)^{\top} \\ L_\lambda'(x, s, \lambda)^{\top} \end{bmatrix} = \begin{bmatrix} \left(f'(x) + \sum_{j=1}^{m} \lambda_j g_j'(x) \right)^{\top} \\ 2\Lambda S e \\ g(x) + s^2 \end{bmatrix}$$

$$= 0_{n+2m}. \tag{28}$$

This system is referred to as the *Lagrange system*. The corresponding Jacobian matrix $F'(x, s, \lambda)$ is given by

$$F'(x, s, \lambda) = \begin{bmatrix} L_{xx}''(x, s, \lambda) & 0_{n \times m} & J^{\top}(x) \\ 0_{m \times n} & 2\Lambda & 2S \\ J(x) & 2S & 0_{m \times m} \end{bmatrix}, \qquad J(x) = \begin{bmatrix} g_1'(x) \\ \vdots \\ g_m'(x) \end{bmatrix}. \tag{29}$$

If the strict complementary condition does not hold at x^*, i.e., there exists an index j such that $\lambda_j^* = 0$ and $s_j^* = 0$, and/or if the active constraint gradients are linearly dependent, and/or a SOSC does not hold at x^*, then the Jacobian matrix (29) is singular at (x^*, s^*, λ^*).

Our goal is to apply the p-factor method described in Sect. 2 to solve system (28). We consider the specific case of $p = 2$ in Sect. 4.1.

4.1   Case of $p = 2$

In the case of $p = 2$ introduce the mapping Φ as following

$$\Phi(z) = F(z) + P_2 F'(z)[h]. \tag{30}$$

To construct Φ we should first identify the functions $F_{i_1}(z), \ldots, F_{i_{m_1}}(z)$ such that the vectors $F'_{i_1}(z^*), \ldots, F'_{i_{m_1}}(z^*)$ are linearly independent. At the same time, we identify the number m_1. Then, if it is necessary we apply the transformation $G(z) = B(z)F(z)$ given by (24) to satisfy the requirement $G'_{m_1+1}(z^*) = 0, \ldots, G'_{n+2m}(z^*) = 0$. We keep the same notation for the system $F(z)$ that satisfies $F'_{m_1+1}(z^*) = 0, \ldots, F'_{n+2m}(z^*) = 0$. Under our assumption P_2 is a diagonal matrix $P_2 = \operatorname{diag}(p_j)_{j=1}^{n+2m}$ that is given by

$$p_j = \begin{cases} 0, & j = 1, \ldots, m_1; \\ 1, & j = m_1 + 1, \ldots, n + 2m. \end{cases}$$

The choice of the vector $h \in \mathbb{R}^{n+2m}$ in (30) is flexible and should be done in such a way that (19) holds. To find z^* we apply 2-factor Newton's method with an initial point z^0 and with the mapping Φ defined by (30):

$$z^{k+1} = z^k - \left[\Phi'(z^k)\right]^{-1}\left[\Phi(z^k)\right], \qquad k = 0, 1, \ldots. \tag{31}$$

The following theorem states the convergence properties of the iterative method (31).

Theorem 2. *Let x^* be a solution of (1). Let $f, g \in C^3(\mathbb{R}^n)$ and let $z^* = (x^*, s^*, \lambda^*)$ satisfy the optimality conditions given in (28). Assume that F defined in (28) be 2-regular at the point z^* along some vector $h \in \mathbb{R}^{n+2m}$, $h \neq 0$. Then there is a neighborhood $V(x^*)$ of z^* in $\mathbb{R}^{n+2m}$ such that for any $z^0 \in V(z^*)$, the iterative sequence $\{z^k\}$ generated by (31) converges to z^* and*

$$\|z^{k+1} - z^k\| \leq C\|z^k - z^*\|^2, \qquad k = 0, 1, \ldots,$$

where $C > 0$ is some constant.

Proof. Let x^* be a solution of (1), and let $\Lambda^*(x^*)$ be a corresponding set of Lagrangian multipliers, i.e., $\Lambda^*(x^*) = \{\lambda^* \in \mathbb{R}^m \,|\, (x^*, s^*, \lambda^*)\}$ is a solution of system (28). Under our assumptions, the set $\Lambda^*(x^*)$ might consist of more than one element. However, condition (19) guarantees that the system $\Phi(z) = 0$ has a unique solution $z^* = (x^*, s^*, \lambda^*)$ corresponding with the solution x^* of the NLP problem. Then the proof of the theorem follows from Theorem 1 and properties of Newton's method [8, Proposition 1.4.1].

For the numerical realization of the method introduced in this section, we use techniques described in Sect. 3. One of the parameters in Sect. 3.1 is function $\mu(z)$. We gave the explicit form of $\mu(z)$ for the case, when F is 2-regular at the point z^*. As we mentioned above, the choice of the function $\mu(z)$ depends on the specific of the problem. At the same time, other techniques might be used for the numerical realization of the proposed method.

5   SOSC and Lagrange Optimality System

In this section, we analyze our assumption made in the paper. We prove that the standard second-order sufficient condition (SOSC) is a sufficient condition for our assumption made in the previous section.

5.1   The SOSC and the 2-Regularity of the Lagrange Optimality System

In this section, we compare the SOSC with our assumption made in the paper. Namely, the following theorem states that the SOSC (3) is a sufficient condition for the 2-regularity of the Lagrange optimality system.

Theorem 3. *Let $f, g \in C^3(\mathbb{R}^n)$. Assume that for $x^* \in \mathbb{R}^n$, there exists a Lagrange multiplier $\lambda^* \in \mathbb{R}^m$ satisfying (2) and a scalar $\nu > 0$ such that SOSC (3) holds. Then Lagrange optimality system (28) is either regular or 2-regular at $z^* = (x^*, s^*, \lambda^*)$ with $s^* \in \mathbb{R}^m$ defined by (7) along some vector $\bar{z} \in \mathbb{R}^{n+2m}$.*

Proof. Because $x^* \in \mathbb{R}^n$ and $\lambda^* \in \mathbb{R}^m$ satisfy (2) and there exists a scalar $\nu > 0$ such that SOSC (3) holds, then Lagrange optimality system (28) has a solution $z^* = (x^*, s^*, \lambda^*)$ with s^* defined by (7).

Let m_A denote the number of the active constraints at x^*, and let $m_N = m - m_A$ denote the number of nonactive constraints. Without loss of generality, assume that the first m_A constraints are active at x^*. Introduce the notation: $\Lambda_A = \text{diag}\,(\lambda_j)_{j=1}^{m_A}$, $\Lambda_N = \text{diag}\,(\lambda_j)_{j=m_A+1}^{m}$, $S_A = \text{diag}\,(s_j)_{j=1}^{m_A}$, $S_N = \text{diag}\,(s_j)_{j=m_A+1}^{m}$,

$$J_A(x) = \begin{bmatrix} g_1'(x) \\ \vdots \\ g_{m_A}'(x) \end{bmatrix}, \qquad J_N(x) = \begin{bmatrix} g_{m_A+1}'(x) \\ \vdots \\ g_m'(x) \end{bmatrix}.$$

Then the Jacobian matrix $F'(x, s, \lambda)$ given in (29) has the form:

$$F'(x, s, \lambda) = \begin{bmatrix} L''_{xx}(x, s, \lambda) & 0_{n \times m_A} & 0_{n \times m_N} & J_A^\top(x) & J_N^\top(x) \\ 0_{m_A \times n} & 2\Lambda_A & 0_{m_A \times m_N} & 2S_A & 0_{m_A \times m_N} \\ 0_{m_N \times n} & 0_{m_N \times m_A} & 2\Lambda_N & 0_{m_N \times m_A} & 2S_N \\ J_A(x) & 2S_A & 0_{m_A \times m_N} & 0_{m_A \times m_A} & 0_{m_A \times m_N} \\ J_N(x) & 0_{m_N \times m_A} & 2S_N & 0_{m_N \times m_A} & 0_{m_N \times m_N} \end{bmatrix}.$$

Let

$$C = L''_{xx}(x^*, s^*, \lambda^*).\tag{32}$$

Consider the following system that defines the vectors $h = (h_1, h_2, h_3, h_4, h_5)$ in the nullspace of $F'(x^*, s^*, \lambda^*)$:

$$\begin{aligned} Ch_1 + J_A(x^*)^\top h_4 + J_N^\top(x^*)h_5 &= 0 \\ 2\Lambda_A^* h_2 + 2S_A^* h4 &= 0 \\ 2\Lambda_N^* h_3 + 2S_N^* h5 &= 0 \\ J_A(x^*)h_1 + 2S_A^* h2 &= 0 \\ J_N(x^*)h_1 + 2S_N^* h3 &= 0 \end{aligned}\tag{33}$$

where $h_1 \in \mathbb{R}^n$, $h_2, h_4 \in \mathbb{R}^{m_A}$, and $h_3, h_5 \in \mathbb{R}^{m_N}$.

Because $\Lambda_N^* = 0_{m_N \times m_N}$ and S_N^* is a full rank diagonal matrix, then $h_5 = 0_{m_N}$. Because $S_A^* = 0_{m_A \times m_A}$, we get from the second and the fourth equation if system (33) that h_2 and h_1 should satisfy the following equalities:

$$\Lambda_A^* h_2 = 0, \qquad J_A(x^*)h_1 = 0.\tag{34}$$

Taking into account that $h_5 = 0_{m_N}$, we obtain from the first equation of system (33) that

$$\begin{aligned} 0 &= \langle Ch_4, h_1 \rangle + \langle J_A(x^*)^\top h_4, h_1 \rangle \\ &= \langle Ch_4, h_1 \rangle + \langle h_4, J_A(x^*)h_1 \rangle. \end{aligned}\tag{35}$$

Now we assume that $h_1 \neq 0$. Then (32), (34), and (35) imply that

$$\begin{aligned} \langle I''_x x(x^*, s^*, \lambda^*)h_1, h_1 \rangle &= 0, \qquad \text{for } h_1 \text{ such that} \\ J_A(x^*)h_1 &= 0, \end{aligned}\tag{36}$$

which contradicts the assumption that SOSC (3) holds. Hence, $h_1 = 0_n$.

Then the fifth equation in (33) yields $h_3 = 0_{m_N}$. Thus, the nullspace of $F'(x^*, s^*, \lambda^*)$ consists of the vectors h in the form

$$h = (0_n, h2, 0_{m_N}, h_1, 0_{m_N}),$$

where

$$\Lambda_A^* h_2 = 0, \qquad J_A(x^*)^\top h_4 = 0.\tag{37}$$

Under assumption of the theorem, NLP (1) might be regular or nonregular at x^*. We consider these two cases separately.

1. First, consider the case when NLP (1) is *regular* at x^*, that is, when the active constraint gradients are linearly independent and the strict complementarity holds at x^*. In this case, system (37) has only trivial solution $(h_2, h_4) = (0, 0)$. Hence, the nullspace of $F'(x^*, s^*, \lambda^*)$ consists of the zero vector only, which means that the Lagrange optimality system (28) is regular at $z^* = (x^*, s^*, \lambda^*)$. This finishes the proof of the theorem in the case when NLP (1) is *regular* at x^*.

2. Otherwise, consider the case when NLP (1) is *nonregular* at x^*, that is, when the strict complementarity does not hold or when the active constraint gradients are linearly dependent at x^*. In this case, equations (37) have a nontrivial solution (h_2, h_4). We will show that in this case there exists an element $\bar{z} \in \mathbb{R}^{n+2m}$ such that the Lagrange system (28) is 2-regular at z^* along $\bar{z}$.

By definition, $F'(x, s, \lambda)$ is a symmetric matrix. Hence,

$$\operatorname{Ker} F'(x^*, s^*, \lambda^*) = \left(\operatorname{Im} F'(x^*, s^*, \lambda^*) \right)^{\perp}.$$

Denote by Λ^0 the set of indices of the *weakly active constraints* as follows:

$$\Lambda^0 = \{ j = 1, \ldots, m \mid \lambda_j^* = 0, \ s_j^* = 0 \},$$

where $\lambda^* = (\lambda_1^*, \ldots, \lambda_m^*)$ is a Lagrange multiplier satisfying the assumption of the theorem. Then the orthoprojector P_2 onto $(\operatorname{Im} F'(x^*, s^*, \lambda^*))^{\perp}$ in $\mathbb{R}^{n+2m}$ is given by

$$P_2 = \begin{pmatrix} 0 & 0 & 0 & 0 & 0 \\ 0 & P_1 & 0 & 0 & 0 \\ 0 & 0 & 0 & 0 & 0 \\ 0 & 0 & 0 & \bar{P}_2 & 0 \\ 0 & 0 & 0 & 0 & 0 \end{pmatrix},$$

where $P_1 = \operatorname{diag}(p_i)$ with $p_i = 1$ for $i \in \Lambda_0$ and $p_i = 0$ for $i \notin \Lambda_0$ and $\bar{P}_2$ is the orthoprojector onto $\operatorname{Ker} J_A^\top$. Note that there might be a set of the Lagrange multipliers $\Lambda^*(x^*)$ corresponding to x^*. However, for every multiplier $\lambda^* \in \Lambda^*(x^*)$, the definition of the matrix P_1 depends on the set Λ_0.

Define the element $\bar{z} \in \mathbb{R}^{n+2m}$ as

$$\bar{z} = (0_n, z_A, 0_{mN}, 0_{mA}, 0_{mN})^\top,$$

where $z_A \in \mathbb{R}^{m_A}$ is the composed of ones, i.e., $z_A = (1, \ldots, 1)^\top$. Then matrix $F''(x^*, s^*, \lambda^*)\bar{z}$ is given as

$$F''(x^*, s^*, \lambda^*)\bar{z} = \begin{bmatrix} 0_{n \times n} & 0_{n \times m_A} & 0_{n \times m_N} & 0_{n \times m_A} & 0_{n \times m_N} \\ 0_{m_A \times n} & 0_{m_A \times m_A} & 0_{m_A \times m_N} & 2E & 0_{m_A \times m_N} \\ 0_{m_N \times n} & 0_{m_N \times m_A} & 0_{m_N \times m_N} & 0_{m_N \times m_A} & 0_{m_N \times m_N} \\ 0_{m_A \times n} & 2E & 0_{m_A \times m_N} & 0_{m_A \times m_A} & 0_{m_A \times m_N} \\ 0_{m_N \times n} & 0_{m_N \times m_A} & 0_{m_N \times m_N} & 0_{m_N \times m_A} & 0_{m_N \times m_N} \end{bmatrix}$$

and for any $y \in \mathbb{R}^{n+2m}$, $y = (y_1, y_2, y_3, y_4, y_5)$, we get

$$
P_2 \left[F''(x^*, s^*, \lambda^*) \bar{z} \right] y = \begin{bmatrix} 0_n \\ 2P_1 E y_4 \\ 0_{m_N} \\ 2\overline{P}_2 E y_2 \\ 0_{m_N} \end{bmatrix},
$$

where E is the identity matrix.

Introduce the matrix $\Phi(\bar{z})$:

$$
\Phi(\bar{z}) = F'(x^*, s^*, \lambda^*) + P_2 \left[F''(x^*, s^*, \lambda^*) \bar{z} \right]. \tag{38}
$$

As follows from our consideration, the definition of matrix $\Phi(\bar{z})$ depends on the vector λ^*. However, for any $\lambda^* \in \Lambda^*(x^*)$, the definition of P_1, $\overline{P}_2$, and $\bar{z}$ yield that

$$
\operatorname{Im} \Phi(\bar{z}) = \mathbb{R}^{n+2m}.
$$

Hence, by Definition 5, the Lagrange optimality system (28) is 2-regular at z^* along $\bar{z}$. This finishes the proof of the theorem.

Acknowledgements. This work was supported by the Russian Foundation for Basic Research (projects no. 17-07-00510, 17-07-00493) and the RAS Presidium Program (program 27).

References

1. Nocedal, J., Wright, S.J.: Numerical Optimization. Springer, New York (1999). https://doi.org/10.1007/978-0-387-40065-5
2. Izmailov, A.F., Solodov, M.V.: Newton-type methods for optimization problems without constraint qualifications. SIAM J. Optim. **15**(1), 210–228 (2004)
3. Wright, S.J.: An algorithm for degenerate nonlinear programming with rapid local convergence. SIAM J. Optim. **15**(3), 673–696 (2005)
4. Tret'yakov, A.A.: Necessary conditions for optimality of pth order. In: Control and Optimization. MSU, Moscow, pp. 28–35 (1983). (in Russian)
5. Tret'yakov, A.A.: Necessary and sufficient conditions for optimality of p-th order. USSR Comput. Math. Math. Phys. **24**(1), 123–127 (1984)
6. Tret'yakov, A.A.: The implicit function theorem in degenerate problems. Russ. Math. Surv. **42**(5), 179–180 (1987)
7. Tret'yakov, A.A., Marsden, J.E.: Factor analysis of nonlinear mappings: p-regularity theory. Commun. Pure Appl. Anal. **2**(4), 425–445 (2003)
8. Bertsekas, D.P.: Nonlinear programming. Athena Scientific, Belmont (1999)
9. Brezhneva, O.A., Tret'yakov, A.A.: The pth-order optimality conditions for inequality constrained optimization problems. Nonlinear Anal. Theor. Methods Appl. **63**(5–7), e1357–e1366 (2005)
10. Tret'yakov, A.A.: Some schemes for solving degenerate optimization problems. In: Optimization and Optimal Control. MSU, Moscow, pp. 45–50 (1985). (in Russian)
11. Belash, K.N., Tret'yakov, A.A.: Methods for solving degenerate problems. USSR Comput. Math. Math. Phys. **28**(4), 90–94 (1988)

12. Izmailov, A.F., Tret'yakov, A.A.: The 2-Regular solutions of Nonregular Problems. Fizmatlit, Moscow (1999). (in Russian)
13. Brezhneva, O.A., Izmailov, A.F., Tret'yakov, A.A., Khmura, A.: An approach to finding singular solutions to a general system of nonlinear equations. Comput. Math. Math. Phys. **40**(3), 365–377 (2000)
14. Brezhneva, O.A., Tret'yakov, A.A.: New Methods for Solving Essentially Nonlinear Problems. Computing Center of the Russian Academy of Sciences, Moscow (2000). (in Russian)

The Nearest Point Theorem for Weakly Convex Sets in Asymmetric Seminormed Spaces

Grigorii E. Ivanov$^{(\boxtimes)}$ [ID], Mariana S. Lopushanski [ID], and Maxim O. Golubev [ID]

Moscow Institute of Physics and Technology (State University), 9 Institutskiy per., Dolgoprudny, Moscow Region 141700, Russian Federation
g.e.ivanov@mail.ru, masha.alexandra@gmail.com, maksimkane@mail.ru

Abstract. Weakly convex sets in asymmetric seminormed spaces are considered. We prove that any point from some neighborhood of such a set has the unique nearest point in the set. The proof of the nearest point theorem is based on the theorem about the diameter of ε-projection which is also important in approximation theory. The notion of weakly convex sets in asymmetric seminormed spaces generalizes known notions of sets with positive reach, proximal smooth sets, and prox-regular sets. By taking the Minkowski functional of the epigraph of some convex function as a seminorm, the results obtained for weakly convex sets can be applied to weakly convex functions whose graphs are weakly convex sets with respect to this seminorm.

Keywords: Weakly convex sets · Asymmetric seminorm Metric projection

1 Introduction

It is well-known that convex analysis plays an important role in optimization and approximation. However, convexity is a qualitative notion. Often we need to compare which set is "more convex". Parametrical convex analysis is a new branch of convex analysis, it considers quantitative characteristics of convexity, which provide more valuable information about geometric and approximation properties of sets and functions. Main objects of parametrical convex analysis are classes of weakly and strongly convex sets and functions.

Weakly convex sets are often considered in the literature under different names. The term "weakly convex sets" was introduced by Vial. However, earlier, in 1959, Federer [1] introduced the notion of sets with positive reach in $\mathbb{R}^n$. Clark, Stern, and Wolenski [2] studied proximally smooth sets in Hilbert space, which are sets with continuously differentiable distance function in some neighborhood of the set. Later Poliquin, Rockafellar and Thibault [3] considered prox-regular

---

Supported by the Russian Foundation for Basic Research, grant 18-01-00209.

Y. Evtushenko et al. (Eds.): OPTIMA 2018, CCIS 974, pp. 21–34, 2019.
https://doi.org/10.1007/978-3-030-10934-9_2

sets in Hilbert spaces and proved that in a Hilbert space the class of uniformly prox-regular sets coincides with the classes of proximally smooth and weakly convex sets. In Banach spaces the prox-regular sets were considered in [4,5] and the equivalence of the uniform prox-regularity and the weak convexity was shown for spaces with convexity and smoothness moduli of power type. In [6] this result was generalized for uniformly smooth and uniformly convex Banach spaces.

Our goal is to develop a unified approach for weakly convex sets and functions. Thus we need to study weakly convex sets in asymmetric seminormed spaces. Then we may apply the results obtained for weakly convex sets to function epigraphs. Asymmetric normed and seminormed spaces were studied in [7–9]. Approximation problems for weakly convex sets in such spaces were considered in [10–17]. In [12] it was shown that weakly convex function can be characterized as a function with a weakly convex epigraph in a seminormed space, where the seminorm is the Minkowski functional of the epigraph of some convex function.

In the present paper, we prove the theorem about the diameter of ε-projection and the nearest point theorem in asymmetric seminormed spaces. In [12] and some other works the seminorm was considered as an additional structure of a normed space, while in the present paper we introduce the seminorm as a primary structure of a real vector space.

2   Definitions and Notation

Let E be a real vector space.

Definition 1. *A function $\mu : E \to \mathbb{R}$ is called* sublinear *if it is* positively homogeneous:

$$\mu(\lambda x) = \lambda \mu(x) \qquad \forall x \in E \quad \forall \lambda \geq 0$$

and subaditive:

$$\mu(x + y) \leq \mu(x) + \mu(y) \qquad \forall x, y \in E.$$

Definition 2. *A sublinear function $\mu : E \to [0, +\infty)$ such that*

$$\max\{\mu(x), \mu(-x)\} > 0 \qquad \forall x \in E \setminus \{0\} \tag{1}$$

is called asymmetric seminorm. *The pair (E, μ) is called* asymmetric seminormed space.

Definition 3. *If asymmetric seminorm μ satisfies additional condition*

$$\mu(x) > 0 \qquad \forall x \in E \setminus \{0\},$$

then it is called asymmetric norm. *In such a case the pair (E, μ) is called* asymmetric normed space.

In some literature (e.g. [8]) a bit different terminology is used: an asymmetric seminorm is called an asymmetric norm and a sublinear function $\mu : E \to [0, +\infty)$ (without axiom (1)) is called an asymmetric seminorm.

Definition 4. *For $\varepsilon > 0$ and $x \in E$ use $U_\varepsilon(x)$ to denote ε-neighborhood of x:*

$$U_\varepsilon(x) = \{y \in E : \; \mu(x - y) < \varepsilon\}.$$

A set $X \subset E$ is called μ-open if for any $x \in X$ there exists $\varepsilon > 0$ such that $U_\varepsilon(x) \subset X$. We shall use τ_μ to denote the family of all μ-open subsets $X \subset E$.

Remark 1. (E, τ_μ) is a topological space.

Remark 2. (i) If μ is an asymmetric norm, then (E, τ_μ) is a Hausdorff space, that is for any different points $x, y \in E$ there exists $\varepsilon > 0$ such that $U_\varepsilon(x) \cap U_\varepsilon(y) = \emptyset$.
(ii) If μ is an asymmetric seminorm, then (E, τ_μ) is not a Hausdorff space in general. Consider, for example, an asymmetric seminormed space (E, μ) with $E = \mathbb{R}$ and

$$\mu(x) = \begin{cases} x, \; x \geq 0, \\ 0, \; x < 0. \end{cases}$$

Then for any $x \in \mathbb{R}$, $\varepsilon > 0$ we have $U_\varepsilon(x) = (x - \varepsilon, +\infty)$ and (E, τ_μ) is not a Hausdorff space.

Definition 5. *Let X be an arbitrary set. A function $\varrho : X \times X \to [0, +\infty)$ is called a* quasi-metric *if*

$$\varrho(x, x) = 0 \quad \forall x \in X,$$

$$\varrho(x, z) \leq \varrho(x, y) + \varrho(y, z) \qquad \forall x, y, z \in X,$$

$$\varrho(x, y) = \varrho(y, x) = 0 \quad \Rightarrow \quad x = y \qquad \forall x, y \in X.$$

Remark 3. Any asymmetric seminormed space (E, μ) possesses a quasi-metric

$$\varrho_\mu(x, y) = \mu(x - y), \qquad x, y \in E,$$

a metric

$$\varrho_\mu^s(x, y) = \max\{\mu(x - y), \mu(y - x)\}, \qquad x, y \in E,$$

and a norm

$$\|x\|_\mu = \max\{\mu(x), \mu(-x)\}, \qquad x \in E.$$

We shall use $\operatorname{cl} X$ to denote the closure of the subset $X \subset E$ with respect the topology induced by metric ϱ_μ^s. Let E^* be the space of linear functionals $p : E \to \mathbb{R}$, which are continuous with respect to metric ϱ_μ^s. The value of a functional $p \in E^*$ at $x \in E$ will be denoted as $\langle p, x \rangle$.

Definition 6. *The* diameter *of a set $X \subset E$ is defined as*

$$\operatorname{diam} X = \sup_{x, y \in X} \mu(x - y).$$

Definition 7. *A sequence $\{x_k\}$ in an asymmetric seminormed space (E, μ) is called*

– ϱ_μ^s-Cauchy *if*

$$\forall \varepsilon > 0 \; \exists N_\varepsilon : \; \forall n, k \geq N_\varepsilon \qquad \varrho_\mu^s(x_n, x_k) < \varepsilon,$$

– convergent to $x \in E$ *(we write $x_k \to x$) if* $\varrho_\mu^s(x_k, x) \to 0$ *as* $k \to \infty$.

Definition 8. *An asymmetric seminormed space* (E, μ) *is called* biBanach space *if any ϱ_μ^s-Cauchy sequence $\{x_k\}$ converges to some $x \in E$.*

Remark 4. If (E, μ) is a biBanach space, then $(E, \|\cdot\|_\mu)$ is a Banach space.

Definition 9. *Let* (E, μ) *be an asymmetric seminormed space.*
   The μ-distance from a point $x \in E$ to a set $A \subset E$ is

$$\varrho_\mu(x, A) = \inf_{a \in A} \mu(x - a).$$

The μ-projection (or μ-nearest point) for a point $x \in E$ on a set $A \subset E$ is

$$P_\mu(x, A) = \{a \in A : \mu(x - a) = \varrho_\mu(x, A)\}.$$

Given $\varepsilon > 0$, the ε-μ-projection of a point $x \in E$ on a set $A \subset E$ is defined as

$$P_\mu^\varepsilon(x, A) = \{a \in A : \mu(x - a) \leq \varrho_\mu(x, A) + \varepsilon\}.$$

The cone of proximal normals *to a set $A \subset E$ at a point $a \in A$ is*

$$N_\mu(a, A) = \{z \in E \mid \exists t > 0 : \; a \in P_\mu(a + tz, A)\}.$$

Definition 10. *A set $A \subset E$ is called μ-weakly convex if for any $a \in A$ and $z \in N_\mu(a, A)$ with $\mu(z) = 1$ one has $a \in P_\mu(a + z, A)$.*

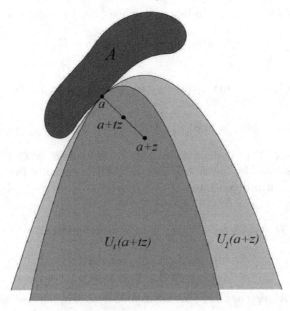

Definition 11. *An asymmetric seminormed space* (E, μ) *is called* parabolic *if for any* $b \in E$

$$\sup_{\substack{x \in E: \ \mu(x) \leq 1, \\ \mu(x+b) > 2}} \mu(-x) < +\infty.$$

Definition 12. *An asymmetric seminormed space* (E, μ) *is called* uniformly convex *if* $\delta_\mu(\varepsilon, R) > 0$ *for any positive* ε *and* R, *where*

$$\delta_\mu(\varepsilon, R) = \inf \left\{ 1 - \mu\left(\frac{x_1 + x_2}{2}\right) \middle| \ x_1, x_2 \in E: \begin{array}{l} \mu(x_1) \leq 1, \ \mu(x_2) \leq 1, \\ \mu(-x_1) \leq R, \ \mu(-x_2) \leq R, \\ \mu(x_1 - x_2) \geq \varepsilon \end{array} \right\}. \tag{2}$$

Remark 5. If (E, μ) is a uniformly convex asymmetric seminormed space, then $(E, \|\cdot\|_\mu)$ is a uniformly convex (and consequently reflexive) normed space.

3 Auxiliary Results

Lemma 1. *Let* (E, μ) *be an asymmetric seminormed space,* $A \subset E$. *Then*

(i) $\varrho_\mu(x_1, A) - \varrho_\mu(x_2, A) \leq \mu(x_1 - x_2) \quad \forall x_1, x_2 \in E$;
(ii) for any $x \in E$ *such that* $\varrho = \varrho_\mu(x, A) > 0$ *one has*

$$U_\varrho(x) \cap A = \emptyset;$$

(iii) function $\varrho_\mu(\cdot, A)$ *is Lipschitz continuous on* E *with Lipschitz constant* 1 *with respect to metric* ϱ_μ^s;
(iv) for any $\varepsilon_1 > 0$, $\varepsilon_2 > \varepsilon_1$ *and* $x_1, x_2 \in E$ *such that* $\mu(x_1 - x_2) + \mu(x_2 - x_1) \leq \varepsilon_2 - \varepsilon_1$ *one has* $P_\mu^{\varepsilon_1}(x_1, A) \subset P_\mu^{\varepsilon_2}(x_2, A)$.

Proof. Assertion

(i) follows from subadditivity of μ.
(ii) Assume the contrary: there exists $a \in U_\varrho(x) \cap A$. Then $\varrho = \varrho_\mu(x, A) \leq \mu(x - a) < \varrho$. Contradiction.
(iii) follows directly from (i);
(iv) Let $a \in P_\mu^{\varepsilon_1}(x_1, A)$. Then $a \in A$ and $\mu(x_1 - a) \leq \varrho_\mu(x_1, A) + \varepsilon_1$. Using item (i), we get

$$\mu(x_2 - a) \leq \mu(x_2 - x_1) + \mu(x_1 - a) < \mu(x_2 - x_1) + \varrho_\mu(x_1, A) + \varepsilon_1$$

$$\leq \mu(x_2 - x_1) + \mu(x_1 - x_2) + \varrho_\mu(x_2, A) + \varepsilon_1 \leq \varrho_\mu(x_2, A) + \varepsilon_2.$$

So, $a \in P_\mu^{\varepsilon_2}(x_2, A)$. □

Lemma 2. *Let* (E, μ) *be an asymmetric seminormed space and* $A \subset E$ *be convex. Then* A *is* μ-*weakly convex.*

Proof. Fix $a \in A$ and $z \in N_\mu(a, A)$ with $\mu(z) = 1$. According to the definition of the cone of proximal normals there exists $t > 0$ such that $a \in P_\mu(a + tz, A)$. It means that $a \in A$ and $\varrho_\mu(a + tz, A) = \mu(tz) = t$. By Lemma 1(ii) we have $U_t(a + tz) \cap A = \emptyset$. According to the Hahn-Banach separation theorem there exists $p \in E^*$ such that

$$\langle p, y \rangle < \langle p, x \rangle \qquad \forall y \in A \quad \forall x \in U_t(a + tz).$$

Consequently,

$$\langle p, y \rangle \leq \langle p, a \rangle \qquad \forall y \in A, \tag{3}$$

$$\langle p, a \rangle < \langle p, x \rangle \qquad \forall x \in U_t(a + tz),$$

that is,

$$-\langle p, z \rangle < \langle p, x \rangle \qquad \forall x \in U_1(0). \tag{4}$$

Adding the inequalities (3) and (4), we obtain

$$\langle p, a + z \rangle > \langle p, y - x \rangle \qquad \forall y \in A \quad \forall x \in U_1(0).$$

It means that $y - a - z \notin U_1(0)$ for any $y \in A$, that is $\mu(a + z - y) \geq 1$ for all $y \in A$. Hence $\varrho_\mu(a + z) \geq 1$. On the other hand, $\varrho_\mu(a + z) \leq \mu(a + z - a) = \mu(z) = 1$. So, $\varrho_\mu(a + z) = \mu(a + z - a)$ and hence $a \in P_\mu(a + z, A)$. □

For any $A \subset E$ we define

$$\gamma_\mu(A) = \sup_{x \in E} \varrho_\mu(x, A),$$

$$S_\mu(A) = \{x \in E : 0 < \varrho_\mu(x, A) < \gamma_\mu(A)\},$$

$$T_\mu(A) = \{x \in E : P_\mu(x, A) \text{ is a singleton}\}.$$

The following proposition is a corollary of [11, Theorem 2.3].

Proposition 1. *Let* (E, μ) *be a parabolic uniformly convex biBanach asymmetric seminormed space and* $A \subset E$ *be* ϱ_μ^s*-closed. Then* $T_\mu(A)$ *is dense in* $S_\mu(A)$, *that is* $S_\mu(A) \subset \mathrm{cl}\, T_\mu(A)$, *where the closure is treated in the sense of topology induced by metric* ϱ_μ^s.

Lemma 3. *Let* (E, μ) *be a parabolic uniformly convex biBanach asymmetric seminormed space. Let* $A \subset E$ *be* ϱ_μ^s*-closed and* μ*-weakly convex set. If* $\gamma_\mu(A) > 0$, *then* $\gamma_\mu(A) \geq 1$.

Proof. Since $\sup_{x \in E} \varrho_\mu(x, A) = \gamma_\mu(A) > 0$ there exists $x_0 \in E$ such that $\varrho_\mu(x_0, A) > 0$. Fix $a_0 \in A$. We have $\varrho_\mu(a_0, A) = 0$. By Lemma 1(iii) the function $\varrho_\mu(\cdot, A)$ is continuous with respect to metric ϱ_μ^s. Therefore the segment $[x_0, a_0]$ includes a point x_1 such that $0 = \varrho_\mu(a_0, A) < \varrho_\mu(x_1, A) < \varrho_\mu(x_0, A) \leq \gamma_\mu(A)$. Hence, $x_1 \in S_\mu(A)$. By Proposition 1 there exists $x \in T_\mu(A) \setminus A$. Consequently, one can find $a \in P_\mu(x, A)$. Denoting $z = \frac{x-a}{\mu(x-a)}$, we have $z \in N_\mu(a, A)$ and $\mu(z) = 1$. Definition 10 implies that $a \in P_\mu(a + z, A)$. So, $\varrho_\mu(a + z, A) = \mu(z) = 1$. Consequently, $\gamma_\mu(A) \geq \varrho_\mu(a + z, A) \geq 1$. □

Lemma 4. *Let (E, μ) be a parabolic asymmetric seminormed space. Then for any $\lambda_1 > 0$, $\lambda_2 > \lambda_1$ and $x_1, x_2 \subset E$ one has*

$$\sup_{\substack{x \in E: \ \mu(x - x_1) \le \lambda_1, \\ \mu(x - x_2) > \lambda_2}} \mu(-x) < +\infty.$$

Proof. Introducing $\lambda = \frac{\lambda_2}{\lambda_1}$, $y_1 = \frac{x_1}{\lambda_1}$, $y_2 = \frac{x_2}{\lambda_1}$, $y = \frac{x}{\lambda_1}$, $z = y - y_1$, $b = y_1 - y_2$, we have

$$\sup_{\substack{x \in E: \ \mu(x - x_1) \le \lambda_1, \\ \mu(x - x_2) > \lambda_2}} \mu(-x) = \sup_{\substack{y \in E: \ \mu(y - y_1) \le 1, \\ \mu(y - y_2) > \lambda}} \lambda_1 \mu(-y) =$$

$$\sup_{\substack{z \in E: \ \mu(z) \le 1, \\ \mu(z + b) > \lambda}} \lambda_1 \mu(-z - y_1) \le \sup_{\substack{z \in E: \ \mu(z) \le 1, \\ \mu(z + b) > \lambda}} \lambda_1 \mu(-z) + \lambda_1 \mu(-y_1).$$

Hence, it suffices to show that

$$\varkappa(b, \lambda) < +\infty \qquad \forall b \in E, \ \lambda > 1, \tag{5}$$

where

$$\varkappa(b, \lambda) = \sup_{\substack{x \in E: \ \mu(x) \le 1, \\ \mu(x + b) > \lambda}} \mu(-x). \tag{6}$$

According to the definition of the parabolic space $\varkappa(b, 2) < +\infty$ for all $b \in E$. It follows from (6) that

$$\varkappa(b, \lambda) \le \varkappa(b, 2) < +\infty \qquad \forall b \in E, \ \lambda \ge 2. \tag{7}$$

Now let us show that

$$\varkappa(b, \lambda) \le \varkappa\left(\frac{b}{\lambda - 1}, 2\right) \qquad \forall b \in E, \ \lambda \in (1, 2). \tag{8}$$

Fix any $b \in E$, $\lambda \in (1, 2)$ and $x \in E$ such that $\mu(x) \le 1$, $\mu(x + b) > \lambda$. Since

$$x + b = (\lambda - 1)\left(x + \frac{b}{\lambda - 1}\right) + (2 - \lambda)x,$$

by sublinearity of μ it follows that

$$\lambda < \mu(x+b) \le (\lambda-1)\mu\left(x + \frac{b}{\lambda - 1}\right) + (2-\lambda)\mu(x) \le (\lambda-1)\mu\left(x + \frac{b}{\lambda - 1}\right) + 2 - \lambda.$$

Hence, $2\lambda - 2 < (\lambda - 1)\mu\left(x + \frac{b}{\lambda-1}\right)$ and $\mu\left(x + \frac{b}{\lambda-1}\right) > 2$. Using (6), we get $\mu(-x) \le \varkappa\left(\frac{b}{\lambda-1}, 2\right)$. This implies (8). It follows by (8) that

$$\varkappa(b, \lambda) \le \varkappa\left(\frac{b}{\lambda - 1}, 2\right) < +\infty \qquad \forall b \in E, \ \lambda \in (1, 2).$$

So, in view of (7) the proof of (5) is completed. $\qquad \square$

Lemma 5. *Let (E, μ) be an asymmetric seminormed space. Let $x, y \in E$ and $R > 0$ be such that $0 < \mu(-x) \leq R \cdot \mu(x), 0 < \mu(-y) \leq R \cdot \mu(y)$. Then*

$$2 \min\{\mu(x), \mu(y)\} \cdot \delta_\mu \left(\mu \left(\frac{x}{\mu(x)} - \frac{y}{\mu(y)} \right), R \right) \leq \mu(x) + \mu(y) - \mu(x + y), \quad (9)$$

where the function $\delta_\mu(\cdot, \cdot)$ is defined in (2).

Proof. Define

$$a = \frac{x}{\mu(x)}, \qquad b = \frac{y}{\mu(y)}, \qquad c = \frac{a + b}{2}, \qquad \delta = \delta_\mu(\mu(a - b), R).$$

Since $\mu(a) = \mu(b) = 1$, $\mu(-a) \leq R$, $\mu(-b) \leq R$, it follows by (2) that

$$\mu(c) = \mu \left(\frac{a + b}{2} \right) \leq 1 - \delta. \tag{10}$$

Denote $\mu_1 = \mu(x)$, $\mu_2 = \mu(y)$. Without loss of generality, we assume that $\mu_2 \leq \mu_1$. By sublinearity of μ in view of $\mu(a) = 1$ we have

$$\mu(x + y) = \mu((\mu_1 - \mu_2)a + \mu_2(a + b)) \leq (\mu_1 - \mu_2) \cdot \mu(a) + \mu_2 \cdot \mu(a + b) = \mu_1 - \mu_2 + 2\mu_2 \cdot \mu(c).$$

Using (10) we get

$$\mu(x + y) \leq \mu_1 + \mu_2 - 2\mu_2\delta = \mu_1 + \mu_2 - 2\delta \min\{\mu_1, \mu_2\}. \qquad \square$$

Lemma 6. *Let (E, μ) be a uniformly convex asymmetric seminormed space. Let $\{x_k\}, \{y_k\}$ be such sequences in E that*

$$\sup_{k \in \mathbb{N}} \mu(-x_k) < +\infty, \qquad \sup_{k \in \mathbb{N}} \mu(-y_k) < +\infty, \tag{11}$$

$$\limsup_{k \to \infty} \mu(x_k) \leq \mu_1, \qquad \limsup_{k \to \infty} \mu(y_k) \leq \mu_2, \qquad \liminf_{k \to \infty} \mu(x_k + y_k) \geq \mu_1 + \mu_2,$$

$$\tag{12}$$

where $\mu_1 > 0$ and $\mu_2 > 0$. Then

$$\lim_{k \to \infty} \left\| \frac{x_k}{\mu_1} - \frac{y_k}{\mu_2} \right\|_\mu = 0.$$

Proof. By sublinearity of μ we have

$$\mu(x_k + y_k) \leq \mu(x_k) + \mu(y_k) \qquad \forall k \in \mathbb{N}.$$

This and (12) imply existence of the following limits:

$$\lim_{k \to \infty} \mu(x_k) = \mu_1, \qquad \lim_{k \to \infty} \mu(y_k) = \mu_2, \qquad \lim_{k \to \infty} \mu(x_k + y_k) = \mu_1 + \mu_2. \tag{13}$$

Without loss of generality, we assume that $\mu(x_k) \geq \frac{\mu_1}{2}$, $\mu(y_k) \geq \frac{\mu_2}{2}$ for any $k \in \mathbb{N}$. Using (11) we conclude that the value

$$R = \sup_{k \in \mathbb{N}} \max \left\{ \frac{\mu(-x_k)}{\mu(x_k)}, \frac{\mu(-y_k)}{\mu(y_k)} \right\}$$

is finite. Lemma 5 implies that

$$2 \min\{\mu_1, \mu_2\} \cdot \delta_\mu \left(\mu \left(\frac{x_k}{\mu(x_k)} - \frac{y_k}{\mu(y_k)}, R \right) \right) \leq \mu(x_k) + \mu(y_k) - \mu(x_k + y_k) \to 0.$$

This and the uniform convexity of the space imply

$$\lim_{k \to \infty} \mu \left(\frac{x_k}{\mu(x_k)} - \frac{y_k}{\mu(y_k)} \right) = 0.$$

So, using (13), we obtain

$$\lim_{k \to \infty} \mu \left(\frac{x_k}{\mu_1} - \frac{y_k}{\mu_2} \right) = 0.$$

Similarly,

$$\lim_{k \to \infty} \mu \left(\frac{y_k}{\mu_2} - \frac{x_k}{\mu_1} \right) = 0.$$

Since

$$\left\| \frac{x_k}{\mu_1} - \frac{y_k}{\mu_2} \right\|_\mu = \max \left\{ \mu \left(\frac{x_k}{\mu_1} - \frac{y_k}{\mu_2} \right), \mu \left(\frac{y_k}{\mu_2} - \frac{x_k}{\mu_1} \right) \right\}$$

the proof is completed.   □

Some results of this section are similar to those obtained previously for normed spaces with seminorm as an additional structure. In particular, Lemma 1 (iii) is an analogue of [12, Lemma 6.1], Lemma 3 is akin to [12, Lemma 5.4] and Lemma 5 is similar to [15, Lemma 2.1].

4   Main Results

Theorem 1. (On the diameter of ε-μ-projection). *Let (E, μ) be a parabolic uniformly convex biBanach asymmetric seminormed space. Let $A \subset E$ be a closed μ-weakly convex set, $r \in (0, 1)$, $b \in E$. Let $\{x_k\}$ be such sequence in E that $\mu(b - x_k) \leq r$ for all $k \in \mathbb{N}$, $\lim_{k \to \infty} \varrho_\mu(x_k, A) = \varrho_0 \in (0, 1 - r)$ and $\{\varepsilon_k\}$ be an infinitesimal sequence of positive reals. Then*

$$\lim_{k \to \infty} \operatorname{diam} P_\mu^{\varepsilon_k}(x_k, A) = 0.$$

Proof. Denote

$$\varepsilon_0 = \frac{1}{8} \min\{\varrho_0, 1 - r - \varrho_0\}. \tag{14}$$

Without loss of generality, we assume that

$$|\varrho_\mu(x_k, A) - \varrho_0| < \varepsilon_k < \varepsilon_0 \qquad \forall k \in \mathbb{N}.$$

Lemma 3 implies that $1 \leq \gamma_\mu(A)$. So, for all $k \in \mathbb{N}$ we have $0 < \varrho_0 - \varepsilon_0 \leq \varrho_\mu(x_k, A) \leq \varrho_0 + \varepsilon_0 < 1 \leq \gamma_\mu(A)$, hence $x_k \in S_\mu(A)$. By Proposition 1 for any $k \in \mathbb{N}$ there exists $z_k \in T_\mu(A)$ such that $\varrho_\mu^s(z_k, x_k) < \varepsilon_k$. Hence, by Lemma 1 (iii) we get

$$|\varrho_\mu(z_k, A) - \varrho_\mu(x_k, A)| < \varepsilon_k \qquad \forall k \in \mathbb{N}. \tag{15}$$

Using Lemma 1(iv) we get

$$P_\mu^{\varepsilon_k}(x_k, A) \subset P_\mu^{3\varepsilon_k}(z_k, A) \qquad \forall k \in \mathbb{N}. \tag{16}$$

In view of (16) it suffices to prove

$$\lim_{k \to \infty} \operatorname{diam} P_\mu^{3\varepsilon_k}(z_k, A) = 0. \tag{17}$$

Since $1 - \varepsilon_0 < 1 \leq \gamma_\mu(A)$ there exists $b_1 \in E$ such that $1 - \varepsilon_0 < \varrho_\mu(b_1, A)$. Hence,

$$\mu(b_1 - a) > 1 - \varepsilon_0 \qquad \forall a \in A. \tag{18}$$

For any $y \in P_\mu^{3\varepsilon_k}(z_k, A)$ we have

$$\mu(z_k - y) \leq \varrho_\mu(z_k, A) + 3\varepsilon_k < \varrho_\mu(x_k, A) + 4\varepsilon_k < \varrho_0 + 5\varepsilon_k.$$

Therefore for any $y \in P_\mu^{3\varepsilon_k}(z_k, A)$

$$\mu(b - y) \leq \mu(b - x_k) + \mu(x_k - z_k) + \mu(z_k - y) \leq r + \varepsilon_k + \varrho_0 + 5\varepsilon_k < r + \varrho_0 + 6\varepsilon_0.$$

In view of (14) we have $r + \varrho_0 + 6\varepsilon_0 < 1 - \varepsilon_0$. Using (18) by Lemma 4 we get

$$\sup_{k \in \mathbb{N}} \sup_{y \in P_\mu^{3\varepsilon_k}(z_k, A)} \mu(-y) < +\infty. \tag{19}$$

Since $z_k \in T_\mu(A)$, for any $k \in \mathbb{N}$ there exists $a_k \in P_\mu(z_k, A)$. We have $\mu(z_k - a_k) = \varrho_\mu(z_k, A) \leq \varrho_0 + 2\varepsilon_0$. Using (18) we obtain

$$1 - \varepsilon_0 < \mu(b_1 - a_k) \leq \mu(b_1 - z_k) + \mu(z_k - a_k) \leq \mu(b_1 - z_k) + \varrho_0 + 2\varepsilon_0$$

and hence

$$\mu(b_1 - z_k) > 1 - \varrho_0 - 3\varepsilon_0.$$

On the other hand

$$\mu(b - z_k) \leq \mu(b - x_k) + \mu(x_k - z_k) < r + \varepsilon_0.$$

Taking into account that $r + \varepsilon_0 < 1 - \varrho_0 - 3\varepsilon_0$ and using Lemma 4 once more, we obtain

$$\sup_{k \in \mathbb{N}} \mu(-z_k) < +\infty. \tag{20}$$

Since $a_k \in P_\mu(z_k, A)$, according to the Definition 10 we have

$$a_k \in P_\mu(c_k, A) \qquad \forall k \in \mathbb{N}, \tag{21}$$

$$c_k = a_k + \frac{z_k - a_k}{\mu(z_k - a_k)}.$$ (22)

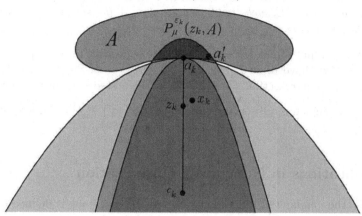

Fix any sequence $\{a_k'\}$ such that

$$a_k' \in P_\mu^{3\varepsilon_k}(z_k, A) \qquad \forall k \in \mathbb{N}.$$ (23)

As $a_k \in P_\mu(z_k, A)$, we have $\mu(z_k - a_k) = \varrho_\mu(z_k, A)$. Taking into account (15) and $\lim\limits_{k \to \infty} \varrho_\mu(x_k, A) = \varrho_0$, we obtain

$$\lim_{k \to \infty} \mu(z_k - a_k) = \varrho_0.$$ (24)

By (21), (22) it follows that $\varrho_\mu(c_k, A) = \mu(c_k - a_k) = 1$. Hence,

$$\mu(c_k - a_k') \geq \varrho_\mu(c_k, A) = 1 \qquad \forall k \in \mathbb{N}.$$ (25)

In view of (15), (23), we get

$$\limsup_{k \to \infty} \mu(z_k - a_k') \leq \lim_{k \to \infty} \varrho_\mu(z_k, A) = \varrho_0.$$ (26)

It follows by (22) that $\mu(c_k - z_k) = 1 - \mu(z_k - a_k)$. So, (24) implies

$$\lim_{k \to \infty} \mu(c_k - z_k) = 1 - \varrho_0.$$ (27)

In view of relations (19)–(27) by Lemma 6 we obtain

$$\lim_{k \to \infty} \left\| \frac{c_k - z_k}{1 - \varrho_0} - \frac{z_k - a_k'}{\varrho_0} \right\|_\mu = 0.$$ (28)

Using (22) we see $\frac{c_k - z_k}{1 - \mu(z_k - a_k)} = \frac{z_k - a_k}{\mu(z_k - a_k)}$. Thus, by (24) we have

$$\lim_{k \to \infty} \left\| \frac{c_k - z_k}{1 - \varrho_0} - \frac{z_k - a_k}{\varrho_0} \right\|_\mu = 0.$$

This and (28) imply that

$$\lim_{k \to \infty} \|a_k - a_k'\|_\mu = 0.$$

Since $\{a_k'\}$ is an arbitrary sequence satisfying (23) we obtain (17). $\qquad \square$

Theorem 2. (On the nearest point). *Let (E, μ) be a parabolic uniformly convex biBanach asymmetric seminormed space. Let $A \subset E$ be a closed μ-weakly convex set. Then for any $x \in E$ such that $0 < \varrho_\mu(x, A) < 1$ the set of μ-nearest points $P_\mu(x, A)$ is a singleton.*

Proof. Fix any infinitesimal decreasing sequence of positive reals $\{\varepsilon_k\}$. According to Theorem 1 we have diam $P_\mu^{\varepsilon_k}(x, A) \to 0$ as $k \to \infty$. Since $(E, \| \cdot \|_\mu)$ is a Banach space, the intersection $\bigcap\limits_{k \in \mathbb{N}} P_\mu^{\varepsilon_k}(x, A)$ of the nested sequence of closed sets $P_\mu^{\varepsilon_k}(x, A)$ is a singleton. Observing that $P_\mu(x, A) = \bigcap\limits_{k \in \mathbb{N}} P_\mu^{\varepsilon_k}(x, A)$, we complete the proof.   □

5   Applications in Nonconvex Optimization

Recall that the *epigraph* of function $f : E \to \mathbb{R} \cup \{-\infty, +\infty\}$ is defined by

$$\text{epi } f = \{(x, y) \in E \times \mathbb{R} \mid x \in E, \ y \geq f(x)\};$$

the *effective domain* of f is

$$\text{dom } f = \{x \in E \mid f(x) \in \mathbb{R}\}.$$

The *infimal convolution* [18] of the functions $f : E \to \mathbb{R} \cup \{+\infty\}$ and $g : E \to \mathbb{R} \cup \{+\infty\}$ is the function $f \boxplus g$, defined by

$$(f \boxplus g)(x) = \inf_{u \in E} \Big(f(x - u) + g(u)\Big), \qquad x \in E. \tag{29}$$

Note that functions f and (or) g in (29) may be nonconvex and in such a case (29) is a problem of nonconvex optimization.

The infimal convolution problem is very important in optimization. In particular, if $f(\cdot) = \| \cdot \|$ and $g(\cdot)$ is the indicator function of a set $A \subset E$ ($g(x) = 0$ if $x \in A$ and $g(x) = +\infty$ otherwise), then $(f \boxplus g)(x)$ is the distance from $x \in E$ to A and the solution of (29) is the metric projection of x onto A. In case $g(\cdot) = \frac{1}{2\lambda}\| \cdot \|^2$ the infimal convolution $f \boxplus g$ is Moreau–Yosida regularization of f [19]. Applications of the infimal convolution to optimal control are considered e.g. in [20–22]. In [23] it is shown that well-posedness properties of optimization problems are very important in subdifferential calculus. Recall that a minimization problem is Tykhonov well posed if it admits a unique solution and any minimizing sequence converges to this solution. Note that the well-posedness property of a minimization problem is needed for the stability of numerical methods to find the solution to the problem. The following proposition provides a necessary and sufficient conditions for the infimal convolution problem to be Tykhonov well posed. This condition will be formulated in terms of μ-weak convexity with respect to the Minkowski functional of the epigraph of some convex function.

The Minkowski functional of a set M is defined as

$$\mu_M(x) = \inf\{t > 0 : \ x \in tM\}.$$

A function $\gamma : E \to \mathbb{R}$ is called *uniformly convex* on a convex set $X \subset E$ if for any sequences $\{x_k\} \subset X$ and $\{x_k'\} \subset X$ the statement $\limsup_{k\to\infty} \left(\gamma(x_k) + \gamma(x_k') - 2\gamma\left(\frac{x_k + x_k'}{2}\right) \right) \leq 0$ implies that $\lim_{k\to\infty} \|x_k - x_k'\| = 0$.

A function $\gamma : E \to \mathbb{R} \cup \{+\infty\}$ is *coercive* if $\lim_{\|x\|\to\infty} \frac{\gamma(x)}{\|x\|} = +\infty$.

Given a function $\gamma : E \to \mathbb{R}$, for any number $r > 0$ and for any function $\gamma : E \to \mathbb{R}$ denote $\gamma_{(r)}(x) = r\gamma\left(\frac{x}{r}\right)$. Note that epi $\gamma_{(r)} = r \cdot$ epi γ.

Proposition 2. *Let E be a Banach space and let $\gamma : E \to \mathbb{R}$ be a supercoercive function, bounded on any bounded set, and uniformly convex on any convex bounded set, $\gamma(0) < 0$. Let $\mu(\cdot)$ be the Minkowski functional of epi γ. Suppose that a function $f : E \to \mathbb{R} \cup \{+\infty\}$ is lower semicontinuous and $\mathrm{dom}\,(f \boxplus \gamma) \neq \emptyset$. The following statements are equivalent:*

(1) epi f is μ-weakly convex;
(2) for any $r \in (0,1)$ and $x_0 \in E$ the problem

$$\min_{u \in E} \left(f(u) + \gamma_{(r)}(x_0 - u) \right)$$

is Tykhonov well posed.

Proposition 2 follows directly from Theorems 3.5–3.7 in [12].

References

1. Federer, H.: Curvature measures. Trans. Amer. Math. Soc. **93**, 418–491 (1959)
2. Clarke, F.H., Stern, R.J., Wolenski, P.R.: Proximal smoothness and lower-C^2 property. J. Convex Anal. **2**(1,2), 117–144 (1995)
3. Poliquin, R.A., Rockafellar, R.T., Thibault, L.: Local differentiability of distance functions. Trans. Amer. Math. Soc. **352**, 5231–5249 (2000)
4. Bernard, F., Thibault, L., Zlateva, N.: Characterization of proximal regular sets in super reflexive Banach spaces. J. Convex Anal. **13**(3,4), 525–559 (2006)
5. Bernard, F., Thibault, L., Zlateva, N.: Prox-regular sets and epigraphs in uniformly convex Banach spaces: various regularities and other properties. Trans. Amer. Math. Soc. **363**, 2211–2247 (2011)
6. Balashov, M.V., Ivanov, G.E.: Weakly convex and proximally smooth sets in Banach spaces. Izv. Math. **73**, 455–499 (2009)
7. Cobzas, S.: Ekeland variational principle in asymmetric locally convex spaces. Topol. Appl. **159**(10,11), 2558–2569 (2012)
8. Cobzas, S.: Functional Analysis in Asymmetric Normed Spaces. Springer, Basel (2013). https://doi.org/10.1007/978-3-0348-0478-3
9. Jordan-Pérez, N., Sánchez-Perez, E.A.: Extreme points and geometric aspects of compact convex sets in asymmetric normed spaces. Topol. Appl. **203**, 15–21 (2016)
10. Borodin, P.A.: On the convexity of N-Chebyshev sets. Izv. Math. **75**(5), 889–914 (2011)
11. Ivanov, G.E.: On well posed best approximation problems for a nonsymmetric seminorm. J. Convex Anal. **20**(2), 501–529 (2013)

12. Ivanov, G.E.: Weak convexity of sets and functions in a Banach space. J. Convex Anal. **22**(2), 365–398 (2015)
13. Ivanov, G.E.: Continuity and selections of the intersection operator applied to nonconvex sets. J. Convex Anal. **22**(4), 939–962 (2015)
14. Ivanov, G.E.: Weak convexity of functions and the infimal convolution. J. Convex Anal. **23**(3), 719–732 (2016)
15. Ivanov, G.E., Lopushanski, M.S.: Well-posedness of approximation and optimization problems for weakly convex sets and functions. J. Math. Sci. **209**(1), 66–87 (2015)
16. Ivanov, G.E., Lopushanski, M.S.: Separation theorems for nonconvex sets in spaces with non-symmetric seminorm. J. Math. Inequal. Appl. **20**(3), 737–754 (2017)
17. Lopushanski, M.S.: Normal regularity of weakly convex sets in asymmetric normed spaces. J. Convex Anal. **25**(3), 737–758 (2018)
18. Fenchel, W.: Convex Cones, Sets and Functions. Mimeographed Lecture Notes, Princeton University (1951)
19. Moreau, J.J.: Proximité et dualité dans un espace hilbertien. Bulletin de la Societe Mathematique de France **93**, 273–299 (1965)
20. Ivanov, G.E., Thibault, L.: Infimal convolution and optimal time control problem I: Frechet and proximal subdifferentials. Set Valued Var. Anal. **26**(3), 581–606 (2018)
21. Ivanov, G.E., Thibault, L.: Infimal convolution and optimal time control problem II: limiting subdifferential. Set Valued Var. Anal. **25**(3), 517–542 (2017)
22. Ivanov, G.E., Thibault, L.: Infimal convolution and optimal time control problem III: minimal time projection set. SIAM J. Optim. **28**(1), 30–44 (2018)
23. Ivanov, G.E., Thibault, L.: Well-posedness and subdifferentials of optimal value and infimal convolution. Set Valued Var. Anal. 1–21 (2018). https://doi.org/10.1007/s11228-018-0493-4

A Modified Duality Method for Solving an Elasticity Problem with a Crack Extending to the Outer Boundary

Robert Namm[1](✉), Georgiy Tsoy[1], and Ellina Vikhtenko[2]

[1] Computing Center of Far Eastern Branch Russian Academy of Sciences,
Kim Yu Chen 65, 680021 Khabarovsk, Russia
rnamm@yandex.ru, tsoy.dv@mail.ru
[2] School of Fundamental and Computer Sciences, Pacific National University,
Tihookeanskaya 136, 680035 Khabarovsk, Russia
vikht.el@gmail.com
http://www.ccfebras.ru/

Abstract. A modified dual method for solving an elasticity problem with a crack extending to the outer boundary is considered. The method is based on a modified Lagrange functional. The convergence of the method is investigated in detail under a natural assumption of H^1-regularity of the solution to the crack problem. Basic duality relation for the primal and dual problems is proposed.

Keywords: Non-penetration condition · Crack · Duality scheme
Modified lagrange functional · Generalized newton method

1 Introduction

The classical statement of an equilibrium problem of an elastic body with a crack is to assume that on the crack faces zero stress conditions are given [1,2]. These conditions do not exclude the possibility of penetration of the crack faces into each other, which is unnatural in terms of mechanics. In recent papers on the crack theory, the models with boundary conditions such as inequalities on the crack faces are considered. These models provide a mutual non-penetration of the crack faces and can be formulated as variational problems of minimization of a convex functional over a closed convex subset of the initial Hilbert space or as variational inequalities. It is important to construct efficient approximate methods for solving nonlinear variational problems of continuum mechanics, in particular, problems with cracks, which can be found in [3–7]. To solve a plane elasticity problem with mutual non-penetration between the crack faces we use a duality scheme based on modified Lagrange functionals. Such functionals for solving variational inequalities of mechanics are considered in [8–13]. In these papers, as a rule, sufficient regularity of the solution to the initial problem is assumed to provide solvability of the duality problem. However, for elasticity

© Springer Nature Switzerland AG 2019
Y. Evtushenko et al. (Eds.): OPTIMA 2018, CCIS 974, pp. 35–48, 2019.
https://doi.org/10.1007/978-3-030-10934-9_3

crack problems regularity of the solution near the edges of the crack can be insufficient for the solvability of the duality problem. Despite this fact, to solve the elasticity problem with a crack we construct and justify a duality scheme and prove a duality relation. Numerical experiments are presented to show the performance of the proposed modified dual method.

2   Problem Statement

In contrast to the [8], we consider the formulation of an elasticity problem with a crack in which one of the edges of the crack extends to the outer boundary of the domain. Let $\Omega \subset R^2$ - be a bounded domain with a regular boundary Γ, and let $\gamma \subset \Omega$ be a cut(crack) with edge lying on the outer boundary. Assume that $\Gamma = \Gamma_0 \bigcup \Gamma_1^+ \bigcup \Gamma_1^- \bigcup \Gamma_1^*$, where Γ_0, Γ_1^+, Γ_1^-, Γ_1^* are nonempty open disjoint subsets of Γ and $\Gamma_1 = \Gamma_1^+ \bigcup \Gamma_1^- \bigcup \Gamma_1^*$ (see Fig. 1).

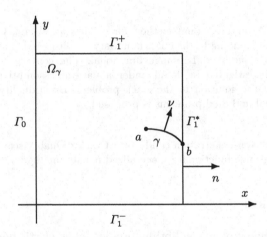

Fig. 1. Elastic body with a crack.

We assume that γ is a smooth curve, without self-intersections, leaving one endpoint of γ under a nonzero angle to Γ. This assumption is important under the study of the duality scheme based on the modified Lagrange functional. Denote $\Omega_\gamma = \Omega \setminus \overline{\gamma}$, where $\overline{\gamma} = \gamma \cup \{a\} \cup \{b\}$. Let ν be the vector of the unit normal on γ. In this case on the crack γ, denote the positive (upper) face by γ^+ and the negative (lower) face γ^-, where the signs $\pm$ correspond to the positive and negative directions of the normal ν to γ. Consider the following elasticity crack problem.

For the displacement vector $v = (v_1, v_2)$, define the deformation tensor

$$\varepsilon_{ij}(v) = \frac{1}{2}\left(\frac{\partial v_i}{\partial x_j} + \frac{\partial v_j}{\partial x_i}\right), \; i,j = 1,2,$$

and the stress tensor

$$\sigma_{ij}(v) = c_{ijkm}\varepsilon_{km}(v),$$

where $C = \{c_{ijkm}\}, i, j, k, m = 1, 2$ is a given elasticity tensor with the usual properties of positive definiteness and symmetry: $c_{ijkm} = c_{jimk} = c_{kmij}$, $c_{ijkm} \in L^\infty(\Omega)$. Summation over repeated indices is assumed.

Let us specify vector-functions of the body and surface forces $f = (f_1, f_2)$ and $p = (p_1, p_2)$, respectively. The boundary value problem is formulated as follows [2,8]:

$$-\frac{\partial \sigma_{ij}}{\partial x_j} = f_i \text{ in } \Omega_\gamma, \ i = 1, 2,$$

$$u = 0 \text{ on } \Gamma_0,$$

$$\sigma_{ij} n_j = p_i \text{ on } \Gamma_1, \ i = 1, 2, \tag{1}$$

where $n = (n_1, n_2)$ is the unit outward normal vector to Γ.

The following conditions are set on γ:

$$[u_\nu] \geq 0, \ [\sigma_\nu(u)] = 0, \ \sigma_\nu(u)[u_\nu] = 0 \text{ on } \gamma,$$

$$\sigma_\nu(u) \leq 0, \ \sigma_\tau(u) = 0 \text{ on } \gamma^\pm. \tag{2}$$

Here $u_\nu = u\nu$, $[u_\nu] = u_\nu^+ - u_\nu^-$, $\sigma_\nu(u) = \sigma_{ij}(u)\nu_i\nu_j$, $[\sigma_\nu(u)] = \sigma_\nu^+(u) - \sigma_\nu^-(u)$, $\sigma_\tau(u) = \sigma(u) - \sigma_\nu\nu$, where $\sigma(u) = (\sigma_1(u), \sigma_2(u))$, $\sigma_i(u) = \sigma_{ij}(u)\nu_j, i = 1, 2$.

Consider a variational problem for a domain with a crack corresponding to the boundary value problem (1), (2). Introduce, as in [2], the set of admissible displacements

$$K = \left\{v \in [H^1(\Omega_\gamma)]^2 : \ [v_\nu] \geq 0 \text{ on } \gamma, \ v = 0 \text{ on } \Gamma_0\right\},$$

where, as before, $[v_\nu] = v_\nu^+ - v_\nu^-$ is a jump of the function $v_\nu = v\nu$ on γ, $v_\nu^\pm \in H^{1/2}(\gamma)$ (see [2, p. 12]). The norm in the space $H^{1/2}(\gamma)$ is defined as

$$\|v_\nu\|_{H^{1/2}(\gamma)}^2 = \|v_\nu\|_{L_2(\gamma)}^2 + \int_\gamma \int_\gamma \frac{|v_\nu(x) - v_\nu(y)|^2}{|x - y|^2} dx dy.$$

The boundary value problem (1), (2) corresponds to the following variational problem [2,8]:

$$\begin{cases} J(v) = \dfrac{1}{2}a(v, v) - \int\limits_{\Omega_\gamma} f_i v_i \, d\Omega - \int\limits_{\Gamma_1} p_i v_i \, d\Gamma \to \min, \\ v \in K, \end{cases} \tag{3}$$

where $a(u, v) = \int\limits_{\Omega_\gamma} c_{ijpm}\varepsilon_{pm}(u)\varepsilon_{ij}(v) \, d\Omega$, $f \in [L_2(\Omega_\gamma)]^2$, $p \in [L_2(\Gamma_1)]^2$.

The problem (3) is equivalent to variational inequality: find $u \in K$ such that for all $v \in K$ inequality

$$a(u, v - u) - \int\limits_{\Omega_\gamma} f_i(v_i - u_i) \, d\Omega - \int\limits_{\Gamma_1} p_i(v_i - u_i) \, d\Gamma \geq 0 \tag{4}$$

is valid.

3 Duality Method for Solving the Elasticity Crack Problem

To solve the problem with a crack extending from one end to the outer boundary, we apply a duality scheme based on the modified Lagrange functional. A similar duality scheme for solving an elasticity problem with an internal crack is investigated in [8]. We recall the assumption made earlier in Sect. 1 that the crack extends to the outer boundary under a nonzero angle to Γ.

Define the space

$$W = \left\{ v \in [H^1(\Omega_\gamma)]^2 : v = 0 \text{ on } \Gamma_0 \right\}.$$

For an arbitrary $m \in L_2(\gamma)$, construct the set

$$K_m = \left\{ v \in W : -[v_\nu] \le m \text{ on } \gamma \right\}.$$

It is easy to show that K_m is a convex set that is closed in the $H^1(\Omega_\gamma)$ norm.

On the space $L_2(\gamma)$, we define a sensitivity functional

$$\chi(m) = \begin{cases} \inf_{v \in K_m} J(v), & \text{if } K_m \ne \varnothing, \\ +\infty & - \text{ otherwise.} \end{cases}$$

Since the crack leaves at the point b on the outer boundary Γ under a nonzero angle to Γ, then the Korn inequality is fulfilled [2]

$$\int_{\Omega_\gamma} \varepsilon_{ij}(v)\varepsilon_{ij}(v)\, d\Omega \ge c\|v\|^2_{H^1_{\Omega_\gamma}} \quad \forall\, v \in W,$$

where $c > 0$ is a constant. The Korn inequality implies the property of strong convexity (and hence coercivity) of the functional $J(v)$ on the space W. Therefore the problem $\inf_{v \in K_m} J(v)$ under the condition $K_m \ne \varnothing$ is solvable. If the function $m \in L_2(\gamma) \setminus H^{1/2}(\gamma)$, the set K_m can be empty [8,12].

The functional $\chi(m)$ is a proper convex functional on $L_2(\gamma)$, but its effective domain $dom\chi = \{m \in L_2(\gamma) : \chi(m) < +\infty\}$ does not coincide with $L_2(\gamma)$. Also note the $dom\chi$ is a convex but not closed in $L_2(\gamma)$, and in our case $\overline{dom\chi} = L_2(\gamma)$.

On the space $W \times L_2(\gamma) \times L_2(\gamma)$, we define the functional

$$Q(v,l,m) = \begin{cases} J(v) + \int_\gamma lm\,d\Gamma + \frac{r}{2}\int_\gamma m^2 d\Gamma, & \text{if } -[v_\nu] \le m \text{ a.e. on } \gamma, \\ +\infty & - \text{ otherwise.} \end{cases}$$

and a modified Lagrange functional $M(v,l)$ on the space $W \times L_2(\gamma)$:

$$M(v,l) = \inf_{m \in L_2(\gamma)} Q(v,l,m) = J(v) + \frac{1}{2r}\int_\gamma \left\{ \left[(l - r[v_\nu])^+ \right]^2 - l^2 \right\} d\Gamma,$$

where $(l - r[v_\nu])^+ \equiv \max\{0, l - r[v_\nu]\}$, $r > 0$ is a constant.

Introduce a modified dual functional

$$\underline{M}(l) = \inf_{v \in W} M(v, l) = \inf_{v \in W} \left\{ J(v) + \frac{1}{2r} \int_\gamma \left\{ [(l - r[v_\nu])^+]^2 - l^2 \right\} d\Gamma \right\}.$$

Since $\inf\limits_{v \in W} \inf\limits_{m \in L_2(\gamma)} Q(v, l, m) = \inf\limits_{m \in L_2(\gamma)} \inf\limits_{v \in W} Q(v, l, m)$, for $\underline{M}(l)$ we also have the following representation [12]:

$$\underline{M}(l) = \inf_{m \in L_2(\gamma)} \left\{ \chi(m) + \int_\gamma lm \, d\Gamma + \frac{r}{2} \int_\gamma m^2 d\Gamma \right\}.$$

It is easy to see that the following estimate is valid for any $l \in L_2(\gamma)$:

$$\underline{M}(l) \le \chi(0) = \inf_{v \in K} J(v). \tag{5}$$

For the functional $\underline{M}(l)$, we define a dual problem:

$$\begin{cases} \underline{M}(l) \to \sup, \\ l \in L_2(\gamma). \end{cases} \tag{6}$$

For a problem with a crack, it is natural to assume only $[H^1(\Omega_\gamma)]^2$ regularity of the solution. In this case, the dual problem (6) may be unsolvable. We consider a dual method for solving a problem with a crack, extending to the outer boundary, in which the solvability of the dual problem (6) is not assumed in advance. A similar study of the dual method for solving an elasticity problem with an internal crack was performed in [8].

Let us investigate the sensitivity functional $\chi(m)$ and the corresponding dual functional $\underline{M}(l)$. We show that the sensitivity functional $\chi(m)$ is weakly lower semicontinuous on $L_2(\gamma)$. Since $\chi(m)$ is a convex functional, it is sufficient to prove that epigraph of the functional is closed. Closure of the epigraph follows from the following two lemmas.

Lemma 1. *Let $\overline{m} \in L_2(\gamma)$ not belong to $\mathrm{dom}\chi$. Then for any sequence $\{m_i\} \subset \mathrm{dom}\chi$ such that $\lim\limits_{i \to \infty} \|m_i - \overline{m}\|_{L_2(\gamma)} = 0$ we have a limit equality $\lim\limits_{i \to \infty} \chi(m_i) = +\infty$.*

Proof. For a function $\overline{m} \notin \mathrm{dom}\chi$, we consider an arbitrary sequence $\{m_i\} \subset \mathrm{dom}\chi$ such that $\lim\limits_{i \to \infty} \|m_i - \overline{m}\|_{L_2(\gamma)} = 0$. Since $K_{m_i} \ne \varnothing$ and the functional $J(v)$ is coercive on W, there exists a unique element $v^i = \arg\min\limits_{v \in K_{m_i}} J(v)$, $(i = 1, 2, ...)$.

Let us show that $\lim\limits_{i \to \infty} \|v^i\|_W = +\infty$.

Assume the contrary, that is, let the sequence $\{v^i\}$ has a bounded subsequence $\{v^{i_j}\}$, $\|v^{i_j}\|_W \le c$ for all i_j, where $c > 0$ is a constant. It follows from the trace theorem that $\left\| [v_\nu^{i_j}] \right\|_{H^{1/2}(\gamma)} \le c_1$, where $c_1 > 0$ is a constant [2]. Then

$\{[v_\nu^{i_j}]\}$ is a compact subsequence in $L_2(\gamma)$. Let $t \in H^{1/2}(\gamma)$ be a weak limit point of this sequence. Without loss of generality t can be considered as a weak limit $\{[v_\nu^{i_j}]\}$ in $H^{1/2}(\gamma)$. Then $\{[v_\nu^{i_j}]\}$ converges to t in the norm in $L_2(\gamma)$. Since $-[v_\nu^{i_j}] \leq m_i$, we have $-t \leq \overline{m}$ on γ. Hence, $K_{\overline{m}} \neq \varnothing$ or $\overline{m} \in dom\chi$. This contradiction shows that $\lim_{i \to \infty} \|v^i\|_W = +\infty$. Since the functional $J(v)$ is coercive on W, we have $\lim_{i \to \infty} \chi(m_i) = \lim_{i \to \infty} J(v^i) = +\infty$.

Lemma 2. *Let $\overline{m} \in L_2(\gamma)$ belong to $dom\chi$. Then for any sequence $\{m_i\} \subset dom\chi$ converging to $\overline{m}$ in $L_2(\gamma)$, the following inequality holds*

$$\varliminf_{i \to \infty} \chi(m_i) \geq \chi(\overline{m}).$$

Proof. Let $\{m_i\} \subset dom\chi$ and $\lim_{i \to \infty} \|m_i - \overline{m}\|_{L_2(\gamma)} = 0$, where $\overline{m} \in dom\chi$. From the sequence $\{m_i\}$, we take a subsequence $\{m_{i_j}\}$ for which

$$\lim_{i \to \infty} \chi(m_{i_j}) = \varliminf_{i \to \infty} \chi(m_i).$$

Consider a subsequence $\{v_{i_j}\}$, where $v^{i_j} = \arg \min_{v \in K_{m_{i_j}}} J(v)$. The sequence $\{v_{i_j}\}$ is bounded in W (otherwise $\lim_{i \to \infty} \chi(m_{i_j}) = +\infty$ and the required inequality is proved). Denote $\widetilde{\Gamma} = \Gamma \cup \gamma^+ \cup \gamma^-$ and let $[H^{1/2}(\widetilde{\Gamma})]^2$ be the space of traces of functions from the space $[H^1(\Omega_\gamma)]^2$ to $\widetilde{\Gamma}$. Let $[H^{1/2}(\Gamma)]^2$ be the space of functions that are the restrictions of functions from $[H^{1/2}(\widetilde{\Gamma})]^2$ to Γ. Since $W \subset [H^1(\Omega_\gamma)]^2 \subset [H^{1/2}(\widetilde{\Gamma})]^2$, we have $\|v_{i_j}\|_{[H^{1/2}(\Gamma)]^2} \leq c$, where $c > 0$ is a constant. In addition, $\{v_{i_j}\}$ is a compact sequence in $L_2(\Gamma)$. Let $\hat{v} \in [H^{1/2}(\Gamma)]^2$ be a weak limit point of this sequence. Without loss of generality $\hat{v}$ may be considered a weak limit $\{v_{i_j}\}$. Then $\{v_{i_j}\}$ converges to $\hat{v}$ in $L_2(\Gamma)$. It follows from the trace theorem [2] that the sequence $\{[v_{i_j}]\}$ is weakly compact in $H^{1/2}(\gamma)$. Let $t \in H^{1/2}(\gamma)$ be a weak limit point of this sequence. Without loss of generality $\{[v_{i_j}]\}$ can be considered a weakly converging sequence, that is, t is a weak limit of $\{[v_{i_j}]\}$ in $H^{1/2}(\gamma)$.

Since the space $H^{1/2}(\gamma)$ is compactly embedded into $L_2(\gamma)$ and $L_2(\gamma)$ is embedded into $H^{-1/2}(\gamma)$, $[v_{i_j}]$ converges to t in the norm $L_2(\gamma)$. Here $H^{-1/2}(\gamma)$ is the dual space to $H^{1/2}(\gamma)$. From the convergence m_{i_j} to $\overline{m}$ in $L_2(\gamma)$, $[v_{i_j}]$ to t in $L_2(\gamma)$, and condition $-[v_{i_j}] \leq m_{i_j}$, we obtain $-t \leq \overline{m}$ on γ.

Let us denote $\widetilde{t} = \arg \min_{v \in W_t} J(v)$, where $W_t = \{v \in W : [v_\nu] = t$ on γ, $v = \hat{v}$ on $\Gamma_1\}$. We have

$$J(v^{i_j}) - J(\widetilde{t}) = a(\widetilde{t}, v^{i_j} - \widetilde{t}) - \int_{\Omega_\gamma} f_s(v_s^{i_j} - \widetilde{t}_s) d\Omega + \frac{1}{2} a(v^{i_j} - \widetilde{t}, v^{i_j} - \widetilde{t}) - \int_{\Gamma_1} p_s(v_s^{i_j} - \hat{v}_s) d\Gamma$$

$$= \langle \mu_1, v^{i_j} - \hat{v} \rangle + \langle \mu_2, [v_\nu^{i_j}] - t \rangle - \int_{\Gamma_1} p_s(v_s^{i_j} - \hat{v}_s) d\Gamma + \frac{1}{2} a(v^{i_j} - \widetilde{t}, v^{i_j} - \widetilde{t}),$$

where $\mu_1 \in [H^{-1/2}(\Gamma)]^2$, $\mu_2 \in H^{-1/2}(\gamma)$. Here

$$\langle \mu_1, v^{i_j} - \hat{v}\rangle + \langle \mu_2, [v^{i_j}_\nu] - t\rangle = a(\tilde{t}, v^{i_j} - \tilde{t}) - \int_{\Omega_\gamma} f_s(v^{i_j}_s - \tilde{t}_s)d\Omega$$

and $\mu_1 + \mu_2 \in ([H^{1/2}(\Gamma)]^2 \times H^{1/2}(\gamma))^*$, where $([H^{1/2}(\Gamma)]^2 \times H^{1/2}(\gamma))^*$ is the space that is dual to $[H^{1/2}(\Gamma)]^2 \times H^{1/2}(\gamma)$.

Since $\{v_{i_j}\}$ weakly converges to $\hat{v}$ in $[H^{1/2}(\Gamma)]^2$ and $\{[v_{i_j}]\}$ weakly converges to t in $H^{1/2}(\gamma)$, owing to the uniqueness of the weak limits we have

$$\lim_{j\to\infty} \langle \mu_1, v^{i_j} - \hat{v}\rangle + \lim_{j\to\infty} \langle \mu_2, [v^{i_j}_\nu] - t\rangle = 0.$$

Therefore, the following estimate is valid:

$$\lim_{j\to\infty} \chi(m_{i_j}) = \lim_{j\to\infty} J(v^{i_j}) \geq J(\tilde{t}) \geq \chi(\overline{m})$$

and, hence,

$$\varliminf_{i\to\infty} \chi(m_i) \geq \chi(\overline{m}).$$

As was noted above, from the lemmas proved follows the weak lower semicontinuity of a convex functional $\chi(m)$ or the closure of its epigraph.

For an arbitrary fixed $l \in L_2(\gamma)$, consider the functional

$$F_l(m) = \chi(m) + \int_\gamma lmd\Gamma + \frac{r}{2}\int_\gamma m^2 d\Gamma , \quad r > 0 - const.$$

It is easy to see that $F_l(m)$ is a functional that is lower semicontinuous on $L_2(\gamma)$. Since $epi\chi$ is a convex closed set in $L_2(\gamma) \times R$, $R = (-\infty, +\infty)$, then according to the Mazur separability theorem [15], there exist $\alpha \in L_2(\gamma)$ and $\beta \in R$ such that

$$\chi(m) + \int_\gamma \alpha md\Gamma + \beta \geq 0 \quad \forall\, m \in dom\chi.$$

Hence, for the functional $F_l(m)$ the following lower estimate holds:

$$F_l(m) \geq -\int_\gamma \alpha md\Gamma + \int_\gamma lmd\Gamma + \frac{r}{2}\int_\gamma m^2 d\Gamma - \beta \geq 0 \quad \forall\, m \in L_2(\gamma).$$

Therefore, $F_l(m) \to +\infty$ as $\|m\|_{L_2(\gamma)} \to +\infty$, that is, $F_l(m)$ is coercive in $L_2(\gamma)$.

It follows from the weak semicontinuity and coercivity of $F_l(m)$ that for any $l \in L_2(\gamma)$ there exist an element $m(l) \in L_2(\gamma)$ such that

$$m(l) = \arg\min_{m\in L_2(\gamma)} F_l(m).$$

It follows from the strong convexity of $F_l(m)$ on $dom\chi$ [16] that for any $l \in L_2(\gamma)$ the element $m(l)$ is unique.

We formulate for the dual functional $\underline{M}(l)$ some characteristic statements that can be proved similarly to Theorems 2–4 in [17].

Theorem 1. *The dual functional $\underline{M}(l)$ is continuous in $L_2(\gamma)$*

Theorem 2. *The dual functional $\underline{M}(l)$ is Gateaux differentiable in $L_2(\gamma)$ and its derivative $\bigtriangledown\underline{\dot{M}}(l)$ satisfies a Lipschitz condition with a constant $\frac{1}{r}$, that is, the following inequality holds:*

$$\| \bigtriangledown \underline{M}(l') - \bigtriangledown \underline{M}(l'') \|_{L_2(\gamma)} \leq \frac{1}{r} \| l' - l'' \|_{L_2(\gamma)} \ \forall\, l', l'' \in L_2(\gamma).$$

It can be shown that $\bigtriangledown\underline{M}(l) = m(l) = \max\{-[u_\nu], -\frac{l}{r}\} \ \forall\, l \in L_2(\gamma)$ [10].

To solve the dual problem (6), consider a gradient method [8]

$$l^{k+1} = l^k + \theta_k m(l^k), \ k = 1, 2, \cdots, \tag{7}$$

with any initial value $l^0 \in L_2(\gamma)$, $\theta_k \in [\tau, 2r - \tau]$, $\tau \in (0, r]$.

Theorem 3. *For the sequence $\{l^k\}$ constructed by the method (7) we have a limit equality*

$$\lim_{k \to \infty} \| m(l^k) \|_{L_2(\gamma)} = 0.$$

The gradient method (7) generates the following algorithm of a Uzawa-type method for solving the problem (3) [8]. At the initial step, $k = 0$, specify an arbitrary function $l^0 \in L_2(\gamma)$ and for every $k = 0, 1, 2, \cdots$ subsequently calculate:

$$(i) \quad u^{k+1} = \arg\min_{v \in W} \left\{ J(v) + \frac{1}{2r} \int_\gamma \left\{ \left[(l^k - r[v_\nu])^+ \right]^2 - (l^k)^2 \right\} d\Gamma \right\}; \tag{8}$$

$$(ii) \quad l^{k+1} = l^k + \theta_k \max\{-[u_\nu^{k+1}], -\frac{l^k}{r}\}, \ \theta_k \in [\tau, 2r - \tau], \ \tau \in (0, r]. \tag{9}$$

Theorem 4. *The following duality relation holds [8]*

$$\sup_{l \in L_2(\gamma)} \underline{M}(l) = \inf_{v \in K} J(v).$$

Note that when the dual problem (6) is solvable, it can be proved that the sequence $\{l^k\}$ is bounded in $L_2(\gamma)$ [8,18]. According to Theorem 3, this means that the following equality is valid:

$$\lim_{k \to \infty} \int_\gamma l^k m(l^k) d\Gamma = 0.$$

Hence, the method (8), (9) converges for the functional of the (3), that is,

$$\lim_{k \to \infty} \chi(m(l^k)) = \lim_{k \to \infty} J(u^{k+1}) = J(u),$$

where u is the solution to the problem (3).

4   Numerical Experiment

The domain Ω is taken as a unit square with crack

$$\gamma = \{(x, y) : 0.75 \leq x \leq 1, y = \sqrt{0.3125^2 - (x - 0.75)^2} + 0.1875\}.$$

To find a solution to the problem (8), we use a finite element method. Figure 2 shows the triangulation of the domain with increasing mesh density near the crack. For triangulation we used Fade2D Delaunay triangulation library (http://www.geom.at/fade2d/html).

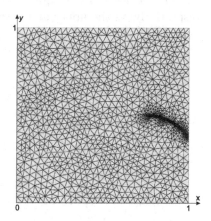

Fig. 2. Delaunay triangulation of domain Ω.

We introduce the following notation: h - edge length on γ, n - number of all triangulation nodes, n_γ - number of triangulation nodes on γ, W_h is the linear shell of the basis functions $\varphi_i(x, y)$, $u_h = (u_1^h, u_2^h)$ is the piecewise linear approximation of the exact solution u:

$$u_1^h(x, y) = \sum_{j=1}^{n} t_j \varphi_j(x, y), \ u_2^h(x, y) = \sum_{j=n+1}^{2n} t_j \varphi_{j-n}(x, y), \ t_j \in R. \qquad (10)$$

Since Ω is a polygon, the embedding $W_h \subset W$ is guaranteed. Thus, the problem (8) is replaced by the finite element problem

$$u^{k+1} = \arg \min_{v \in W_h} \left\{ J(v) + \frac{1}{2r} \int_{\gamma} \left\{ \left[(l^k - r[v_\nu])^+ \right]^2 - (l^k)^2 \right\} d\Gamma \right\}. \qquad (11)$$

We approximate the boundary integral of γ using the trapezium quadrature rule. Let $t = (t_1, t_2, \ldots, t_n, t_{n+1}, \ldots, t_{2n})$, then the minimization problem (11) is reduced to finding the optimal values of t_i. For this, we use the generalized Newton method [19].

Let us introduce the stiffness matrix $A = (a_{ij}) \in R^{2n \times 2n}$ and column vector of right side $F = (f_i) \in R^{2n}$. Define the vector $\alpha^k = (\alpha_1^k, \alpha_2^k, \ldots, \alpha_{n_\gamma}^k)$ - approximate value of the dual variable l^k and gradient $g(t)$ of the corresponding finite-dimensional functional

$$g(t) = At - F + \beta(t), \tag{12}$$

where $\beta(t) = (b_i) \in R^{2n}$. For convenience, we denote by $\{i_j^+\}, \{i_j^-\}_{j=1,n_\gamma}$ - the numbers of nodes lying respectively on the upper and lower sides of the crack faces. Then

$$b_i = 0, \quad \text{for all } i \neq \{i_j^+, i_j^-, i_j^+ + n, i_j^- + n\}.$$

Otherwise, if

$$\lambda_j = \alpha_j^k - r(t_{i_j^+} - t_{i_j^-})\nu_j^1 - r(t_{i_j^+ + n} - t_{i_j^- + n})\nu_j^2 > 0, \quad j = \overline{1, n_\gamma} ,$$

where (ν_j^1, ν_j^2) - normal vector in the j-th node, then

$$b_{i_j^+} = -h\nu_j^1 \lambda_j^+,$$
$$b_{i_j^-} = h\nu_j^1 \lambda_j^+,$$
$$b_{i_j^+ + n} = -h\nu_j^2 \lambda_j^+,$$
$$b_{i_j^- + n} = h\nu_j^2 \lambda_j^+.$$

The generalized Newton method looks in the following way:

(1) At the initial step, set t^0
(2) For every $m = 0, 1, 2, \cdots$ sequentially calculate

$$t^{m+1} = t^m - (\partial g(t^m))^{-1} g(t^m). \tag{13}$$

(3) Check

$$\|t^{m+1} - t^m\|_\infty < \varepsilon_t, \varepsilon_t = 10^{-12}.$$

Here $\partial g(t)$ is a generalized Jacobian $g(t)$:

$$\partial g(t) = A + D(t) ,$$

where $D(t) = (d_{ij}) \in R^{2n \times 2n}$ is a symmetric sparse matrix. If $\lambda_j > 0$, then matrix $D(t)$ will have nonzero elements:

$$d_{i_j^+, i_j^+} = rh(\nu_j^1)^2, \ d_{i_j^+, i_j^-} = -rh(\nu_j^1)^2, \ d_{i_j^+, i_j^+ + n} = rh\nu_j^1\nu_j^2, \ d_{i_j^+, i_j^- + n} = -rh\nu_j^1\nu_j^2,$$

$$d_{i_j^-, i_j^-} = rh(\nu_j^1)^2, \ d_{i_j^-, i_j^+} = -rh(\nu_j^1)^2, \ d_{i_j^-, i_j^- + n} = rh\nu_j^1\nu_j^2, \ d_{i_j^-, i_j^+ + n} = -rh\nu_j^1\nu_j^2,$$

$$d_{i_j^+ + n, i_j^+ + n} = rh(\nu_j^2)^2, d_{i_j^+ + n, i_j^- + n} = -rh(\nu_j^2)^2, d_{i_j^+ + n, i_j^+} = rh\nu_j^1\nu_j^2, d_{i_j^+ + n, i_j^-} = -rh\nu_j^1\nu_j^2,$$

$$d_{i_j^- + n, i_j^- + n} = rh(\nu_j^2)^2, d_{i_j^- + n, i_j^+ + n} = -rh(\nu_j^2)^2, d_{i_j^- + n, i_j^-} = rh\nu_j^1\nu_j^2, d_{i_j^- + n, i_j^+} = -rh\nu_j^1\nu_j^2.$$

At step (ii), we find the new value of the dual variable:

$$\alpha_j^{k+1} = (\lambda_j)^+, \; j = \overline{1, n_\gamma}. \tag{14}$$

The stop criterion for the Uzawa method has the following form:

$$\|\alpha^{k+1} - \alpha^k\|_\infty < \varepsilon_\alpha, \varepsilon_\alpha = 10^{-8}.$$

Let us present the results of numerically solving the problem. The parameter values are as follows: $f = (f_1, f_2) = (0,0)$, the right side surface force $p_1|_{\Gamma_1^*} = -27$ MPa, $p_2|_{\Gamma_1^*} = 0$ MPa, on the upper side $p_1|_{\Gamma_1^+} = 0$ MPa, $p_2|_{\Gamma_1^+} = -1$ MPa and on the lower side $p_1|_{\Gamma_1^-} = 0$ MPa, $p_2|_{\Gamma_1^-} = 1$ MPa, Youngs elasticity modulus $E = 73000$ MPa, the Poisson coefficient $\mu = 0.34$, the constant $r = 10^{10}$, $h = 0.003$.

Numerical experiments were conducted on a hybrid computing cluster based on the OpenPOWER architecture. It should be noted that the generalized Newton method is easy and well parallelized and its main computational complexity consists in finding the inverse matrix. Therefore, the calculations were performed on NVIDIA Tesla P100 GPU using the cuBLAS library. This allows to significantly accelerate the speed of execution compared to the version on the CPU.

The results of the numerical solution are presented graphically in Figs. 3 and 4. The graphs show that the jump $[u_\nu] \geq 0$ on the crack, so there is no mutual penetration between the crack faces into each other. In addition, it can be seen from the Fig. 4 that the value of the dual variable is greater than zero at points where crack faces are stuck together. This indicates the presence of a normal stress in these nodes.

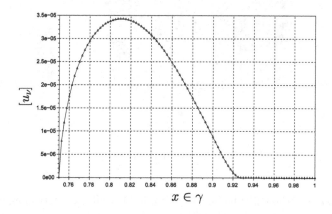

Fig. 3. Jump of the function u_ν.

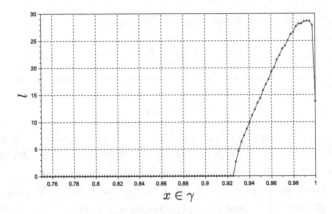

Fig. 4. Value of the dual variable l.

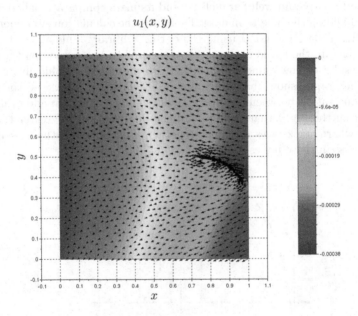

Fig. 5. Value of function $u_1(x, y)$.

Figures 5 and 6 show displacement vector field u with an increasing coefficient of 100 in both axes and value of functions $u_1(x, y)$, $u_2(x, y)$ using a heatmap.

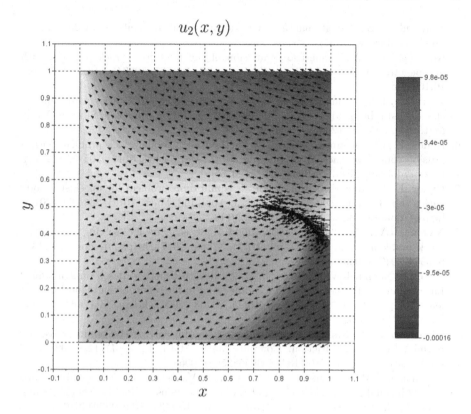

Fig. 6. Value of function $u_2(x, y)$.

5   Conclusions

A small number of steps (ii) provides fast convergence of the Uzawa method by dual variable l. In the above example, only 5 iterations by dual variable are performed. It should be noted that the number of iterations (13) for the generalized Newton method at step (i) also turned out to be relatively small. On the first step of the Uzawa method, 9 iterations of type (13) are executed, and on the next steps only 2 iterations. Thus, the numerical calculations confirm that modified Lagrange functionals make it possible to efficiently solve mathematical models of an elastic body with a crack with nonlinear boundary conditions in the form of inequalities.

Acknowledgments. This study was supported by the Russian Foundation for Basic Research (Project 17-01-00682 A). Numerical experiments were performed on a computational cluster of the Shared Facility Center "Data Center of FEB RAS" [20].

References

1. Morozov, N.F.: Mathematical Problems of Crack Theory. Nauka, Moscow (1982)
2. Khludnev, A.M.: Problems of Elasticity Theory in Non-Smooth Domains. Fizmatlit, Moscow (2010)

3. Kravchuk, A.S.: Variational and Quasi-Variational Inequalities in Mechanics. MGAPI, Moscow (1997)
4. Kovtunenko, V.A.: Numerical simulation of the non-linear crack problem with non-penetration. Math. Methods Appl. Sci. **27**, 163–179 (2004). https://doi.org/10.1002/mma.449
5. Vtorushin, E.V.: A numerical investigation of a model problem for deforming an elasto-plastic body with a crack under non-penetration condition. Sib. Zh. Vych. Mat. **9**(4), 335–344 (2006)
6. Rudoy, E.M.: Domain decomposition method for a model crack problem with a possible contact of crack edges. Comput. Math. Math. Phys. **55**(2), 305–316 (2015). https://doi.org/10.1134/S0965542515020165
7. Rudoy, E.M.: Numerical solution of an equilibrium problem for an elastic body with a thin delaminated rigid inclusion. J. Appl. Ind. Math. **10**(2), 264–276 (2016). https://doi.org/10.1134/S1990478916020113
8. Namm, R.V., Tsoy, G.I.: A modified dual scheme for solving an elastic crack problem. Num. Anal. Appl. **10**(1), 37–46 (2017). https://doi.org/10.1134/S1995423917010050
9. Woo, G., Namm, R.V., Sachkov, S.A.: An iterative method based on a modified Lagrangian functional for finding a saddle point in the semicoercive signorini problem. Comput. Math. Math. Phys. **46**(1), 23–33 (2006). https://doi.org/10.1134/S0965542506010052
10. Vikhtenko, E.M., Namm, R.V.: Duality scheme for solving the semicoercive signorini problem with friction. Comput. Math. Math. Phys. **47**(12), 2023–2036 (2007). https://doi.org/10.1134/S0965542507120068
11. Vikhtenko, E.M., Maksimova, N.N., Namm, R.V.: Sensitivity functionals in variational inequalities of mechanics and their applications to duality schemes. Num. Anal. Appl. **7**(1), 36–44 (2014). https://doi.org/10.1134/S1995423914010042
12. Vikhtenko, E.M., Woo, G., Namm, R.V.: Sensitivity functionals in contact problems of elasticity theory. Comput. Math. Math. Phys. **54**(7), 1218–1228 (2014). https://doi.org/10.1134/S0965542514070112
13. Vikhtenko, E.M., Namm, R.V.: On duality method for solving model crack problem. Tr. IMM UrO RAN **22**, 36–43 (2016)
14. Hlaváček, I., Haslinger, Ya., Nechas, I., Lovišhek, Ya.: Numerical Solution of Variational Inequalities. Springer, New York (1988)
15. Kufner, A., Fuchik, S.: Nonlinear Differential Equations. Nauka, Moscokw (1988)
16. Vasiliev, F.P.: Methods for Solving Extremal Problems. Nauka, Moscow (1981)
17. Vikhtenko, E.M., Woo, G., Namm, R.V.: The methods for solution semi-coercive variational inequalities of mechanics on the basis of modified Lagrangian functionals. Dalnevos. Mat. Zh. **14**, 6–17 (2014)
18. Vikhtenko, E.M., Woo, G., Namm, R.V.: Modified dual scheme for finite-dimensional and infinite-dimensional convex optimization problems. Dalnevos. Mat. Zh. **17**, 158–169 (2017)
19. Mangasarian, O.L.: A generalized Newton method for absolute value equations. Optim. Lett. **3**(1), 101–108 (2009). https://doi.org/10.1007/s11590-008-0094-5
20. Shared Facility Center "Data Center of FEB RAS" (Khabarovsk). http://lits.ccfebras.ru

Subgradient Method with Polyak's Step in Transformed Space

Petro Stetsyuk[1] , Viktor Stovba[1(✉)] , and Zhanna Chernousova[2]

[1] V.M. Glushkov Institute of Cybernetics of NAS of Ukraine,
40 Glushkov ave., Kyiv 03187, Ukraine
stetsyukp@gmail.com, vik.stovba@gmail.com
[2] National Technical University of Ukraine "Igor Sikorsky Kyiv Polytechnic
Institute", 37 Prosp. Peremohy, Kyiv 03056, Ukraine
chernjant@ukr.net

Abstract. We consider two subgradient methods (methods A and B) for finding the minimum point of a convex function for the known optimal value of the function. Method A is a subgradient method, which uses the Polyak's step in the original space of variables. Method B is a subgradient method in the transformed space of variables, which uses Polyak's step in the transformed space. For both methods a proof of the convergence of finding the minimum point with a given accuracy by the value of the function was performed. Examples of ravine convex (smooth and non-smooth) functions are given, for which convergence of method A is slow. It is shown that with a suitable choice of the space transformation matrix method B can be significantly accelerated in comparison with method A for ravine convex functions.

Keywords: Subgradient method · Polyak's step
Space transformation

1 Introduction

Let $f(x)$ be a convex function, $x \in R^n$. Denote its minimal value as $f^* = f(x^*)$ and without loss of generality assume, that the point x^* is the unique minimum point. A subgradient $g_f(x)$ satisfies the following condition:

$$(x - x^*, g_f(x)) \geq f(x) - f^*, \quad \forall x \in R^n. \tag{1}$$

Here (x, y) is the scalar product of vectors $x \in R^n$ and $y \in R^n$.

If $f(x)$ is continuously differentiable at the point $\overline{x}$, then the subgradient $g_f(\overline{x})$ is determined uniquely and coincides with $\nabla f(\overline{x})$ – the gradient of the function $f(x)$ at the point $\overline{x}$. At non-smoothness points of function $f(x)$, the subgradient $g_f(\overline{x})$ is not uniquely determined.

Supported by Volkswagen Foundation (grant No. 90 306).

Y. Evtushenko et al. (Eds.): OPTIMA 2018, CCIS 974, pp. 49–63, 2019.
https://doi.org/10.1007/978-3-030-10934-9_4

The inequality (1) follows from the definition of the subgradient $g_f(x)$ for the convex function $f(x)$. Indeed, the subgradient $g_f(x)$ at the point x satisfies the inequality

$$f(y) - f(x) \geq (g_f(x), y - x) \quad \text{for all} \quad y \in R^n. \tag{2}$$

The inequality (2) also holds for x^* – the minimum point. For x^* it turns into inequality $f(x^*) - f(x) \geq (g_f(x), x^* - x)$ which, taking into account the fact that $f(x^*) = f^*$, can be written as inequality (1).

If f^* is known, then to find an approximation to the point $x^* \in R^n$ one can use the Polyak's subgradient method [1]:

$$x_{k+1} = x_k - h_k \frac{g_f(x_k)}{\|g_f(x_k)\|}, \quad h_k = \frac{f(x_k) - f^*}{\|g_f(x_k)\|}, \quad k = 0, 1, 2, \ldots. \tag{3}$$

The step h_k is called the Polyak's step (or the Agmon-Motzkin-Schoenberg step). For the first time, Polyak's step was used for minimization of piecewise linear convex functions. In 1954 Agmon [2] and Motzkin, Schoenberg [3] used this step in relaxation method for finding at least one of the solutions of feasible system of linear inequalities. In 1965 Eremin [4] generalized this relaxation method for the systems of convex inequalities.

The geometric sense of the method (3) is as follows. The function $f(x)$ is approximated by a linear function $\tilde{f}(x) = f(x_k) + (g_f(x_k), x - x_k)$ and the step is selected so that this approximation function becomes equal to f^* (i.e. $\tilde{f}(x_{k+1}) = f^*$). For convex function $f(x)$ the step h_k determines the value of the maximum shift from the point x_k in the direction of the normalized anti-subgradient, for which the condition (1) guarantees that the angle between the antisubgradient at the point x_k and the direction from the point x_{k+1} to the minimum point x^* will not be obtuse. It means the following fact. Subgradient at the point x_k defines the hyperplane localizing the point x^* in the half-space in the direction of the antisubgradient. If it is moved by the value of the maximal shift in the direction of the normalized antisubgradient, new hyperplane localizes x^* in the half-space in regard to the point x_{k+1}.

Consider the Polyak's subgradient method (method A, Sect. 2) and its accelerated version (method B, Sect. 3) for finding an approximation to the minimum point of ravine convex functions. As a stopping criterion we use condition $f(x_k) - f^* \leq \varepsilon$; for an arbitrarily small $\varepsilon > 0$ it allows us to find the point $x_\varepsilon^* = x_k$ such that $f(x_\varepsilon^*) \leq f^* + \varepsilon$. We consider methods A and B for more general case of a convex function $f(x)$, when its subgradient $g_f(x)$ satisfies the following condition:

$$(x - x^*, g_f(x)) \geq m(f(x) - f^*), \quad \forall x \in R^n, \tag{4}$$

where parameter $m \geq 1$. The parameter m is introduced to take into account special classes of convex functions: for example, for a piecewise linear non-smooth function $m = 1$, for quadratic smooth function $m = 2$, for functions $f(x) = \sum_{i=1}^{k} \left| \sum_{j=1}^{n} a_{ij} x_j - b_i \right|^p$, where $p > 1$, $m = p$. Parameter m can be actively used for differentiable homogeneous of degree σ functions. The equality $\sigma(f(x)-f^*) = (x - x^*, g_f(x))$ is true for them, so the parameter m can be chosen $m = \sigma > 1$.

2   Method A: The Polyak's Subgradient Method

2.1   Description of Method A

If the convex function $f(x)$ satisfies the condition (4) and f^* is known, then to find the point $x_\varepsilon^* \in R^n$ such that $f(x_\varepsilon^*) \leq f^* + \varepsilon$, one can use the following iterative method.

Initialization. Let f^* and $m \geq 1$ be given. Let's select the starting point $x_0 \in R^n$, the value $\varepsilon > 0$ and go to the next iteration with the value x_0.

Iterative Process. Let the point $x_k \in R^n$ be found on the k-th iteration. To proceed to the $(k + 1)$-th iteration, we perform the following actions.

A1. Calculate $f(x_k)$ and $g_f(x_k)$. If $f(x_k)-f^* \leq \varepsilon$, then STOP ($k^* = k, x_\varepsilon^* = x_k$).
A2. Calculate the next point

$$x_{k+1} = x_k - h_k \frac{g_f(x_k)}{\|g_f(x_k)\|}, \qquad h_k = \frac{m(f(x_k) - f^*)}{\|g_f(x_k)\|},$$

A3. Go to the $(k + 1)$-th iteration with x_{k+1}.

Theorem 1. *The sequence $\{x_k\}_{k=0}^{k^*-1}$ generated by the method A satisfies the inequalities*

$$\|x_{k+1} - x^*\|^2 \leq \|x_k - x^*\|^2 - \frac{m^2(f(x_k)-f^*)^2}{\|g_f(x_k)\|^2}, \qquad k = 0, 1, 2, \ldots \qquad (5)$$

Proof. From A2 for an arbitrary k $(0 \leq k \leq k^* - 1)$ we have

$$\|x_{k+1} - x^*\|^2 = \left\| x_k - x^* - h_k \frac{g_f(x_k)}{\|g_f(x_k)\|} \right\|^2$$

$$= \|x_k - x^*\|^2 - 2h_k \frac{(x_k - x^*, g_f(x_k))}{\|g_f(x_k)\|} + h_k^2.$$

Taking into account that from (4) it follows

$$\frac{(x_k - x^*, g_f(x_k))}{\|g_f(x_k)\|} \geq \frac{m(f(x_k) - f^*)}{\|g_f(x_k)\|} = h_k,$$

we have

$$\|x_{k+1} - x^*\|^2 \leq \|x_k - x^*\|^2 - h_k^2 = \|x_k - x^*\|^2 - \left(\frac{m(f(x_k) - f^*)}{\|g_f(x_k)\|} \right)^2,$$

that gives the inequalities (5). Theorem is proved.

Theorem 1 guarantees that in Polyak's subgradient method the distance to the minimum point decreases monotonically. Besides, the following inequalities are satisfied:

$$(x^* - x_{k+1}, -g_f(x_k)) \geq 0, \quad k = 0, 1, \ldots, \tag{6}$$

Really, the inequalities (6) follow from the fact, that using (4) we have

$$(x^* - x_{k+1}, -g_f(x_k)) = (x_{k+1} - x^*, g_f(x_k))$$
$$= (x_k - x^* - h_k \frac{g_f(x_k)}{\|g_f(x_k)\|}, g_f(x_k)) = (x_k - x^*, g_f(x_k)) - h_k\|g_f(x_k)\|$$
$$= (x_k - x^*, g_f(x_k)) - m(f(x_k) - f^*) \geq 0.$$

The inequalities (6) mean that for convex function $f(x)$, satisfying the condition (4), h_k determines the value of the maximum shift in the direction of the normalized antisubgradient. It is hereby guaranteed that angle between the antisubgradient and the direction from point x_{k+1} to the minimum point x^* will not be obtuse.

2.2   Octave Function PolyakA

The Octave program **PolyakA** finds an approximation x_ε^* to the minimum point of a convex function $f(x)$ of n variables, which for the method A is determined by the following input data: the starting point x_0; f^* - value of function at the minimum point; $m \geq 1$ – parameter, defining length of shift along antisubgradient, i.e. satisfying the condition (4); stop parameters ε_f and **maxitn**.

The program uses the octave-function having form **function [f, g] = calcfg (x)**, which calculates the value of the function $f = f(x)$ and its subgradient $g_f(x)$ at the point x, $g_f(x) \in \partial f(x)$. The name of the function **calcfg (x)** can be arbitrary, which octave syntax permits. The program **PolyakA** uses the following input and output parameters.

```
# Program PolyakA for the method A (P.Stetsyuk, June 23, 2018)
# Input parameters:
# calcfg - reference to function for calculation of f(x) and g(x)
# x0 - the starting point, x0(1:n)
# fstar - value of the function at the minimum point
# m - length of shift along anti-subgradient (m>=1)
# epsf, maxitn - stop parameters
# intp - print information every intp iteration
# Output parameters:
# x - the minimum point, which was found by the program, x(1:n)
# f - the value of the function f at the point x
# itn - the number of iterations used by the program
# info - exit code (0 = epsf, 4 = maxitn)
```

The status of the point x is defined by the return code **info** at iteration **itn** = **k**: **info**=**0** – stop if the point x_k was found, for which $f(x_k) - f^* \leq \varepsilon_f$; **info**=**0** – stop if **itn** > **maxitn** (the process was exceeded the maximum number of iterations).

The program **PolyakA** is an octave-function and its code is presented below.

```
function [x,f,itn,info] = PolyakA(calcfg,x0,fstar,m,      #row01
                          epsf,maxitn,intp);              #.....
itn=0; x=x0; [f,g] = calcfg(x); dg=norm(g);              #row02
if(intp>0)                                                #row03
    printf("itn %4d f %14.6e \n", itn, f); # xprint = x', #.....
endif                                                     #.....
for(itn = 1:maxitn)                                       #row04
    if(f-fstar < epsf) info = 0; return; endif            #row05
    g1=g/dg; hs=m*(f-fstar)/dg;                           #row06
    x -= hs * g1;                                         #row07
    [f,g] = calcfg(x); dg=norm(g);                        #row08
    if(mod(itn,intp)==0)                                  #row09
        printf("itn %4d f %14.6e \n",itn,f); # xprint = x', #.....
    endif                                                 #.....
endfor                                                    #row10
info = 4;                                                 #row11
endfunction                                               #row12
```

The code consists of **14** lines and most of them contain more than one **octave** operator. The iterative process is executed in a loop **for** (lines **4–10**), where for k-th iteration subgradient $g_f(x_k)$ is stored as a column vector **g**, and normalized subgradient $g_f(x_k)$ is stored as a column vector **g1**. In the loop **for** stop condition is: deviation of function value from optimal value is less than ε_f (line **5**).

Using the program **PolyakA**, one can find reasonably accurate approximations for the minimum point of smooth convex function, surface levels of which are characterized by a small degree of elongation. The method A converges fast to the minimum point of a convex function if angles between successive subgradients on every iteration are acute, or obtuse being close to the right angle.

The disadvantage of the method A is its slow convergence for ravine functions. Below we consider in more detail the convergence of the method A for functions $f_1(x_1, x_2) = |x_1| + t|x_2|$, $t > 1$, and $f_2(x_1, x_2) = x_1^2 + tx_2^2$, $t \gg 1$.

2.3   Computational Experiments for Ravine Functions

In this section we describe the results of computational experiments using the program **PolyakA**. The experiments are connected with the study of the method A convergence rate for ravine convex functions of two variables – piecewise linear function $f_1(x_1, x_2) = |x_1| + t|x_2|$, $t > 1$, $f^* = 0$, $x^* = (0, 0)^T$ (problem **sabs**) and quadratic function $f_2(x_1, x_2) = x_1^2 + tx_2^2$, $t \gg 1$, $f^* = 0$, $x^* = (0, 0)^T$ (problem **quad**). The calculations were conducted on a Pentium 2.5 GHz computer with system WindowsXP/32 using GNU Octave version 3.0.0. For the problems **sabs** and **quad** we used the octave-functions for calculation of $f(x)$ and $g_f(x)$ in the next form

```
function [f,g] = squad(x)          |   function [f,g] = sabs(x)
global t;                          |   global t;
f=x(1,1)*x(1,1)+t*x(2,1)*x(2,1);   |   f=abs(x(1,1))+t*abs(x(2,1));
g(1,1) = 2*x(1,1);                 |   g(1,1) = sign(x(1,1));
g(2,1) = 2*t*x(2,1);               |   g(2,1) = t*sign(x(2,1));
endfunction                        |   endfunction
```

In case of non-smooth function in two variables $f_1(x_1, x_2) = |x_1| + t|x_2|$, $t > 1$, the rate of convergence of the method A is determined by the geometric progression with the common ratio

$$q = \sqrt{1 - 1/t^2} \qquad (7)$$

and will be very slow for large values of t. Thus, for example, the zigzag-shaped trajectory of the method A for finding the minimum point of the slightly ravine function $f_1(x_1, x_2) = |x_1| + 5|x_2|$ is illustrated in Fig. 1.

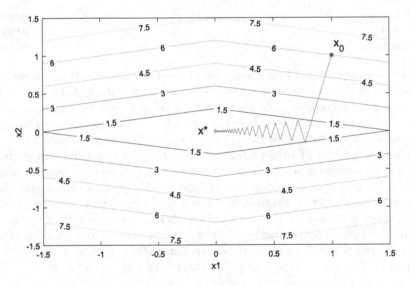

Fig. 1. The trajectory of the method A searching the approximation to the minimum point of function $f_1(x_1, x_2) = |x_1| + 5|x_2|$: $m = 1$, $f^* = 0$, $x_0 = (1, 1)^T$

An analogous situation appears when minimizing the ravine quadratic function $f(x_1, x_2) = x_1^2 + tx_2^2$ (see Fig. 2). For instance, if $t = 6$, $m = 2$ and $\varepsilon = 0.01$, then the method A generates the sequence $x_0 = (1.00, 1.00)^T$, $x_1 = (0.811, -0.135)^T$, $x_2 = (0.338, 0.338)^T$, $x_3 = (0.274, -0.046)^T$, $x_4 = (0.114, 0.114)^T$, $x_5 = (0.093, -0.015)^T$ (see Fig. 2).

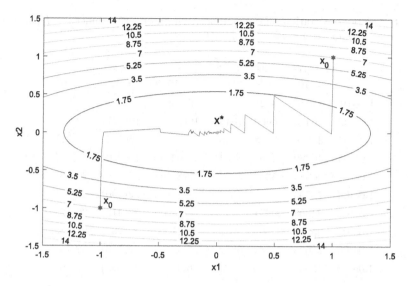

Fig. 2. Two trajectories of the method A searching the approximation to the minimum point of the quadratic function $f_2(x_1, x_2) = x_1^2 + 6x_2^2$: (1) $m = 2$, $f^* = 0$, $x_0 = (1, 1)^T$, (2) $m = 1$, $f^* = 0$, $x_0 = (-1, -1)^T$

Table 1 shows the results of the method A for finding the minimum points of functions $f_1(x_1, x_2) = |x_1| + t|x_2|$ and $f_2(x_1, x_2) = x_1^2 + tx_2^2$ at different elongation degrees, which are determined by the parameter t. The first column contains the accuracies from 10^{-1} to 10^{-10}, with which the approximations to the minimum point are calculated. The columns 2, 3, 4 contain the number of iterations needed for minimizing piecewise linear function $f_1(x_1, x_2) = |x_1| + t|x_2|$ with $t = 100$, 50, 25 respectively. The columns 5, 6, 7 show the numbers of iterations for the quadratic function $f_2(x_1, x_2) = x_1^2 + tx_2^2$ with $t = 10000, 1000, 100$, where the number of iterations is given in brackets if the parameter $m = 1$ is used for the method A, instead of $m = 2$. From the columns 5–7 it is easy to see that the number of iterations with $m = 1$ is much greater than the number of iterations with $m = 2$. The situation is similar for the convex functions $f(x_1, x_2) = x_1^4 + 10000x_2^4$ and $f(x_1, x_2) = (x_1 + 1.001x_2)^4 + (1.001x_1 + x_2)^4$, for which $m = 4$. Thus, with the accuracy $\varepsilon_f = 10^{-20}$ for the first function to find the minimum point it is needed 4 iterations if $m = 4$ and 50 iterations if $m = 1$. For the second function it is needed 2 iterations if $m = 4$ and 45 iterations if $m = 1$.

From Table 1 we see the slow convergence of the method A if the functions $f_1(x_1, x_2)$ and $f_2(x_1, x_2)$ are ravine. The greater the elongation of the level surfaces to which the larger values of the parameter t correspond, the slower the convergence of the method A. As the accuracy for finding an approximation to the minimum point increases, the number of iterations increases significantly. Moreover, the situation will be worse for a piecewise linear function $f_1(x_1, x_2)$. So, for example, to achieve an accuracy of 10^{-10} at $t = 100$ it took 118547 iterations. However, when the elongation of the level surfaces was reduced by the

Table 1. Number of iterations of the method A for finding x_ε^* of functions $f_1(x_1, x_2) = |x_1| + t|x_2|$ and $f_2(x_1, x_2) = x_1^2 + tx_2^2$: $f^* = 0$, $x_0 = (1, 1)^T$

ε_f	$f_1(x_1, x_2) = |x_1| + t|x_2|$			$f_2(x_1, x_2) = x_1^2 + tx_2^2$		
	$t = 100$	$t = 50$	$t = 25$	$t = 10000$	$t = 1000$	$t = 100$
1.0e-01	14930	3721	925	6(139)	6(97)	6(9)
1.0e-02	26443	6600	1645	10(295)	10(127)	10(29)
1.0e-03	37956	9478	2365	12(551)	12(194)	12(51)
1.0e-04	49469	12356	3084	16(646)	16(230)	16(68)
1.0e-05	60982	15234	3804	20(885)	20(324)	20(86)
1.0e-06	72495	18113	4523	22(1003)	22(390)	22(98)
1.0e-07	84008	20991	5243	26(1135)	26(418)	26(121)
1.0e-08	95521	23869	5962	30(1332)	30(485)	28(135)
1.0e-09	107034	26747	6682	32(1518)	32(513)	32(150)
1.0e-10	118547	29625	7401	36(1947)	36(543)	36(169)

factor of 4, in order to achieve the same accuracy it was necessary to use substantially fewer iterations: a total of 7401. In case of minimizing the quadratic function, the situation is somewhat different: for $m = 2$, the number of iterations does not practically depend on the degree of elongation. For example, to achieve the same accuracy of 10^{-10}, it took only 36 iterations. If we use the parameter $m = 1$ for quadratic functions, then the number of iterations increases as the parameter t increases.

Below we consider the modification of the subgradient method with Polyak's step, where the accelerated convergence with respect to the method A can be provided by choosing the space transformation matrix.

3 Method B: Subgradient Method with Polyak's Step in the Transformed Space of Variables

3.1 Polyak's Step in the Transformed Space of Variables

We will consider the subgradient Polyak's step in the transformed space of variables, following the Shor's second idea [5] for accelerating the methods by using linear space transformations. This idea gave ellipsoid method and r-algorithms (subgradient methods with space dilation in the direction of successive subgradients) for convex optimization [6–8].

Let's make the substitution of variables $x = By$, where B is a nonsingular $n \times n$-matrix (that is, there exists an inverse matrix $A = B^{-1}$). The subgradient $g_\varphi(y)$ of the convex function $\varphi(y) = f(By)$ at the point $y = Ax$ of the transformed space of variables satisfies the inequality

$$(y - y^*, g_\varphi(y)) \geq m(\varphi(y) - \varphi^*), \quad \forall y \in R^n, \tag{8}$$

where $g_\varphi(y) = B^T g_f(x)$, $\varphi^* = \varphi(y^*) = f(By^*)$, $y^* = Ax^*$. Indeed, since $A = B^{-1}$ and $x = By$, the inequality (4) can be rewritten in the form

$$(A(x - x^*), B^T g_f(x)) \geq m(f(By) - f(By^*)), \quad \forall By \in R^n,$$

whence we obtain the inequality (8).

To find the point x^*, the subgradient method with the Polyak's step in the transformed space (defined by the nonsingular matrix B) has the following form:

$$x_{k+1} = x_k - h_k B \frac{B^T g_f(x_k)}{\|B^T g_f(x_k)\|}, \quad h_k = \frac{m(f(x_k) - f^*)}{\|B^T g_f(x_k)\|}, \quad k = 0, 1, 2, \ldots. \quad (9)$$

Here h_k is the Polyak's step (the Agmon-Motzkin-Schoenberg step), but in the transformed space of variables $y = Ax$. This follows from the fact that in the transformed space of variables the method (9) is written as a subgradient process

$$y_{k+1} = y_k - h_k \frac{g_\varphi(y_k)}{\|g_\varphi(y_k)\|}, \quad h_k = \frac{m(\varphi(y_k) - \varphi^*)}{\|g_\varphi(y_k)\|}, \quad k = 0, 1, 2, \ldots. \quad (10)$$

The Polyak's step in the transformed space of variables has the same properties as the Polyak's step in the original space. They are defined by the minimum value of the function and the inequality (8) associated with φ^*.

3.2   Description of Method B

To find the point $x_\varepsilon^* \in R^n$ for which $f(x_\varepsilon^*) \leq f^* + \varepsilon$, the subgradient method with the Polyak's step in the transformed space is represented by the following iterative procedure.

Initialization. We have f^* and $m \geq 1$. We choose the point $x_0 \in R^n$, the nonsingular $n \times n$ matrix B, and the value $\varepsilon > 0$. We move to the next iteration with the value x_0.

Iterative Process. Let $x_k \in R^n$ be found on the k-th iteration. To proceed to the $(k + 1)$-th iteration, we perform the following actions.

B1. Calculate $f(x_k)$ and $g_f(x_k)$. If $f(x_k) - f^* \leq \varepsilon$, then STOP ($k^* = k$, $x_\varepsilon^* = x_k$).
B2. Calculate the next point

$$x_{k+1} = x_k - h_k B \frac{B^T g_f(x_k)}{\|B^T g_f(x_k)\|}, \quad h_k = \frac{m(f(x_k) - f^*)}{\|B^T g_f(x_k)\|},$$

B3. Go to the $(k + 1)$-th iteration with x_{k+1}.

Theorem 2. *The sequence $\{x_k\}_{k=0}^{k^*-1}$ generated by the method B satisfies the inequalities*

$$\|A(x_{k+1} - x^*)\|^2 \leq \|A(x_k - x^*)\|^2 - \frac{m^2(f(x_k) - f^*)^2}{\|B^T g_f(x_k)\|^2}, \quad k = 0, 1, \ldots \quad (11)$$

Proof. From B2 for an arbitrary k $(0 \leq k \leq k^* - 1)$ we have

$$\|A(x_{k+1} - x^*)\|^2 = \left\| A(x_k - x^*) - h_k \frac{B^T g_f(x_k)}{\|B^T g_f(x_k)\|} \right\|^2$$

$$= \|A(x_k - x^*)\|^2 - 2h_k \frac{(x_k - x^*, g_f(x_k))}{\|B^T g_f(x_k)\|} + h_k^2.$$

Taking into account that from (4) it follows the inequality

$$\frac{\left(x_k - x^*, g_f(x_k)\right)}{\|B^T g_f(x_k)\|} \geq \frac{m(f(x_k) - f^*)}{\|B^T g_f(x_k)\|} = h_k,$$

we have

$$\|A(x_{k+1} - x^*)\|^2 \leq \|A(x_k - x^*)\|^2 - h_k^2 = \|A(x_k - x^*)\|^2 - \left(\frac{m(f(x_k) - f^*)}{\|B^T g_f(x_k)\|} \right)^2,$$

which gives the inequalities (11). Theorem is proved.

Theorem 2 guarantees that in the subgradient method with Polyak's step in the transformed space of variables the distance to the minimum point decreases monotonically in the transformed space. Besides, the following inequalities are satisfied:

$$(A(x^* - x_{k+1}), -B^T g_f(x_k)) \geq 0, \quad k = 0, 1, \dots, \tag{12}$$

which can be written as inequalities

$$(y^* - y_{k+1}, g_\varphi(y_k)) \geq 0, \quad k = 0, 1, \dots . \tag{13}$$

Really, the inequalities (12) follow from the fact, that using (8) and (10) we have

$$(x^* - x_{k+1}, -g_f(x_k)) = (x_{k+1} - x^*, g_f(x_k))$$

$$= (x_k - x^* - h_k B \frac{B^T g_f(x_k)}{\|B^T g_f(x_k)\|}, g_f(x_k)) = (x_k - x^*, g_f(x_k)) - h_k \|B^T g_f(x_k)\|$$

$$= (x_k - x^*, g_f(x_k)) - m(f(x_k) - f^*) \geq 0.$$

The inequalities (13) mean that for convex function $\varphi(x)$, satisfying the condition (8), the step h_k determines the value of the maximum shift in the direction of the normalized antisubgradient. It is hereby guaranteed, that the angle between the antisubgradient and the direction from the point y_{k+1} to the minimum point y^* will not be obtuse in the transformed space of variables.

3.3   Octave Function PolyakB and Computational Experiments

The Octave function **PolyakB** finds an approximation x_ε^* of a convex function $f(x)$, which for the method B is determined by the following input data: B – $n \times n$-matrix for transformation of space; x_0 – the starting point; f^* – value of the function at the minimum point; $m \geq 1$ – parameter, satisfying the condition (4); stop parameters ε_f and **maxitn**. Its code is presented below.

```
# Program PolyakB for the method B (P.Stetsyuk, June 23, 2018)
# Input parameters:
# calcfg - name of the function for calculation of f(x) and g(x)
# B - n*n-matrix for transformation of space
# x0 - the starting point, x0(1:n)
# fstar - value of the function at the minimum point
# m - length of shift along anti-subgradient (m>=1)
# epsf, maxitn - stop parameters
# intp - print information every intp iteration
# Output parameters:
# x - the minimum point, which was found by the program, x(1:n)
# f - the value of the function f at the point x
# itn - the number of iterations used by the program
# info - exit code (0 = epsf, 4 = maxitn)
function [x,f,itn,info] = PolyakB(calcfg,B,x0,fstar,m,      #row01
                                  epsf,maxitn,intp);        #.....
itn=0; x=x0; [f,g] = calcfg(x);                             #row02
if(intp>0)                                                  #row03
    printf("itn %4d f %14.6e \n", itn, f);  # xprint = x',  #.....
endif                                                       #.....
for(itn = 1:maxitn)                                         #row04
    if(f-fstar < epsf) info = 0; return; endif             #row05
    g1=B'*g; dg1=norm(g1);                                  #row06
    g2=g1/dg1; hs=m*(f-fstar)/dg1;                          #row07
    x -= hs * B * g2;                                       #row08
    [f,g] = calcfg(x); dg=norm(g);                          #row09
    if(mod(itn,intp)==0)                                    #row10
        printf("itn %4d f %14.6e \n",itn,f);  # xprint = x',#.....
    endif                                                   #.....
endfor                                                      #row11
info = 4;                                                   #row12
endfunction                                                 #row13
```

Using the program **PolyakB** one can find reasonably accurate approximations for the minimum point of a convex function. If the matrix B is chosen such that in the transformed space of variables the level surfaces of ravine functions are elongated less than in the original space of variables, then the method B will converge faster than the method A. For example, if the matrix $B = \mathrm{diag}(1; 0.5)$, then for function $f_1(x_1, x_2) = |x_1| + t|x_2|$, $t > 1$, the convergence rate of the method B is determined by the geometric progression with the common ratio $q' = \sqrt{1 - 4/t^2}$, which is less than $q = \sqrt{1 - 1/t^2}$ for the method A (see the formula (7)). If $m = 2$, $B = \mathrm{diag}(1; 0.7)$ and $\varepsilon = 0.01$, then while minimizing function $f_2(x_1, x_2) = x_1^2 + 6x_2^2$ the method B generates sequence $x_0 = (1.00, 1.00)^T$, $x_1 = (0.624, -0.104)^T$, $x_2 = (0.136, 0.136)^T$, $x_3 = (0.085, -0.014)^T$.

For ravine functions the subgradient Polyak's method with the space transformation (method B) will be more effective than the subgradient Polyak's method without space transformation (method A). This is consistent with the number of iterations of method B for finding ten successively refined approximations to the minimum point of the function $f_1(x_1, x_2) = |x_1| + 10|x_2|$ for six different matrices B, each of which corresponds to its column in Table 2. The matrices B are obtained as a result of dilation of space of variables in the direction x_2 with dilation coefficients $\alpha = 1$; 1.5; 2; 3; 4; 5. The matrix B has the form

$$B = \begin{pmatrix} 1 & 0 \\ 0 & \frac{1}{\alpha} \end{pmatrix}$$

and if $\alpha = 1$, then it coincides with the identity matrix. The case $\alpha = 1$ corresponds to method A.

Table 2. Number of iterations of the method B for minimization of the function $f_1(x_1, x_2) = |x_1| + 10|x_2|$, $x_0 = (1, 1)^T$

ε_f	$\alpha = 1$	$\alpha = 1.5$	$\alpha = 2$	$\alpha = 3$	$\alpha = 4$	$\alpha = 5$
1.0e-01	147	63	33	6	10	9
1.0e-02	262	114	62	19	17	13
1.0e-03	377	165	91	31	24	18
1.0e-04	492	216	119	44	31	22
1.0e-05	607	268	148	57	38	27
1.0e-06	722	319	177	70	45	31
1.0e-07	837	370	206	82	53	36
1.0e-08	952	421	234	95	60	40
1.0e-09	1068	472	263	108	67	45
1.0e-10	1183	523	292	121	74	49

From Table 2 we see that the number of iterations of the Polyak's method with space transformation decreases monotonically when the degree of ravine of function $\varphi_1(y_1, y_2) = |y_1| + \frac{10}{\alpha}|y_2|$ decreases in the transformed space of variables (corresponds to increasing dilation coefficient α).

The situation with speed of convergence will be more complex for essentially ravine function

$$f_3(x_1, x_2) = \max \left\{ x_1^2 + (2x_2 - 2)^2 - 3, x_1^2 + (x_2 + 1)^2 \right\}.$$

From Table 3 we see that there is no monotonous decrease in the number of iterations when the value of α, space dilation coefficient in the direction x_2, increases. But there is some gap that occurs at the coefficient value $\alpha = 2$ (corresponds to column with $\alpha = 2$). Figure 3 shows the trajectories of the method A ($\alpha = 1$) and the method B ($\alpha = 2$) using solid and dashed lines, respectively. On the trajectory of the method B all the points obtained by the

Table 3. Number of iterations of the method B for minimization of the function $f_3(x_1, x_2) = \max\{x_1^2 + (2x_2 - 2)^2 - 3, x_1^2 + (x_2 + 1)^2\}$, $x_0 = (1,1)^T$, **maxitn** = 100000

ε_f	$\alpha = 1$	$\alpha = 1.5$	$\alpha = 2$	$\alpha = 3$	$\alpha = 4$	$\alpha = 5$
1.0e-01	16	4	4	5	7	8
1.0e-02	162	37	4	6	7	9
1.0e-03	1604	679	5	6	8	9
1.0e-04	16004	7079	5	6	9	446
1.0e-05	–	71079	6	8	8061	6206
1.0e-06	–	–	6	–	98061	63806
1.0e-07	–	–	6	–	–	–

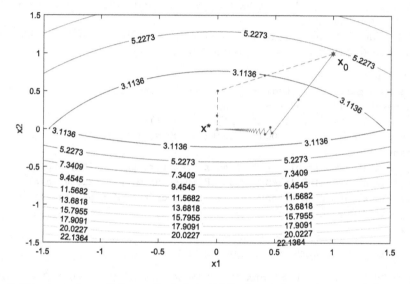

Fig. 3. The trajectory of the method A (solid line) and the trajectory of the method B (dashed line) for finding the approximation to the minimum point of piecewise quadratic function $f_3(x_1, x_2) = \max\{x_1^2 + (2x_2 - 2)^2 - 3, x_1^2 + (x_2 + 1)^2\}$: $m = 1$, $f^* = 1$, $x_0 = (1,1)^T$, $B = \mathrm{diag}(1; 0.5)$, $x^* = (0,0)$

algorithm are drawn, on the trajectory of the method A the first four points are drawn.

Consider the use of the PolyakA and PolyakB programs for minimizing convex quadratic function $f_4(x) = \|Ax - b\|^2$ from variables $x \in R^n$, where $A = \|a_{ij}\|_{i,j=1}^{l,n}$ is an arbitrary $l \times n$-matrix and l-dimensional vector b is such that its components $b_i = \sum_{j=1}^n a_{ij}$, $i = 1, \ldots, l$. For the function, $f^* = 0$, and if the matrix A has full rank, then the function has a unique minimum point $x^* = (1, 1, \ldots, 1)^T$. The octave-function for calculation of the values of the function $f_4(x)$ and its gradient has the form

```
function [f,g] = fgfun4(x)
global A b;
temp=A*x- b;
f = temp'*temp;
g=2*A'*temp;
endfunction
```

Let us consider the test example with 500×100-matrix having the following form

$$A = \left\| \begin{array}{cccc} 100 & 0 & 0...0 \\ 0 & 100 & 0...0 \\ & A_1 & \end{array} \right\|,$$

where A_1 is the 498×100-matrix that is generated using a random number generator from the range $[0,3]$.

The code of octave-program for comparing methods A and B where the matrix B is $diag(0.1; 0.1; 1; ...1)$ is the following

```
global A b;
n=100; x0 = zeros(n,1); B=diag([0.1 0.1 ones(1,n-2)]);
rand("seed", 2018); A1=3*rand(498,n);
A=[diag([100 100]) zeros(2,n-2); A1]; b=sum(A')';
m=2; fstar=0.d0; epsf = 0.0001; maxitn = 50000; intp=1000;
for(itn = 1:6)
  [xA,fA,itnA,infoA]=PolyakA(@fgfun4,x0,fstar,m,epsf,maxitn,intp);
  [xB,fB,itnB,infoB]=PolyakB(@fgfun4,B,x0,fstar,m,epsf,maxitn,intp);
  epsf, itnA, dnA=norm(xA-ones(n,1)), itnB, dnB=norm(xB-ones(n,1)),
  epsf=epsf/100;
endfor
```

The result of its work is the following table

epsf=1.0e-04	itnA= 695	dnA=8.0612e-04	itnB= 86	dnB=1.2175e-04
epsf=1.0e-06	itnA=1085	dnA=8.2861e-05	itnB=108	dnB=1.1773e-05
epsf=1.0e-08	itnA=1491	dnA=8.3470e-06	itnB=130	dnB=1.1353e-06
epsf=1.0e-10	itnA=1901	dnA=8.3881e-07	itnB=150	dnB=1.3526e-07
epsf=1.0e-12	itnA=2313	dnA=8.3994e-08	itnB=172	dnB=1.3018e-08
epsf=1.0e-14	itnA=2725	dnA=8.4440e-09	itnB=194	dnB=1.2523e-09

from which we see that method B requires significantly fewer iterations than method A. For example, to achieve the accuracy 10^{-14}, it is necessary to perform 2725 iterations for method A, while for method B it is necessary only 194 iterations.

4   Conclusion

In the paper, two subgradient methods for finding the minimum point of a convex function for the known optimal value of the function have been considered. For both methods proof of the convergence to the minimum point with a given accuracy by the value of the function was presented. It has been shown that for ravine convex functions one can significantly accelerate the second method in comparison with the first method.

The considered methods can be used to find the feasible point of a consistent system of convex inequalities

$$f_i(x) \le 0, \quad i = 1, \ldots, l, \quad x \in R^n.$$

This problem is equivalent to the minimization of the non-smooth convex function $\psi(x) = \max\{0, \max_{1 \le i \le l} f_i(x)\}$, for which $\psi^* = 0$. A particular case of the system of convex inequalities is the system of linear inequalities. The problems of linear programming can be reduced to this case using constraints of the primal and dual problems. For minimizing $\psi(x)$ by the methods A and B the parameter $m = 1$ should be used. The parameter $m = 2$ can be used to find a solution to a consistent system of linear equations (either full rank or underdetermined or overdetermined).

The authors would like to thank Oleksii Lykhovyd and Volodymyr Zhydkov for their help in preparing this paper.

References

1. Polyak, B.T.: Minimization of unsmooth functionals. USSR Comput. Math. Math. Phys. **9**(3), 14–29 (1969)
2. Agmon, S.: The relaxation method for linear inequalities. Can. J. Math. **6**(3), 382–392 (1954)
3. Motzkin, T., Schoenberg, I.J.: The relaxation method for linear inequalities. Can. J. Math. **6**(3), 393–404 (1954)
4. Eremin, I.I.: Generalization of the Motzkin-Agmon relaxational method. Uspekhi Mat. Nauk. **20**(2), 183–187 (1965)
5. Sergienko, I.V., Stetsyuk, P.I.: On N.Z. Shor's three scientific ideas. Cybern. Syst. Anal. **48**(1), 2–16 (2012)
6. Shor, N.Z.: Minimization Methods for Non-Differentiable Functions. SSCM, vol. 3. Springer, Berlin (1985). https://doi.org/10.1007/978-3-642-82118-9
7. Shor, N.Z.: Nondifferentiable Optimization and Polynomial Problems. Kluwer Academic Publishers, Boston (1998)
8. Stetsyuk, P.I.: Methods of Ellipsoids and r-Algorithms. Eureka, Chisinau (2014)

Mirror Descent and Constrained Online Optimization Problems

Alexander A. Titov[1]([⊠]), Fedor S. Stonyakin[1,2], Alexander V. Gasnikov[1,3], and Mohammad S. Alkousa[1]

[1] Moscow Institute of Physics and Technologies, Moscow, Russia
{a.a.titov,mohammad.alkousa}@phystech.edu, gasnikov@yandex.ru
[2] V. I. Vernadsky Crimean Federal University, Simferopol, Russia
fedyor@mail.ru
[3] Adyghe State University, Caucasus Mathematical Center, Maykop, Russia

Abstract. We consider the following class of online optimization problems with functional constraints. Assume, that a finite set of convex Lipschitz-continuous non-smooth functionals are given on a closed set of n-dimensional vector space. The problem is to minimize the arithmetic mean of functionals with a convex Lipschitz-continuous non-smooth constraint. In addition, it is allowed to calculate the (sub)gradient of each functional only once. Using some recently proposed adaptive methods of Mirror Descent the method is suggested to solve the mentioned constrained online optimization problem with an optimal estimate of accuracy. For the corresponding non-Euclidean prox-structure, the case of a set of n-dimensional vectors lying on the standard n-dimensional simplex is considered.

Keywords: Online convex optimization
Non-smooth constrained optimization
Adaptive mirror descent · Non-euclidean prox-structure · Unit simplex

1 Introduction

Online convex optimization plays a key role in solving the problems, where statistical information is being updated [12,13]. There are a lot of examples of such problems, concerning internet network, consumer data sets or financial market. Quite a few branches of science also face the above-mentioned problems, for example, machine learning applications [14]. The important example is the decision-making problem [13,15]. Suppose, we are given N experts and range of admissible solutions lie on the unit simplex. Every expert gives his estimates of losses with the possible solution and the problem is to minimize total losses

The research by Alexander A. Titov (Sect. 5) and Fedor S. Stonyakin (Sect. 4) was supported by the Russian Science Foundation according to the research project 18-71-00048. The research by Alexander V. Gasnikov (Sect. 3) was supported by the Russian Foundation for Basic Research according to the research project 18-29- 03071 mk.

Y. Evtushenko et al. (Eds.): OPTIMA 2018, CCIS 974, pp. 64–78, 2019.
https://doi.org/10.1007/978-3-030-10934-9_5

from the point view of all experts (the arithmetic mean). Therefore, in recent years, methods for solving online optimization problems have been actively developed [8–14, 16].

In problems of online convex optimization, it is required to minimize the sum (or the arithmetic mean) of several convex Lipschitz functionals f_i ($i = \overline{1, N}$) given on some closed set $Q \subset \mathbb{R}^n$. It should be noted that it is possible to calculate the (sub)gradient $\nabla f_i(x)$ of each functional f_i only once. Our paper is devoted to some optimal methods for the following type of problems

$$
\begin{cases}
\frac{1}{N} \sum_{i=1}^{N} f_i(x) \to \min\limits_{x \in Q} \\
s.t. \quad g(x) \leq 0
\end{cases}
\tag{1}
$$

We assume that the functionals f_i and g satisfy the Lipschitz property, i.e. there exists a number $M > 0$, such that

$$
|g(x) - g(y)| \leq M \|x - y\|,
\tag{2}
$$

$$
|f_i(x) - f_i(y)| \leq M \|x - y\| \quad \forall i = \overline{1, N}.
\tag{3}
$$

We can explain the meaning of such formulation of the problem in the following situation. Suppose that we are engaged in some kind of activity during the fixed number of days. Each day can be productive or non-productive. We want to live out N productive days (not necessarily in a row, there can be some non-productive days within this period) so that the total nerve costs (characterized by $f_i(x)$) would be minimal. Note that we pay nervous expenses only in productive days when we try to do something. In non-productive days we do nothing, our aim is to return to the productive state, but we do not pay any costs. The productivity of the day is determined by the condition $g(x^k) \leq \varepsilon$. Let's define index i as the number of the productive day. This day we receive feedback from the outside world in the next form: $\nabla f_i(x^k)$ and using this information we build a strategy for the next day x^{k+1}. In non-productive days, we get information about how far have we gone out of the functional constraint and we try to return to this framework. There is no point in arranging unnecessary non-productive days. Therefore, it is also desirable to minimize the number of non-productive days for a given N. The proposed algorithm provides a small amount of costs simultaneously, ensuring that the number of non-productive days will be no more than $O(N)$.

The optimization problems of non-smooth functionals with constraints attract widespread interest in large-scale optimization and its applications [6, 23]. There are various methods of solving this kind of optimization problems. Some examples of these methods are: bundle-level method [19], penalty method [24], Lagrange multipliers method [7]. Among them, Mirror Descent (MD) [1, 4, 18] is viewed as a simple method for non-smooth convex optimization.

Note that a functional constraint, generally, can be non-smooth. That is why we consider subgradient methods. These methods have a long history starting from the method for deterministic unconstrained problems and Euclidean setting in [21] and the generalization for constrained problems in [20], where the idea of steps switching between the direction of subgradient of the objective and the direction of subgradient of the constraint was suggested. Non-Euclidean extension usually referred to as Mirror Descent, originated in [17,18] and was later analyzed in [4]. An extension for constrained problems was proposed in [18], see also a recent version in [3].

Usually, the step size and stopping rule for Mirror Descent requires to know the Lipschitz constant of the objective function and constraint, if any. Adaptive stepsizes, which do not require this information, are considered in [5] unconstrained problems, and in [3] for constrained problems. Recently, in [2] optimal algorithms of Mirror Descent for convex programming problems with Lipschitz functional constraints with both adaptive step selection and adaptive stopping criteria were proposed for a number of classes of problems. Also, there were considered some modifications of these methods for the case of problems with many functional constraints in [22]. In [14] authors considered adaptive algorithms for online convex optimization problem with constraints, but with only standard Euclidean prox-structure.

In this paper, we propose adaptive and non-adaptive algorithms for solving the problem (1). Note that we consider arbitrary proximal structure, which seems essential for the problem of experts [10–13]. The paper consists of an Introduction and five main sections. In Sect. 2 we give some basic notation concerning convex optimization problems with functional constraints and online optimization problems. In Sect. 3 we propose a non-adaptive algorithm of Mirror Descent for the considered online optimization problem (1). Section 4 is devoted to an adaptive analog of this method (Algorithm 2).

Also in Sect. 4, by analogy with [22], we propose a modification of Algorithm 2 for problems with several functional constraints (Algorithm 3). It is shown that Algorithms 1, 2 and 3 are optimal accurate to multiplication by constants under the condition of nonnegativity of the regret (see Theorems 1 and 2). In Sect. 5 the condition of negative regret is considered. In this case, we get the optimal quality of estimation by the objective function, but the estimation of the number of non-productive steps is worse than (19). In the last section, we consider some numerical experiments that allow us to compare the work of Algorithms 1, 2, and 3 for certain examples.

Summing up, contributions of this paper are as follows:

- two methods (adaptive and non-adaptive) were proposed to solve the online optimization problem for an arbitrary prox-structure;
- the number of non-productive steps is $O(N)$ in the case of nonnegative regret;
- the number of non-productive steps is $O(N^2)$, but the accuracy by regret is better.

2 Problem Statement and Standard Mirror Descent Basics

Let $(E, ||\cdot||)$ be a normed finite-dimensional vector space and E^* be the conjugate space of E with the norm:

$$||y||_* = \max_x \{\langle y, x \rangle, ||x|| \leq 1\},$$

where $\langle y, x \rangle$ is the value of the continuous linear functional y at $x \in E$.

Let $Q \subset E$ be a (simple) closed convex set, $d : Q \to \mathbb{R}$ be a distance generating function (d.g.f) which is continuously differentiable and 1-strongly convex w.r.t. the norm $||\cdot||$, i.e.

$$\forall x, y \in Q \quad \langle \nabla d(x) - \nabla d(y), x - y \rangle \geq ||x - y||^2,$$

and assume that $\min_{x \in Q} d(x) = d(0)$. Suppose, we have a constant Θ_0 such that $d(x_*) \leq \Theta_0^2$, where x_* is a solution of (1).

Note that if there is a set of optimal points for (1) $X_* \subset Q$, we may assume that

$$\min_{x_* \in X_*} d(x_*) \leq \Theta_0^2.$$

For all $x, y \in Q \subset E$ consider the corresponding Bregman divergence

$$V(x, y) = d(y) - d(x) - \langle \nabla d(x), y - x \rangle.$$

Standard proximal setups, i.e. Euclidean, entropy, ℓ_1/ℓ_2, simplex, nuclear norm, spectahedron can be found, e.g. in [5]. Let us define the proximal mapping operator standardly

$$\text{Mirr}_x(p) = \arg\min_{u \in Q} \{\langle p, u \rangle + V(x, u)\} \quad \text{for each} \quad x \in Q \quad \text{and} \quad p \in E^*.$$

We make the simplicity assumption, which means that $\text{Mirr}_x(p)$ is easily computable. There are well-known examples of distance generating function, let us denote ℓ_p norm by $||x||_p$, and the unit simplex in $\mathbb{R}^n$ by

$$S_n(1) = \left\{ x \in \mathbb{R}_+^n \mid \sum_{i=1}^n x_i = 1 \right\}.$$

Consider two cases:

– if $p = 1$, then

$$d(x) = \ln n + \sum_{k=1}^n x_k \ln x_k, \quad V(x, y) = \sum_{k=1}^n x_k \ln \left(\frac{x_k}{y_k} \right); \qquad (4)$$

– if $p = 2$, then $d(x) = \frac{1}{2}||x||_2^2$, $V(x, y) = \frac{1}{2}||x - y||_2^2$.

Let $Q = \mathbb{B}_p^n(1) = \{x \in \mathbb{R}^n; \|x\|_p \leq 1\}$ be the unit ball with l_p norm. One can note the following: if $p \geq 2$, then it is optimal to choose the l_2-norm and the Euclidean prox-structure.

Define q by $\frac{1}{p} + \frac{1}{q} = 1$ and consider $1 \leq p \leq 2$, then $q \geq 2$. If in this case $q = O(\ln n)$, then it is optimal to choose l_p-norm and prox-structure with distance generating function

$$d(x) = \frac{1}{2(p-1)} \|x\|_p^2.$$

In all these cases $R^2 = \max_{x \in Q} d(x) \geqslant \Theta_0^2$.

For $q > \Omega(\ln n)$, we choose l_a-norm, where

$$a = \frac{2 \ln n}{2 \ln n - 1}$$

and prox-structure with distance generating function

$$d(x) = \frac{1}{2(a-1)} \|x\|_a^2.$$

In this case

$$R^2 = O(\ln n) \geqslant \Theta_0^2 \quad \text{and} \quad \Theta_0 \leqslant O(\sqrt{\ln n}). \tag{5}$$

Let us remind one well-known statement (see, e.g. [5]).

Lemma 1. *Let $f : Q \to \mathbb{R}$ be a convex subdifferentiable function over the convex set Q and $z = Mirr_y(h\nabla f(y))$ for some $h > 0$, $y, z \in Q$. Then for each $x \in Q$*

$$h\langle \nabla f(y), y - x \rangle \leq \frac{h^2}{2} \|\nabla f(y)\|_*^2 + V(y, x) - V(z, x). \tag{6}$$

3   Online Optimization for the Case of Non-negative Regret: Non-adaptive Algorithm

Assume that the method produces N productive steps and each step the (sub)gradient of exactly one functional of the objectives is calculated. Denote the number of non-productive steps by N_J. Let's consider the non-adaptive method for the problem (1) with a constant step, which depends on the Lipschitz constant M. As a result, we get a sequence $\{x^k\}_{k \in I}$ (on productive steps), which can be considered as a solution to the problem (1) with accuracy δ (see (7)).

By Lemma 1

$$f_i(x^k) - f_i(x) \leq \frac{h}{2}M^2 + \frac{V(x^k, x)}{h} - \frac{V(x^{k+1}, x)}{h} = \frac{\varepsilon}{2} + \frac{V(x^k, x)}{h} - \frac{V(x^{k+1}, x)}{h}$$

$$g(x^k) - g(x) \leq \frac{h}{2}M^2 + \frac{V(x^k, x)}{h} - \frac{V(x^{k+1}, x)}{h} = \frac{\varepsilon}{2} + \frac{V(x^k, x)}{h} - \frac{V(x^{k+1}, x)}{h}$$

Algorithm 1. Constrained Online Optimization: Non-Adaptive Mirror Descent
Algorithm

Require: $\varepsilon, N, \Theta_0^2, Q, d(\cdot), x^0$
1: $i := 1, \ k := 0$;
2: **repeat**
3:   **if** $g(x^k) \leqslant \varepsilon$ **then**
4:     $h = \frac{\varepsilon}{M^2}$;
5:     $x^{k+1} := Mirr[x^k](h\nabla f_i(x^k))$;
6:     $i := i + 1$;
7:     $k := k + 1$;
8:   **else**
9:     $h = \frac{\varepsilon}{M^2}$;
10:     $x^{k+1} := Mirr[x^k](h\nabla g(x^k))$;
11:     $k := k + 1$;
12:   **end if**
13: **until** $i = N + 1$
14: Guaranteed accuracy:

$$\delta := \frac{\varepsilon}{2} + \frac{M^2\Theta_0^2}{\varepsilon N} - \frac{\varepsilon N_J}{2N} \tag{7}$$

Taking summation over productive and non-productive steps, we get

$$\sum_{i=1}^{N}(f_i(x^k) - f_i(x^*)) + \sum_{k \in J}(g(x^k) - g(x^*))$$

$$\leq \frac{\varepsilon}{2}(N + N_J) + \frac{1}{h}\sum_{k=0}^{N+N_J-1}\left(V(x^k, x^*) - V(x^{k+1}, x^*)\right)$$

$$= \frac{\varepsilon}{2}(N + N_J) + \frac{M^2}{\varepsilon}\sum_{k=0}^{N+N_J-1}\left(V(x^k, x^*) - V(x^{k+1}, x^*)\right),$$

then

$$\sum_{i=1}^{N}(f_i(x^k) - f_i(x^*)) \leq \frac{\varepsilon}{2}N + \frac{M^2\Theta_0^2}{\varepsilon} - \frac{\varepsilon}{2}N_J \tag{8}$$

and by virtue of (7)

$$\frac{1}{N}\sum_{i=1}^{N}f_i(x^k) - \min_{x \in Q}\frac{1}{N}\sum_{i=1}^{N}f_i(x) \leq \delta. \tag{9}$$

If we assume the nonnegativity of the regret (i.e. the left side in (8)) and

$$\delta \leq \varepsilon = \frac{C}{\sqrt{N}} \quad \text{for some } C > 0, \tag{10}$$

then we get

$$0 \leq N + \frac{2M^2\Theta_0^2}{\varepsilon^2} - N_J = N + \frac{2M^2\Theta_0^2}{C^2}N - N_J,$$

then

$$N_J \leq N \cdot \left(1 + \frac{2M^2\Theta_0^2}{C^2}\right) \sim O(N).$$

Thus, we have the following result

Theorem 1. *Suppose Algorithm 1 works exactly N productive steps. After the stopping of the Algorithm 1, the following inequality holds:*

$$\frac{1}{N}\sum_{i=1}^{N} f_i(x^k) - \min_{x \in Q} \frac{1}{N}\sum_{i=1}^{N} f_i(x) \leq \delta.$$

For the case (10) and

$$\frac{1}{N}\sum_{i=1}^{N} f_i(x^k) - \min_{x \in Q} \frac{1}{N}\sum_{i=1}^{N} f_i(x) \geq 0$$

there will be no more than

$$N \cdot \left(1 + \frac{2M^2\Theta_0^2}{C^2}\right) \sim O(N). \tag{11}$$

non-productive steps.

Remark 1. The estimate (11) is optimal for the considered class of problems [12].

Corollary 1. *If $Q = S_n(1)$ and the corresponding prox-structure is chosen as (4), then by (5) the estimate (11) modifies into*

$$N_J \leqslant N \cdot \left(1 + \frac{2M^2 \ln n}{C^2}\right).$$

4 Adaptive Mirror Descent for the Case of Non-negative Regret

Now, let us consider the adaptive analog of Algorithm 1 for problem (1). The main feature is a nondecreasing stepsize with consideration of the norm of (sub)gradient of the objective function or the constraints in a particular step. Therefore, the proposed algorithm will work until there are exactly N productive steps. As a result, we get a sequence $\{x^k\}_{k \in I}$ on productive steps, which can be considered as a solution to the problem (1) with accuracy δ (see (12)).

By Lemma 1

$$f_i(x^k) - f_i(x) \leq \frac{h_k}{2}\|\nabla f_i(x^k)\|_*^2 + \frac{V(x^k, x)}{h_k} - \frac{V(x^{k+1}, x)}{h_k}$$

Algorithm 2. Constrained Online Optimization: Adaptive Mirror Descent Algorithm

---

Require: $\varepsilon, N, \Theta_0^2, Q, d(\cdot), x^0$
1: $i := 1, \ k := 0$;
2: **repeat**
3:    **if** $g(x^k) \leqslant \varepsilon$ **then**
4:       $M_k := \|\nabla f_i(x^k)\|_*$;
5:       $h_k = \Theta_0 \left(\sum\limits_{t=0}^{k} M_t^2 \right)^{-1/2}$;
6:       $x^{k+1} := Mirr[x^k](h_k \nabla f_i(x^k))$;
7:       $i := i + 1$;
8:       $k := k + 1$;
9:    **else**
10:      $M_k := \|\nabla g(x^k)\|_*$;
11:      $h_k = \Theta_0 \left(\sum\limits_{t=0}^{k} M_t^2 \right)^{-1/2}$;
12:      $x^{k+1} := Mirr[x^k](h_k \nabla g(x^k))$;
13:      $k := k + 1$;
14:    **end if**
15: **until** $i = N + 1$
16: Guaranteed accuracy:

$$\delta := \frac{2\Theta_0}{N} \left(\sum_{i=0}^{N+N_J-1} M_i^2 \right)^{1/2} - \varepsilon \cdot \frac{N_J}{N}. \tag{12}$$

$$g(x^k) - g(x) \leq \frac{h_k}{2} \|\nabla g(x^k)\|_*^2 + \frac{V(x^k, x)}{h_k} - \frac{V(x^{k+1}, x)}{h_k}$$

Dividing each inequality by h_k and summing up for k from 0 to $N + N_J - 1$, and by using the definition of h_k, we obtain

$$\sum_{k \in I} \left(f(x^k) - f(x_*) \right) + \sum_{k \in J} \left(g(x^k) - g(x_*) \right) \leq \sum_{k=0}^{N+N_J-1} \frac{h_k M_k^2}{2}$$

$$+ \sum_{k=0}^{N+N_J-1} \frac{1}{h_k} \left(V(x^k, x_*) - V(x^{k+1}, x_*) \right) \text{ and}$$

$$\sum_{k=0}^{N+N_J-1} \frac{1}{h_k} \left(V(x^k, x_*) - V(x^{k+1}, x_*) \right) = \frac{1}{h_0} V(x^0, x_*) + \sum_{k=0}^{N+N_J-2} \left(\frac{1}{h_{k+1}} - \frac{1}{h_k} \right) V(x^{k+1}, x_*)$$

$$- \frac{1}{h_{N+N_J-1}} V(x^k, x_*) \leq \frac{\Theta_0^2}{h_0} + \Theta_0^2 \sum_{k=0}^{N+N_J-2} \left(\frac{1}{h_{k+1}} - \frac{1}{h_k} \right) = \frac{\Theta_0^2}{h_{N+N_J-1}}.$$

Whence, by the definition of step sizes h_k,

$$\sum_{i=1}^{N}\left(f_i(x^k)-f(x_*)\right)+\sum_{k\in J}\left(g(x^k)-g(x_*)\right)\le\sum_{k=0}^{N+N_J-1}\frac{h_kM_k^2}{2}+\frac{\Theta_0^2}{h_{N+N_J-1}}$$

$$\le\sum_{k=0}^{N+N_J-1}\frac{\Theta_0}{2}\frac{M_k^2}{\left(\sum_{j=0}^{k}M_j^2\right)^{1/2}}+\Theta_0\left(\sum_{k=0}^{N+N_J-1}M_k^2\right)^{1/2}\le2\Theta_0\left(\sum_{k=0}^{N+N_J-1}M_k^2\right)^{1/2}$$

$$(13)$$

where we used inequality

$$\sum_{i=0}^{N+N_J-1}\frac{M_i^2}{\left(\sum_{j=0}^{i}M_j^2\right)^{1/2}}\le2\left(\sum_{i=0}^{N+N_J-1}M_i^2\right)^{1/2},$$

which can be proved by induction. Since, for $k\in J$, $g(x^k)-g(x_*)\ge g(x^k)>\varepsilon$, we get

$$\sum_{i=1}^{N}(f_i(x^k)-f_i(x^*))<\varepsilon N-\varepsilon(N+N_J)+2\Theta_0\left(\sum_{i=0}^{N+N_J-1}M_i^2\right)^{1/2}.\qquad(14)$$

and by (12)

$$\frac{1}{N}\sum_{i=1}^{N}f_i(x^k)-\min_{x\in Q}\frac{1}{N}\sum_{i=1}^{N}f_i(x)\le\delta.\qquad(15)$$

If we assume the nonnegativity of the regret (i.e. the left side in (14)) and the accuracy is given by (10), one can get

$$\varepsilon(N+N_J)\le\varepsilon N+2\Theta_0\left(\sum_{i=0}^{N+N_J-1}M_i^2\right)^{1/2}\le\varepsilon N+2M\Theta_0\cdot\sqrt{N+N_J},$$

$$N_J^2\le\frac{4M^2\Theta_0^2(N+N_J)}{\varepsilon^2}=\frac{4M^2\Theta_0^2(N+N_J)N}{C^2}.$$

Further,

$$\frac{N_J^2}{N^2+NN_J}=\frac{\left(\frac{N_J}{N}\right)^2}{1+\frac{N_J}{N}}\le\frac{4M^2\Theta_0^2}{C^2}$$

and $N_J=O(N)$. Thus, we have come to the following result.

Theorem 2. *Suppose Algorithm 2 works exactly N productive steps. After the stopping of the Algorithm 2, the following inequality holds:*

$$\frac{1}{N}\sum_{i=1}^{N}f_i(x^k)-\min_{x\in Q}\frac{1}{N}\sum_{i=1}^{N}f_i(x)\le\delta.$$

For the case of (10) *and*

$$\frac{1}{N}\sum_{i=1}^{N} f_i(x^k) - \min_{x \in Q} \frac{1}{N}\sum_{i=1}^{N} f_i(x) \geq 0$$

there will be no more than $O(N)$ *non-productive steps.*

Remark 2. Algorithm 2 is optimal for the considered class of problems [12].

Remark 3. Let's consider a modification of the proposed Algorithm 2 for the case of a set of functional constraints $g_m : Q \rightarrow \mathbb{R}$ ($m = \overline{1, K}$). We assume, that all the functionals g_m satisfy the Lipschitz condition:

$$|g_m(x) - g_m(y)| \leq M||x - y|| \ \forall x, y \in Q, \ m = \overline{1, K}. \tag{16}$$

In this case, instead of a set of convex functional constraints $\{g_m(\cdot)\}_{m=1}^{K}$ we can consider one constraint, given as $g : Q \rightarrow \mathbb{R}$, where

$$g(x) = \max_{m=\overline{1,K}} g_m(x), \quad |g(x) - g(y)| \leq M||x - y|| \ \forall x, y \in Q.$$

This method will be also optimal, but in practice it can give better accuracy (see Remark 4 below).

5   The Case of Negative Regret

Now we consider the situation when after the stopping of any of the above algorithms, it turns out that the regret is negative. In this case the following inequality

$$\frac{1}{N}\sum_{i=1}^{N} f_i(x^k) - \min_{x \in Q} \frac{1}{N}\sum_{i=1}^{N} f_i(x) \leq 0 \tag{18}$$

holds. It is already impossible to justify the optimality of the number of non-productive steps in view of the right-hand side of inequality (18).

Note that the set of productive steps is not empty, because for arbitrary p steps when the inequality

$$\sum_{k=1}^{p} \frac{1}{M_k^2} \geq \frac{2\Theta_0^2}{\varepsilon^2}$$

is satisfied, one of these p steps will necessarily be productive (see [2,22]). If all the other $p-1$ steps are non-productive (without loss of generality let the last step be productive), then

$$\sum_{k=1}^{p-1} \frac{1}{M_k^2} < \frac{2\Theta_0^2}{\varepsilon^2}$$

and

$$p - 1 < \frac{2M^2\Theta_0^2}{\varepsilon^2}.$$

Algorithm 3. Online Optimization: Adaptive Mirror Descent Algorithm Modification for the Case of Many Constraints

---

Require: $\varepsilon, N, \Theta_0^2, Q, d(\cdot), x^0$
1: $i := 1, \ k := 0$;
2: **repeat**
3: **if** $g(x^k) \leqslant \varepsilon$ **then**
4:      $M_k := \|\nabla f_i(x^k)\|_*$;
5:      $h_k = \Theta_0 \left(\sum_{t=0}^{k} M_t^2 \right)^{-1/2}$;
6:      $x^{k+1} := Mirr[x^k](h_k \nabla f_i(x^k))$;
7:      $i := i + 1$;
8:      $k := k + 1$;
9: **else**
10:     $M_k := \|\nabla g_{m(k)}(x^k)\|_*$ for some $g_{m(k)}(\cdot)$: $g_{m(k)}(x^k) > \varepsilon$
11:     $h_k = \Theta_0 \left(\sum_{t=0}^{k} M_t^2 \right)^{-1/2}$;
12:     $x^{k+1} := Mirr[x^k](h_k \nabla g_{m(k)}(x^k))$;
13:     $k := k + 1$;
14: **end if**
15: **until** $i = N + 1$
16: Guaranteed accuracy:

$$\delta := \frac{2\Theta_0}{N} \left(\sum_{i=0}^{N+N_J-1} M_i^2 \right)^{1/2} - \varepsilon \cdot \frac{N_J}{N}. \tag{17}$$

It is clear, that running the method for a sufficiently long time, it is possible to achieve N productive steps. At the same time between each two successive productive steps there will be no more than $\frac{2M^2\Theta_0^2}{\varepsilon^2}$ non-productive steps, i.e. the number of all non-productive steps will be no more than

$$\frac{2M^2\Theta_0^2}{\varepsilon^2} N.$$

In comparison with the previous items, for $\varepsilon = \frac{C}{\sqrt{N}}$ there will be no more than

$$\frac{2M^2\Theta_0^2}{\varepsilon^2} N = O(N^2) \tag{19}$$

non-productive steps.

6   Numerical Experiments

To compare of Algorithms 1, 2 and 3, some numerical tests were carried out. Consider four different examples with objective function

$$f(x) = \frac{1}{N} \sum_{i=1}^{N} |\langle a_i, x \rangle - b_i|.$$

For the coefficients a_i and constants b_i for $i = 1, \ldots, N$, with different values of N. Let $A \in \mathbb{R}^{N \times 11}$ be a matrix with entries drawn from different random distributions. Then a_i^T are rows in the matrix $A' \in \mathbb{R}^{N \times 10}$, which is introduced from A, by eliminating the last column, and b_i are the entries of the last column in the matrix A. In details, entries of A drawn

- In example 1, from a normal distribution with mean (center) equalling 0 and standard deviation (width) equalling 1.
- In example 2, from a uniform distribution over $[0, 1)$.
- In example 3, from the standard exponential distribution with a scale parameter of 1.
- In example 4, from a Gumbel distribution with the location of the mode equalling 1 and the scale parameter equalling 2.

For the function of constraints $g(x) = \max\limits_{i \in \overline{1,m}} g_i(x)$, we take $m = 3$ and the functionals $g_i(x) = \langle \alpha_i, x \rangle$, where α_i^T are the rows of the matrix

$$\begin{pmatrix} 1 & 1 & 1 & 1 & 1 & 1 & 1 & 1 & 1 & 1 \\ 1 & 2 & 3 & 4 & 5 & 6 & 7 & 8 & 9 & 10 \\ 1 & 2 & 4 & 6 & 8 & 10 & 12 & 14 & 16 & 18 \end{pmatrix}$$

We choose standard Euclidean proximal setup as a prox-function, starting point $x^0 = \dfrac{(1, 1, \ldots, 1)}{\sqrt{10}}$, $\Theta_0 = 3$, $\varepsilon = \frac{1}{\sqrt{N}}$ and

$$Q = \{x = (x_1, x_2, \ldots, x_{10}) \in \mathbb{R}^{10} \mid x_1^2 + x_2^2 + \ldots + x_{10}^2 \le 1\}.$$

The results of the work of Algorithms 1, 2 and 3 are represented in Tables 1, 2 and 3 below, respectively, demonstrate the comparison between these algorithms. The number of non-productive steps are denoted by *nonprod.*, time is given in seconds and parts of the second, δ is guaranteed accuracy of the solution approximation found (sequence $\{x^k\}_{k \in I}$ on productive steps).

All experiments were implemented in Python 3.4, on computer fitted with Intel(R) Core(TM) i7-8550U CPU @ 1.80 GHz, 1992 Mhz, 4 Core(s), 8 Logical Processor(s). RAM of the computer is 8 GB.

From Tables 1 and 2 one can see, that the adaptive Algorithm 2 always works better than non-adaptive Algorithm 1. It is clearly shown in all the examples by

Table 1. Results of Algorithm 1.

	nonprod.	time	δ
ex. 1, $N = 3000$	7041	00.444	187.473
ex. 2, $N = 6000$	12645	00.812	132.565
ex. 3, $N = 7000$	15814	00.958	122.730
ex. 4, $N = 10000$	24971	01.523	102.682

Table 2. Results of Algorithm 2.

	nonprod.	time	δ
ex. 1, $N = 3000$	39	00.149	0.426
ex. 2, $N = 6000$	2821	00.404	0.223
ex. 3, $N = 7000$	5543	00.586	0.405
ex. 4, $N = 10000$	12576	01.104	0.692

the number of non-productive steps, running time of the algorithms and guaranteed accuracy δ. Where the number of non-productive steps and δ produced by Algorithm 2 is very small compared to the Algorithm 1.

From Table 3, we can see, that there is a difference between the number of non-productive steps produced by Algorithms 2 and 3, but the guaranteed accuracy δ and the running time produced by Algorithm 3 is smaller compared to Algorithm 2.

Table 3. Results of Algorithm 3.

	nonprod.	time	δ
ex. 1, $N = 3000$	47	00.121	0.414
ex. 2, $N = 6000$	2835	00.333	0.220
ex. 3, $N = 7000$	5563	00.454	0.394
ex. 4, $N = 10000$	12885	00.807	0.680

Remark 4. To show the advantages of Algorithm 3, as compared to Algorithm 2, one additional numerical test was carried out. Let's now take the functionals of constraints g_i, $i = 1, 2, 3$ as follows

$$g_1(x) = \sum_{i=1}^{10} i \cdot x_i + 1, \quad g_2(x) = \sum_{i=1}^{10} 10i \cdot x_i, \quad g_3(x) = \sum_{i=1}^{10} 50i \cdot x_i.$$

with the same all previous parameters: starting point $x^0 = \dfrac{(1, 1, \ldots, 1)}{\sqrt{10}}$, $\Theta_0 = 3$,

$$Q = \{x = (x_1, x_2, \ldots, x_{10}) \in \mathbb{R}^{10} \mid x_1^2 + x_2^2 + \ldots + x_{10}^2 \leq 1\},$$

but with $\varepsilon = 0.5$. Table 4 below demonstrate the comparison between Algorithms 2 and 3, for the objective function $f(x) = \frac{1}{3} \sum_{i=1}^{3} f_i(x)$, where

$$f_1(x) = \sqrt{\sum_{i=1}^{9}(x_i + x_{i+1})^2}, \quad f_2(x) = \sqrt{0.1 \left(\sum_{i=1}^{10} x_i^2 + \sum_{i=1}^{9} x_i x_{i+1}\right)}, \quad f_3(x) = \sqrt{\sum_{i=1}^{10} x_i^2}.$$

Table 4. Results of Algorithms 2 and 3.

ex. 5, $N = 3$	nonprod.	time	δ
Algorithm 2	1	00.044	1961.954
Algorithm 3	2	00.030	9.608

From Table 4, one can see, that Algorithm 3 works better than Algorithm 2, since the difference between the non-productive steps is very small, equalling only one, and the guaranteed accuracy δ produced by Algorithm 3 is very small compared to the precision produced by Algorithm 2.

References

1. Bayandina, A., Gasnikov, A., Gasnikova, E., Matsievsky, S.: Primal-dual mirror descent for the stochastic programming problems with functional constraints. Comput. Math. Math. Phys. (2018, accepted). https://arxiv.org/pdf/1604.08194.pdf. (in Russian)
2. Bayandina, A., Dvurechensky, P., Gasnikov, A., Stonyakin, F., Titov, A.: Mirror descent and convex optimization problems with non-smooth inequality constraints. In: Large-Scale and Distributed Optimization, pp. 181–213. Springer, Cham (2018)
3. Beck, A., Ben-Tal, A., Guttmann-Beck, N., Tetruashvili, L.: The comirror algorithm for solving nonsmooth constrained convex problems. Oper. Res. Lett. **38**(6), 493–498 (2010)
4. Beck, A., Teboulle, M.: Mirror descent and nonlinear projected subgradient methods for convex optimization. Oper. Res. Lett. **31**(3), 167–175 (2003)
5. Ben-Tal, A., Nemirovski, A.: Lectures on Modern Convex Optimization. Society for Industrial and Applied Mathematics, Philadelphia (2001)
6. Ben-Tal, A., Nemirovski, A.: Robust Truss Topology Design via semidefinite programming. SIAM J. Optim. **7**(4), 991–1016 (1997)
7. Boyd, S., Vandenberghe, L.: Convex Optimization. Cambridge University Press, New York (2004)
8. Bubeck, S., Eldan, R.: Multi-Scale Exploration of Convex Functions and Bandit Convex Optimization. e-print (2015). http://research.microsoft.com/en-us/um/people/sebubeck/ConvexBandits.pdf
9. Bubeck, S., Cesa-Bianchi, N.: Regret analysis of stochastic and nonstochastic multi-armed bandit problems. Found. Trends Mach. Learn. **5**(1), 1–122 (2012)
10. Gasnikov, A.V., Lagunovskaya, A.A., Morozova, L.E.: On the relationship between simulation logit dynamics in the population game theory and a mirror descent method in online optimization using the example of the shortest path problem. Proc. MIPT **7**(4), 104–113 (2015). (in Russian)
11. Gasnikov, A.V., Lagunovskaya, A.A., Usmanova, I.N., Fedorenko, F.A., Krymova, E.A.: Stochastic online optimization. Single-point and multi-point non-linear multi-armed bandits. Convex and strongly-convex case. Autom. Remote Control **78**(2), 224–234 (2017)
12. Hazan, E., Kale, S.: Beyond the regret minimization barrier: optimal algorithms for stochastic strongly-convex optimization. JMLR **15**, 2489–2512 (2014)

13. Hazan, E.: Introduction to online convex optimization. Found. Trends Optim. **2**(3–4), 157–325 (2015)
14. Jenatton, R., Huang, J., Archambeau, C.: Adaptive Algorithms for Online Convex Optimization with Long-term Constraints (2015). https://arxiv.org/abs/1512.07422
15. Kalai, A., Vempala, S.: Efficient algorithms for online decision problems. J. Comput. Syst. Sci. **71**, 291–307 (2005)
16. Lugosi, G., Cesa-Bianchi, N.: Prediction, Learning and Games. Cambridge University Press, New York (2006)
17. Nemirovskii, A.: Efficient methods for large-scale convex optimization problems. Ekonomika i Matematicheskie Metody (1979). (in Russian)
18. Nemirovsky, A., Yudin, D.: Problem Complexity and Method Efficiency in Optimization. Wiley, New York (1983)
19. Nesterov, Y.: Introductory Lectures on Convex Optimization: A Basic Course. Kluwer Academic Publishers, Massachusetts (2004)
20. Polyak, B.: A general method of solving extremum problems. Sov. Math. Dokl. **8**(3), 593–597 (1967). (in Russian)
21. Shor, N.Z.: Generalized gradient descent with application to block programming. Kibernetika **3**(3), 53–55 (1967). (in Russian)
22. Stonyakin, F.S., Alkousa, M.S., Stepanov, A.N., Barinov, M.A.: Adaptive mirror descent algorithms in convex programming problems with Lipschitz constraints. Trudy Instituta Matematiki i Mekhaniki URO RAN **24**(2), 266–279 (2018)
23. Shpirko, S., Nesterov, Y.: Primal-dual subgradient methods for huge-scale linear conic problem. SIAM J. Optim. **24**(3), 1444–1457 (2014)
24. Vasilyev, F.: Optimization Methods. Fizmatlit, Moscow (2002). (in Russian)

Primal-Dual Newton's Method with Steepest Descent for Linear Programming

Vitaly Zhadan$^{(\boxtimes)}$ (iD)

Dorodnicyn Computing Centre, FRC "Computer Science and Control" of RAS,
Vaviliva st., 40, 119333 Moscow, Russia
zhadan@ccas.ru

Abstract. The primal-dual method for solving linear programming problems is considered. In order to determine the search directions the non-perturbed system of optimality conditions is solved by Newton's method. If this system is degenerate, then an auxiliary linear complementarity problem is solved for obtained unique directions. Starting points and all consequent points are feasible. The step-lengths are chosen from the steepest descent approach based on minimization of the dual gap. The safety factor is not introduced, and trajectories are allowed to move along the boundaries of the feasible sets. The convergence of the method at a finite number of iterations is proved.

Keywords: Linear programming · Primal-dual method
Newton's method · Steepest descent · Finite convergence

1 Introduction

Newton's method is one of the main numerical tools for solving systems of non-linear equations and optimization problems. Last decades Newton's method was widely used for solving linear programming (LP) problems [1–3]. On the basis of Newton's method, numerous interior point algorithms were developed [4–7]. The primal-dual methods are most efficient from this class of methods. In all these methods the pure or perturbed system of optimality conditions is solved by Newton's method. These methods have excellent theoretical properties and good practical performance.

The first variant of the primal-dual method based on the solution of the perturbed system of optimality conditions was proposed by Kojima et al. [8]. Afterward, many other variants of this method were developed in which very skillful rules for choosing step-lengths and perturbation coefficients were suggested [9–14]. The exhaustive review of the primal-dual methods together with

This work was supported by the Russian Foundation for Basic Research (project no. 17-07-00510).

Y. Evtushenko et al. (Eds.): OPTIMA 2018, CCIS 974, pp. 79–94, 2019.
https://doi.org/10.1007/978-3-030-10934-9_6

another path following methods had been given in [5–7,13]. The numerical experiments carried out in [15,16] showed their good practical performance even if the perturbation coefficients are taken very small.

In this paper, we will consider the limit variant of the primal-dual method in which the pure non-perturbed system of optimality conditions is solved by Newton's method. We suppose that starting points and all consequent points are feasible. The step-lengths in primal and dual spaces are chosen from the steepest descent approach basing on minimization of the dual gap. We have not use the safety factor and allow trajectories to move along the boundaries of the feasible sets. Similar to the simplex method the proposed variant of Newton's method finds the exact solutions of primal and dual problems at a finite number of iterations.

The more labor-consuming operation in primal-dual methods is solving of the linear system of equations that determines the search directions. The standard approach is based on the solution of the normal linear system with the symmetric positive definite matrix. The important feature of the proposed variant of the Newton method is that the dimension of this linear system decreases during iterations, tending to zero. Such an approach had been proposed firstly in [14]. In this paper, we applied it for another pair of primal and dual problems. It gives us the possibility to use different step-lengths in primal and dual spaces.

The paper is organized as follows. In Sect. 2 we formally stand the linear programming problem and its dual. We also introduce required notations and give some definitions. In Sect. 3 and 4 we introduce the linear system of equations for finding the search directions and describe the solutions of this system. Since current points may belong to the boundaries of the feasible sets it is possible that the linear system of equations is undetermined and therefore has many solutions. In order to choose the unique solution, we suggest to fix some of the unknown variables and after that to solve the system with respect to remaining variables. We suggest determining the fixed variables from a solution of auxiliary linear complementarity problem.

In Sect. 5 we describe the iterative process of the primal-dual Newton's method with the steepest descent. We show that, if all points of the trajectory are not vertices of the feasible sets, then method solves both LP problems at the number of iterations that does not exceed the number of variables.

2   Problem Statement and Definitions

Consider the linear programming problem in standard formulation

$$c_* = \min \, \langle c, x \rangle, \quad Ax = b, \quad x \geq 0_n, \tag{1}$$

where $c \in \mathbb{R}^n$, $b \in \mathbb{R}^m$, and A is a matrix of the size $m \times n$ with $m < n$. By 0_n is denoted the zero n-dimensional vector, and by $\langle \cdot, \cdot \rangle$ — the eucledean inner product. The dual problem to (1) has the following form

$$b^* = \max \, \langle b, u \rangle, \quad y = c - A^T u, \quad y \geq 0_n. \tag{2}$$

We assume that both problems (1) and (2) have solutions. Then $c_* = b^*$. We assume also that the matrix A has full rank equal to m.

Multiplying the equality $y = c - A^T u$ from (2) by the matrix A, we obtain the equation $AA^T u = A(c - y)$. Since AA^T is a nonsingular matrix, we get

$$u = (AA^T)^{-1} A(c - y). \tag{3}$$

Observe that this expression for u is valid if and only if the vector $c - y$ belongs to the range-space of the matrix A^T.

Let K be a full rank matrix such that its rows generate the null-space of A. The matrix K has the size $l \times n$, where $l = n - m$. The range-space of the matrix A^T coincides with the null-space of the matrix K.

After substitution u from (3) in the objective function of the dual problem (2) we have

$$\langle b, u \rangle = \langle A^T(AA^T)^{-1}b, c \rangle - \langle A^T(AA^T)^{-1}b, y \rangle.$$

Therefore, denoting $f = A^T(AA^T)^{-1}b \in \mathbb{R}^n$ and $d = Kc \in \mathbb{R}^l$, we obtain that solution of the dual problem can be derived from the solution of the following problem

$$f_* = \min \langle f, y \rangle, \quad Ky = d, \quad y \geq 0_n. \tag{4}$$

The optimal value f_* of the objective function in problem (4) is connected with the optimal value b^* in dual problem (2) by equality $b^* + f_* = b^T(AA^T)^{-1}Ac$. Below we will denote by X and Y the feasible sets in problems (2) and (4), respectively, that is $X = \{x \in \mathbb{R}^n_+ : Ax = b\}$, $Y = \{y \in \mathbb{R}^n_+ : Ky = d\}$. Together with (2) we will call the problem (4) by *dual problem*.

Compose from m linearly independent columns of the matrix A the nonsingular sub-matrix B_A of order m. The remaining columns of A we place in the sub-matrix N_A. We call $\bar{x}$ by *basic solution* of the equation $Ax = b$, if $A\bar{x} = b$ and all components of $\bar{x}$, corresponding to columns from the sub-matrix N_A, are equal to zero. The point $\bar{x} \in \mathbb{R}^n$ we call *basic*, if $\bar{x}$ is a basic solution of the equation $Ax = b$. The basic point $\bar{x}$ is called feasible, if $\bar{x} \in X$. The matrix B_A is a basis of the basic point $\bar{x}$.

Definition 1. *The basic point $\bar{x}$ is called non-degenerate, if all components $\bar{x}^i$, corresponding to columns from the matrix B_A, are nonzero.*

Definition 2. *The problem (1) is called non-degenerate, if all the feasible basic points are non-degenerate. The problem (1) is called strictly non-degenerate, if all basic points (not only feasible basic points) are non-degenerate.*

In similar way we define the non-degenerate solutions of the equation $Ky = d$ and feasible *basic points* $\bar{y} \in Y$. We say that the problem (4) is *non-degenerate*, if all feasible basic points are non-degenerate. In the case, when all basic points (not only feasible basic points) are non-degenerate, we say that the problem (4) is *strictly non-degenerate*.

If the problem (1) is strictly non-degenerate, then the vector b does not belong to any linear sub-space, generated by less than m columns of the matrix

A. Similarly, if the problem (4) is strictly non-degenerate, then the vector d does not belong to any linear sub-space, generated by less than l columns of the matrix K. In what follows, we suppose that both problems (1) and (4) are at least non-degenerate.

Denote by $D(x)$ the diagonal matrix with a vector x on its diagonal. In order that both problems (1) and (4) have solutions it is necessary and sufficient that the following system of equalities and inequalities

$$D(x)y = 0_n, \quad Ax = b, \quad Ky = d, \quad x \geq 0_n, \quad y \geq 0_n \qquad (5)$$

has solution.

Definition 3. *The pair $[x, y]$ is called interior, if $x \geq 0_n$ and $y \geq 0_n$. The interior pair $[x, y]$ is called feasible, if $Ax = b$ and $Ky = d$. The pair $[x, y]$ is called complementary, if $x^i y^i = 0$, $1 \leq i \leq n$.*

The feasible complementary pair $[x, y]$ is *optimal*, i.e. x is a solution of problem (1), and y is a solution of problem (4).

Let $[x, y]$ be an interior pair. We split the index set $J^n = [1 : n]$ onto four subsets, depending from x and y (some of these subsets may be empty):

$$J_B(x,y) = \left\{ i \in J^n : x^i > 0, \ y^i = 0 \right\}, \quad J_N(x,y) = \left\{ i \in J^n : x^i = 0, \ y^i > 0 \right\}, \qquad (6)$$

$$J_P(x,y) = \left\{ i \in J^n : x^i > 0, \ y^i > 0 \right\}, \quad J_Z(x,y) = \left\{ i \in J^n : x^i = 0, \ y^i = 0 \right\}. \qquad (7)$$

In order that the feasible pair $[x, y]$ be complementary and hence optimal, it is necessary and sufficient that $J_P(x, y) = \emptyset$.

Denote by $|J|$ the number of elements in the index set J. It follows from the assumption about the non-degeneracy of problem (1) that at every feasible pair $[x, y]$ the following inequality

$$|J_P(x,y)| + |J_B(x,y)| \geq m \qquad (8)$$

holds. Similarly, from the assumption about the non-degeneracy of problem (4) we obtain that

$$|J_P(x,y)| + |J_N(x,y)| \geq l. \qquad (9)$$

Therefore, due to (8) and (9)

$$|J_B(x,y)| + |J_Z(x,y)| \leq m, \qquad |J_N(x,y)| + |J_Z(x,y)| \leq l. \qquad (10)$$

We conclude from these inequalities that $|J_Z(x,y)| \leq \min\{m, l\}$.

Definition 4. *The interior pair $[x, y]$ is called regular, if $J_Z(x, y) = \emptyset$. Otherwise, when $J_Z(x, y) \neq \emptyset$, the interior pair $[x, y]$ is called irregular.*

Proposition 1. *Let problems (1) and (4) be non-degenerate. Then the optimal pair $[x_*, y_*]$ is regular.*

Proof. In optimal pair $[x_*, y_*]$ necessarily $J_P(x_*, y_*) = \emptyset$. Hence, it follows from (8) and (9) that $|J_B(x_*, y_*)| \geq m$ and $|J_N(x_*, y_*)| \geq l$. Therefore, if $|J_Z(x_*, y_*)| > 0$, we get the contradiction with the inequalities (10). $\qquad \square$

In accordance with partition of the index set J^n on subsets (6), (7) we partition also the matrices A and K on sub-matrices. Moreover, without loss of generality, we suppose that at the pair $[x, y]$ these sub-matrices are placed in the following order

$$A = [A_P, A_B, A_N, A_Z], \qquad K = [K_P, K_B, K_N, K_Z].$$

The same partition will be applied for components of vectors x and y:

$$x = \left[x^P;\ x^B;\ x^N;\ x^Z \right], \qquad y = \left[y^P;\ y^B;\ y^N;\ y^Z \right]. \tag{11}$$

Here we use ";" for adjoining vectors or components of a vector in a column. In the case of regular pair $[x, y]$ the last sub-matrices in the matrices A and K, as well the last sub-vectors in the vectors x and y are absent.

Lemma 1. *At any feasible pair $[x, y]$ matrices $A_{BZ} = [A_B, A_Z]$ and $K_{NZ} = [K_N, K_Z]$ have full ranks.*

Proof. Suppose at first, that y is a vertex of the set Y. Then $K_P y^P + K_N y^N = d$. Moreover, $K_{PN} = [K_P, K_N]$ is a square matrix of the order l. It follows from assumption about non-degeneracy of problem (4) that K_{PN} is non-singular matrix. The matrix A_{BZ} is also square of the order m. Show by contradiction that it is non-singular. Indeed, if the matrix A_{BZ} is singular, then there exists a nonzero vector $z \in \mathbb{R}^m$ such that $A_{BZ}^T z = 0_m$. But in this case for the remaining sub-matrix $A_{PN} = [A_P, A_N]$ the following inequality $A_{PN}^T z \neq 0_m$ must hold, otherwise, we should have $A^T z = 0_m$. The last equality contradicts to the assumption about the linear independence of rows of the matrix A.

According to the definition of the matrix K:

$$K_P A_P^T + K_B A_B^T + K_N A_N^T + K_Z A_Z^T = 0_{lm}.$$

After multiplying this equality from the right by z we obtain $K_{PN} A_{PN}^T z = 0_l$. Since $A_{PN}^T z \neq 0$, it follows from here the linear dependence of columns of the matrix K_{PN}. This contradicts to the assumption about the non-degeneracy of the vertex $y \in Y$. Therefore, the matrix A_{BZ} is nonsingular.

Now, consider the case, where y is not a vertex of the set Y. Then the number of columns in the matrix A_{BZ} is less than m. Since Y is a convex set, there exists at least one vertex $\bar{y}$ of the set Y such that $\bar{y}^i = 0$ for $i \in J_B(x, y) \bigcup J_Z(x, v)$. As it is established just now, the matrix $\bar{A}_{BZ}$, corresponding to the pair $[x, \bar{y}]$, is non-singular. Hence, the columns of $\bar{A}_{BZ}$ are linearly independent. Because the matrix A_{BZ} is a sub-matrix of $\bar{A}_{BZ}$, the matrix A_{BZ} has a full rank.

By analogy, it can be established that the matrix K_{NZ} also has full rank. $\quad \square$

Corollary 1. *For any feasible pair $[x, y]$ the matrices A_B and A_Z side by side with the matrices K_N and K_Z have full ranks.*

3 Newton's Directions at Regular Pairs

The Newton method can be applied for solving the system of equalities from (5). Let $[x, y]$ be a feasible pair, and let Δx and Δy be the search directions. Making one iteration by Newton's method, we obtain the updated pair $[\bar{x}, \bar{y}]$ such that

$$\bar{x} = x + \alpha \Delta x, \qquad \bar{y} = y + \beta \Delta y. \tag{12}$$

Here $\alpha > 0$ and $\beta > 0$ are step-lengths, choosing in such a way that updated pair $[\bar{x}, \bar{y}]$ is also feasible. The search directions Δx and Δy satisfy the following system of linear equations

$$D(y)\Delta x + D(x)\Delta y = -D(x)y, \qquad A\Delta x = 0_m, \qquad K\Delta y = 0_l, \tag{13}$$

which is a system of $2n$ equations with respect to $2n$ variables — components of the directions Δx and Δy. The system (13) is nonhomogeneous, if $[x, y]$ is not an optimal pair. The matrix of the linear system (13) has the following block form

$$W(x, y) = \begin{bmatrix} D(y) & D(x) \\ A & 0_{mn} \\ 0_{ln} & K \end{bmatrix}. \tag{14}$$

Consider at first the case, when the feasible non-optimal pair $[x, y]$ is regular. Then (14) can be rewritten in more detailed form as

$$W(x, y) = \begin{bmatrix} D(y^P) & 0 & 0 & D(x^P) & 0 & 0 \\ 0 & 0 & 0 & 0 & D(x^B) & 0 \\ 0 & 0 & D(y^N) & 0 & 0 & 0 \\ A_P & A_B & A_N & 0 & 0 & 0 \\ 0 & 0 & 0 & K_P & K_B & K_N \end{bmatrix}. \tag{15}$$

The following result is valid.

Proposition 2. *Let $[x, y]$ be a regular pair. Then the matrix (15) is nonsingular.*

Partitioning the directions Δx and Δy on components $\Delta x = [\Delta x^P; \Delta x^B; \Delta x^N]$ and $\Delta y = [\Delta y^P; \Delta y^B; \Delta y^N]$, we get that the system (13) can be rewritten as

$$D(y^P)\Delta x^P + D(x^P)\Delta y^P = -D(x^P)y^P,$$
$$D(x^B)\Delta y^B = 0, \qquad D(y^N)\Delta x^N = 0,$$
$$A_P\Delta x^P + A_B\Delta x^B + A_N\Delta x^N = 0, \qquad K_P\Delta y^P + K_B\Delta y^B + K_N\Delta y^N = 0. \tag{16}$$

It follows from here that $\Delta x^N = 0$, $\Delta y^B = 0$. Thus, the system (16) is simplified

$$D(y^P)\Delta x^P + D(x^P)\Delta y^P = -D(x^P)y^P,$$
$$A_P\Delta x^P + A_B\Delta x^B = 0, \qquad K_P\Delta y^P + K_N\Delta y^N = 0. \tag{17}$$

Let us give the solution of system (17), which is depended from a number of indices in the set $J_B(x, y)$. From the first inequality (10) follows that this number does not exceed m. Consider separately three possible cases.

(a) $|J_B(x,y)| = m$. Under this assumption, A_B is a square nonsingular matrix. Since $K\Delta y = 0$, we have that $\Delta y = A^T \Delta u$ for some vector $\Delta u \in \mathbb{R}^m$. In particular, $\Delta y^B = A_B^T \Delta u = 0$. It follows from here, that $\Delta u = 0$. Then $\Delta y = 0$, and due to the first equality (17) $\Delta x^P = -x^P$. Because of the second equality (17) we derive also that $\Delta x^B = A_B^{-1} A_P x^P$. Thus,

$$\Delta x^P = -x^P, \quad \Delta x^B = A_B^{-1} A_P x^P, \quad \Delta x^N = 0, \quad \Delta y = 0. \tag{18}$$

If the equality $|J_B(x,y)| = m$ is valid, then y is a vertex of the set Y.

(b) $0 < |J_B(x,y)| < m$. Let $\mathcal{R}(A_B)$ be a rangespace of the matrix A_B, and let $\mathcal{R}^\perp(A_B)$ be its orthogonal complement. We take an arbitrary basis in $\mathcal{R}^\perp(A_B)$ and introduce into consideration a full rank matrix H_B, which columns are the vectors of this basis. From the equality $\Delta y^B = A_B^T \Delta u = 0$ follows that $\Delta u = H_B p$ for some $p \in \mathbb{R}^{m_B}$, where $m_B = m - |J_B(x,y)|$. Therefore, $\Delta y^P = A_P^T H_B p$ and $\Delta y^N = A_N^T H_B p$.

Introduce also the matrix $S_P = D(x^P)D^{-1}(y^P)$. Then after multiplying the first equation (17) by the matrix $D^{-1}(y^P)$ and after substituting Δy^P, we obtain

$$\Delta x^P = -\left(S_P A_P^T H_B p + x^P\right). \tag{19}$$

Further, we substitute Δx^P in the second equation (17) and multiply the obtained equality from the left by H_B^T. Taking into account that $H_B^T A_B = 0$, we derive the linear system of equations

$$\left(H_B^T A_P S_P A_P^T H_B\right) p = -H_B^T A_P x^P$$

with the solution $p = -\left(H_B^T A_P S_P A_P^T H_B\right)^{-1} H_B^T A_P x^P$. Taking into account the equality $A_P x^P + A_B x^B = b$, we obtain also $p = -\left(H_B^T A_P S_P A_P^T H_B\right)^{-1} H_B^T b$.

Let

$$\Gamma_P = H_B \left(H_B^T A_P S_P A_P^T H_B\right)^{-1} H_B^T. \tag{20}$$

We have from (17) and (19), that

$$\begin{aligned}\Delta x^P &= S_P A_P^T \Gamma_P b - x^P, \qquad \Delta x^N = 0 \\ \Delta x^B &= \left(A_B^T A_B\right)^{-1} A_B^T \left(I - A_P S_P A_P^T \Gamma_P\right) b - x^B.\end{aligned} \tag{21}$$

Here and in what follows I is an identity matrix of the corresponding order.

Using the equality $\Delta y = A^T H_B p$, we get directions in the y-space:

$$\Delta y^P = -A_P^T \Gamma_P b, \quad \Delta y^B = 0, \quad \Delta y^N = -A_N^T \Gamma_P b. \tag{22}$$

(c) $|J_B(x,y)| = 0$. Under this assumption $A_P x^P = b$, and we obtain from the second equation (17), that $A_P \Delta x^P = 0$. Moreover, because of (8) the inequality $|J_P(x,y)| \geq m$ holds. Thus, columns of the matrix A_P^T are linearly independent.

We have $\Delta y^P = A_P^T \Delta u$ for some m-dimensional vector Δu. Therefore, the first equation (17) is reduced to

$$\Delta x^P + S_P A_P^T \Delta u = -x^P. \tag{23}$$

Multiplying this equation by A_P, we obtain $\Delta u = -\Gamma_P A_P x^P = -\Gamma_P b$. Here, unlike to (20)

$$\Gamma_P = \left(A_P S_P A_P^T\right)^{-1}. \tag{24}$$

Observe that the matrix (24) coincides with the matrix (20) in the case, where H_B is the matrix of a basis in the whole space $\mathbb{R}^m$. Therefore, H_B is a square nonsingular matrix of order m.

According to (17) and (23) $\Delta x^P = S_P A_P^T \Gamma_P b - x^P$, $\Delta x^N = 0$. Using the relation $\Delta y = A^T \Delta u$, we obtain also: $\Delta y^P = -A_P^T \Gamma_P b$, $\Delta y^N = -A_N^T \Gamma_P b$.

The solution of the system (16) in x-space in all three cases can be represented in the form (21). However, the matrix Γ_P is a zero matrix in the case (a), and it coincides with the matrix (24) in the case (c). The same remark, concerning directions in the y-space, also can be done.

4   Newton's Directions at Irregular Pairs

At irregular pairs $[x, y]$ the set $J_Z(x, y)$ is not empty, i.e. $n_Z = |J_Z(x, y)| > 0$. At these pairs there exists at least one zero row in the matrix $W(x, y)$. Therefore, the matrix $W(x, y)$ is necessarily singular at such pairs. However, all components of the right-hand side in (13), corresponding to zero rows of $W(x, y)$, are also zeros. Therefore, the system (13) at irregular pairs $[x, y]$ is undetermined, and thus, (13) has the whole set of solutions.

Suppose that the direction Δx^Z is chosen by some way and is fixed. Then, carrying over the last item in the equality

$$A_P \Delta x^P + A_B \Delta x^B + A_N \Delta x^N + A_Z \Delta x^Z = 0$$

to the right hand side, we get from (16) the system of $2n - n_Z$ equations with respect to $2n - n_Z$ unknowns

$$\begin{aligned}
D(y^P)\Delta x^P + D(x^P)\Delta y^P &= -D(x^P)y^P, \\
D(x^B)\Delta y^B = 0, \qquad D(y^N)\Delta y^N &= 0, \\
A_P \Delta x^P + A_B \Delta x^B + A_N \Delta x^N &= -A_Z \Delta x^Z, \\
K_P \Delta y^P + K_B \Delta y^B + K_N \Delta y^N + K_Z \Delta y^Z &= 0.
\end{aligned} \tag{25}$$

The matrix of this system has the form

$$W(x, y) = \begin{bmatrix}
D(y^P) & 0 & 0 & D(x^P) & 0 & 0 & 0 \\
0 & 0 & 0 & 0 & D(x^B) & 0 & 0 \\
0 & 0 & D(y^N) & 0 & 0 & 0 & 0 \\
A_P & A_B & A_N & 0 & 0 & 0 & 0 \\
0 & 0 & 0 & K_P & K_B & K_N & K_Z
\end{bmatrix}. \tag{26}$$

This matrix is derived from the general matrix (14) by deleting from upper and left halves of (14) rows and columns with numbers corresponding to indices from the set $J_Z(x, y)$. The matrix (26) is square of order $2n - n_Z$.

Proposition 3. *Let $[x, y]$ be an irregular pair. Then matrix (26) is nonsingular.*

By Proposition 3 the system (25) has unique solution. Moreover, since $\Delta x^N = 0$ and $\Delta y^B = 0$, the system is reduced to the following one

$$
\begin{aligned}
D(y^P)\Delta x^P + D(x^P)\Delta y^P &= -D(y^P)x^P, \\
A_P\Delta x^P + A_B\Delta x^B &= -A_Z\Delta x^Z, \\
K_P\Delta y^P + K_N\Delta y^N + K_Z\Delta y^Z &= 0.
\end{aligned}
\tag{27}
$$

Solution of system (27) is similar to solution of system (17) with the exception of the case (a). At irregular pair $[x, y]$ the case (a) is impossible, since due to first inequality (10) necessarily $|J_B(x, y)| < m$. Solutions in the cases (b) and (c) are obtained by analogy with these cases at regular pairs. After carrying out of the corresponding calculations we obtain instead of (21)

$$
\begin{aligned}
\Delta x^P &= \left(S_P A_P^T \Gamma_P A_P x^P - A_Z \Delta x^Z\right) - x^P, \quad \Delta x^N = 0, \\
\Delta x^B &= \left(A_B^T A_B\right)^{-1} A_B^T \left(I - A_P S_P A_P^T \Gamma_P\right)\left(A_P x^P - A_Z \Delta x^Z\right).
\end{aligned}
\tag{28}
$$

For directions in y-space we get

$$
\begin{aligned}
\Delta y^P &= -A_P^T \Gamma_P \left(A_P x^P - A_Z \Delta x_Z\right), \quad \Delta y^B = 0, \\
\Delta y^N = -A_N^T \Gamma_P \left(A_P x^P - A_Z \Delta x^Z\right), \quad \Delta y^Z &= -A_Z^T \Gamma_P \left(A_P x^P - A_Z \Delta x^Z\right).
\end{aligned}
\tag{29}
$$

Consider now the problem of choosing Δx^Z in irregular pairs. First of all, in order to preserve the feasibility in problem (1), it is necessary that $\Delta x^Z \geq 0$. Moreover, in order to preserve the feasibility in problem (4), it is necessary that $\Delta y^Z \geq 0$, i.e.

$$
\Delta y^Z = -A_Z^T \Gamma_P \left(A_P x^P - A_Z \Delta x^Z\right) \geq 0.
\tag{30}
$$

Since $H_B^T A_B$ is a zero matrix, the matrix $\Gamma_P A_B$ is also zero. Thus, if we take into account the equality $A_P x^P + A_B x^B = b$, then the inequality (30), can be written as $\Delta y^Z = Q\Delta x^Z + q \geq 0$, where $Q = A_Z^T \Gamma_P A_Z$, $q = -A_Z^T \Gamma_P b$.

As Q is the Gram matrix composed of linearly independent vectors, it is non-negative definite. In fact, the more strong assertion holds.

Proposition 4. *The symmetric matrix Q is positive definite at any feasible irregular pair $[x, y]$.*

Proof. First of all note, that at irregular pair $[x, y]$ due to non-degeneracy of the problem (1) the inequality $|J_B(x, y)| < m$ holds. Note also, that according to Lemma 1 and corollary to it the matrices A_{BZ} and A_Z have full ranks.

Let $|J_B(x, y)| > 0$. We take some nonzero vector $u \in \mathbb{R}^{n_Z}$ and denote by w the vector $w = H_B A_Z u$. This vector w is also nonzero since otherwise, the nonzero vector $A_Z u$ would be orthogonal to all columns of the matrix H_B. But these columns form a basis in the nullspace of the matrix A_B. Therefore, it means that the vector $A_Z u$ belongs to the rangespace of the matrix A_B. Hence, the columns of the matrix A_{BZ} are linear dependent. Since the matrix A_{BZ}

has full rank, this is impossible. From here, taking into account the positive definiteness of the matrix $H_B^T A_P S_P A_P^T H^B$, we get

$$\langle u, Qu \rangle = \langle w, \left(H_B^T A_P S_P A_P^T H^B \right)^{-1} w \rangle > 0.$$

Thus, the matrix Q is positive definite.

In the case, where $J_B(x, y) = \emptyset$, the matrix Q has the form $A_Z^T (A_P S_P A_P^T)^{-1} A_Z$. Therefore, the inequality

$$\langle u, Qu \rangle = \langle A_Z u, (A_P S_P A_P^T)^{-1} A_Z u \rangle > 0$$

follows from the fact that the matrix A_Z has full rank.   □

At the irregular pair $[x, y]$ the complementarity condition, i.e. equalities $x^i y^i = 0$, are carried out for all indices $i \in J_Z(x, y)$. Moreover, it is desirable to choose the direction Δx^Z in such a way that at the updated pair these equalities do not violate. We derive from here the auxiliary linear complementarity problem (LCP):

$$\Delta y^Z = Q \Delta x^Z + q, \quad \Delta x^Z \geq 0, \quad \Delta y^Z \geq 0, \quad \langle \Delta x^Z, \Delta y^Z \rangle = 0. \quad (31)$$

Since the matrix Q is positive definite, the solution of problem (31) always exists, and what is more, this solution is unique [17]. Hence, taking at irregular pairs the solutions of problem (31) as the directions Δx^Z and Δy^Z, we obtain that the Newton's directions Δx and Δy are defined by unique manner. Moreover, $\Delta x^i \Delta y^i = 0$ for all indices i from the set

$$J_{BNZ}(x, y) = J_B(x, y) \cup J_N(x, y) \cup J_Z(x, y), \quad (32)$$

Therefore, the complementarity conditions for these indices are preserved at the updated pair.

It follows from the inequalities (8)–(10) that $|J_Z(x, y)| \leq |J_P(x, y)|$. Thus, the dimension of the auxiliary LCP does not exceed the number of indices in the set $J_P(x, y)$. Moreover, as it had been mentioned above, this dimension does not more than any of two numbers m or $n - m$.

Observe, that Newton's directions both at regular and irregular pairs $[x, y]$ can be defined also, using the approach, in which the main role play the number of indices in the set $J_N(x, y)$ instead of $J_B(x, y)$. In this case we need in other basis in the sub-space $\mathcal{R}^{\perp}(K_N)$, defined by columns of another matrix H_N.

5   The Newton Steepest Descent Algorithm

At any feasible pair $[x, y]$ the direction Δx belongs to the nullspace of the matrix A and the direction Δy belongs to the nullspace of the matrix K. Since these two sub-spaces are orthogonal to each other, the equality $\langle \Delta x, \Delta y \rangle = 0$ holds. Moreover, because of $\Delta y^B = 0$ and $\Delta x^N = 0$, and because of $\langle \Delta x^Z, \Delta y^Z \rangle = 0$, if the pair $[x, y]$ is irregular, we obtain that $\langle \Delta x^P, \Delta y^P \rangle = 0$. At optimal pair $[x, y]$ due to homogeneity of system (13) and non-singularity of its matrix we derive, that $\Delta x = 0_n$, $\Delta y = 0_n$.

Proposition 5. *Let the feasible non-optimal pair $[x, y]$ be regular. If the point x is a vertex of the set X, then $\Delta x = 0_n$. Similarly, if the point y is a vertex of the set Y, then $\Delta y = 0_n$.*

Proof. We will prove only the first assertion. Since x is a vertex of the set X, the following equality $|J_P(x, y)| + |J_B(x, y)| = m$ holds, and, what is more, the square sub-matrix $A_{PB} = [A_P, A_B]$ of the matrix A is nonsingular. Furthermore, according to (17) the equality $A_P \Delta x^P + A_B \Delta x^B = 0_m$ takes place. From here, taking into account the linear independence of the columns of the matrix A_{PB}, we get that $\Delta x^P = 0$, $\Delta x^B = 0$. As $\Delta x^N = 0$, we obtain that $\Delta x = 0_n$. □

It can be shown that, if both problems (1) and (4) are strictly non-degenerate, then at any feasible pair $[x, y]$ (regular or irregular) $\Delta x = 0_n$ only when x is a vertex of the feasible set X. Similarly, $\Delta y = 0_n$ only when y is a vertex of the feasible set Y. The following assertion is also takes place

Proposition 6. *Let problems (1) and (4) be strictly non-degenerate. Let also $[x, y]$ be a feasible pair. Then $\langle c, \Delta x \rangle \le 0$, $\langle f, \Delta y \rangle \le 0$. Moreover, the equalities are possible if and only if x is a vertex of the set X or y is a vertex of the set Y.*

Assume that the starting feasible pair $[x_0, y_0]$ is given. Assume also that after k steps of the iterative process the feasible pair $[x_k, y_k]$ is obtained. On the next $k + 1$ iteration we set

$$x_{k+1} = x_k + \alpha_k \Delta x_k, \quad y_{k+1} = y_k + \beta_k \Delta y_k, \tag{33}$$

where the search directions Δx_k and Δy_k are calculated by formulas (28) and (29) at $x = x_k$, $y = y_k$. Moreover, if the pair $[x_k, y_k]$ is regular, the components Δx_k^Z and Δy_k^Z are determined from the solution of LCP (31). The step-lengths α_k and β_k are taken in such a way that the updated pair $[x_{k+1}, y_{k+1}]$ is feasible.

Introduce into consideration the function $V(x, y) = \langle x, y \rangle$, defined on the set $X \times Y$. If x_* and y_* are the solutions of the problems (1) and (4), respectively, then $V(x_*, y_*) = 0$. For any other feasible pair $[x, y]$ the inequality $V(x, y) > 0$ is fulfilled. Let us compute the variation of the value $V(x, y)$ under the passage from the pair $[x_k, y_k]$ to the next pair $[x_{k+1}, y_{k+1}]$. Taking into account the orthogonality of directions Δx_k and Δy_k, we have:

$$\begin{aligned} V(x_k + \alpha \Delta x_k, y_k + \beta \Delta y_k) &= \langle x_k + \alpha \Delta x_k, y_k + \beta \Delta y_k \rangle \\ &= V(x_k, y_k) + \alpha \sigma_1(x_k, y_k) + \beta \sigma_2(x_k, y_k), \end{aligned} \tag{34}$$

where $\sigma_1(x, y) = \langle y, \Delta x \rangle$, $\sigma_2(x, y) = \langle x, \Delta y \rangle$.

Denote by e the vector with all components equal to one. Multiplying by e^T the first equation from (17) or the first equation from (27), we get

$$\sigma_1(x, y) + \sigma_2(x, y) = -V(x, y). \tag{35}$$

Moreover, by results, stated above: $\sigma_1(x, y) = \langle y^P, \Delta x^P \rangle$, $\sigma_2(x, y) = \langle x^P, \Delta y^P \rangle$.

Let us verify that $\sigma_1(x, y) \le 0$, $\sigma_2(x, y) \le 0$ at any feasible pair $[x, y]$, which does not coincide with the solutions of problems (1), and (4). Moreover, at least one of these inequalities is strict.

Proposition 7. *Let the feasible pair* $[x, y]$ *do not coincide with the solutions of problems (1) and (4). Then* $\sigma_1(x, y) \leq 0$, $\sigma_2(x, y) \leq 0$ *and*

$$\min\{\sigma_1(x, y), \sigma_2(x, y)\} < 0. \tag{36}$$

Proof. Since the feasible pair $[x, y]$ is non-optimal, the set $J_P(x, y)$ is not empty and $V(x, y) > 0$. Then after multiplication of first equations in (17) or (27) from the left by $(\Delta x^P)^T D^{-1}(x^P)$ we get $\sigma_1(x, y) = -\langle \Delta x^P, S_P^{-1} \Delta x^P \rangle \leq 0$. Moreover, the equality takes place if and only if $\Delta x^P = 0$. Similarly, we obtain that $\sigma_2(x, y) \leq 0$.

The inequality (36) follows from (35), since $V(x, y) > 0$ and both terms in left hand side are non-negative. $\square$

It follows from (34) and Proposition 7 that, in order to obtain the maximal decrease of the value of the function $V(x, y)$ at iteration, the step-lengths α_k and β_k must be chosen maximal possible under the condition that the updated points x_{k+1}, y_{k+1} do not abandon the feasible sets, that is

$$\alpha_k = \arg\max\left\{\alpha \geq 0: \ x_k + \alpha \Delta x_k \in \mathbb{R}_+^n\right\},$$
$$\beta_k = \arg\max\left\{\beta \geq 0: \ y_k + \beta \Delta y_k \in \mathbb{R}_+^n\right\}.$$

The iterations are carried out until the condition $V(x_k, y_k) = 0$ will be fulfilled.

Since $\Delta x^N = 0$ and $\Delta y^B = 0$, not all components of the vectors x_k and y_k are changed at each iteration, but only part of them, to be exact,

$$x_{k+1}^P = x_k^P + \alpha_k \Delta x_k^P, \quad x_{k+1}^B = x_k^B + \alpha_k \Delta x_k^B, \tag{37}$$
$$y_{k+1}^P = y_k^P + \beta_k \Delta y_k^P, \quad y_{k+1}^N = y_k^N + \beta_k \Delta y_k^N.$$

Aside from these components, at irregular pair $[x_k, y_k]$ the components x_k^Z and y_k^Z also may be changed: $x_{k+1}^Z = \alpha_k \Delta x_k^Z$, $y_{k+1}^Z = \beta_k \Delta y_k^Z$. Moreover, these components may only increase. Thus, only the components Δx_k^P, Δx_k^B and Δy_k^P, Δy_k^N are influenced on the choose of step-lengths.

By analogy with [18] introduce the indicator vectors μ^P, μ^B and μ^N, setting

$$\mu^P = e + D^{-1}(x^P)\Delta x^P = -D^{-1}(y^P)\Delta y^P,$$
$$\mu^B = e + D^{-1}(x^B)\Delta x^B, \quad \mu^N = -D^{-1}(y^N)\Delta y^N. \tag{38}$$

Denote also

$$\mu_{\min}^P = \min_{i \in J_P(x_k, y_k)} \mu^i, \quad \mu_{\min}^B = \min_{i \in J_B(x_k, y_k)} \mu^i, \quad \mu_{\min}^{PB} = \min\left[\mu_{\min}^P, \mu_{\min}^B\right], \tag{39}$$

$$\mu_{\max}^P = \max_{i \in J_P(x_k, y_k)} \mu^i, \quad \mu_{\max}^N = \max_{i \in J_N(x_k, y_k)} \mu^i, \quad \mu_{\max}^{PN} = \max\left[\mu_{\max}^P, \mu_{\max}^N\right].$$

Then

$$\alpha_k = \left[1 - \mu_{\min}^{PB}\right]_+^{-1}, \quad \beta_k = \left[\mu_{\max}^{PN}\right]_+^{-1}, \tag{40}$$

where $a_+ = \max[0, a]$.

Lemma 2. *At any feasible non-optimal pair* $[x_k, y_k]$ *the following inequalities* $\mu_{\min}^P \leq 1$ *and* $\mu_{\max}^P \geq 0$ *hold. Moreover, from two inequalities* $\mu_{\min}^P \leq 0$ *and* $\mu_{\max}^P \geq 1$ *at least one is always takes place.*

Proof. Let the matrix Γ_P has the general form (20). Then the expression for μ^P from (38) may be written as $\mu^P = D^{-1}(\psi^P)\mathcal{P}\psi^P$, where $\mathcal{P} = S_P^{1/2} A_P^T \Gamma_P A_P S_P^{1/2}$ and $\psi_P = D^{1/2}(x^P)D^{1/2}(y^P)e$.

Assume that $\mu_{\min}^P > 1$. Then $\mu^P > e$, and, thus, $\mathcal{P}\psi^P > \psi^P > 0$. Therefore, $\langle \mathcal{P}\psi^P, \mathcal{P}\psi^P \rangle > \langle \mathcal{P}\psi^P, \psi^P \rangle$. However, this inequality is impossible, since for the orthogonal projector $\mathcal{P}$ the equality

$$\langle \mathcal{P}\psi^P, \mathcal{P}\psi^P \rangle = \langle \mathcal{P}^2\psi^P, \psi^P \rangle = \langle \mathcal{P}\psi^P, \psi^P \rangle \tag{41}$$

holds. Hence, $\mu_{\min}^P \leq 1$.

Similarly, if all components of the vector μ^P negative, then $\mathcal{P}\psi^P < 0$. Thus, $\langle \mathcal{P}\psi^P, \psi^P \rangle < 0$ for $\psi^P > 0$. But

$$\langle \mathcal{P}\psi^P, \psi^P \rangle = \langle \mathcal{P}^2\psi^P, \psi^P \rangle = \langle \mathcal{P}\psi^P, \mathcal{P}\psi^P \rangle = \|\mathcal{P}\psi^P\|^2 \geq 0.$$

This inequality contradicts to the previous one.

At last, let us show, that from two inequalities $\mu_{\min}^P \leq 0$ and $\mu_{\max}^P \geq 1$ at least one is always takes place. Otherwise, we have $0 < \mathcal{P}\psi^P < \psi^P$. Hence, $\langle \mathcal{P}\psi^P, \mathcal{P}\psi^P \rangle < \langle \mathcal{P}\psi^P, \psi^P \rangle$. This inequality contradicts (41). □

Proposition 8. *From two step-lengths* α_k *and* β_k *at least one step-length is not more than one.*

Proof. This assertion follows from (40) and from Lemma 2, since $\mu_{\min}^{PB} \leq \mu_{\min}^P$ and $\mu_{\max}^{PN} \geq \mu_{\max}^P$. □

Denote

$$J_{BZ}(x,y) = J_B(x,y) \cup J_Z(x,y), \quad J_{NZ}(x,y) = J_N(x,y) \cup J_Z(x,y).$$

According to (28), (29) and (40) at every k^{th} iteration the following inclusions

$$J_B(x_k, y_k) \subseteq J_{BZ}(x_{k+1}, y_{k+1}), \quad J_N(x_k, y_k) \subseteq J_{NZ}(x_{k+1}, y_{k+1}). \tag{42}$$

are valid. Moreover, $J_Z(x_k, y_k) \subseteq J_{BNZ}(x_{k+1}, y_{k+1})$, where the set $J_{BNZ}(x,y)$ is defined in (32).

Lemma 3. *Let* $J_{PBN}(x,y) = J_P(x,y) \cup J_B(x,y) \cup J_N(x,y)$. *Then at every* k^{th} *iteration* $J_P(x_k, y_k) \subseteq J_{PBN}(x_{k+1}, y_{k+1})$.

Proof. It is sufficient to show that for any index $i \in J_P(x_k, y_k)$ the inclusion $i \in J_Z(x_{k+1}, y_{k+1})$ is impossible. By contradiction, assume that such index $i \in J_P(x_k, y_k)$ exists. Since

$$x_{k+1}^P = D(x_k^P)\left[e + \alpha_k\left(\mu_k^P - e\right)\right], \qquad y_{k+1}^P = D(y_k^P)\left[e - \beta_k\mu_k^P\right],$$

we have $1 + \alpha_k(\mu^i - 1) = 0$ and $1 - \beta_k\mu^i = 0$. These equalities are possible if and only if $0 < \mu^i < 1$. Hence, $\alpha_k = (1 - \mu^i)^{-1} > 1$ and $\beta_k = (\mu^i)^{-1} > 1$. These two inequalities contradict to Proposition 8. □

Proposition 9. *At every k^{th} iteration the inclusion*

$$J_P(x_{k+1}, y_{k+1}) \subseteq J_P(x_k, y_k) \qquad (43)$$

holds. Thus, if at some k^{th} iteration the index i leaves the set $J_P(x_k, y_k)$, that is $i \notin J_P(x_{k+1}, y_{k+1})$, then it can not return into the index set J_P conversely on consequent iterations.

Proof. The inclusion (43) follows from (42) and assertion of Lemma 3. □

Definition 5. *By non-complementarity degree of the feasible pair $[x, y]$ we will call the number of indices in the set $J_P(x, y)$.*

If at feasible pair $[x, y]$ the non-complementarity degree is equal to zero, then this pair is optimal. The irregular pair $[x, y]$ with the non-complementarity degree equal to one consists from vertices of the sets X and Y.

According to Proposition 9 the non-complementarity degree does not increase during the iterations. In what follows, we will say that the kth iteration is *active*, if the non-complementarity degree of the pair $[x_{k+1}, y_{k+1}]$ is strictly less than the non-complementarity degree of the pair $[x_k, y_k]$. Observe, that if the iteration is not active, then the consequent pair $[x_{k+1}, y_{k+1}]$ without fail will be irregular. Hence, if the pair $[x_{k+1}, y_{k+1}]$ is regular, then the active iteration occurs. Clearly, if the method (33) solves problems (1) and (4), then it solves these problems at most n active iterations.

Theorem 1. *Let the method (33) solve the problems (1) and (2) at k_* iterations. Let also all pairs $[x_k, y_k]$, $0 \le k \le k_*$ be regular. Then $k_* \le n$.*

Consider now the case, where there are non-active iterations among all iterations. In other words, consider the case, where there are irregular pairs $[x_{k+1}, y_{k+1}]$ among all pairs. Observe, that it is possible only when at $[x_k, y_k]$ at least one point x_k or y_k is a vertex of the feasible set.

Proposition 10. *Let $[x_k, y_k]$ be a regular pair, and let the point x_k be a non-optimal vertex of the set X. Then the updated pair $[x_{k+1}, y_{k+1}]$ is irregular and*

$$J_Z(x_{k+1}, y_{k+1}) \subseteq J_N(x_k, y_k), \qquad J_P(x_{k+1}, y_{k+1}) = J_P(x_k, y_k).$$

Moreover, an active iteration will occur at finite number of iterations or the process (33) will leave this vertex.

The similar assertions are valid for vertices of the set Y.

Theorem 2. *Let the starting pair $[x_0, y_0]$ be feasible. Then the optimal pair will be obtained by the method (33) at finite number of iterations.*

Proof. Proof follows from Proposition 10 and arguments given above, since the number of vertices is finite. □

By Theorem 2 the method solves both problems (1) and (4) at finite number of iterations. However, the behavior of the method essentially depends on a number of vertices, which belong to the trajectory. The best situation is when the trajectory does not contain such vertices. By Theorem 1 the solutions of (1) and (4) can be obtained in this case at number of iterations which does not exceed the number of variables in (1) and (4).

6   Conclusion

We presented the feasible primal-dual algorithm for solving linear programming problems which can be considered as the special variant of the known interior-point methods. In our variant of the method, we do not introduce the safety factor and allow to trajectories move along the boundaries of the feasible sets. The combinatorial structure of the boundaries is used in algorithms in order to obtain the optimal solutions at a finite number of iterations. Therefore, the algorithm can be recommended for application at the final stage of computations, when the points of trajectories are in some neighborhoods of the optimal solutions.

References

1. Luenberger, D.: Linear and Nonlinear Programming. Addison-Wesle, London (1984)
2. Vanderbei, R.J.: Linear Programming. Foundations and Extensions. Kluwer Academic Publishers, Boston (1997)
3. Vasilyev, F.P., Ivanitskiy, A.Y.: In-Depth Analysis of Linear Programming. Springer, Nethelands (2001)
4. Nesterov, Y.E., Nemirovski, A.: Interior Point Polynomial Algorithms in Convex Programming. SIAM Publications, Philadelphia (1994)
5. Wright, S.J.: Primal-Dual Interior-Point Methods. SIAM, Philadelphia (1997)
6. Roos, C., Terlaky, T., Vial, J-Ph: Theory and Algorithms for Linear Optimization an Interior Point Approach. Wiley, Chichester (1997)
7. Gondzio, J.: Interior point methods 25 years later. Eur. J. Oper. Res. **218**, 587–601 (2012)
8. Kojima, M., Mizuno, S., Yoshise, A.: A primal-dual interior point method for linear programming. In: Megiddo, N. (ed.) Progress in Mathematical Programming. Interior Point and Related Methods, pp. 29–47. Springer, Berlin (1989). https://doi.org/10.1007/978-1-4613-9617-8_2
9. Kojima, M., Megiddo, N., Mizuno, S.: A primal-dual infeasible-interior point method for linear programming. Math. Program. **61**, 263–280 (1993)
10. Monteiro, R.C., Adler, I.: Interior path-following primal-dual algorithms. Part I: linear programming. Math. Program. **44**, 27–41 (1989)
11. Todd, M.J., Ye, Y.: A centered projective algorithm for linear programming. Math. Oper. Res. **15**, 508–529 (1990)
12. Ye, Y., Guler, O., Tapia, R., Zhang, Y.: A quadratically convergent $O(\sqrt{n}L)$-iteration algorithm for linear programming. Math. Program. **59**, 151–162 (1993)
13. Gonzaga, C.: Path-following methods for linear programming. SIAM Rev. **34**, 167–224 (1992)
14. Zhadan, V.G.: Primal-dual Newton method for linear programming problems. Comp. Maths. Math. Phys. **39**, 14–28 (1999)
15. McShane, K., Monma, C., Shanno, D.: On implementation of primal-dual interior point method for linear programming. ORSA J. Comput. **1**, 70–89 (1989)
16. Lustig, I.J., Monma, C., Shanno, D.: On implementing Mehrotra's predictor-corrector interior point method for linear programming. SIAM J. Optim. **2**, 435–449 (1992)

17. Cottle, R.W., Pang, J.-S., Stone, R.E.: The Linear Complementarity Problem. Academic press Inc., Boston (1992)
18. El-Bakry, A.S., Tapia, R.A., Zhang, Y.: A study of indicators for identifying zero variables in interior point methods. SIAM Rev. **36**, 45–72 (1994)

Combinatorial and Discrete Optimization

Sufficient Conditions of Polynomial Solvability of the Two-Machine Preemptive Routing Open Shop on a Tree

Ilya Chernykh[1,2,3(✉)]

[1] Sobolev Institute of Mathematics, Novosibirsk, Russia
`idchern@math.nsc.ru`
[2] Novosibirsk State University, Novosibirsk, Russia
[3] Novosibirsk State Technical University, Novosibirsk, Russia

Abstract. The routing open shop problem with preemption allowed is a natural combination of the metric TSP problem and the classical preemptive open shop scheduling problem. While metric TSP is strongly NP-hard, the preemptive open shop is polynomially solvable for any (even unbounded) number of machines. The previous research on the preemptive routing open shop is mostly focused on the case with just two nodes of the transportation network (problem on a link). It is known to be strongly NP-hard in the case of an unbounded number of machines and polynomially solvable for the two-machine case. The algorithmic complexity of both two-machine problem on a triangular network and a three-machine problem with two nodes are still unknown. The problem with a general transportation network is a generalization of the metric TSP and therefore is strongly NP-hard.

We describe a wide polynomially solvable subclass of the preemptive routing open shop on a tree. This class allows building an optimal schedule with at most one preemption in linear time. For any instance from that class optimal makespan coincides with the standard lower bound. Therefore, the result, previously known for the problem on a link, is generalized on a special case on an arbitrary tree. The algorithmic complexity of the general case of the two-machine problem on a tree remains unknown.

Keywords: Scheduling · Routing open shop
Preemption · Polynomially solvable subclass
Overloaded node · Overloaded edge

1 Introduction

The open shop scheduling problem, introduced in [8], can be described as follows. Sets $\mathcal{M} = \{M_1, \ldots, M_m\}$ of machines and $\mathcal{J} = \{J_1, \ldots, J_n\}$ of jobs are given

This research was supported by the program of fundamental scientific researches of the SB RAS No I.5.1., project No 0314-2016-0014, and by the Russian Foundation for Basic Research, projects 17-01-00170, 17-07-00513 and 18-01-00747.

© Springer Nature Switzerland AG 2019
Y. Evtushenko et al. (Eds.): OPTIMA 2018, CCIS 974, pp. 97–110, 2019.
https://doi.org/10.1007/978-3-030-10934-9_7

and each machine M_i has to process an operation on each job J_j, this operation O_{ji} requires $p_{ji} \geqslant 0$ time units to complete. Operations of each job have to be processed in an arbitrary order (which has to be chosen by a scheduler). Any machine cannot process more than one operation simultaneously. The goal is to minimize the *makespan* $C_{\max}$, i.e. maximal completion time of the operation. Following the standard three-field notation (see [10] for example) the open shop problem with m machines is denoted by $Om||C_{\max}$ (or $O||C_{\max}$ if the number of machines is not a constant). It is known ([8]) to be polynomially solvable in the case of two machines and is NP-hard for $m \geqslant 3$. The problem $O||C_{\max}$ with the unbounded m is strongly NP-hard and for any $\rho < \frac{5}{4}$ no ρ-approximation algorithm for $O||C_{\max}$ exists (unless $P = NP$) [14].

On the other hand, the open shop problem becomes polynomially solvable (even with unbounded m) if we allow preemption of the operations [8]. In this case each operation O_{ji} can be separated into arbitrary number of *fragments* with total processing time equals p_{ji}. The preemptive open shop problem is denoted by $O|pmtn|C_{\max}$. It was shown in [8] that the optimal makespan for $O|pmtn|C_{\max}$ always coincides with the standard lower bound

$$\bar{C} \doteq \max\{\max_i \sum_{j=1}^{n} p_{ji}, \max_j \sum_{i=1}^{m} p_{ji}\}.$$ Note that the last property also holds for

the $O2||C_{\max}$ problem.

Most of the classical scheduling models (open shop included) share the following disadvantage. It is supposed that each machine is able to start any operation at the same time moment it completed the previous one. In a real-life environment that's not always the case. Usually, jobs represent some material objects, therefore some delays between processing operations of two subsequent jobs may be unavoidable. Such delays can be machine-dependent, job- or sequence-dependent, and taking them into account can make the problem harder to investigate. Still, there is a number of papers considering such *transportation delays* (see [3,11,13] for example). However, the problem we are considering in this paper uses a different approach to model transportation delays.

We consider the *routing open shop* problem [2] which can be described as the open shop meeting the metric traveling salesman problem (TSP). Let the input of the TSP be given by an edge-weighted graph G. Jobs from $\mathcal{J}$ are distributed between the nodes of G, each node contains at least one job. Machines are mobile and are initially located at the predefined node referred to as *the depot*. Machines have to travel over the edges of G, weights of the edges represent travel times for each machine and satisfy the triangle inequality. Any number of machines can travel over the same edge at the same time. All the machines have to visit each node of G, process all the respective operations (under the feasibility constraints from the open shop problem), and to return back to the depot. The makespan $R_{\max}$ is the *return time moment* of chronologically last machine and has to be minimized. The problem with m machines is denoted by $ROm||R_{\max}$, or $ROm|G = X|R_{\max}$ if we want to specify the structure X of the graph G. In the latter case we use either classic notation from graph theory, like

K_p for the *complete graph* with p vertices, or graph's typical names, like *tree* or *chain*.

The general routing open shop problem contains the metric TSP as a special case, moreover, the problem with the single machine is equivalent to the metric TSP and therefore is strongly NP-hard. On the other hand, the problem with zero travel times (or with $G = K_1$) is equivalent to the open shop problem and is NP-hard for $m \geqslant 3$. However, the problem remains NP-hard even in the simplest problem case $RO2|G = K_2|R_{max}$ [2]. An FPTAS for that case is described in [9].

The preemptive variant of the routing open shop problem $ROm|pmtn|R_{max}$ is still poorly investigated. The paper [12] focuses on the research of the problem with $G = K_2$. It is shown that $RO2|pmtn, G = K_2|R_{max}$ is polynomially solvable (and in this case the optimal makespan coincides with the standard lower bound, see Sect. 2 for details), while the problem $RO|pmtn, G = K_2|R_{max}$ is strongly NP-hard. The algorithmic complexity of the problem $ROm|pmtn, G = K_2|R_{max}$ is still an open question for any constant $m \geqslant 3$. One more polynomially solvable special case of the $RO2|pmtn|R_{max}$ is described in [5]. This case is based on the existence and location of so-called *overloaded node* in graph G (see Sect. 2.) The algorithmic complexity of $RO2|pmtn, G = K_3|R_{max}$ is unknown for the time being.

In this paper, we consider the problem $RO2|pmtn, G = tree|R_{max}$, and describe several special cases which are solvable to the optimum in linear time, with the optimal makespan equals the standard lower bound. The special cases are formulated in terms of load distribution between the nodes, the formulation involves the definitions of the overloaded node and *overloaded edge* (Sect. 2). We describe an algorithm which builds an optimal schedule with at most one preemption, therefore generalizing the polynomially solvable case from [12] to a problem where graph G is an arbitrary tree with one of our special conditions. The algorithmic complexity of the general problem $RO2|pmtn, G = tree|R_{max}$ remains unknown.

The structure of the remainder of the paper is as follows. Section 2 contains a detailed problem description, necessary notation and the formulation of known results we use. In Sect. 3 we describe the procedure of instance reduction, which is the main part of our algorithm. Special conditions of the polynomial solvability are described in Sect. 4, followed by the description of the main algorithm and the proof of its optimality. Concluding remarks and some open questions are given in Sect. 5.

2   Preliminary Notes

Let us give a formal description of the routing open shop problem with preemption allowed.

A problem instance combines inputs from the metric TSP and the open shop problem in the following manner. A connected graph $G = \langle V, E \rangle$ is given, a non-negative weight function $\tau : E \to \mathbb{Z}_{\geqslant 0}$ is defined. The function τ satisfies the triangle inequality. One of the nodes $v_0 \in V$ is chosen to be the *depot*. Jobs from the set $\mathcal{J} = \{J_1, \ldots, J_n\}$ are distributed among the nodes from V. A set of jobs located at $v \in V$ is denoted by $\mathcal{J}(v)$ and is non-empty for any node with the possible exclusion of the depot. Machines from the given set $\mathcal{M} = \{M_1, \ldots, M_m\}$ are initially located at the depot and each machine can travel over the edges of G, the travel time of each machine for an edge $e \in E$ equals $\tau(e)$. Any number of machines can travel over the same edge in any direction at the same time. Each machine M_i has to perform an *operation* O_{ji} on every job J_j. This operation takes $p_{ji} \in \mathbb{Z}_{\geqslant 0}$ time units and requires the machine to be at the location of J_j: while the machine is in the node v, it can only process operations of jobs from $\mathcal{J}(v)$. Different operations of the same job cannot be processed simultaneously, and each machine can process at most one operation at a time. Machines are allowed to interrupt the processing of any operations but required to resume (and complete) the operation later. It means that any operation O_{ji} can be divided into any number of suboperations referred to as the *fragments* with total processing time equals to p_{ji}. Machines have to return to the depot after processing all the operations. We use notation $p_{ji}(I)$, $G(I)$, $\tau(I; e)$ and $\mathcal{J}(I; v)$, if we want to specify a problem instance I.

A *schedule* S can be described by specifying the processing interval for all fragments of each operation:

$$S = \left\{ \{[s_1(O_{ji}), c_1(O_{ji})], \ldots, [s_{\xi_{ji}}(O_{ji}), c_{\xi_{ji}}(O_{ji})]\} \,\middle|\, i = 1, \ldots, m,\ j = 1, \ldots, n \right\}.$$

Notation $s(O_{ji}) \doteq s_1(O_{ji})$ and $c(O_{ji}) \doteq c_{\xi_{ji}}(O_{ji})$ specifies the starting time and the completion time of the operation O_{ji} in some current schedule. We also use $s(O)$ and $c(O)$ to denote the starting and the completion times, respectively, for some fragment O. We also use the following notation to indicate the total number of fragments in schedule S:

$$\xi(S) \doteq \sum_{i,j} (\xi_{ji} - 1).$$

Note that $\xi(S) = 0$ means that schedule S is nonpreemptive.

Let $\mathrm{dist}(v, u)$ denote the weighted distance between the nodes v and u, i.e. the minimal total weight of edges belonging to some chain connecting v and u. So $\mathrm{dist}(v, u)$ is the shortest time needed for a machine to reach u from v. We also use notation $\mathrm{dist}(I; v, u)$ for a specific instance I.

Definition 1. *A schedule S for an instance I is referred to as* feasible *if it satisfies the following conditions:*

1. *For each operation O_{ji} inequalities $s_1(O_{ji}) \leqslant c_1(O_{ji}) \leqslant \cdots \leqslant s_{\xi_{ji}}(O_{ji}) \leqslant c_{\xi_{ji}}(O_{ji})$ hold and*

$$\sum_{k=1}^{\xi_{ji}} (c_k(O_{ji}) - s_k(O_{ji})) = p_{ji}(I).$$

2. *Let fragments O and O' belong to the same job or are processed by the same machine. Then*

$$\big(s(O), c(O)\big) \cap \big(s(O'), c(O')\big) = \emptyset.$$

3. *If operation of job $J_j \in \mathcal{J}(v)$ is the first to start by machine M_i then*

$$s(O_{ji}) \geqslant \mathrm{dist}(I; v_0, v).$$

4. *If machine M_i processes a fragment O of operation O_{ji} before the processing of a fragment O' of operation $O_{j'i}$, $J_j \in \mathcal{J}(v)$, and $J_{j'} \in \mathcal{J}(v')$, then*

$$s(O') \geqslant c(O) + \mathrm{dist}(I; v, v').$$

Suppose an operation of job $J_j \in \mathcal{J}(v)$ is the last to be processed by machine M_i in a schedule S. Then we define the *return time* of machine M_i as

$$R_i(S) \doteq c(O_{ji}) + \mathrm{dist}(v, v_0).$$

The *makespan* of a schedule S is $R_{\max}(S) \doteq \max_i R_i(S)$. The goal is to find a feasible schedule minimizing the makespan.

We use the following notation for some problem instance I.

- $\ell_i(I) \doteq \sum_{j=1}^{n} p_{ji}(I)$—the *load* of machine M_i;
- $\ell_{\max}(I) \doteq \max_i \ell_i(I)$—the *maximal machine load*;
- $d_j(I) \doteq \sum_{i=1}^{m} p_{ji}(I)$—the *length* of job J_j;
- $d_{\max}(I; v) \doteq \max_{J_j \in \mathcal{J}(v)} d_j(I)$ — the maximal job length at node v;
- $\Delta(I; v) \doteq \sum_{J_j \in \mathcal{J}(v)} d_j(I)$—the *load* of node v;
- $\Delta(I) \doteq \sum_{v \in V} \Delta(I; v)$—the *total load* of instance I;
- $T^*(I)$—the optimum for an underlying TSP, i.e. the length of the shortest cyclic route visiting each node at least once;
- $R^*_{\max}(I)$—the optimal makespan.

We omit I from the notation in case when it does not lead to a confusion.

The following *standard lower bound* on the optimum for the routing open shop problem was introduced in [1]:

$$\bar{R}(I) \doteq \max \left\{ \ell_{\max}(I) + T^*(I), \max_{v \in V}\big(d_{\max}(I; v) + 2\mathrm{dist}(I; v_0, v)\big) \right\}. \quad (1)$$

Note that $\bar{R}$ is still a lower bound for the preemptive version of the routing open shop problem. It also coincides with $\bar{C}$ in case all edges have zero weight (in this case our problem is reduced to the classical open shop problem).

Our study is focused on the case of two machines. In this case we use simplified notation for the operations of each job J_j: a_j and b_j instead of O_{j1} and O_{j2}, respectively. Moreover, we use the same notation (a_j and b_j) for operations' processing times whenever it does not lead to a confusion.

3   The Procedure of Instance Reduction

In this section, we study some general properties of an instance of $RO2||R_{\max}$ and describe the reduction procedure which helps to reduce the number of jobs and to simplify the graph structure preserving the standard lower bound $\bar{R}$. One of the important properties of the procedure is its *reversibility*: any feasible schedule for a reduced instance can be treated as a feasible schedule with the same makespan for the initial instance. In general case, this procedure can increase the optimal makespan. However, in the next section, we prove that for our special case of $RO2|pmtn, G = tree|R_{\max}$ the instance reduction procedure also preserves the optimum. Therefore, it can be used as a part of an exact algorithm for solving the initial instance to the optimum. The procedure is based on two types of instance transformation: *job aggregation* and *terminal edge contraction*. The first one is described in detail in [6], while the second was used in [4] for a certain generalization of the routing open shop problem. We provide all the necessary details below.

Definition 2. *Let I be an instance of the problem $ROm||R_{\max}$ with graph $G = \langle V; E \rangle$, and $\mathcal{K} \subseteq \mathcal{J}(I; v)$ for some $v \in V$. Then we say that instance I' is obtained from I by* aggregation *of jobs from $\mathcal{K}$ if*

$$\mathcal{J}(I'; v) \doteq \mathcal{J}(I; v) \setminus \mathcal{K} \cup \{J_{j_{\mathcal{K}}}\}, \ \forall i = 1, \ldots, m \ p_{j_{\mathcal{K}} i}(I') \doteq \sum_{J_j \in \mathcal{K}} p_{ji}(I),$$

$$\forall u \neq v \ \mathcal{J}(I'; u) = \mathcal{J}(I; u).$$

(Here $j_{\mathcal{K}}$ is some new job index. A job $J_{j_{\mathcal{K}}}$ is to replace the set of jobs $\mathcal{K}$.) An instance $\tilde{I}$ obtained from I by a series of job aggregations is referred to as an aggregation *of I.*

It is easy to observe that any feasible schedule (preemptive or non-preemptive) for an aggregation $\tilde{I}$ can be treated as a feasible schedule for the initial instance I: one just need to replace an aggregated operation with a sequence of operations of jobs from $\mathcal{K}$ to be processed in any order with no idle time. Therefore, $R_{\max}^*(\tilde{I}) \geqslant R_{\max}^*(I)$. Also as soon as we obtained a new job $J_{j_{\mathcal{K}}}$ in $\mathcal{J}(I'; v)$, it is possible that $d_{j_{\mathcal{K}}} > d_{\max}(I; v)$, so job aggregation can lead to the growth of the standard lower bound. Specifically, (1) implies

$$\bar{R}(I') > \bar{R}(I) \text{ if and only if } d_{j_{\mathcal{K}}} > \bar{R}(I) - 2\mathrm{dist}(v_0, v). \tag{2}$$

We use job aggregation to simplify the instance preserving the standard lower bound. Such an aggregation is referred to as a *valid* one. A natural question arises, is it possible to perform a valid job aggregation of a whole set $\mathcal{J}(I; v)$ for some $v \in V$. To answer that question, we use the following definition from [5].

Definition 3. *A node $v \in V$ of an instance I of the problem $ROm||R_{\max}$ is referred to as* overloaded *if*

$$\Delta(I; v) > \bar{R}(I) - 2\mathrm{dist}(I; v_0, v).$$

Otherwise the node is called underloaded.

The job aggregation of the set $\mathcal{J}(I; v)$ is valid if and only if the node v is underloaded. Therefore, any node containing single job is an underloaded one.

By $L_V(I)$ we denote the number of overloaded nodes in an instance I. It was proved in [6] that for every instance I of the $RO2||R_{\max}$ problem $L_V(I) \leqslant 1$. Further in this section we prove a more general result (Proposition 1).

Now let us describe the terminal edge contraction operation.

Definition 4. *Let $v \in V \setminus \{v_0\}$ be some terminal node in graph G, containing a single job J_j in an instance I of the $ROm||R_{\max}$ problem. Let $e = [v, u] \in E$ be the edge incident to v. By the* contraction *of the edge e we understand the following instance transformation:*

$$\mathcal{J}(I'; u) \doteq \mathcal{J}(I; u) \cup \{J_{j'}\};\ p_{j'i}(I') \doteq p_{ji}(I) + 2\tau(e);\ G(I') \doteq G(I) \setminus \{v\}.$$

Consider instance I', obtained from I by the contraction of edge e. Any feasible schedule for I' with no preemption of operations of the transformed job $J_{j'}$ can be treated as a feasible schedule for the initial instance I. One just need to replace a scheduled interval of an operation $O_{j'i}$ with three consecutive intervals: traveling of the machine M_i over the edge e to the node v, performing of the operation O_{ji}, and traveling back to the node u.

Consider the two-machine case of our problem. Again, we want to perform an edge contraction operation only if it does not lead to the growth of the standard lower bound. Otherwise, the edge is called *overloaded*. The following definition describes the exact condition, under which an edge is overloaded.

Definition 5. *Let $v \in V \setminus \{v_0\}$ be some terminal node in graph G, containing a single job J_j in instance I of the $RO2||R_{\max}$ problem. Let $e = [v, u] \in E$ be the edge incident to v. The edge e is referred to as* overloaded *if*

$$d_j(I) + 4\tau(e) > \bar{R}(I) - 2\mathrm{dist}(I; v_0, u). \tag{3}$$

For any problem instance I, we denote the number of overloaded edges by $L_E(I)$. The following property of any instance of $RO2||R_{\max}$ is fundamental for the procedure of instance reduction.

Proposition 1. *Let I be an instance of the problem $RO2||R_{\max}$. Then $L_V(I) + L_E(I) \leqslant 1$.*

Proof. As proved in [6], any instance of $RO2||R_{\max}$ contains at most one overloaded node, so $L_V(I) \leqslant 1$. Let us prove, that I contains at most one overloaded edge. Note that (1) implies

$$\Delta(I) = \ell_1(I) + \ell_2(I) \leqslant 2(\bar{R}(I) - T^*(I)). \tag{4}$$

Let v and v' be two different terminal nodes with single job in each, J_j and $J_{j'}$ respectively; $e = [u, v]$ and $e' = [u', v']$ be the edges, incident to v and v', respectively, and both edges are overloaded. (Note that there is a possibility that $u = u'$.) From (3) we have

$$d_j + 4\tau(e) > \bar{R} - 2\mathrm{dist}(v_0, u);\ d_{j'} + 4\tau(e') > \bar{R} - 2\mathrm{dist}(v_0, u'),$$

and therefore

$$\Delta \geqslant d_j + d_{j'} > 2\bar{R} - 2\mathrm{dist}(v_0, u) - 2\mathrm{dist}(v_0, u') - 4\tau(e) - 4\tau(e'). \qquad (5)$$

Consider a graph $G' = G \setminus \{v, v'\}$. Let T'^* be the optimum of the TSP on G'. Then due to the metric property of distances we have

$$T'^* \geqslant \mathrm{dist}(v_0, u) + \mathrm{dist}(v_0, u'). \qquad (6)$$

Also, due to the fact that edges e and e' are terminal,

$$T^* = T'^* + 2\tau(e) + 2\tau(e'). \qquad (7)$$

Indeed, in order to visit terminal nodes, one needs to travel twice over their respective incident edges. Combining (5), (6) and (7), we obtain the inequality

$$\Delta > 2\bar{R} - 2T'^* - 4\tau(e) - 4\tau(e') \geqslant 2\bar{R} - 2T^*.$$

By contradiction with (4) we have $L_E(I) \leqslant 1$.

Now suppose $L_V(I) = L_E(I) = 1$. Let $e = [u, v]$ be the overloaded edge, $v \neq v_0$ is terminal node with single job J_j. Note that node v is underloaded as it contains a single job. Let $v' \neq v$ be the overloaded node. Then, by Definitions 3 and 5 we have

$$\Delta(v') > \bar{R} - 2\mathrm{dist}(v_0, v'), \qquad (8)$$

$$d_j + 4\tau(e) > \bar{R} - 2\mathrm{dist}(v_0, u). \qquad (9)$$

By using reasoning similar to that of (6) and (7), we deduce

$$T^* \geqslant \mathrm{dist}(v_0, v') + \mathrm{dist}(v_0, u) + 2\tau(e).$$

Using this inequality, together with (8) and (9), we obtain

$$\Delta \geqslant \Delta(v') + d_j > 2\bar{R} - 2\mathrm{dist}(v_0, v') - 2\mathrm{dist}(v_0, u) - 4\tau(e) \geqslant 2\bar{R} - 2T^*,$$

contradicting (4). This concludes the proof of the Proposition. $\qquad \square$

As it was shown in [6, Statement 2], one can, by means of job aggregation, easily (in linear time) transform any instance of $RO2||R_{\max}$ into an instance with the same value of $\bar{R}$, which has exactly one job in each underloaded node and at most three in the overloaded one (if such node exists). This transformation is *reversible* in the sense, that any feasible schedule for the transformed instance can be interpreted as a feasible schedule for the initial instance. We can also simplify the transformed instance, using terminal node contraction operations for each underloaded terminal edge. This operation is also reversible if we consider nonpreemptive schedule for the simplified instance. Specifically, we lose the reversibility, if a schedule built for the simplified instance is preemptive for at least one operation of any job, transformed by any edge contraction operation. Therefore, in order to maintain reversibility under our instance reduction, we need to flag some of the transformed jobs as *nonpreemtive*. Such jobs have to be scheduled in nonpreemptive mode only. Let us describe such the simplification procedure in detail.

The instance reduction procedure.

INPUT: An instance I of the problem $RO2|pmtn|R_{\max}$.

OUTPUT: A simplified instance $\tilde{I}$, in which some of the jobs are flagged as nonpreemptive.

1. **For each** $v \in V$
   1.1. **If** v is underloaded **then** perform the job aggregation of $\mathcal{J}(v)$
   1.2. **Else**
      1.2.1. Perform the job aggregation at v, as described in [6, Statement 2].
      1.2.2. Choose jobs from $\mathcal{J}(v)$, J_α and J_β, such that

$$d_\alpha + d_\beta > \bar{R}(I) - 2\text{dist}(v_0, v). \tag{10}$$

2. **For each** *terminal node* $v \neq v_0$:
   2.1. $e \doteq [u, v]$ (e is incident to v, u is adjacent to v),
   2.2. **If** e *is underloaded*
      2.2.1. Let J_j is the only job in $\mathcal{J}(v)$,
      2.2.2. Perform the contraction of e: $J_j \rightarrow J_{j'}$, flag $J_{j'}$ as nonpreemptive.
      2.2.3. **If** u is underloaded **then** perform the job aggregation of $\mathcal{J}(u)$ and flag the resulting job as nonpreemptive.
3. **End.**

Note that the running time of the procedure is $O(n)$. The existence of jobs necessary at step 1.2.2 follows from the proof of Statement 2 from [6]. All the new jobs, obtained by means of the edge contraction operation, are flagged as nonpreemptive, therefore, any feasible schedule for the instance obtained is reversible.

The following Lemma describes all possible variants of the reduced instance for the problem $RO2|G = tree|R_{\max}$.

Lemma 1. *Let I be an instance of $RO2|G = tree|R_{\max}$ and $\tilde{I}$ is obtained from I using the instance reduction procedure. Then $\bar{R}(\tilde{I}) = \bar{R}(I)$ and the graph $G(\tilde{I})$ satisfies exactly one of the following conditions:*

1. *$G(\tilde{I})$ has a single node v_0;*
2. *$G(\tilde{I})$ is a chain connecting v_0 with an overloaded node v and each node except v contains only one job;*
3. *$G(\tilde{I})$ is a chain connecting v_0 with a node v with single job at each node, and the edge incident to v is overloaded.*

Proof. Each job aggregation used in the Procedure is valid and, therefore, does not grow the standard lower bound. Terminal edge contractions are applied only to underloaded edges, therefore, $\bar{R}(\tilde{I}) = \bar{R}(I)$.

Consider the case $G(\tilde{I}) \neq K_1$. Note that steps 1.1 and 2.2.3 guarantee that each underloaded node in $\tilde{I}$ contains exactly one job. Therefore, each terminal node in $G(\tilde{I})$ is either v_0, or overloaded, or incident to an overloaded edge. By Proposition 1 the graph $G(\tilde{I})$ contains at most two terminal nodes and hence is a chain. Step 2 of the procedure continues until we have no more underloaded terminal edges. Therefore, a terminal edge is contracted unless it is overloaded, or incident to the depot, or incident to an overloaded node. Note that first and third options are mutually exclusive, which implies the Lemma.   □

Note that if the initial instance I contains an overloaded node, then it remains to be overloaded during the instance reduction procedure, leading to a case 2 of Lemma 1. Due to the Step 1.2.2 of the procedure, in the resulting (reduced) instance that node contains jobs J_α and J_β, allowed to be preempted and satisfying the condition (10). We use that property to prove the main result in the next Section.

4   The Main Result

The main result of the paper is Theorem 1, describing sufficient conditions of the polynomial solvability of the problem $RO2|pmtn, G = tree|R_{\max}$. It is based on the following two Lemmas.

Lemma 2. *Let I be an instance of $RO2||R_{\max}$, with $G(I)$ being a chain connecting v_0 and v_k, $k \geqslant 1$, each node v_p contains a single job J_p and the edge $[v_{k-1}, v_k]$ is overloaded. Then one can build a feasible nonpreemptive schedule S for I, such that $R_{\max}(S) = \bar{R}(I)$, in linear time.*

Proof. Let $T \doteq \mathrm{dist}(v_0, v_{k-1})$ and $\mu \doteq \tau([v_{k-1}, v_k])$. Then $T^* = 2(T + \mu)$. As soon as the edge $[v_{k-1}, v_k]$ is overloaded, we have

$$d_k + 4\mu > \bar{R} - 2T.$$

Therefore, (4) implies

$$\sum_{j=0}^{k-1} d_j + 2T = \Delta - d_k + 2T < 2\bar{R} - 2T^* - \bar{R} + 2T + 4\mu + 2T = \bar{R}. \qquad (11)$$

The schedule S can be built in the following manner. Let the first machine process the operations in an order $a_0 \to a_1 \to \cdots \to a_k$, and the second machine uses the following sequence: $b_k \to b_{k-1} \to \cdots \to b_0$, such that job J_k is processed in order $b_k \to a_k$, and all other jobs are proceeded in order $a_j \to b_j$ (see Fig. 1).

According to a well-known fact from calendar planning, the makespan of the schedule S coincides with the length of a critical path in graph from Fig. 1:

$$R_{\max}(S) = \max \left\{ \ell_1 + T^*, \ell_2 + T^*, d_k + 2\mathrm{dist}(v_0, v_k), \sum_{j=0}^{k-1} d_j + 2T \right\}.$$

From (1) and (11) we obtain

$$R_{\max}(S) = \bar{R}(I),$$

which concludes the proof.   □

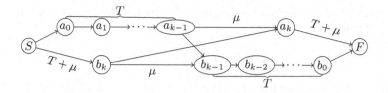

Fig. 1. A scheme of an optimal schedule for an instance with overloaded edge.

Lemma 3. *Let I be an instance of $RO2|pmtn|R_{\max}$, with $G(I)$ being a chain connecting v_0 and an overloaded node $v_k = v$, $k \geqslant 1$, each underloaded node v_p contains a single job J_p, and $\mathcal{J}(v_k)$ contains a number of jobs, including J_α and J_β, which are allowed to be preempted and satisfy the condition (10). Then one can build a feasible schedule S for I with $\xi(S) \leqslant 1$, such that $R_{\max}(S) = \bar{R}(I)$, in linear time.*

Proof. Let $\mathcal{J}(v_k) \setminus \{J_\alpha, J_\beta\} = \{J_{\gamma_1}, \ldots, J_{\gamma_l}\}$. Let $T \doteq \sum_{p=1}^{k} \tau([v_{p-1}, v_p])$, then we have $T^* = 2T$. Without loss of generality we may assume

$$a_\alpha \geqslant b_\beta \tag{12}$$

(this can be achieved by renumeration of machines and/or jobs J_α, J_β). Note that (4) and (10) imply

$$\sum_{j=0}^{k-1} d_j + \sum_{t=1}^{l} d_{\gamma_t} + T^* < \bar{R}. \tag{13}$$

Consider the schedule S_1 built in the following manner. The first machine processes the operations in order $a_0 \to \cdots \to a_{k-1} \to a_{\gamma_1} \to \cdots \to a_{\gamma_l} \to a_\alpha \to a_\beta$, and the second machine uses the order $b_\alpha \to b_\beta \to b_{\gamma_l} \to \cdots \to b_{\gamma_1} \to b_{k-1} \to \cdots \to b_0$. All the jobs except J_α and J_β a processed first by the machine M_1, then M_2 (see Fig. 2).

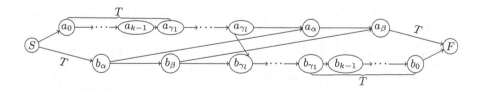

Fig. 2. A scheme of the schedule S_1.

By the reasoning similar to that of the proof of Lemma 2, using (12) we have

$$R_{\max}(S_1) = \max\left\{ \ell_1 + T^*, \ell_2 + T^*, \sum_{j=0}^{k-1} d_j + \sum_{t=1}^{l} d_{\gamma_t} + T^*, T^* + b_\alpha + a_\alpha + a_\beta \right\}.$$

Let $R_{\max}(S_1) > \bar{R}$ (otherwise the schedule S_1 satisfies the Lemma and the proof is done). Then by (1) and (13),

$$R_{\max}(S_1) = T^* + b_\alpha + a_\alpha + a_\beta.$$

Note that in this case the first machine idles before the processing of the operation a_α for the time $b_\alpha - \sum_{j=0}^{k-1} a_j - \sum_{t=1}^{l} a_{\gamma_t} = \varepsilon > 0$.

Now let us compare ε and a_β and consider two cases.

Case 1. $\varepsilon \geqslant a_\beta$.

In this case, the early schedule S_2 built according to the scheme from Fig. 3 satisfies the Lemma.

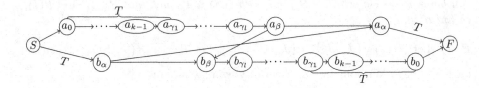

Fig. 3. A scheme of the schedule S_2

Indeed, the second machine works without idle times as soon as $c(b_\alpha) \geqslant c(a_\beta)$ due to the assumption of the case considered. Therefore, by (1)

$$R_{\max}(S_2) = \max\{\ell_2 + T^*, d_\alpha + T^*\} \leqslant \bar{R}.$$

Case 2. $\varepsilon < a_\beta$.

In this case we divide the operation a_β into two fragments, a'_β and a''_β, such that $a_\beta = \varepsilon$, and build the early schedule S_3 according to the scheme from the Fig. 4. Note that $c(a'_\beta) = c(b_\alpha)$, and by (12) $c(a_\alpha) \geqslant c(b_\beta)$, hence both machines do not idle and $R_{\max}(S_3) = \bar{R}$.

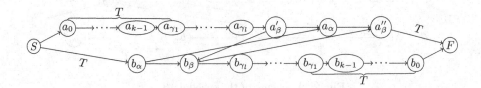

Fig. 4. A scheme of the schedule S_3.

The construction of each of the schedules considered obviously requires linear time. This observation concludes the proof of the Lemma.   □

Now we are ready to declare the main result of the paper.

Theorem 1. *Let I be an instance of the $RO2|pmtn, G = tree|R_{\max}$ problem, and one of the following conditions is true:*

1. *$G(I)$ contains an overloaded node,*
2. *the instance reduction procedure transforms $G(I)$ into a single node,*
3. *$G(I)$ contains an overloaded edge or such an edge occurs during the application of the instance reduction procedure.*

Then one can build a feasible schedule S for I with $\xi(S) \leqslant 1$, such that $R_{\max}(S) = \bar{R}(I)$, in linear time.

Proof. Straightforward from Lemmas 1, 2 and 3. □

5 Conclusion

From Lemma 1 we know, that any instance with $G = tree$ can be reduced into another instance with $G = chain$ preserving the standard lower bound. The resulting chain is *irreducible*, as it has either an overloaded terminal node or overloaded terminal edge (for now the trivial case of a single-node chain is excluded from consideration). Lemmas 2 and 3 provide ways to build an optimal schedule for both of those cases in linear time. Unfortunately, in the case with an overloaded node, we cannot guarantee the reversibility of the reduction, if the initial instance does not have an overloaded node: we need a preemption in the overloaded node, but we cannot apply preemption to jobs, transformed with edge contraction operation. From the other hand, we can avoid the necessity of preemption in case the terminal node is *superoverloaded*. Such a property was introduced in [7] and basically means that we obtain exactly three jobs in the overloaded node under the reduction procedure, and no further aggregation is valid. It was proved in [7] that any two-node chain with superoverloaded node admits construction of a feasible nonpreemptive schedule with makespan equal to the standard lower bound. The proof can be easily extended to the case of a chain of arbitrary length, providing that the superoverloaded node is terminal. This gives us one more sufficient condition of polynomial solvability of the problem under consideration.

We would like to suggest the following open questions for future investigation.

Open Question 1. Is the problem $RO2|pmtn, G = tree|R_{\max}$ polynomially solvable?

Open Question 2. Is it true that for any instance of $RO2|pmtn, G = tree|R_{\max}$ optimal makespan coincides with the standard lower bound?

Open Question 3. What is the minimal value ξ such that for any instance of $RO2|pmtn, G = tree|R_{\max}$ there exists an optimal schedule S with $\xi(S) \leqslant \xi$?

References

1. Averbakh, I., Berman, O., Chernykh, I.: A 6/5-approximation algorithm for the two-machine routing open-shop problem on a two-node network. Eur. J. Oper. Res. **166**(1), 3–24 (2005). https://doi.org/10.1016/j.ejor.2003.06.050
2. Averbakh, I., Berman, O., Chernykh, I.: The routing open-shop problem on a network: complexity and approximation. Eur. J. Oper. Res. **173**(2), 531–539 (2006). https://doi.org/10.1016/j.ejor.2005.01.034
3. Brucker, P., Knust, S., Edwin Cheng, T., Shakhlevich, N.: Complexity results for flow-shop and open-shop scheduling problems with transportation delays. Ann. Oper. Res. **129**, 81–106 (2004). https://doi.org/10.1023/b:anor.0000030683.64615.c8
4. Chernykh, I.: Routing open shop with unrelated travel times. In: Kochetov, Y., Khachay, M., Beresnev, V., Nurminski, E., Pardalos, P. (eds.) DOOR 2016. LNCS, vol. 9869, pp. 272–283. Springer, Cham (2016). https://doi.org/10.1007/978-3-319-44914-2_22
5. Chernykh, I., Kuzevanov, M.: Sufficient condition of polynomial solvability of two-machine routing open shop with preemption allowed. Intellektual'nye sistemy **17**(1–4), 552–556 (2013). (in Russian)
6. Chernykh, I., Lgotina, E.: The 2-machine routing open shop on a triangular transportation network. In: Kochetov, Y., Khachay, M., Beresnev, V., Nurminski, E., Pardalos, P. (eds.) DOOR 2016. LNCS, vol. 9869, pp. 284–297. Springer, Cham (2016). https://doi.org/10.1007/978-3-319-44914-2_23
7. Chernykh, I., Pyatkin, A.: Refinement of the optima localization for the two-machine routing open shop. In: Proceedings of the 8th International Conference on Optimization and Applications (OPTIMA 2017). CEUR Workshop Proceedings (1987), vol. 1987, pp. 131–138 (2017)
8. Gonzalez, T.F., Sahni, S.: Open shop scheduling to minimize finish time. J. ACM **23**(4), 665–679 (1976). https://doi.org/10.1145/321978.321985
9. Kononov, A.: On the routing open shop problem with two machines on a two-vertex network. J. Appl. Ind. Math. **6**(3), 318–331 (2012). https://doi.org/10.1134/s1990478912030064
10. Lawler, E.L., Lenstra, J.K., Rinnooy Kan, A.H.G., Shmoys, G.B.: Sequencing and scheduling: algorithms and complexity. In: Logistics of Production and Inventory. Elsevier (1993)
11. Lushchakova, I., Soper, A., Strusevich, V.: Transporting jobs through a two-machine open shop. Nav. Res. Logist. **56**, 1–18 (2009). https://doi.org/10.1002/nav.20323
12. Pyatkin, A., Chernykh, I.: The open shop problem with routing at a two-node network and allowed preemption. J. Appl. Ind. Math. **6**(3), 346–354 (2012). https://doi.org/10.1134/s199047891203009x
13. Strusevich, V.: A heuristic for the two-machine open-shop scheduling problem with transportation times. Discret. Appl. Math. **93**(2), 287–304 (1999). https://doi.org/10.1016/S0166-218X(99)00115-8
14. Williamson, D.P., Hall, L.A., Hoogeveen, J.A., Hurkens, C.A.J., Lenstra, J.K., Sevast'janov, S.V., Shmoys, D.B.: Short shop schedules. Oper. Res. **45**(2), 288–294 (1997). https://doi.org/10.1287/opre.45.2.288

On Complexity and Exact Solution
of Production Groups Formation Problem

Anton Eremeev[1(✉)] [iD], Alexander Kononov[2] [iD], and Igor Ziegler[1]

[1] Omsk Department, Sobolev Institute of Mathematics SB RAS, Omsk, Russia
eremeev@ofim.oscsbras.ru, icygler@hwdtech.ru
[2] Sobolev Institute of Mathematics SB RAS, Novosibirsk, Russia
alvenko@math.nsc.ru

Abstract. The success of a modern enterprize is substantially determined by the effectiveness of staff selection and formation of various kinds of functional groups. Creation of such groups requires consideration of different factors depending on the activity of the groups. The problem of production groups formation, considered in this paper, asks for an assignment of workers to jobs taking into account the implicational constraints. The first result of the paper states the NP-hardness of the problem under consideration. The second result is a branch and bound method, which uses supplementary assignment problems for computing bounds. A software implementation of the algorithm is made, and a computational experiment is carried out, comparing the proposed algorithm with the CPLEX solver on randomly generated input data.

Keywords: Integer programming · Optimization on graphs
Production groups · Branch and bound algorithm

1 Introduction

The success of a modern enterprize is substantially determined by the effectiveness of staff selection and formation of various kinds of functional groups. The specifics of these groups substantially depends on the area where the enterprize is functioning: these may be production brigades, art groups, expert boards etc. Construction of such groups requires consideration of numerous factors. For example, when forming production groups, it is necessary to make an assignment to positions ensuring quality and timely performance of work, observance of working conditions, accounting for interpersonal and hierarchical relations in the team and other requirements. To solve the problems of this kind, models and methods of discrete optimization, in particular, the methods of Integer Linear Programming (ILP). Integrality of variables in these models allows taking into account the alternative choice of applicants for positions in groups.

Presently, the research on optimization in personnel management extends in several directions. Development and use of models based on the well-known assignment problems and their generalizations proved to be fruitful in many

Y. Evtushenko et al. (Eds.): OPTIMA 2018, CCIS 974, pp. 111–122, 2019.
https://doi.org/10.1007/978-3-030-10934-9_8

circumstances [2, 10–12, 15]. If different qualities of each candidate are characterized numerically, a sufficiently balanced team of maximum size may be found using the dynamic programming or quadratic programming as proposed in [6]. Alternatively, it may be required to find a sufficiently large team, which is as balanced as possible [5]. Other optimization and game-theoretic methods are also proposed for personnel management [4, 8, 9, 13, 14].

In many cases, the production group formation problem may be formulated as follows. Suppose that an enterprize creates a production group, given a certain number of candidates in the labor market, not less than the number of the vacant positions. Each of the candidates may be assigned to at most one position and each position should be occupied by one person. For each of the candidates, the cost of assignment of the worker in any vacant position is known. It is required to form a production group with respect to the above-mentioned conditions so that the total expenses are minimized [2]. Analogous problem with a *maximization* criterion may be considered in case we have the information about the *efficiency* $s_{ij} > 0$ of assignment of each worker i to any of the jobs j, and the overall efficiency of the team is the sum of efficiencies taken over all jobs. In the latter case, the production group formation problem may be considered as a variation of the maximization version of the assignment problem.

The assignment problem is known to be solvable in polynomial time (see e.g. [4]). This problem, however, ignores many factors, thus requiring other problem formulations to be considered. For example in [2, 9, 10], a personnel management problem is considered, accounting for the so-called *strained* interpersonal relations, forbidding collaboration of workers with the pairwise strained relationship.

Implicational Constraints. Suppose a worker i_1 is considered for job j_1 and a worker i_2 is considered for job j_2. If the jobs j_1 and j_2 require an interaction of the workers and if the relations of workers i_1 and i_2 are strained, then such assignment is considered *undesirable*. In what follows, W denotes the set of all tuples $\langle (i_1, j_1), (i_2, j_2) \rangle$ that imply undesirable assignments in the sense described above. Each tuple $\langle (i_1, j_1), (i_2, j_2) \rangle \in W$ defines an implicational constraint, stating that if i_1 is assigned on job j_1, then i_2 *should not* be assigned on job j_2. Interaction of workers performing two jobs is a symmetric relation, therefore we will assume that if $\langle (i_1, j_1), (i_2, j_2) \rangle \in W$ then $\langle (i_1, j_2), (i_2, j_1) \rangle \in W$ as well.

A detailed treatment of production groups formation problems with respect to strained interpersonal relations was made in [1, 2, 9, 10].

Let us consider another type of implicational constraints. Suppose that some workers i_1 and i_2 have developed some important skills of working in a pair on jobs j_1 and j_2. If assignment of worker i_1 on position j_1 and i_2 on position j_2 will benefit from their joint skills, then this pair of assignments will be called consistent. In what follows, S denotes the set of all tuples $\langle (i_1, j_1), (i_2, j_2) \rangle$ that imply consistent assignments. Each tuple $\langle (i_1, j_1), (i_2, j_2) \rangle \in S$ defines an implicational constraint, stating that if i_1 is assigned on job j_1, then i_2 *should* be assigned on job j_2. Note that for the set of consistent assignments, the symmetry observed in the previous case is not necessarily assumed. Indeed, a pair of workers i, i' may have developed joint skills while i performed specific jobs and i' performed other specific jobs related to the jobs of i. Swapping the jobs of these two workers can make their joint skills irrelevant.

Contribution of the Paper. In the present paper, we focus on the problems with respect to consistent assignments. In particular, we prove the strong NP-hardness of finding a feasible solution to the problem of production groups formation, accounting for consistent assignments. An exact branch and bound algorithm are proposed for solving the problem and results of computational experiments with this algorithm are compared to the performance of two other known algorithms. Throughout the paper, we consider the maximization version of the problem (maximization of the team efficiency), although analogous results hold for the minimization problem (minimization of assignment cost) as well. A brief description of the proposed branch and bound algorithm and some preliminary computational results were published in [15].

Boolean Linear Programming Problem Formulation. Let I be the set of workers, J be the set of jobs, $|I| = |J|$. Let $n := |I|$ and $m := |J|$. Assume that two sets of tuples are given W and S, describing the implicational constraints as defined above. For all $i \in I$, $j \in J$ we will assume that the variable x_{ij} equals 1 if a worker i is assigned on job j, otherwise $x_{ij} = 0$. Denote a solution by $x = (x_{ij})$. Then the production groups formation problem may be formulated in terms of Boolean linear programming as follows.

$$F(x) := \sum_{i \in I} \sum_{j \in J} s_{ij} x_{ij} \to \max, \tag{1}$$

$$\sum_{i \in I} x_{ij} = 1 \quad \forall j \in J, \tag{2}$$

$$\sum_{j \in J} x_{ij} \le 1 \quad \forall i \in I, \tag{3}$$

$$x_{ij} + x_{i'j'} \le 1 \quad \forall \langle (i,j), (i',j') \rangle \in W, \tag{4}$$

$$x_{ij} \le x_{i'j'} \quad \forall \langle (i,j), (i',j') \rangle \in S, \tag{5}$$

$$x_{ij} \in \{0,1\} \quad \forall i \in I, j \in J. \tag{6}$$

Here the objective function (1) shows the total outcome the enterprize receives from the assignment of workers on the jobs. Equalities (2) indicate that one worker is assigned to each job. Inequalities (3) ensure that each worker is assigned on at most one job. Inequalities (4) and (5) are the implicational constraints taking into account the strained relationships and consistent assignments, respectively. Note that expressions (1)–(3), (6) define the well-known assignment problem, which is efficiently solvable. Addition of (4) turns it into an NP-hard problem [2,9]. The authors of [2,9] have also proposed a branch and bound algorithm for this NP-hard problem, showing competitive results in comparison to CPLEX.

The production groups formation with respect to consistent assignments is defined by the expressions (1)–(3), (5) and (6). This special case is also NP-hard as it will be shown below.

The rest of the paper is structured as follows. Section 2 demonstrates the hardness of the problem under consideration. In Sect. 3, we propose an exact branch and bound method, which uses supplementary assignment problems for computing bounds. This algorithm is applicable to the general case of the problem (1)–(6), where both strained relationships and consistent assignments are taken into account. Section 4 presents the results of a computational experiment, comparing the proposed algorithm with the CPLEX solver on randomly generated input data. Conclusions are provided in Sect. 5.

2   Analysis of Complexity

2.1   Strained Relationships

First, let us consider the special case where only strained relationships are considered, i.e. $S = \emptyset$. Kolokolov and Afanasyeva [1] prove the NP-hardness of a problem of this type by reducing the NP-complete problem INDEPENDENT SET [7] to it. Their proof is based on the following idea. Given a graph $G = (V, E)$ and an integer m, the set V may be considered as the set of candidates, and each edge $e \in E$ may be modeled by a strained relationship between the candidates corresponding to its endpoints. Now, if all m jobs are defined as mutually interacting, then finding a feasible solution to this production groups formation problem is equivalent to finding an independent set of the given size m in G, which provides an answer the NP-complete decision problem INDEPENDENT SET. Therefore the reasoning from [1] implies.

Theorem 1. *The problem of finding a feasible solution to the production groups formation problem is NP-hard in the strong sense, even in the case of $S = \emptyset$.*

2.2   Consistent Assignments

Now we show that the special case, where the implicational constraints demand only the assignments consistency, is also NP-hard.

Theorem 2. *The problem of finding a feasible solution to the production groups formation problem is NP-hard in the strong sense, even in the case of $W = \emptyset$.*

Proof. Let us reduce the NP-complete 3-SATISFIABILITY problem [7] to the problem of finding a feasible production group. Suppose that a 3-SATISFIABILITY instance with n logical variables $z_1, \ldots, z_n$ is given by m clauses $C_k = (\ell_k^1 \vee \ell_k^2 \vee \ell_k^3)$, $k = 1, \ldots, m$, where the literals $\ell_k^1, \ell_k^2, \ell_k^3$ are associated to logical variables

$z_{r(k,1)}, z_{r(k,2)}$ and $z_{r(k,3)}$ respectively, i.e. $\ell_k^1 = z_{r(k,1)}$ or $\ell_k^1 = \bar{z}_{r(k,1)}$, $\ell_k^2 = z_{r(k,2)}$ or $\ell_k^2 = \bar{z}_{r(k,2)}$, and $\ell_k^3 = z_{r(k,3)}$ or $\ell_k^3 = \bar{z}_{r(k,3)}$.

For each logical variable z_r, we define a worker i_r, two jobs j_r, j_r' and tuples $\langle (i_r, j), (i_0, j_0) \rangle$ for all $j \notin \{j_r, j_r'\}$, where the supplementary worker i_0 and the job j_0 are introduced specifically to rule out any solution with the "forbidden" assignment (i_0, j_0). To this end, we also define a supplementary worker i_{-1} and jobs j_{-1}, j_{-2}, and introduce two tuples $\langle (i_0, j_0), (i_{-1}, j_{-1}) \rangle$ and $\langle (i_{-1}, j_{-1}), (i_0, j_{-2}) \rangle$. Clearly, an assignment of the worker i_0 to the job j_0 leads to infeasibility. We will also introduce two jobs j_0' and j_{-1}', not involved into any tuple, so that the supplementary workers i_0 and i_{-1} may be assigned to these jobs without violation of feasibility.

For each clause C_k, we construct a worker i_k^{lit}, three jobs $j_{k1}^{\text{lit}}, j_{k2}^{\text{lit}}$ and j_{k3}^{lit}, the tuples

$$\langle (i_k^{\text{lit}}, j), (i_0, j_0) \rangle \quad \forall j \notin \{j_{k1}^{\text{lit}}, j_{k2}^{\text{lit}}, j_{k3}^{\text{lit}}\},$$

that allow to assign i_k^{lit} only to jobs $j_{k1}^{\text{lit}}, j_{k2}^{\text{lit}}$ and j_{k3}^{lit}. Besides that, for each of the three positions $p = 1, 2, 3$ of clause C_k, we define the tuples

$$\langle (i_k^{\text{lit}}, j_{kp}^{\text{lit}}), (i_{r(k,p)}, j_{r(k,p)}) \rangle \quad \text{if variable } z_{r(k,p)} \text{ is unnegated in clause } C_k,$$

$$\langle (i_k^{\text{lit}}, j_{kp}^{\text{lit}}), (i_{r(k,p)}, j_{r(k,p)}') \rangle \quad \text{if variable } z_{r(k,p)} \text{ is negated in clause } C_k,$$

associating satisfaction of a clause C_k to the values of its variables $z_{r(k,1)}, z_{r(k,2)}$ and $z_{r(k,3)}$. The choice of a job for the worker i_k^{lit} selects a literal that should ensure the clause C_k is satisfied (if it is possible).

Construction of additional "dummy" workers, not involved in any tuples, ensures the equality $|J| = |I|$, implying that each worker should be assigned on some job.

Now in case a satisfying assignment $z_1^*, \ldots, z_n^*$ of variables exists, we first of all assign each worker i_r, $r = 1, 2, \ldots, n$ a job j_r, if $z_r^* = $ "true" or a job j_r', if $z_r^* = $ "false" otherwise. In each clause C_k one can choose a position $p \in \{1, 2, 3\}$ where $z_{r(k,p)}^* = 1$ and this variable is unnegated in C_k, or a position $\bar{p} \in \{1, 2, 3\}$, where $z_{r(k,\bar{p})}^* = 0$ and this variable is negated in C_k. Then we can assign a worker i_k^{lit} on job j_{kp}^{lit} in the first case or on job $j_{k\bar{p}}^{\text{lit}}$ in the second case. To ensure the assignment is feasible, in view of the tuple $\langle (i_k^{\text{lit}}, j_{kp}^{\text{lit}}), (i_{r(k,p)}, j_{r(k,p)}) \rangle$, it is necessary to assign $i_{r(k,p)}$ on job $j_{r(k,p)}$ in the first case. In the second case, in view of the tuple $\langle (i_k^{\text{lit}}, j_{k\bar{p}}^{\text{lit}}), (i_{r(k,\bar{p})}, j_{r(k,\bar{p})}') \rangle$, it is necessary to assign $i_{r(k,\bar{p})}$ on job $j_{r(k,\bar{p})}'$.

The supplementary worker i_0 is assigned on the job j_0', the supplementary worker i_{-1} is assigned on the job j_{-1}'.

To those jobs j_r or j_r', which are not assigned a worker yet, we can assign arbitrary dummy workers (such workers will be available since $|J| = |I|$).

The obtained solution will be feasible because all clauses are satisfied by the assignment $z_1^*, \ldots, z_n^*$ and therefore no conflicts in assignments of the workers i_r will turn out.

At the same time, given a feasible solution (x_{ij}), by looking at assignments of the workers i_r, one can easily construct an assignment of logical variables for the 3-SATISFIABILITY instance, which will be satisfying because each clause C_k will be satisfied by a literal, corresponding to the job of the worker i_k^{lit}.

Thus we have constructed a polynomial reduction of 3-SATISFIABILITY to the problem of finding a feasible solution to the production groups formation problem, involving no numerical parameters, and the statement of the theorem holds.   □

3   Branch and Bound Algorithm

In this section, we propose a branch and bound algorithm for the problem (1)–(6), where the lower bounds are obtained from solutions of supplementary assignment problems. The Hungarian algorithm is employed to solve these problems. The branching process consists in iterative partitioning the set of feasible solutions defined by (2)–(6) and removing those subsets which probably do not contain the optimal solution. Each branching decision focuses on the satisfaction of one of the constraints violated by a solution to the current assignment subproblem, thus resembling the technique used in [2] and other papers. Each subset in the partitions emerging in this process is considered as a node of the branching tree and the original set defined by (2)–(6) is the root of the tree.

Let Rec denote the so far best-found value of the objective function. Initially, we assume $Rec := 0$. The algorithm has the following outline.

Step 0. Solve the root assignment problem (1)–(3), (6) and denote the obtained solution by x'. If all constraints (4), (5) are satisfied, then the solution x' is optimal for (1)–(6). Otherwise go to Step 1.

Step 1.
1.1. Branching 1. If for some $\langle (i,j), (i',j') \rangle \in W$ holds $x'_{ij} + x'_{i'j'} > 1$ or $x'_{ij'} + x'_{i'j} > 1$, then the problem splits into two subproblems. These subproblems correspond to the left-hand decedent node where worker i is not assigned on jobs j, j', i.e. $x_{ij} = x_{ij'} = 0$, and the right-hand decedent node where worker i' is not assigned on jobs j, j', i.e. $x_{i'j} = x_{i'j'} = 0$. We first choose the left-hand decedent as the current node and go to Step 2.
1.2. Branching 2. If for some $\langle (i,j), (i',j') \rangle \in S$ holds $x_{i'j'} - x_{ij'} < 0$, then the problem splits into two subproblems. These subproblems correspond to the left-hand decedent node where $x_{ij} = x_{i'j'} = 1$, and the right-hand decedent node where $x_{ij} = 1$. We first choose the left-hand decedent as the current node and go to Step 2.
1.3. Transition to the Right-Hand Decedent or to the Parent
– If the current node is the tree root and all nodes of the right-hand subtree are fathomed, then go to Step 3.
– If the current node is not the tree root and all nodes of the right-hand subtree are fathomed, then mark the current node as fathomed and go to Step 1.3 with the parent as the current node.
– If the current node is not the tree root and not all nodes of the right-hand subtree are fathomed, then go to Step 2 with the right-hand decedent as the current node.

Step 2. Subproblem Evaluation and Comparison to the Record. If the current assignment problem is infeasible, then go to Step 1.3. Otherwise, find an optimal solution x' to the current assignment problem and compute $F(x')$.

- If $F(x') \leq Rec$ then further processing of the subtree starting in this node can not improve the best incumbent. Mark the current node as fathomed and go to Step 1.3.
- If $F(x') > Rec$ and x' satisfies all constraints (4) but violates at least one of the constraints (5) then go to Step 1.2 for further branching.
- If $F(x') > Rec$ and x' satisfies all constraints (4) and (5) then update the best incumbent, put $Rec := F(x')$, and go to Step 1.3.
- If $F(x') > Rec$ and x' violates at least one of the constraints (4) then go to Step 1.1 for further branching.

Step 3. Completion. If $Rec > 0$ then the best incumbent is an optimal solution. Otherwise, the instance is infeasible.

Note that the branching into two subproblems on Step 1 does not exclude any feasible solutions to problem (1)–(6). The algorithm terminates after a finite number of steps because of the number of constraints (4) and (5) is finite and the depth of the branching the tree does not exceed this number. Therefore the proposed algorithm outputs an optimal solution if such solution exists.

4   Computational Experiments

We developed a software package and carried out a computational experiment to investigate the proposed branch and bound algorithm (BB) in comparison with two other known algorithms.

Constraint-Generation Algorithm. The Boolean linear programming problem (1)–(3), (5), (6) may also be solved by a modification [15] of constraint-generation method used in [2,3,10] and other works. This algorithm starts with the assignment problem (1)–(3) and iteratively adds the inequalities (4) or (5) if an optimal solution to the current problem, found by a MIP-solver, violates some of the implicational constraints. The method terminates when all constraints are satisfied in the current solution. This algorithm is implemented using CPLEX MIP solver. It is denoted as CUT in what follows. A diagram of the algorithm is given in (Fig. 1). A more detailed algorithm outline is provided in the appendix.

Experimental Setup. The calculations were carried out on a computer with an Intel (R) Core (TM) 2 processor i7-4770 CPU 3.4 GHz. The algorithm BB was compared to a stand-alone CPLEX MIP-solver (version 12) with the default settings and to the CUT algorithm using the same CPLEX version. The outcomes of all three algorithms on all instances were identical in terms of optimal objective value or infeasibility conclusion. We measured the CPU run time of each algorithm regardless whether the outcome of computations gave an optimal solution or a conclusion that an instance is infeasible. Let A be an algorithm, then $\overline{T}(A)$ will denote the average computing time of algorithm A (in seconds).

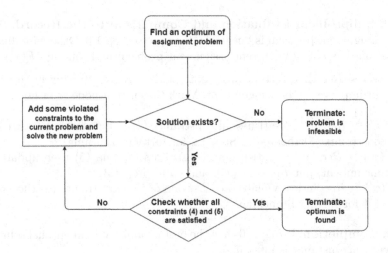

Fig. 1. Diagram of the CUT algorithm

Problem Instances. As tests cases, problems with random initial data were used. The sets of consistent assignments were generated in a symmetric way as follows. First of all, we generate two random graphs $G_1 = (V_1, E_1)$ and $G_2 = (V_2, E_2)$ where $V_1 = I$, $V_2 = J$ and in the graph G_1 each vertex has a degree at most 1. Each edge in G_1 indicates a *coherent relationship* of the corresponding pair of workers. Each edge in G_2 indicates that the corresponding pair of jobs requires an interaction. Then for each pair of workers $\{i, i'\} \in E_1$ and each pair of jobs $\{j, j'\} \in E_2$ we create two tuples $\langle (i, j), (i', j') \rangle$ and $\langle (i, j'), (i', j) \rangle$ and add them to S. In total, the resulting set S has a cardinality of $2|E_1| \cdot |E_2|$. This problem generation procedure was executed with the following input parameters:

– the number of workers n,
– the number of jobs m,
– the number of coherent interpersonal relations $|E_1|$,
– the number of jobs that require interaction $|E_2|$,
– lower and upper limits on performance values s_{ij}.

Each series of testing problems consists of randomly generated instances with fixed values of the above-mentioned parameters. In total, 75 problem series were solved, each of which contained 50 testing instances. Parameters of the input data generator are presented in Table 1. The values $s_{ij} \in \{1, 2, \ldots, 100\}$ were chosen with the uniform distribution.

Computational Results. The average running time of the CPLEX package turned out to be much longer than that of the CUT and BB algorithms for small-dimension tasks, which is especially noticeable in the $S1 - S13$ series (see Fig. 2).

Computational experiments on the larger problems showed that the CUT algorithm had an advantage in computing time, compared to CPLEX on series

Table 1. Parameters for Random Instances Generator

Series	n	m	$|E_1|$	$|E_2|$	Series	n	m	$|E_1|$	$|E_2|$	Series	n	m	$|E_1|$	$|E_2|$
S1	25	25	10	10	S26	500	500	100	800	S51	500	500	400	900
S2	25	25	10	20	S27	500	500	100	900	S52	500	500	400	1000
S3	25	25	10	30	S28	500	500	100	1000	S53	500	500	500	500
S4	25	25	10	40	S29	500	500	200	200	S54	500	500	500	600
S5	25	25	10	50	S30	500	500	200	300	S55	500	500	500	700
S6	25	25	10	60	S31	500	500	200	400	S56	500	500	500	800
S7	25	25	10	70	S32	500	500	200	500	S57	500	500	500	900
S8	25	25	10	80	S33	500	500	200	600	S58	500	500	500	1000
S9	25	25	10	90	S34	500	500	200	700	S59	500	500	600	600
S10	25	25	10	100	S35	500	500	200	800	S60	500	500	600	700
S11	25	25	10	110	S36	500	500	200	900	S61	500	500	600	800
S12	25	25	10	120	S37	500	500	200	1000	S62	500	500	600	900
S13	25	25	10	150	S38	500	500	300	300	S63	500	500	600	1000
S14	50	50	50	50	S39	500	500	300	400	S64	500	500	700	700
S15	50	50	50	100	S40	500	500	300	500	S65	500	500	700	800
S16	50	50	50	200	S41	500	500	300	600	S66	500	500	700	900
S17	50	50	100	100	S42	500	500	300	700	S67	500	500	700	1000
S18	100	100	100	100	S43	500	500	300	800	S68	500	500	800	800
S19	100	100	100	200	S44	500	500	300	900	S69	500	500	800	900
S20	100	100	100	500	S45	500	500	300	1000	S70	500	500	800	1000
S21	500	500	100	300	S46	500	500	400	400	S71	500	500	900	900
S22	500	500	100	400	S47	500	500	400	500	S72	500	500	900	1000
S23	500	500	100	500	S48	500	500	400	600	S73	1000	1000	1000	1000
S24	500	500	100	600	S49	500	500	400	700	S74	1000	1000	1000	2000
S25	500	500	100	700	S50	500	500	400	800	S75	1000	1000	1000	3000

$S14 - S75$ in 60% of the cases and the BB algorithm had an advantage in 79% (see Fig. 3). In this figure, the problem series are sorted by increasing value of $|G_1|$, and in the case of equal $|G_1|$, the series are sorted by increasing value of $|G_2|$ (displayed by grey columns). It is interesting that the value of $|G_2|$ influenced the CPU time much less than the value of $|G_1|$. When we increased the number of jobs and the number of workers to 1000 on series $S72 - S75$, CPLEX solver was faster.

Experimental comparison of CUT, BB, and CPLEX on random instances combining constraints (4) and (5) was carried out in [15] and demonstrated a similar behavior of the algorithms.

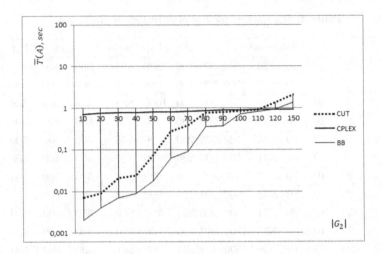

Fig. 2. The average execution time of BB, CUT and CPLEX on Series 1–13

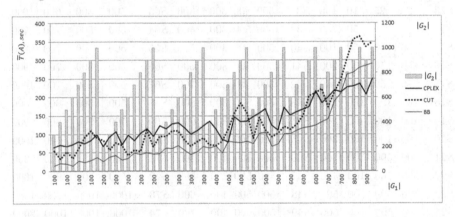

Fig. 3. The average execution time of BB, CUT and CPLEX on Series 17–72

5   Conclusion

The production group formation problems with respect to consistent assignments are considered. The strong NP-hardness of the problem is established. An exact branch and bound algorithm are proposed for solving the problem with implicational constraints accounting for consistent assignments as well as strained interpersonal relations. A software implementation of the branch and bound method is compared with the CPLEX solver and a constraint generation method on randomly generated data. The computational experiment shows that the proposed branch and bound method outperforms the CPLEX solver on problems with practically suitable dimension, while some large-scale series are faster solved by the CPLEX.

Further research might address improvements of the proposed exact algorithm by adaptive branching direction and heuristic search for the initial incumbent. The branch and bound algorithm may also be extended to the more general problem formulation where each tuple instead of a single consistent assignment implies a set of possible workers for the related position. Testing of the method on real-life data is another direction for further research.

Acknowledgement. This research is supported by RFBR projects 16-01-00740 and 17-07-00513.

Appendix

This appendix contains a detailed description of the algorithm CUT from [15]. The number of constraints defined by (4) and (5) can be too large for a straightforward application of a standard ILP algorithm. However, the inequalities may be added gradually, as it is done in cutting plane algorithms. This idea was implemented in the CUT procedure based on the CPLEX package.

Step 0. Solve the root assignment problem (1)–(3), (6) and go to Step 1.

Step 1. If a solution exists then go to Step 2. Otherwise, go to Step 5.

Step 2. If the obtained solution does not satisfy some of the constraints (4), (5) then we add one violated constraint to the current problem and go to Step 3. (The tie-breaking rule is described below.) Otherwise, go to Step 4.

Step 3. Solve the current problem and go to Step 1.

Step 4. The obtained solution is optimal. Stop.

Step 5. Problem (1)–(6) is infeasible. Stop

If several violated constraints (4), (5) are identified in Step 2, we need to choose one constraint to add to the current problem. To this end, for each tuple $\langle (i, j), (i', j') \rangle$, associated to a violated constraint, we count how many times the index i occurs in other violated constraints (4), (5). Denote this number by N_i. Analogous values N_j, $N_{i'}$, and $N_{j'}$ are calculated for j, i', and j'. Then we assign the *weight* to each violated inequality, defined as $N_i + N_j + N_{i'} + N_{j'}$. Finally, we choose the inequality with the maximum weight and add it to the problem.

References

1. Afanasyeva, L.D., Kolokolov, A.A.: Design and analysis of algorithm for solving some formation of production groups problems. Omsk Sci. Bull. **2**(110), 39–42 (2012). (in Russian)
2. Afanasyeva, L.D., Kolokolov, A.A.: Study and solution of a production groups formation problem. Vestnik UGATU. **5**, 20–25 (2013). (in Russian)
3. Borisovsky, P.A., Delorme, X., Dolgui, A.: Balancing reconfigurable machining lines via a set partitioning model. Int. J. Prod. Res. **52**(13), 4026–4036 (2013). https://doi.org/10.1080/00207543.2013.849857

4. Burkard, R.E., Dell'Amico, M., Martello, S.: Assignment problems. SIAM, Philadelphia (2009). https://doi.org/10.1137/1.9781611972238
5. Eremeev, A.V., Kel'manov, A.V., Pyatkin, A.V.: On the complexity of some Euclidean optimal summing problems. Doklady Math. **93**(3), 286–288 (2016). https://doi.org/10.1134/S1064562416030157
6. Eremeev, A.V., Kel'manov, A.V., Pyatkin, A.V.: On complexity of searching a subset of vectors with shortest average under a cardinality restriction. In: Ignatov, D.I., et al. (eds.) AIST 2016. CCIS, vol. 661, pp. 51–57. Springer, Cham (2017). https://doi.org/10.1007/978-3-319-52920-2_5
7. Garey, M.R., Johnson, D.S.: Computers and Intractability. A Guide to the Theory of NP-Completeness. W.H. Freeman and Company, San Francisco (1979)
8. Il'ev, V., Il'eva, S., Kononov, A.: Short survey on graph correlation clustering with minimization criteria. In: Kochetov, Y., Khachay, M., Beresnev, V., Nurminski, E., Pardalos, P. (eds.) DOOR 2016. LNCS, vol. 9869, pp. 25–36. Springer, Cham (2016). https://doi.org/10.1007/978-3-319-44914-2_3
9. Kolokolov, A.A., Afanasyeva, L.D.: Research of production groups formation problem subject to logical restrictions. J. Siberian Federal Univ. Math. Phys. **6**(2), 145–149 (2013)
10. Kolokolov, A.A., Rubanova, N.A., Tsygler, I.A.: Research and solution of some small groups formation problems based on discrete optimization. OMSK Sci. Bull. **4**(145), 139–142 (2016). (in Russian)
11. Kolokolov, A.A., Rubanova, N.A., Ziegler, I.A.: Solution of personnel management problems with respect to some binary relations. In: Proceedings of 12th International School-Seminar "Optimization Problems of Complex Systems", Novosibirsk, pp. 278–284 (2016). (in Russian)
12. Kolokolov, A.A., Rubanova, N.A., Ziegler, I.A.: Solution of small production groups formation problems using discrete optimization. In: Proceedings of the 4th International Conference "Information Technologies for Intelligent Decision Making Support", (ITIDS 2016), vol. 1, pp. 215–218. UGATU, Ufa (2016). (in Russian)
13. Novikov, D.A.: Mathematical Models of Teams Building and Functioning. Phismathlit, Moscow (2008). (in Russian)
14. Sigal, I.C., Ivanova, A.P.: Introduction to Applied Discrete Programming. Phismathlit, Moscow (2007). (in Russian)
15. Ziegler, I.A.: On some algorithms of solving goal-oriented groups formation problems. Young Russia Adv. Technol. Ind. **2**, 51–53 (2017). OmSTU, Omsk, (in Russian)

Time Complexity of the Ageev's Algorithm to Solve the Uniform Hard Capacities Facility Location Problem

Edward Kh. Gimadi[1,2]([✉]) [iD] and Anna A. Kurochkina[3]

[1] Sobolev Institute of Mathematics, pr. Koptyuga, 4, 630090 Novosibirsk, Russia
`gimadi@math.nsc.ru`
[2] Novosibirsk State University, 2 Pirogova Str., 630090 Novosibirsk, Russia
[3] Siberian State University of Telecommunications and Information Sciences, Kirova 86, 630102 Novosibirsk, Russia

Abstract. We show that the facility location problem with uniform hard capacities can be solved by the Ageev's algorithm in $O(m^3n^2)$ time, where m is the number of facilities and n is the number of clients. This improves the results $O(m^5n^2)$ of Ageev in 2004 and $O(m^4n^2)$ of Ageev, Gimadi, and Kurochkin in 2009.

Keywords: Facility location problem · Capacitated · Uniform
Network · Path graph · Exact algorithm
Dynamic programming technique · Polynomial · Time complexity

1 Introduction

In the facility location problem with hard capacities (CFLP) we are given a set of facilities $\mathcal{F}$ and a set of clients $\mathcal{C}$. Each facility $i \in \mathcal{F}$ has a nonnegative integral capacity u_i and can be closed or opened. In the case of opened facility $i \in \mathcal{F}$, a non-negative cost g_i^0 is imposed. Each client $j \in J$ has a nonnegative integral demand b_j that must be serviced by opened facilities. Servicing a unit of demand of client j by facility i costs c_{ij}. The objective is to open a subset of facilities and assign the demand to facilities so that all client demands satisfied and the total cost of opening facilities and servicing demands is minimized.

CFLP can be formulated as follows:

$$W(S,x) = \sum_{i \in S} f_i + \sum_{i \in S} \sum_{j \in \mathcal{C}} c_{ij} x_{ij} \rightarrow \min_{(S \subseteq \mathcal{F},\, x)}, \tag{1}$$

$$\sum_{i \in S} x_{ij} = b_j,\ j \in \mathcal{C}, \tag{2}$$

Y. Evtushenko et al. (Eds.): OPTIMA 2018, CCIS 974, pp. 123–130, 2019.
https://doi.org/10.1007/978-3-030-10934-9_9

$$\sum_{j \in \mathcal{C}} x_{ij} \leq a_i, \; i \in \mathcal{S}, \tag{3}$$

$$x_{ij} \geq 0, \; \text{integer}, \; i \in \mathcal{F}, j \in \mathcal{C}, \tag{4}$$

where

$S \subseteq \mathcal{F}$ means a set of open facilities;

f_i is a cost of opening the facility i;

a_i is a capacity of the facility i;

b_j is amount of product required by the client j;

c_{ij} equal to the transportation cost of a unit production from the facility i to the client j;

x_{ij} equal to the amount of the demand of client j that is serviced by facility i; the constraints (2) guarantee that each client demand is satisfied while the constraints (3) mean that the facility capacities are not exceeded.

It is necessary to choose a set of open facilities S and determine the transportation plan $x = (x_{ij})$ so that meet the demand of each client and minimize total costs of opening the facilities and delivering a product. An admissible solution of the problem with the set S and the quantities $x = (x_{ij})$ will be denoted by (S, x).

The special case of CFLP, when all u_i are sufficiently large, is the well-known uncapacitated facility location problem (see, e.g., [4]). In particular, this means that CFLP is NP-hard [5].

In practical applications the service costs (c_{ij}) often satisfy the triangle inequality. This case of the problem is called a Metric CFLP and has been shown to be constant-factor approximable [3,6,7,9,10].

The Metric CFLP includes an important network case when service costs g_{ij} are induced by the shortest distances between vertices of a connected graph. More precisely, $\mathcal{F}$ and $\mathcal{C}$ are assumed to be vertex subsets of an undirected connected graph $G = (V, E)$ with nonnegative edge lengths $l(e)$, and g_{ij} shortest path between vertices i and j. In this setting, the input contains the edge lengths $l(e)$ instead of the matrix (c_{ij}). We will refer to this special case of CFLP as a Network CFLP.

The Network CFLP remains weakly NP-hard on any class of graphs as it contains as a special case the knapsack problem (consider the case when all edges have zero lengths). In particular, the Network CFLP is weakly NP-hard even on path graphs.

Mirchandani et al. [8] consider CFLP on path graphs with clients having only unit demands and show that it is polynomially solvable under some additional assumption.

Ageev in [1] (2004) consider a different special case of CFLP when the clients have arbitrary demands but the facilities have equal capacities, i.e., $a_i \equiv a$ (commonly referred to as the facility location problem with uniform hard capacities [5], henceforth UCFLP. For the UCFLP on path graphs (Path UCFP) in [1] a polynomial-time algorithm $\mathcal{A}$ was constructed with time complexity $O(m^5 n^2)$, where $m = |\mathcal{F}|$ and $n = |\mathcal{C}|$. Later (in 2008) by Ageev et al. [2], the time complexity of the algorithm $\mathcal{A}$ was improved in m times to the value $O(m^4 n^2)$.

In this paper, we show that the Path UCFP can be solved by the Ageev's algorithm $\mathcal{A}$ in $O(m^3 n^2)$ time.

This paper is organized as follows. In the next section we give definitions and notation from [1] that are used throughout the paper, and describe known facts on the existence of optimal solutions for the Path UCFLP having special properties. In Sect. 3 we apply the dynamic programming technique, presented in [1,2] to develop a polynomial time algorithm $\mathcal{A}$, and refine the recurrent relations for finding such a solution with improved time complexity.

2   Basic Definitions, Notation, and Known Facts

It is convenient to use the obviously equivalent formulation of UCFLP as the problem on a line. Accordingly, we assume that the set of facilities $\mathcal{F} = \{i_1, \ldots, i_m\}$ and the set of clients $\mathcal{C} = \{j_1, \ldots, j_n\}$ are point subsets on a line such that $i_1 \leq \ldots \leq i_m$ and $j_1 \leq \ldots \leq j_n$. Then the distance c_{ij} between $i \in \mathcal{F}$ and $j \in \mathcal{C}$ is defined as the distance between these points on the line. As before, the cost of opening the facility i, $i = 1, \ldots, m$, is f_i.

For any $i' \leq i''$, call a subset $\sigma \subseteq \mathcal{F}$ a *facility segment* and denote it by $[i' \leq i'']$ if $\sigma = \{i', \ldots, i''\}$. Similarly, for any $j' \leq j''$, we call a subset $\tau \subseteq \mathcal{C}$ a *client* segment and denote it by $[j' \leq j'']$ if $\tau = \{j', \ldots, j''\}$.

Let (S, x) be a feasible solution of 1–4. We say that $i \in \mathcal{F}$ services $j \in \mathcal{C}$ if $x_{ij} > 0$. A facility i is called *consumed* if $\sum_{j \in \mathcal{C}} x_{ij} = a$ and *non-consumed* if $0 < \sum_{j \in \mathcal{C}} x_{ij} < a$. Note that consumed and non-consumed facilities are necessarily open. It is clear that we may confine our considerations to the instances of UCFLP satisfying the obvious assumption:

$$\sum_{j=1}^{n} b_j \leq m a. \tag{5}$$

Let us remind the results of the analysis problem obtained in the paper [1]. The main result shows that the set of optimal solutions of the Path UCFLP has solutions with special properties that can be found by a polynomial time dynamic programming algorithm.

Consider an arbitrary instance $\mathcal{I}$ of the UCFLP problem on a line. It is convenient to imagine a set of facilities and clients as a set of points lying on two parallel copies of the original line.

On all subsequent drawings, the upper line will correspond to the set $\mathcal{F}$, and the lower set is $\mathcal{C}$. The order of points on each of the straight lines coincides with the order of the original (see Fig. 1). To each a possible solution (S, x) of a problem with input $\mathcal{I}$ we associate a bipartite graph in which between the vertices $i \in \mathcal{F}$ and $j \in \mathcal{C}$ there exists an edge if and only if $x_{ij} > 0$, i.e. the facility i serves the client j (see Fig. 2).

Lemma 1 *[1]. The instance $\mathcal{I}$ has an optimal solution (S, x) satisfying the following property:*

(1^0) *for any $i', i'' \in \mathcal{F}$ and $j', j'' \in \mathcal{C}$ such that $i' < i''$ and $j' < j''$ either $x_{i'j''} = 0$, or $x_{i''j'} = 0$.*

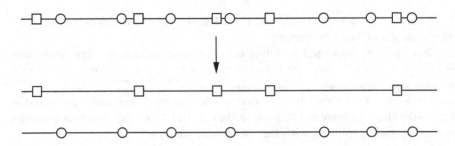

Fig. 1. Possible facilities are represented by squares on the upper line and the clients — by circles on the lower one.

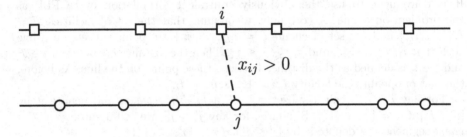

Fig. 2. The dot line indicates that facility i services client j.

Remark 1. *In the graphical terms Lemma 1 asserts that $\mathcal{I}$ has an optimal solution whose graph is a plane graph without cycles (a plane tree) (see Fig. 3 for illustration).*

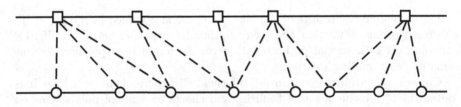

Fig. 3. The solution in the figure has property (1^0).

From Lemma 1 follows the statement that describes structural properties of a feasible solution satisfying (1^0).

Corollary 1. *Let (S, x) be a feasible solution of $\mathcal{I}$ satisfying (1^0). Then*

(i) *for any $j', j'', j''' \in C$ such that $j' < j'' < j'''$ and $i \in \mathcal{F}$, $x_{ij'} > 0$ and $x_{ij'''} > 0$ implies $x_{ij''} = b_{j''}$;*

(ii) *for any facility segment $\sigma \subseteq \mathcal{F}$, the set of clients serviced by facilities in σ is a segment (we refer to it as the client segment serviced by σ);*

(iii) for any disjoint facility segments σ_1 and σ_2, the client segments serviced by σ_1 and σ_2 can share at most one client.

Remark 2. *Lemma 1 and its Corollary hold in the case of arbitrary capacities. Property (1^0) is not sufficient for constructing a polynomial-time algorithm. However, it can be shown that Lemma 1 leads to a pseudo-polynomial algorithm for the Path CFLP, which runs in polynomial time in the case of unit demands.*

Let (S, x) be a feasible solution of $\mathcal{I}$. Call a facility segment σ *simple* if it contains open facilities and at most one facility in σ is non-consumed. Call two open facilities i', i'', $i' < i''$, *neighboring* if the segment $[i' + 1, i'' - 1]$ does not contain open facilities. We say that open facilities i', i'' are *contiguous* if the client segments serviced by i' and i'' share a client (i.e., have a common endpoint) and *discontiguous* otherwise (see Fig. 4).

The following statement is a crucial result of this section.

Lemma 2 *[1]. The instance $\mathcal{I}$ has an optimal solution (S, x) satisfying (1^0) and the following property (2^0): for any non-consumed facilities i' and i'', the facility segment $[i', i'']$ contains open facilities i^* and i^{**} that are neighboring and discontiguous.*

Remark 3. In the graphical terms Lemma 2 can be equivalently formulated as follows: *The instance $\mathcal{I}$ has an optimal solution (S, x) whose graph is a plane graph (property (1^0)) and such that no two non-consumed facilities of (S, x) lie in the same connected component of the graph (property (2^0)).*

See Fig. 4 for an example of a solution with properties (1^0) and (2^0).

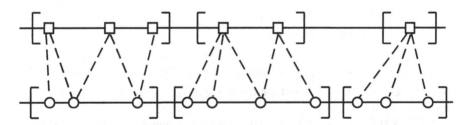

Fig. 4. Graph representation of a solution with properties (1^0) and (2^0).

A straightforward corollary of Lemma 2 is the following statement:

Lemma 3 *[1]. If an optimal solution of $\mathcal{I}$ satisfies properties (1^0) and (2^0), then the set of facilities $\mathcal{F}$ is the union of disjoint simple facility segments, of which no two share a client.*

3   Recurrence Relations and Computational Complexity

In this section we apply the results of the previous section to develop a polynomial-time algorithm for the UCFLP on a line. The algorithm $\mathcal{A}$ finds an optimal solution to the problem satisfying properties (1^0) and (2^0).

For any $i \in \{1, \ldots, m\}$ and $j \in \{1, \ldots, n\}$, let $R(i, j)$ denote the cost of an optimal solution to the modification of $\mathcal{I}$ with the set of facilities $\{1, \ldots, i\}$ and the set of clients $\{1, \ldots, j\}$ subject to the additional constraints (1^0) and (2^0). We formally set $R(i, 0) = 0$ for all $i \in \{0, \ldots, m\}$ and $R(0, j) = +\infty$ for all $j \in \{1, \ldots, n\}$.

As follows from Lemma 2,

$$R(m, n) = W(S^*, x^*),$$

where (S^*, x^*) is an optimal solution of the original problem.

For any $1 \leq i' \leq i'' \leq m$ and $1 \leq j' \leq j'' \leq n$, denote by $Q_{i',i''}(j', j'')$ the cost of an optimal solution to the instance with the set of facilities $[i', i'']$ and the set of clients $[j', j'']$ subject to the additional constraints (1^0) and that the set of facilities $[i', i'']$ is a simple segment, i.e., at most one of open facilities in $[i', i'']$ is non-consumed. If the instance does not satisfy condition 5, then we set $Q_{i',i''}(j', j'') = +\infty$.

It is shown in [1] that by Lemma 3, for any $1 \leq i' \leq m$ and $1 \leq j' \leq n$, the following recurrent relations hold

$$R(i, j) = \min_{1 \leq i' \leq i,\, 1 \leq j' \leq j} \left\{ R(i' - 1, j' - 1) + Q_{i',i}(j', j) \right\}. \tag{6}$$

In [2], to calculate $Q_{i',i''}(j', j'')$, for any $1 \leq i' \leq i'' \leq m$ and $1 \leq j' \leq j'' \leq n$, the following formula is proposed

$$Q_{i',i''}(j', j'') = \min_{1 \leq t \leq r,\, i' \leq i \leq i''} \left\{ \widetilde{G}^L_{i-1,t-1}(j') + S_{it}(j', j'') + \widetilde{G}^R_{i+1,r-t}(j'') \right\},$$

where
$r = r_{j'j''} = \left\lceil \dfrac{b_{j'} + \ldots + b_{j''}}{a} \right\rceil$ is the number open facilities in the segment $[i', i'']$,
to meet of the aggregate demand for the segment $[j', j'']$, provided that $[i', i'']$ is a simple segment;

$\widetilde{G}^L_{i,t}(j')$ is the optimal costs associated with the opening first t of consumed facilities in the segment $[i', i]$, when facility i need not be open;

$\widetilde{G}^R_{i+1,r-t}(j'')$ is the similar costs, associated with the opening of t consumed facilities in segment $[i, i'']$;

$S_{it}(j', j'')$ denote costs associated with the discovery of only non-consumed facilities, including the costs of servicing by this facility its area of demand. This non-consumed facility has the number t among open ones from i', and the number $r - t + 1$ among open if counted from i''.

Remark 4. *As shown in [2], the calculation of all the necessary values $\widetilde{G}^L_{i,t}(j')$, $\widetilde{G}^R_{i,t}(j')$ and $S_{it}(j', j'')$ is performed in a time $O(m^2 n), O(m^2)$ and $O(m^2 n^2)$, accordingly.*

The main result of the paper [2] is the justification of the time complexity of the algorithm $\mathcal{A}$, equal to the value of $O(m^4 n^2)$.

To improve this result we introduce the following notation for all $1 \leq i \leq m$, $1 \leq j' \leq j'' \leq n$:

$$\widetilde{Q}_i(j',j'') = \min_{1 \leq t \leq r} \left\{ \widetilde{G}^L_{i-1,t-1,j'} + S_{it}(j',j'') + \widetilde{G}^R_{i+1,r-t,j''} \right\}. \tag{7}$$

Taking into account this notation and Lemma 3, we obtain the final formula for counting $Q_{i',i''}(j',j'')$:

$$Q_{i',i''}(j',j'') = \min_{i' \leq i \leq i''} \widetilde{Q}_i(j',j''). \tag{8}$$

Theorem 1. *The computation of all the quantities necessary for solving the problem Path UCFLP can be performed in time $O(m^3 n^2)$.*

Proof. By Remark 2 the calculation of all the necessary values $\widetilde{G}^L_{i,t}(j')$, $\widetilde{G}^R_{i,t}(j')$ and $S_{it}(j',j'')$ is performed in a time no more $O(m^2 n^2)$.

Calculation of $\widetilde{Q}_i(j',j'')$ from the formula (7) for all i and for fixed j',j'' will require $O(m^2)$ operations. Then to calculate these values, for all parameter values, no more than $O(m^2 n^2)$ are required.

The complexity of computing one value of $Q_{i',i''}(j',j'')$ as can be seen from the formula (8), does not exceed $O(m)$-time. Calculation these values for all parameter values require no more than $O(m^3 n^2)$ operations.

From the form of the recurrence formula (6), we conclude that with the found $Q_{i',i''}(j',j'')$ the calculation of all the values of $R(i,j)$ takes time $O(m^2 n^2)$.

Thus, as a result of the work of our algorithm, we get a solution with an optimum equal to $R(m,n)$ for a time limited to the value $O(m^3 n^2)$.   □

The Theorem proved.

4   Conclusion

In this paper using the definition and useful properties of the previous papers in [1,2] we improve the time complexity of solving the Path Uniform Capacitated FLP by Ageev's algorithm $\mathcal{A}$ in m^2 time respective to [1] and in m time compared to [2].

For the further research it is interesting to apply this approach and ideas to the Path CFLP with different facility capacities, and maybe to achieve more improvement of time complexity for solving the Path UCFLP.

Acknowledgments. Sections 1 and 2 of this research were supported by the Russian Foundation for Basic Research (project 16-07-00168), by the Russian Ministry of Science and Education under the 5–100 Excellence Programme and by the program of fundamental scientific researches of the SB RAS I.5.1. Section 3 is supported by the Russian Science Foundation (project 16-11-10041).

References

1. Ageev, A.A.: A polynomial-time algorithm for the facility location problem with uniform hard capacities on path graph. In: Proceedings of the 2nd International Workshop on Discrete Optimization Methods in Production and Logistics, DOM 2004, Omsk, pp. 28–32 (2004)
2. Ageev, A.A., Gimadi, E.Kh., Kurochkin, A.A.: A polynomial-time algorithm for the facility location problem on the path with uniform hard capacities. In: Discrete Analysis and Operation Research. IM SO RAN, Novosibirsk, vol. 16, no. 5, pp. 3–17 (2009)
3. Chudak, F.A., Williamson, D.P.: Improved approximation algorithms for capacitated facility location problems. In: Cornuéjols, G., Burkard, R.E., Woeginger, G.J. (eds.) IPCO 1999. LNCS, vol. 1610, pp. 99–113. Springer, Heidelberg (1999). https://doi.org/10.1007/3-540-48777-8_8
4. Cornuéjols, G., Nemhauser, G.L., Wolsey, L.A.: The uncapacitated facility location problem. In: Mirchandani, P., Francis, R. (eds.) Discrete Location Theory, pp. 119–171. Wiley, New York (1990)
5. Garey, M.R., Johnson, D.S.: Computers and Intractablity. Freeman, San Francisco (1979)
6. Korupolu, M.R., Plaxton, C.G., Rajaraman, R.: Analysis of a local search heuristic Algorithms, pp. 1–10 (1998)
7. Mahdian, M., Pál, M.: Universal facility location. In: Di Battista, G., Zwick, U. (eds.) ESA 2003. LNCS, vol. 2832, pp. 409–421. Springer, Heidelberg (2003). https://doi.org/10.1007/978-3-540-39658-1_38
8. Mirchandani, P., Kohli, R., Tamir, A.: Capacitated location problems on a line. Transp. Sci. **30**(1), 75–80 (1996)
9. Pál, M., Tardos, E., Wexler, T.: Facility location with hard capacities. In: Proceeding of the 42nd Annual Symposium on Foundations of Computer Science (2001)
10. Zhang, J., Chen, B., Ye, Y.: A multi-exchange local search algorithm for the capacitated facility location problem. Manuscript (2003)

An Exact Algorithm of Searching for the Largest Size Cluster in an Integer Sequence 2-Clustering Problem

Alexander Kel'manov[1,2], Sergey Khamidullin[1],
Vladimir Khandeev[1,2]([envelope]), and Artem Pyatkin[1,2]

[1] Sobolev Institute of Mathematics, 4 Koptyug Ave., 630090 Novosibirsk, Russia
{kelm,kham,khandeev,artem}@math.nsc.ru
[2] Novosibirsk State University, 2 Pirogova St., 630090 Novosibirsk, Russia

Abstract. A problem of partitioning a finite sequence of points in Euclidean space into two subsequences (clusters) maximizing the size of the first cluster subject to two constraints is considered. The first constraint deals with every two consecutive indices of elements of the first cluster: the difference between them is bounded from above and below by some constants. The second one restricts the value of a quadratic clustering function that is the sum of the intracluster sums over both clusters. The intracluster sum is the sum of squared distances between cluster elements and the cluster center. The center of the first cluster is unknown and determined as the centroid (i.e. as the mean value of its elements), while the center of the second one is zero.

The strong NP-hardness of the problem is shown and an exact algorithm is suggested for the case of integer coordinates of input points. If the space dimension is bounded by some constant this algorithm runs in a pseudopolynomial time.

Keywords: Euclidean space · Sequence · 2-partition
Longest subsequence · Quadratic variance · NP-hard problem
Integer coordinates · Exact algorithm · Fixed space dimension
Pseudopolynomial running time

1 Introduction

We consider a problem of 2-clustering a finite sequence of points in Euclidean space into two disjoint clusters maximizing the size of the first cluster under some constraints. The problem, in particular, models the noise-prove search for the maximum subset of objects close to each other in the set of time-ordered measurement results. Our goal is to find out the problem complexity and to propose an efficient algorithm for this problem.

The study is motivated, on the one hand, by the absence of published theoretical results on the problem complexity status and algorithms with guaranteed

Y. Evtushenko et al. (Eds.): OPTIMA 2018, CCIS 974, pp. 131–143, 2019.
https://doi.org/10.1007/978-3-030-10934-9_10

performance bounds, and on the other hand, by its importance in some applications (see next section).

The paper is organized as follows. Section 2 contains the problem formulation, its interpretation, related problems, their distinctive features, and known algorithmic results. In Sect. 3 we analyze the computational complexity of the considered problem and show that it is strongly NP-hard. Section 4 contains some auxiliary results used later for the algorithm analysis. Finally, in Sect. 5, we present our algorithm.

2   Problem Statement and Related Results

Denote by $\mathbb{R}$ the set of real numbers, by $\| \cdot \|$ the Euclidean norm, and by $\langle \cdot, \cdot \rangle$ the scalar product.

All of the problems considered below are close to the well-known *Minimum Sum-of-Squares Clustering* (MSSC) problem, also known as the *k-Means* problem [1–5]. Complexity status of MSSC problem was studied in [6–9]. Even its basic two-cluster variant, i.e., 2-MSSC (or 2-Means) problem, is strongly NP-hard [6]. Recall that in 2-MSSC it is required to find a 2-partition minimizing over all $\mathcal{C} \subset \mathcal{Y}$ the value of

$$\sum_{y \in \mathcal{C}} \|y - \overline{y}(\mathcal{C})\|^2 + \sum_{y \in \mathcal{Y} \setminus \mathcal{C}} \|y - \overline{y}(\mathcal{Y} \setminus \mathcal{C})\|^2 ,$$

where $\overline{y}(\mathcal{C}) = \frac{1}{|\mathcal{C}|} \sum_{y \in \mathcal{C}} y$ and $\overline{y}(\mathcal{Y} \setminus \mathcal{C}) = \frac{1}{|\mathcal{Y} \setminus \mathcal{C}|} \sum_{y \in \mathcal{Y} \setminus \mathcal{C}} y$.

In 2-MSSC, the objective function is the sum over both clusters of the intra-cluster sums of squared distances between its elements and center. The unknown center of each cluster is defined as a centroid (a geometrical center), i.e. as the mean value of its elements.

In all problems considered below, in contrast to 2-MSSC, only the center of the first cluster is unknown and determined as the centroid, while the center of the second one is given at some point in Euclidean space (without loss of generality, at the origin).

Our study focuses on the following problem.

Problem 1. Given a sequence $\mathcal{Y} = (y_1, \ldots, y_N)$ of points in d-dimensional Euclidean space, positive integers $T_{\min}$, $T_{\max}$, and a number $\alpha \in (0, 1)$. *Find* a subset $\mathcal{M} = \{n_1, \ldots, n_M\} \subseteq \mathcal{N} = \{1, \ldots, N\}$ of largest size such that

$$T_{\min} \leq n_m - n_{m-1} \leq T_{\max} \leq N, \quad m = 2, \ldots, M , \tag{1}$$

and

$$F(\mathcal{M}) = \sum_{j \in \mathcal{M}} \|y_j - \overline{y}(\mathcal{M})\|^2 + \sum_{j \in \mathcal{N} \setminus \mathcal{M}} \|y_j\|^2 \leq \alpha \sum_{j \in \mathcal{N}} \|y_j - \overline{y}(\mathcal{N})\|^2 , \tag{2}$$

where $\overline{y}(\mathcal{M}) = \frac{1}{|\mathcal{M}|} \sum_{i \in \mathcal{M}} y_i$ and $\overline{y}(\mathcal{N}) = \frac{1}{N} \sum_{i \in \mathcal{N}} y_i$ are the centroids of the multisets $\{y_i \in \mathcal{Y} \mid i \in \mathcal{M}\}$ and $\{y_i \in \mathcal{Y} \mid i \in \mathcal{N}\}$ respectively.

There are some clustering problems related to Problem 1. The closest among them is the following

Problem 2. Given a sequence $\mathcal{Y} = (y_1, \ldots, y_N)$ of points in d-dimensional Euclidean space and some positive integers $T_{\min}, T_{\max}$. Find a subset $\mathcal{M} = \{n_1, \ldots, n_M\} \subseteq \mathcal{N} = \{1, \ldots, N\}$ of indices of the sequence elements minimizing $F(\mathcal{M})$, subject to constraints (1).

So, in Problem 2 a partition of $\mathcal{Y}$ into two clusters $\{y_i \in \mathcal{Y} \mid i \in \mathcal{M}\}$ and $\{y_i \in \mathcal{Y} \mid i \in \mathcal{N} \setminus \mathcal{M}\}$ minimizing the objective function $F(\mathcal{M})$ is looked for, while in Problem 1, $F(\mathcal{M})$ defines one of constraints. If $\mathcal{M}^*$ is an optimal solution of Problem 2, then $F(\mathcal{M}^*) \leq F(\mathcal{M})$ for any $\mathcal{M} \subseteq \mathcal{N}$. Therefore, in Problem 1, the inequality (2) defines a subset of admissible solutions of Problem 2, and a subsequence of largest size is searched for in this subset.

Problem 2 has the following interpretation (see, for example, [10–14]). There is a time series containing N measurements $y_1, \ldots, y_N$ of d numerical characteristics of two objects. Each measurement result contains an error and the matching between the elements of the time series and the objects are unknown. One object has the known characteristics (without the loss of generality all of them equal to 0), but the characteristics of the other one are unknown and should be found. It is known that the time interval between every two successive measurings of the second object is bounded from above and below by some constants $T_{\max}$ and $T_{\min}$. The number M of measurements of the second object is also unknown. It is required to find two disjoint subsequences (i.e. two clusters) using the criterion of minimum-sum-of-squared distances. The first subsequence (with indices in $\mathcal{M}$) corresponds to the object with unknown characteristics and the second one (with indices in $\mathcal{N} \setminus \mathcal{M}$) is the complementary subsequence whose elements correspond to the object having known characteristics. In addition, it is required to estimate the unknown characteristics (taking into account the measuring errors in the data), i.e. to find $\overline{y}(\mathcal{M})$.

Problem 1 has a similar interpretation. Namely, we need to find in a given sequence $\mathcal{Y}$ a multisubset $\{y_i \in \mathcal{Y} \mid i \in \mathcal{M}\}$ of largest size whose elements correspond to the object with unknown characteristics subject to the constraint on the sum of two above mentioned intracluster sums. This sum should be at most the threshold (the right part of (2)) that is equal to α times the total quadratic scatter of the input sequence $\mathcal{Y}$ of points relative to its centroid $\overline{y}(\mathcal{N})$. The values of α, $T_{\max}$ and $T_{\min}$ define the set of admissible solutions in Problem 1.

The next Problems 3 and 4 are closely related with Problems 1 and 2 respectively. This relation is caused by the presence of the cluster with given (at the origin) center in the problems formulations.

Problem 3. Given an N-element set $\mathcal{Y}$ of points in d-dimensional Euclidean space and a number $\alpha \in (0, 1)$. Find a subset $\mathcal{C} \subset \mathcal{Y}$ of largest cardinality such that

$$f(\mathcal{C}) = \sum_{y \in \mathcal{C}} \|y - \overline{y}(\mathcal{C})\|^2 + \sum_{y \in \mathcal{Y} \setminus \mathcal{C}} \|y\|^2 \leq \alpha \sum_{y \in \mathcal{Y}} \|y - \overline{y}(\mathcal{Y})\|^2 , \qquad (3)$$

where $\overline{y}(\mathcal{C}) = \frac{1}{|\mathcal{C}|} \sum_{y \in \mathcal{C}} y$ and $\overline{y}(\mathcal{Y}) = \frac{1}{|\mathcal{Y}|} \sum_{y \in \mathcal{Y}} y$ are the centroids of the subset $\mathcal{C}$ and the given set $\mathcal{Y}$ respectively.

Problem 4 (2-MSSC with given center). Given an N-element set $\mathcal{Y}$ of points in Euclidean space of dimension d. *Find* a 2-partition of $\mathcal{Y}$ into clusters $\mathcal{C}$ and $\mathcal{Y} \backslash \mathcal{C}$ minimizing the value of $f(\mathcal{C})$.

These problems differ from Problems 1 and 2 only by the absence of the constraints on the indices of the elements of the first cluster. The function $f(\mathcal{C})$ has the same interpretation as $F(\mathcal{M})$.

All mentioned 2-clustering problems with a given center, as well as known MSSC problem, have a simple geometric interpretation. They are closely related to Combinatorial Geometry, Statistics, Data Science, Data approximation, Data mining, Machine learning. In these research areas, the data can be either sets (tables) or sequences (time series or signals), and clustering algorithms are the key tools for processing them (see, for example, [15–25] and papers cited therein). So, developing the tools for new clustering models is of great interest for these areas.

By this time, there are no available algorithmic results for Problem 1, so below we present known results only for the closest problems where one cluster has a center at the origin, namely, for Problems 2, 3 and 4.

Strong NP-hardness of Problem 4 was proved in [26,27]. In [28], a 2-approximation polynomial algorithm with time complexity $\mathcal{O}(dN^2)$ was proposed for the problem.

The following results were obtained for the variant of Problem 4 with given cluster cardinalities. Denote by M the size of the cluster with the unknown center and by D the maximum absolute value of coordinates over all of the points in the input set.

Strong NP-hardness of this variant of Problem 4 follows directly from [26,27] while its NP-hardness was proved earlier in [29–31].

It was shown in [32,33] that the problem is solvable in $\mathcal{O}(d^2 N^{2d})$ time, which is polynomial when the space dimension d is fixed. A faster exact algorithm with $\mathcal{O}(dN^{d+1})$ running time was proposed in [34]. Additionally, an exact algorithm for the case of integer inputs was presented in [32]. The time complexity of the algorithm is $\mathcal{O}(dN(2MD + 1)^d)$. If the space dimension is fixed, the algorithm is pseudopolynomial.

In [35], a 2-approximation polynomial algorithm with time complexity $\mathcal{O}(dN^2)$ was proposed for the problem.

A PTAS having $\mathcal{O}(dN^{2/\varepsilon+1}(9/\varepsilon)^{3/\varepsilon})$ time complexity, where ε is a relative error, was proposed in [36].

It was shown in [37] that there is no FPTAS for the problem unless $P = NP$. In the same paper, the algorithm finding a $(1 + \varepsilon)$-approximate solution in $\mathcal{O}(dN^2(\sqrt{2q/\varepsilon} + 2)^d)$ time for given $\varepsilon \in (0, 1)$ was proposed. For the fixed space dimension d, the algorithm runs in $\mathcal{O}(N^2(1/\varepsilon)^{d/2})$ time and implements an FPTAS. Moreover, in [38], the modification of this algorithm with improved running time $\mathcal{O}\left(\sqrt{d}N^2\left(\frac{\pi e}{2}\right)^{d/2}\left(\sqrt{\frac{2}{\varepsilon}}+2\right)^d\right)$ was suggested. The algorithm implements an FPTAS with $\mathcal{O}(N^2(1/\varepsilon)^{d/2})$ running time in the case of the fixed space dimension and remains polynomial for instances of dimension $d = \mathcal{O}(\log N)$. In this case it implements a PTAS with $\mathcal{O}\left(N^{C(1.05+\log(2+\sqrt{\frac{2}{\varepsilon}}))}\right)$ time, where C is a positive constant.

A randomized algorithm was proposed in [39]. Under assumption $M \geq \beta N$, where $\beta \in (0, 1)$ is some constant, and given $\varepsilon > 0$ and $\gamma \in (0, 1)$, the algorithm finds a $(1+\varepsilon)$-approximate solution of the problem with probability at least $1-\gamma$ in $\mathcal{O}(dN)$ time that is linear for both d and N. In the same paper, the conditions were found under which the algorithm is asymptotically exact. Namely, it finds a $(1+\varepsilon_N)$-approximate solution of the problem in $\mathcal{O}(dN^2)$ time with probability at least $1 - \gamma_N$, where $\varepsilon_N \to 0$ and $\gamma_N \to 0$ as $N \to \infty$. Note that it is the best algorithmic result now.

As for Problem 3, it remains poorly-studied one up to now. Strong NP-hardness of this problem was recently shown in [40, 41]. The only known algorithmical result for Problem 3 is an exact algorithm for the case of this problem with integer input points coordinates [40, 41]. The time complexity of the algorithm is $\mathcal{O}(dN^2(2MD+1)^d)$. If the space dimension is bounded by some constant, the algorithm is pseudopolynomial.

Finally, let us recall the known results for Problem 2 where the input is not a set but a sequence. The strong NP-hardness follows from the strong NP-hardness of Problem 4 since Problem 4 is a particular case of Problem 2 with $T_{\min} = 1$ and $T_{\max} = N$.

In [42], the variant of Problem 2 with $T_{\min}$ and $T_{\max}$ as parameters was studied and it was established that Problem 2 is strongly NP-hard for every $T_{\min} < T_{\max}$. In the trivial case when $T_{\min} = T_{\max}$ Problem 2 is solvable in a polynomial time.

The only known algorithmical result for the general variant of Problem 2 is a 2-approximation polynomial algorithm with $\mathcal{O}(N^2(T_{\max} - T_{\min} + d))$ running time [43]. One can estimate this running time by $\mathcal{O}(dN^3)$ since $T_{\max} - T_{\min} < N$. But if $T_{\min}$ and $T_{\max}$ are bounded by some constants, then the algorithm running time is $\mathcal{O}(dN^2)$.

The following results were obtained for the variant of Problem 2 with fixed cluster sizes. It was proved in [44] that the problem is strongly NP-hard for any $T_{\min} < T_{\max}$ and polynomially solvable for $T_{\min} = T_{\max}$.

A particular case of Problem 2 of partitioning a sequence with $T_{\min} = 1$ and $T_{\max} = N$ is equivalent to strongly NP-hard Problem 4 of partitioning a set, which does not admit [37] FPTAS unless P $=$ NP. In other words, Problem 2 of partitioning a sequence is the generalization of strongly NP-hard Problem 4 of partitioning a set. Therefore, according to [45], Problem 2 also admits neither exact polynomial, nor exact pseudopolynomial algorithms, nor FPTAS unless P $=$ NP. A 2-approximation polynomial algorithm with complexity $\mathcal{O}(N^2(M(T_{\max} - T_{\min} + 1) + d))$ was proposed in [10, 46]. In [11], an exact algorithm was introduced for the case where the components of the points are integers. The running time of the algorithm is $\mathcal{O}(N(M(T_{\max} - T_{\min} + 1) + d)(2MD + 1)^d)$ which is pseudopolynomial in the case of fixed space dimension.

An approximation algorithm was presented in [12]. This algorithm finds a $(1 + \varepsilon)$-approximate solution of Problem 2 in

$$\mathcal{O}(N^2(M(T_{\max} - T_{\min} + 1) + d)(\sqrt{\frac{2d}{\varepsilon}} + 2)^d)$$

time, and it implements an FPTAS if the space dimension is fixed.

In [13,14], a randomized algorithm was presented. For an established parameter value, given $\varepsilon > 0$ and fixed $\gamma \in (0,1)$, this algorithm finds a $(1 + \varepsilon)$-approximate solution of the problem with a probability of at least $1 - \gamma$ in $\mathcal{O}(N(M(T_{\max} - T_{\min} + 1) + d))$ time. The conditions are established under which the algorithm is asymptotically exact and its time complexity is $\mathcal{O}(N^2(M(T_{\max} - T_{\min} + 1) + d))$.

For the considered Problem 1, no algorithms with guaranteed performance bounds are known; its complexity status is also open.

In the current paper, we show that Problem 1 is strongly NP-hard and propose an exact algorithm for the special case of the problem with integer coordinates of the input points. Our algorithm finds an optimal solution in $\mathcal{O}(N^2(N(T_{\max} - T_{\min} + 1) + d)(2ND + 1)^d)$ time. If the space dimension is bounded by some constant, our algorithm runs in a pseudopolynomial time.

3   Computational Complexity

The following statement is true since Problem 3 is a special case of Problem 1 when $T_{\min} = 1$ and $T_{\max} = N$.

Proposition 1. *Problem 1 is strongly NP-hard.*

Moreover, the following statement is true.

Proposition 2. *If $T_{\min}$ and $T_{\max}$ are parameters (are not inputs), Problem 1 is still strongly NP-hard.*

Proof. First, note that the right-hand side of (2) does not depend on $\mathcal{M}$, and for a given input is a constant. Therefore, Problem 1 in a verification form can be formulated in the following way.

Problem 5. Given a sequence $\mathcal{Y} = (y_1, \ldots, y_N)$ of points in d-dimensional Euclidean space, some positive integers $T_{\min}$, $T_{\max}$, K, and a number $A > 0$. *Question:* is there a subset $\mathcal{M} = \{n_1, \ldots, n_M\} \subseteq \mathcal{N}$ of cardinality $M \geq K$ such that

$$F(\mathcal{M}) \leq A,$$

while the elements of the tuple $(n_1, \ldots, n_M)$ satisfy the constraints (1).

Next, a verification form of Problem 2 is as follows.

Problem 6. Given a sequence $\mathcal{Y} = (y_1, \ldots, y_N)$ of points in d-dimensional Euclidean space, some positive integers $T_{\min}$, $T_{\max}$ and a number $A > 0$. *Question:* is there a subset $\mathcal{M} = \{n_1, \ldots, n_M\} \subseteq \mathcal{N}$ such that

$$F(\mathcal{M}) \leq A,$$

while the elements of the tuple $(n_1, \ldots, n_M)$ satisfy the constraints (1).

Note that Problem 6 is the special case of Problem 5 with $M = 1$. Thus, the proposition follows from the fact that Problem 6 is strongly NP-complete even if $T_{\min}$ and $T_{\max}$ are parameters. $\qquad\square$

4   Algorithm Foundations

At first, we present an algorithm for the following auxiliary

Problem 7. Given a sequence $\mathcal{Y} = (y_1, \ldots, y_N)$ of points in d-dimensional Euclidean space, a point $x \in \mathbb{R}^d$, and positive integers $T_{\min}$, $T_{\max}$ and $M > 1$. *Find* a subset $\mathcal{M} = \{n_1, \ldots, n_M\} \subseteq \mathcal{N}$ of indices of the sequence elements such that

$$f^x(\mathcal{M}) = \sum_{i \in \mathcal{M}} \|y_i - x\|^2 + \sum_{j \in \mathcal{N} \setminus \mathcal{M}} \|y_j\|^2 \to \min, \qquad (4)$$

while the elements of the tuple $(n_1, \ldots, n_M)$ satisfy the constraints (1).

Remark 1. According to the statement of Problem 7, we have $M \in \{2, \ldots, M_{\max}\}$, where

$$M_{\max} = \left\lfloor \frac{N-1}{T_{\min}} \right\rfloor + 1.$$

The following dynamic programming scheme finds a solution $\mathcal{M}^x$ of Problem 7. Here we use the fact that minimizing $f^x(\mathcal{M})$ in (4) is equivalent to finding a subset $\mathcal{M}$ of vectors having the maximum sum $G^x_{\max}$ of scalar products with x. Below, ω_m denotes the set of possible positions for m-th vector (i.e., there must be sufficient space for $m - 1$ vectors with lesser indices and $M - m$ vectors with larger indices) and $\gamma^-_{m-1}(n)$ is the set of possible positions for $(m - 1)$-th vector assuming that m-th one has index n.

Algorithm $\mathcal{A}_1$.

Input: a sequence $\mathcal{Y}$, a point x, and numbers $T_{\min}$, $T_{\max}$ and M.

Step 1. Compute $g(n) = \langle y_n, x \rangle$ for each $n \in \mathcal{N}$.

Step 2. Calculate the values of $G^x_m(n)$ for each $n \in \omega_m$ using the following recurrences

$$G^x_m(n) = \begin{cases} g(n), & \text{if } n \in \omega_1, \ m = 1; \\ g(n) + \max\limits_{j \in \gamma^-_{m-1}(n)} G^x_{m-1}(j), & \text{if } n \in \omega_m, \ m = 2, \ldots, M, \end{cases}$$

where

$$\omega_m = \left\{ n \mid 1 + (m-1)T_{\min} \le n \le N - (M-m)T_{\min} \right\},$$

$$\gamma^-_{m-1}(n) = \omega_{m-1} \cap \left\{ j \mid n - T_{\max} \le j \le n - T_{\min} \right\},$$

and $m = 1, \ldots, M$.

Step 3. Compute $G^x_{\max} = \max\limits_{n \in \omega_M} G^x_M(n)$, find the tuple $\mathcal{M}^x = (n^x_1, \ldots, n^x_M)$ by formulas

$$n^x_M = \arg \max\limits_{n \in \omega_M} G^x_M(n),$$

$$n^x_{m-1} = \arg \max\limits_{n \in \gamma^-_m(n^x_m)} G^x_m(n), \quad m = M, M-1, \ldots, 2,$$

and calculate $f^x_{\min} = f(\mathcal{M}^x)$ by (4).

Output: the tuple $\mathcal{M}^x = (n^x_1, \ldots, n^x_M)$ and the value of $f^x_{\min}$.

Remark 2. It was proved in [10,46] that algorithm $\mathcal{A}_1$ finds an optimal solution of Problem 7 in $\mathcal{O}(N(M(T_{\max} - T_{\min} + 1) + d))$ time.

Remark 3. If $T_{\max}$ and $T_{\min}$ are the part of input, one can estimate the time complexity of $\mathcal{A}_1$ by $\mathcal{O}(N(MN + d))$ since the value of $(T_{\max} - T_{\min} + 1)$ is at most N.

Remark 4. Algorithm $\mathcal{A}_1$ finds an optimal solution of Problem 7 in $\mathcal{O}(N(M+d))$ time in the case when $T_{\max}$ and $T_{\min}$ are some constants (this case is typical for applications).

We also need the following variant of Problem 2 in which cluster sizes are fixed.

Problem 8. Given a sequence $\mathcal{Y} = (y_1, \ldots, y_N)$ of points in d-dimensional Euclidean space and some positive integers $T_{\min}, T_{\max}$ and $M > 1$. *Find* a subset $\mathcal{M} = \{n_1, \ldots, n_M\} \subseteq \mathcal{N} = \{1, \ldots, N\}$ of indices of the sequence elements minimizing $F(\mathcal{M})$, subject to constraints (1).

Finally, we need the following algorithm based on the algorithm $\mathcal{A}_1$ and finding a solution of Problem 8 in the case when the coordinates of all points in the sequence $\mathcal{Y}$ are integers from the interval $[-D, D]$.

Algorithm $\mathcal{A}_2$.
Input: a sequence $\mathcal{Y}$, numbers $T_{\min}, T_{\max}$ and M.
Step 1. Find the value of D by the formula

$$D = \max_{y \in \mathcal{Y}} \max_{j \in \{1, \ldots, d\}} |(y)^j|,$$

where $(y)^j$ is the j-th coordinate of y. Construct the set $\mathcal{D}$ by formula

$$\mathcal{D} = \{x \mid (x)^j = \frac{1}{M}(v)^j, \ (v)^j \in \mathbb{Z}, \ |(v)^j| \le MD, \ j = 1, \ldots, d\}.$$

Step 2. Using algorithm $\mathcal{A}_1$, for each $x \in \mathcal{D}$, find the value of $f^x_{\min}$ and a solution $\mathcal{M}^x = \{n^x_1, \ldots, n^x_M\}$ of Problem 7.
Step 3. Find the point $x_A = \arg\min_{x \in \mathcal{D}} f^x_{\min}$ and the corresponding subset $\mathcal{M}^{x_A}$ and put $\mathcal{M}_A(M) = \mathcal{M}^{x_A}$
Output: the tuple $\mathcal{M}_A(M)$.

Remark 5. It was proved in [11] that if components of all points in the sequence $\mathcal{Y}$ are integers from the interval $[-D, D]$ then the algorithm $\mathcal{A}_2$ finds an optimal solution of Problem 8 in $\mathcal{O}(N(M(T_{\max} - T_{\min} + 1) + d)(2MD + 1)^d)$ time.

Remark 6. In the case when $T_{\max}$ and $T_{\min}$ are the part of input, one can estimate the time complexity of $\mathcal{A}_2$ by $\mathcal{O}(N(MN+d)(2MD+1)^d)$ since the value of $(T_{\max} - T_{\min} + 1)$ is at most N. In this case, if the space dimension d is bounded by some constant, then the running time of $\mathcal{A}_2$ is $\mathcal{O}(MN^2(MD)^d)$, which is pseudopolynomial.

Remark 7. In the typical for applications case when T_{max} and T_{min} are some constants, one can estimate the time complexity of $\mathcal{A}_2$ by

$$\mathcal{O}(N(M+d)(2MD+1)^d).$$

In this case, if the space dimension d is bounded by some constant (that is also typical for applications), then the running time of $\mathcal{A}_2$ is $\mathcal{O}(MN(MD)^d)$, which is pseudopolynomial.

5   Exact Algorithm

We suggest the following approach for finding a solution of Problem 1. For each $M = 2, \ldots, N$, an auxiliary Problem 8 is solved by the algorithm $\mathcal{A}_2$. In the family of the found solutions, we choose the admissible (i.e. satisfying the inequality (2)) solution of the maximum length. The following algorithm implements this approach.

Algorithm $\mathcal{A}$.

Input: a sequence $\mathcal{Y}$ and numbers T_{min}, T_{max}, and α.

Step 1. Compute $A = \alpha \sum_{j \in \mathcal{N}} \|y_j - \overline{y}(\mathcal{N})\|^2$.

Step 2. For every $M = 2, \ldots, M_{max}$ using the algorithm $\mathcal{A}_2$ find an exact solution $\mathcal{M}_A(M) = (n_1, \ldots, n_M)$ of Problem 8 and calculate for this solution the value of the objective function $F(\mathcal{M}_A(M))$.

Step 3. In the family $\{\mathcal{M}_A(M), M = 2, \ldots, M_{max}\}$ of the sets obtained in Step 2 find a set $\mathcal{M}_A$ of maximum cardinality for which $F(\mathcal{M}_A) \leq A$.

Output: the tuple $\mathcal{M}_A$.

The following theorem establishes the main result of this paper.

Theorem 1. *If the points of $\mathcal{Y}$ have integer components lying in $[-D, D]$, then algorithm $\mathcal{A}$ finds an optimal solution of Problem 1 in*

$$\mathcal{O}(N^2(N(T_{max} - T_{min} + 1) + d)(2ND+1)^d)$$

time.

Proof. Let $\mathcal{M}^*$ be the optimal solution of Problem 1, $M^* = |\mathcal{M}^*|$. Note that $\mathcal{M}^*$ is an admissible solution of Problem 8 with $M = M^*$. Since $\mathcal{M}_A(M)$ is an optimal solution of the same problem,

$$F(\mathcal{M}_A(M)) \leq F(\mathcal{M}^*) \leq A.$$

Therefore, $\mathcal{M}_A(M)$ was considered at Step 3 of the algorithm, and the family $\{\{\mathcal{M}_A(M), M = 2, \ldots, M_{max}\} \mid F(\mathcal{M}_A(M)) \leq A\}$ is not empty. From the definition of Step 3 we get

$$|\mathcal{M}_A| \geq |\mathcal{M}_A(M)| = M = |\mathcal{M}^*|.$$

On the other hand, since $|\mathcal{M}_A|$ is an admissible solution of Problem 1, $|\mathcal{M}_A| \leq |\mathcal{M}^*|$. Therefore, $|\mathcal{M}_A| = |\mathcal{M}^*|$.

Evaluate the time complexity of the algorithm. Step 1 is done in qN operations. Step 2 requires at most $\mathcal{O}(N^2(N(T_{\max} - T_{\min} + 1) + d)(2MD + 1)^d)$ operations due to Remark 5. Step 3 is fulfilled in $\mathcal{O}(N)$ time. Summing up the costs required at all steps yields the time complexity bound for algorithm $\mathcal{A}$. $\square$

Remark 8. If the space dimension d is bounded by some constant, then the algorithm's time complexity is $\mathcal{O}(N^4(ND)^d)$ since $T_{\max} - T_{\min} + 1 \leq N$.

Remark 9. Algorithm $\mathcal{A}$ finds an optimal solution of Problem 1 in

$$\mathcal{O}(dN^3(2ND + 1)^d)$$

time in the case when $T_{\max}$ and $T_{\min}$ are some constants. If the space dimension d is bounded by some constant, then the algorithm's time complexity is $\mathcal{O}(N^3(ND)^d)$.

In the case of fixed space dimension d, the algorithm is pseudopolynomial since D is a numerical value given at input.

6   Conclusion

In this paper, we have considered one hard to solve clustering problem of searching for the longest subsequence (the subsequence of largest size) in a finite sequence of points in Euclidean space. Firstly, we have shown the strong NP-hardness of the problem. Secondly, an exact algorithm for the cases of a problem with integer coordinates of the input points has been constructed. This algorithm is pseudopolynomial if the space dimension is bounded by some constant.

In our opinion, the presented algorithm would be useful as a tool for solving problems in applications related to Data science, Data mining, Machine learning, and time-ordered Data processing.

It is clear that the algorithm can be used to solve practical problems having integer instances of small dimensions. Therefore, the development of approximation algorithms with guaranteed accuracy for these problems is of considerable interest.

Acknowledgments. The study presented in Sects. 3 and 5 was supported by the Russian Science Foundation, project 16-11-10041. The study presented in Sects. 2 and 4 was supported by the Russian Foundation for Basic Research, projects 16-07-00168 and 18-31-00398, by the Russian Academy of Science (the Program of basic research), project 0314-2016-0015, and by the Russian Ministry of Science and Education under the 5-100 Excellence Programme.

References

1. MacQueen, J.B.: Some methods for classification and analysis of multivariate observations. In: Proceedings of the 5th Berkeley Symposium on Mathematical Statistics and Probability, vol. 1, pp. 281–297. University of California Press, Berkeley (1967)
2. Rao, M.: Cluster analysis and mathematical programming. J. Am. Stat. Assoc. **66**, 622–626 (1971)
3. Hansen, P., Jaumard, B., Mladenovich, N.: Minimum sum of squares clustering in a low dimensional space. J. Classifi. **15**, 37–55 (1998)
4. Hansen, P., Jaumard, B.: Cluster analysis and mathematical programming. Math. Program. **79**, 191–215 (1997)
5. Fisher, R.A.: Statistical Methods and Scientific Inference. Hafner, New York (1956)
6. Aloise, D., Deshpande, A., Hansen, P., Popat, P.: NP-hardness of Euclidean sum-of-squares clustering. Mach. Learn. **75**(2), 245–248 (2009)
7. Drineas, P., Frieze, A., Kannan, R., Vempala, S., Vinay, V.: Clustering large graphs via the singular value decomposition. Mach. Learn. **56**, 9–33 (2004)
8. Dolgushev, A.V., Kel'manov, A.V.: On the algorithmic complexity of a problem in cluster analysis. J. Appl. Ind. Math. **5**(2), 191–194 (2011)
9. Mahajan, M., Nimbhorkar, P., Varadarajan, K.: The planar k-means problem is NP-hard. Theor. Comput. Sci. **442**, 13–21 (2012)
10. Kel'manov, A.V., Khamidullin, S.A.: An approximating polynomial algorithm for a sequence partitioning problem. J. Appl. Ind. Math. **8**(2), 236–244 (2014)
11. Kel'manov, A.V., Khamidullin, S.A., Khandeev, V.I.: Exact pseudopolynomial algorithm for one sequence partitioning problem. Autom. Remote Control. **78**(1), 66–73 (2017)
12. Kel'manov, A.V., Khamidullin, S.A., Khandeev, V.I.: A fully polynomial-time approximation scheme for a sequence 2-cluster partitioning problem. J. Appl. Ind. Math. **10**(2), 209–219 (2016)
13. Kel'manov, A.V., Khamidullin, S.A., Khandeev, V.I.: A randomized algorithm for a sequence 2-clustering problem. Comput. Math. Math. Phys. **58**(12) (2018, in publishing)
14. Kel'manov, A., Khamidullin, S., Khandeev, V.: A randomized algorithm for 2-partition of a sequence. In: van der Aalst, W.M.P., et al. (eds.) AIST 2017. LNCS, vol. 10716, pp. 313–322. Springer, Cham (2018). https://doi.org/10.1007/978-3-319-73013-4_29
15. Bishop, C.M.: Pattern Recognition and Machine Learning. Springer, New York (2006)
16. James, G., Witten, D., Hastie, T., Tibshirani, R.: An Introduction to Statistical Learning. Springer, New York (2013). https://doi.org/10.1007/978-1-4614-7138-7
17. Hastie, T., Tibshirani, R., Friedman, J.: The Elements of Statistical Learning, 2nd edn. Springer, New York (2009). https://doi.org/10.1007/978-0-387-84858-7
18. Aggarwal, C.C.: Data Mining: The Textbook. Springer, Cham (2015). https://doi.org/10.1007/978-3-319-14142-8
19. Goodfellow, I., Bengio, Y., Courville, A.: Deep Learning (Adaptive Computation and Machine Learning series). The MIT Press, Cambridge (2017)
20. Shirkhorshidi, A.S., Aghabozorgi, S., Wah, T.Y., Herawan, T.: Big data clustering: a review. In: Murgante, B., Misra, S., Rocha, A.M.A.C., Torre, C., Rocha, J.G., Falcão, M.I., Taniar, D., Apduhan, B.O., Gervasi, O. (eds.) ICCSA 2014. LNCS, vol. 8583, pp. 707–720. Springer, Cham (2014). https://doi.org/10.1007/978-3-319-09156-3_49

21. Jain, A.K.: Data clustering: 50 years beyond k-means. Pattern Recognit. Lett. **31**(8), 651–666 (2010)
22. Pach, J., Agarwal, P.K.: Combinatorial Geometry. Wiley, New York (1995)
23. Fu, T.-C.: A review on time series data mining. Eng. Appl. Artif. Intell. **24**(1), 164–181 (2011)
24. Kuenzer, C., Dech, S., Wagner, W. (eds.): Remote Sensing Time Series. Remote Sensing and Digital Image Processing, vol. 22. Springer, Cham (2015). https://doi.org/10.1007/978-3-319-15967-6
25. Liao, T.W.: Clustering of time series data – a survey. Pattern Recognit. **38**(11), 1857–1874 (2005)
26. Kel'manov, A.V., Pyatkin, A.V.: On the complexity of a search for a subset of "similar" vectors. Dokl. Math. **78**(1), 574–575 (2008)
27. Kel'manov, A.V., Pyatkin, A.V.: On a version of the problem of choosing a vector subset. J. Appl. Ind. Math. **3**(4), 447–455 (2009)
28. Kel'manov, A.V., Khandeev, V.I.: A 2-approximation polynomial algorithm for a clustering problem. J. Appl. Ind. Math. **7**(4), 515–521 (2013)
29. Gimadi, E.Kh., Kel'manov, A.V., Kel'manova, M.A., Khamidullin, S.A.: A posteriori detection of a quasi periodic fragment in numerical sequences with given number of recurrences. Sib. J. Ind. Math. **9** (1(25)), 55–74 (2006). (in Russian)
30. Gimadi, E.Kh., Kel'manov, A.V., Kel'manova, M.A., Khamidullin, S.A.: A posteriori detecting a quasiperiodic fragment in a numerical sequence. Pattern Recognit. Image Anal. **18**(1), 30–42 (2008)
31. Baburin, A.E., Gimadi, E.Kh., Glebov, N.I., Pyatkin, A.V.: The problem of finding a subset of vectors with the maximum total weight. J. Appl. Ind. Math. **2**(1), 32–38 (2008)
32. Kel'manov, A.V., Khandeev, V.I.: An exact pseudopolynomial algorithm for a problem of the two-cluster partitioning of a set of vectors. J. Appl. Ind. Math. **9**(4), 497–502 (2015)
33. Gimadi, E.Kh., Pyatkin, A.V., Rykov, I.A.: On polynomial solvability of some problems of a vector subset choice in a Euclidean space of fixed dimension. J. Appl. Ind. Math. **4**(1), 48–53 (2010)
34. Shenmaier, V.V.: Solving some vector subset problems by Voronoi diagrams. J. Appl. Ind. Math. **10**(4), 560–566 (2016)
35. Dolgushev, A.V., Kel'manov, A.V.: An approximation algorithm for solving a problem of cluster analysis. J. Appl. Ind. Math. **5**(4), 551–558 (2011)
36. Dolgushev, A.V., Kel'manov, A.V., Shenmaier, V.V.: Polynomial-time approximation scheme for a problem of partitioning a finite set into two clusters. Proc. Steklov Inst. Math. **295**(Suppl. 1), 47–56 (2016)
37. Kel'manov, A.V., Khandeev, V.I.: Fully polynomial-time approximation scheme for a special case of a quadratic Euclidean 2-clustering problem. J. Appl. Ind. Math. **56**(2), 334–341 (2016)
38. Kel'manov, A., Motkova, A., Shenmaier, V.: An approximation scheme for a weighted two-cluster partition problem. In: van der Aalst, W.M.P., et al. (eds.) AIST 2017. LNCS, vol. 10716, pp. 323–333. Springer, Cham (2018). https://doi.org/10.1007/978-3-319-73013-4_30
39. Kel'manov, A.V., Khandeev, V.I.: A randomized algorithm for two-cluster partition of a set of vectors. Comput. Math. Math. Phys. **55**(2), 330–339 (2015)
40. Kel'manov, A.V., Khandeev, V.I., Panasenko A.V.: Exact algorithms for the special cases of two hard to solve problems of searching for the largest subset. In: van der Aalst, W.M.P., et al. (eds.) AIST 2018. LNCS, vol. 11179, pp. 294–304. Springer, Cham (2018)

41. Kel'manov, A.V., Khandeev, V.I.: Panasenko A.V.: Exact algorithms for two hard to solve 2-clustering problems. Pattern Recognit. Image Anal. **27**(4) (2018, in publishing)
42. Kel'manov, A.V., Pyatkin, A.V.: On complexity of some problems of cluster analysis of vector sequences. J. Appl. Ind. Math. **7**(3), 363–369 (2013)
43. Kel'manov, A.V., Khamidullin, S.A.: An approximation polynomial-time algorithm for a sequence Bi-clustering problem. Comput. Math. Math. Phys. **55**(6), 1068–1076 (2015)
44. Kel'manov, A.V., Pyatkin, A.V.: NP-completeness of some problems of choosing a vector subset. J. Appl. Ind. Math. **5**(3), 352–357 (2011)
45. Garey, M.R., Johnson, D.S.: Computers and Intractability: A Guide to the Theory of NP-Completeness. Freeman, San Francisco (1979)
46. Kel'manov, A.V., Khamidullin, S.A.: Posterior detection of a given number of identical subsequences in a quasi-periodic sequence. Comput. Math. Math. Phys. **41**(5), 762–774 (2001)

NP-hardness of Some Max-Min Clustering Problems

Alexander Kel'manov[1,2] (iD), Vladimir Khandeev[1,2(✉)] (iD),
and Artem Pyatkin[1,2] (iD)

[1] Sobolev Institute of Mathematics, 4 Koptyug Ave., 630090 Novosibirsk, Russia
[2] Novosibirsk State University, 2 Pirogova St., 630090 Novosibirsk, Russia
{kelm,khandeev,artem}@math.nsc.ru

Abstract. We consider some consimilar problems of searching for disjoint clusters in the finite set of points in Euclidean space. The goal is to maximize the minimum subset size so that the value of each intracluster quadratic variation would not exceed a given constant. We prove that all considered problems are NP-hard even on a line.

Keywords: Euclidean space · Clustering · Max-Min problem
NP-hardness · Quadratic variation

1 Introduction

The subject of this study includes some discrete optimization problems of searching for disjoint subsets in a finite set of Euclidean points. These optimization problems model a fundamental clustering problem, which is typical for many applications (such as Data analysis, Machine learning, Data mining, Statistics, Pattern recognition). The aim of our research is finding out the complexity status of these problems. The research is motivated by the importance of the problems both for theory and applications and also by the absence of any published results for them (see the next section).

The paper is organized as follows. In Sect. 2, the problems statement, motivation and applications are given. In Sect. 3, the complexity of the problems is analyzed.

2 Problem Statement and Related Problems

The considered problems are motivated as follows. Given a nonhomogeneous set of products containing two different groups of similar (by some characteristics) products and some waste products (or outliers). The characteristics of the products in each group have deviations from a known standard due to production reasons. The admissible deviations are determined by a given threshold. Products whose characteristics exceed the threshold are outliers. The matching

Y. Evtushenko et al. (Eds.): OPTIMA 2018, CCIS 974, pp. 144–154, 2019.
https://doi.org/10.1007/978-3-030-10934-9_11

between products and groups is unknown. The goal is to find in the set two groups of homogeneous products containing the maximum number of similar by the characteristics elements and to determine the outliers.

One of the possible models for this applied problem is the following optimization problem.

Problem 1. *Given* an N-element set $\mathcal{Y}$ of points in d-dimensional Euclidean space and a number $\alpha \in (0,1)$. *Find* non-empty disjoint subsets $\mathcal{C}_1, \mathcal{C}_2$ and points y_1, y_2 in the set $\mathcal{Y}$ such that

$$\min\{|\mathcal{C}_1|, |\mathcal{C}_2|\} \to \max \tag{1}$$

under constraints

$$\sum_{y \in \mathcal{C}_i} \|y - y_i\|^2 \le \alpha \sum_{y \in \mathcal{Y}} \|y - \bar{y}(\mathcal{Y})\|^2, \quad i = 1, 2, \tag{2}$$

where

$$\bar{y}(\mathcal{Y}) = \frac{1}{|\mathcal{Y}|} \sum_{y \in \mathcal{Y}} y$$

is a centroid (geometrical center) of the set $\mathcal{Y}$.

The applied problem given above can be also modeled by the following

Problem 2. *Given* an N-element set $\mathcal{Y}$ of points in d-dimensional Euclidean space, a number $\alpha \in (0,1)$, and points $z_1, z_2 \in \mathbb{R}^d$. *Find* non-empty disjoint subsets $\mathcal{C}_1, \mathcal{C}_2$ such that (1) holds under constraints

$$\sum_{y \in \mathcal{C}_i} \|y - z_i\|^2 \le \alpha \sum_{y \in \mathcal{Y}} \|y - \bar{y}(\mathcal{Y})\|^2, \quad i = 1, 2.$$

Problem 1 can be interpreted as a search for two disjoint subsets (clusters) of points concentrated near two unknown points in the input set. The union of these subsets may not cover the whole set. The right-hand side of the inequalities (2) equals to the α-proportion of the total quadratic variation of the points of the input set with respect to its centroid. The left-hand side of the inequalities (2) for each $i = 1, 2$ determines the total intracluster spread of points from the cluster $\mathcal{C}_i$ with respect to the sought point y_i in $\mathcal{Y}$. The level of concentration of points in the clusters is determined by the α-proportion of the total quadratic spread of the points of the input set. The minimum size of the clusters should be maximized under the upper bound on the level of the intracluster concentration of points.

The interpretation of Problem 2 is similar. The only difference is that the points z_1 and z_2 are additionally specified at the input.

Examples of input sets on the plane are given in Figs. 1, 2 and 3.

On the left part of Fig. 1, two clusters (subsets) are clearly visible as concentrations of points, and remaining points can be treated as outliers (elements not belonging to the clusters).

An example of an input set with five concentrated clusters and outliers on the plane is given in the right part of Fig. 1.

Examples of input sets with clusters whose centers are close and convex hulls overlap are given in Fig. 2.

On the left part of Fig. 3, an example of an input set with two clusters and outliers on the plane is given. In this example, one cluster has greater variation and it's convex hull contains the second cluster. A similar example with three clusters is given in the right part of Fig. 3.

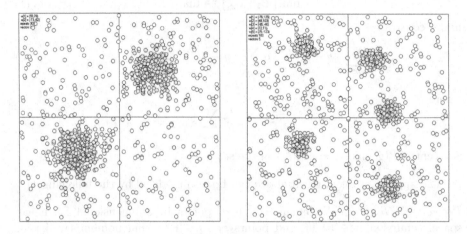

Fig. 1. Instances with two (left part) and five (right part) distant clusters of concentrated points

Note that considered Problems 1 and 2 are not equivalent to any similar in formulation well known clustering problems — k-$MSSC$ (or k-$means$), k-$median$, k-$medoid$, k-$center$, k-$center$ $clustering$ $with$ $outliers$, etc. (see, e.g., [1–17]). Here we emphasize also that the key issue in discrete optimization is the complexity status of recently arisen optimization problems.

As an additional motivation for the research consider some other possible applications. Remind that the data clustering, i.e. search for the subsets in data and partitioning the data into the subsets by various criteria of intercluster homogeneity (similarity of the elements of the same cluster) and separability (nonsimilarity) of the clusters is one of the main problems in Machine learning and Data mining, in Pattern recognition and Big-scaling data (see [18–27] and papers cited there).

The search for homogeneous subsets is typical for Data editing and Data cleaning from unpredictable random elements, so-called outliers, that could be caused by possible failures of measuring devices (see, e.g., [28–30] and papers cited there). In Figs. 1, 2 and 3 the points not belonging to the concentrated clusters can be treated as outliers. The data cleaning from outliers is a necessary procedure in the mentioned problems of Machine learning, Data mining and

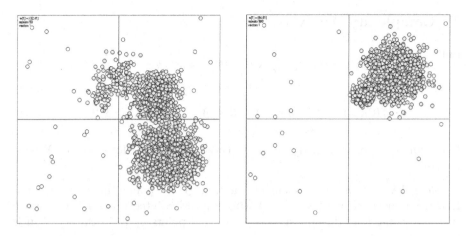

Fig. 2. Instances with four (left part) and two (right part) clusters whose centers are close and convex hulls overlap

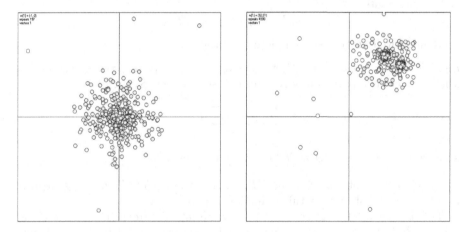

Fig. 3. Instances with two (left part) and three (right part) clusters whose centers almost coincide. The convex hull of the cluster with the greatest variation contains the other cluster(s)

Pattern recognition since these extraneous data can significantly worsen the quality of the machine learning. As an evident consequence of such worsening, the probability of false decisions on the recognizable objects increases.

Note also that the similar in formulation problems of searching for subsets with restrictions on the total intracluster quadratic variation of points were considered in [31–35]. However, these problems are not equivalent to the problems under consideration.

The main result of this paper is the proof of NP-hardness of Problem 1 and Problem 2 and their generalizations for the case of an arbitrary number of clusters.

3   Complexity Analysis

Put

$$A = \alpha \sum_{y \in \mathcal{Y}} \|y - \overline{y}(\mathcal{Y})\|^2,$$

$$f(\mathcal{C}_i, z_i) = \sum_{y \in \mathcal{C}_i} \|y - z_i\|^2, \quad \mathcal{C}_i \subseteq \mathcal{Y}, \quad z_i \in \mathbb{R}^d, \quad i = 1, 2.$$

Reformulate Problem 1 in the form of decision problem. Moreover, let $\mathcal{Y}$ be a multiset.

Problem 1A. *Given* an N-element multiset $\mathcal{Y}$ of points in d-dimensional Euclidean space, a number $A > 0$, and a positive integer M. *Find* whether there are non-empty disjoint subsets $\mathcal{C}_1, \mathcal{C}_2$ and points y_1, y_2 in the set $\mathcal{Y}$ such that

$$\min\{|\mathcal{C}_1|, |\mathcal{C}_2|\} \geq M, \tag{3}$$

under constraints

$$f(\mathcal{C}_i, y_i) \leq A, \quad i = 1, 2.$$

Remind well-known NP-complete [36] problem

PARTITION 1. *Given* a multiset $\mathcal{X} = \{x_1, \ldots, x_K\}$ of positive integers. *Find* whether there is a partition of $\mathcal{X}$ into multisets $\mathcal{S}_1$ and $\mathcal{S}_2$ such that

$$\sum_{x \in \mathcal{S}_1} x = \sum_{x \in \mathcal{S}_2} x.$$

Theorem 1. *Problem 1A is NP-complete even on a line.*

Proof. Clearly, Problem 1A is in NP. Reduce the *PARTITION* 1 problem to Problem 1A and show that this reduction is polynomial.

Let $\mathcal{X}$ be a K-element multiset of positive integers from an instance of the *PARTITION* 1 problem. Construct the following instance of Problem 1A with $d = 1$. Put

$$N = 2K + 2, \quad M = K + 1,$$

$$\mathcal{Y} = \{b_1, b_2, a_1, \ldots, a_K, \delta_1, \ldots, \delta_K\},$$

where

$$b_1 = b_2 = -B\,,$$

$$B > \max\left\{\frac{1}{4}\sum_{x \in \mathcal{X}} x, \sqrt{\frac{1}{2}\sum_{x \in \mathcal{X}} x}\right\},$$

$$a_k = 0, \quad \delta_k = \sqrt{x_k}, \quad k = 1, \ldots, K$$

and

$$A = B^2 + \frac{1}{2}\sum_{x \in \mathcal{X}} x.$$

Without loss of generality, $K \geq 3$. Show that the required subsets $\mathcal{C}_1$, $\mathcal{C}_2$ and points y_1, y_2 exist in Problem 1A if and only if the subsets $\mathcal{S}_1$ and $\mathcal{S}_2$ exist in *PARTITION* 1 problem.

Assume, the subsets $\mathcal{S}_1$ and $\mathcal{S}_2$, where

$$S_i = \{x_k \mid k \in I_i\}, \quad i = 1, 2,$$

$$I_1 \cup I_2 = \{1, \ldots, K\}, \quad I_1 \cap I_2 = \emptyset,$$

exist in *PARTITION* 1 problem. Put

$$y_i = 0, \quad C_i = \{b_i\} \cup \{\delta_k \mid k \in I_i\} \cup \{a_k \mid k \in I_{3-i}\}, \quad i = 1, 2.$$

Then

$$f(\mathcal{C}_i, y_i) = \|b_i - y_i\|^2 + \sum_{k \in I_i} \|\delta_k - y_i\|^2 = B^2 + \sum_{k \in I_i} x_k$$

$$= B^2 + \frac{1}{2} \sum_{k \in \{1, \ldots, K\}} x_k = A, \quad i = 1, 2,$$

i.e., the required subsets $\mathcal{C}_1$, $\mathcal{C}_2$ and points y_1, y_2 exist in Problem 1A.

Let now the required subsets $\mathcal{C}_1$, $\mathcal{C}_2$ and points y_1, y_2 exist in Problem 1A. Note that

$$|\mathcal{C}_1| = |\mathcal{C}_2| = K + 1.$$

Show that the points b_1 and b_2 lie in different subsets. Indeed, if, for example, $b_1, b_2 \in \mathcal{C}_1$, then since

$$|\mathcal{C}_1| = K + 1 \geq 4$$

for every choice of y_1 we have

$$f(\mathcal{C}_1, y_1) \geq 2B^2 > A,$$

contradicting $f(\mathcal{C}_1, y_1) \leq A$.

Show that no subset can contain all points δ_k, $k = 1, \ldots, K$. Indeed, otherwise assume $\delta_k \in \mathcal{C}_1$, $k = 1, \ldots, K$. If $y_1 = -B$ or $y_1 = \delta_r$ for some $r \in \{1, \ldots, K\}$, then

$$f(\mathcal{C}_1, y_1) \geq (B + \min_{k=1,\ldots,K} \delta_k)^2 > B^2 + 2B > A,$$

a contradiction with $f(\mathcal{C}_1, y_1) \leq A$. So, $y_1 = 0$. But then

$$f(\mathcal{C}_1, y_1) = B^2 + \sum_{k=1,\ldots,K} \delta_k^2 = B^2 + \sum_{x \in \mathcal{X}} x > A,$$

a contradiction.

Since both subsets $\mathcal{C}_1$ and $\mathcal{C}_2$ contain at least one point δ_k, $k \in \{1, \ldots, K\}$, and one of the points b_1, b_2, the same arguments as above imply that $y_1 = y_2 = 0$.

Since the points b_1 and b_2 are in different subsets and both C_1, C_2 contain points from $\{\delta_1, \ldots, \delta_K\}$ and $\{a_1, \ldots, a_K\}$, we may suppose that

$$C_i = \{b_i\} \cup \{\delta_k \mid k \in I_i\} \cup \{a_k \mid k \in I_{3-i}\}, \quad i = 1, 2,$$

where

$$I_1 \cup I_2 = \{1, \ldots, K\}, \quad I_1 \cap I_2 = \emptyset.$$

Using $y_i = 0$, $i = 1, 2$, we get

$$f(C_i, y_i) = \|b_i - y_i\|^2 + \sum_{k \in I_i} \|\delta_k - y_i\|^2 = B^2 + \sum_{k \in I_i} x_k$$

$$\leq A = B^2 + \frac{1}{2} \sum_{k \in \{1, \ldots, K\}} x_k, \quad i = 1, 2.$$

Hence,

$$\sum_{k \in I_1} x_k = \sum_{k \in I_2} x_k = \frac{1}{2} \sum_{k \in \{1, \ldots, K\}} x_k.$$

So, *PARTITION* 1 problem also has the required subsets S_1 and S_2.

For the convenience, in the proof we deal with $\delta_k = \sqrt{x_k}$ that could be irrational. However, it is easy to see that the same arguments are valid for rational $\delta_k = \sqrt{x_k} - \varepsilon_k$ where each $\varepsilon_k < 1/(2KB)$ because all x_k are integers. Therefore the constructed reduction is polynomial. □

It follows from Theorem 1 that Problem 1 is NP-hard even on a line. As a corollary, the following generalization of Problem 1 is also NP-hard.

Problem 3. Given an N-element set $\mathcal{Y}$ of points in d-dimensional Euclidean space, a positive integer J, and a number $\alpha \in (0, 1)$. *Find* non-empty disjoint subsets $C_1, \ldots, C_J$ and points $y_1, \ldots, y_J$ in the set $\mathcal{Y}$ such that

$$\min\{|C_1|, \ldots, |C_J|\} \to \max, \tag{4}$$

under constraints

$$\sum_{y \in C_i} \|y - y_i\|^2 \leq \alpha \sum_{y \in \mathcal{Y}} \|y - \overline{y}(\mathcal{Y})\|^2, \quad i = 1, \ldots, J.$$

In this problem, the number of clusters is a part of the input. For the case when the number of clusters is not a part of the input, the following result holds.

Theorem 2. *If the number of clusters is not a part of input then for every fixed $J \geq 2$ Problem 3 is NP-hard even on a line.*

The proof uses an easy induction on J, and its details are omitted here.

Now let us analyze the complexity of the problem 2. Restate Problem 2 in the form of decision problem with $\mathcal{Y}$ as a multiset.

Problem 2A. *Given* an N-element multiset $\mathcal{Y}$ of points in d-dimensional Euclidean space, a number $A > 0$, points $z_1, z_2 \in \mathbb{R}^d$, and a positive integer M. *Find* whether there are non-empty disjoint subsets $\mathcal{C}_1, \mathcal{C}_2$ in the set $\mathcal{Y}$ such that inequality (3) holds under constraints

$$f(\mathcal{C}_i, z_i) \leq A, \quad i = 1, 2.$$

Remind the *PARTITION* 1 problem with the additional requirement that the sought subset should have the same cardinalities remains NP-complete [36].

PARTITION 2. Given a $2K$-element multiset $\mathcal{X} = \{x_1, \ldots, x_{2K}\}$ of positive integers. *Find* whether there is a partition of $\mathcal{X}$ into multisets $\mathcal{S}_1$ and $\mathcal{S}_2$ such that $|\mathcal{S}_1| = |\mathcal{S}_2|$ and

$$\sum_{x \in \mathcal{S}_1} x = \sum_{x \in \mathcal{S}_2} x$$

The following theorem is true.

Theorem 3. *Problem 2 A is NP-complete even on a line.*

Problem 2 A is evidently in NP. Reduce *PARTITION* 2 problem to Problem 2A and show that this reduction is polynomial.

By this way, we construct the following instance of Problem 2A with $d = 1$. Put

$$N = 2K, \quad M = K,$$

$$\mathcal{Y} = \{\delta_1, \ldots, \delta_{2K}\},$$

$$z_1 = z_2 = 0,$$

where

$$\delta_k = \sqrt{x_k}, \quad k = 1, \ldots, 2K$$

and

$$A = \frac{1}{2} \sum_{x \in \mathcal{X}} x.$$

Now, it is easy to see that *PARTITION* 2 problem has the required subsets $\mathcal{S}_1$ and $\mathcal{S}_2$, if and only if Problem 2 has the required subsets $\mathcal{C}_1$ and $\mathcal{C}_2$.

As a corollary, the following generalization of Problem 1 is also NP-hard.

Problem 4. Given an N-element set $\mathcal{Y}$ of points in d-dimensional Euclidean space, a positive integer J, a number $\alpha \in (0, 1)$, and points $z_1, \ldots, z_J \in \mathbb{R}^d$. *Find* non-empty disjoint subsets $\mathcal{C}_1, \ldots, \mathcal{C}_J$ in the set $\mathcal{Y}$ such that (4) holds under the constraints

$$\sum_{y \in \mathcal{C}_i} \|y - z_i\|^2 \leq \alpha \sum_{y \in \mathcal{Y}} \|y - \overline{y}(\mathcal{Y})\|^2, \quad i = 1, \ldots, J.$$

In this problem, the number of clusters is a part of the input. For the case when the number of clusters is not a part of the input, the following result holds.

Theorem 4. *If the number of clusters is not a part of input then for every fixed $J \geq 2$ Problem 4 is NP-hard even on a line.*

The proof is based on an easy induction and is omitted here.

Thus, it follows from the theorems that all considered problems are NP-hard even on a line.

4   Conclusion

NP-hardness of some problems of searching for disjoint subsets in a finite set of Euclidean points is proved in the paper. We have shown that all these problems are NP-hard even on a line. Constructing algorithms with guaranteed performance for these problems is a matter of immediate prospects.

Acknowledgments. The study presented was supported by the Russian Science Foundation, project 16-11-10041.

References

1. MacQueen, J.B.: Some methods for classification and analysis of multivariate observations. In: Proceedings of the 5th Berkeley Symposium on Mathematical Statistics and Probability, vol. 1, pp. 281–297. University of California Press, Berkeley (1967)
2. Rao, M.: Cluster analysis and mathematical programming. J. Amer. Stat. Assoc. **66**, 622–626 (1971)
3. Hansen, P., Jaumard, B., Mladenovich, N.: Minimum sum of squares clustering in a low dimensional space. J. Classif. **15**, 37–55 (1998)
4. Hansen, P., Jaumard, B.: Cluster analysis and mathematical programming. Math. Program. **79**, 191–215 (1997)
5. Aloise, D., Deshpande, A., Hansen, P., Popat, P.: NP-hardness of Euclidean sum-of-squares clustering. Mach. Learn. **75**(2), 245–248 (2009)
6. Drineas, P., Frieze, A., Kannan, R., Vempala, S., Vinay, V.: Clustering large graphs via the singular value decomposition. Mach. Learn. **56**, 9–33 (2004)
7. Arora, S., Raghavan, P., Rao, S.: Approximation schemes for Euclidean k-medians and related problems. In: Proceedings of the 30th Annual ACM Symposium on Theory of Computing, pp. 106–113 (1998)
8. Kariv, O., Hakimi, S.: An algorithmic approach to network location problems. Part 1: the p-centers. SIAM J. Appl. Math. **37**, 513–538 (1979)
9. Feder, T., Greene, D.: Optimal algorithms for approximate clustering. In: Proceedings of 20th Annual ACM Symposium on Theory of Computing, pp. 434–444 (1988)
10. Hochbaum, D.S., Shmoys, D.B.: A best possible heuristic for the k-center problem. Math. Oper. Res. **10**(2), 180–184 (1985)
11. Kaufman, L., Rousseeuw, P.J.: Clustering by means of Medoids, in statistical data analysis based on the L_1-Norm and related methods. Edited by Dodge, Y., pp. 405–416. Dodge, North-Holland (1987)
12. Krarup, J., Pruzan, P.: The simple plant location problem: survey and synthesis. Eur. J. Oper. Res. **12**(1), 36–81 (1983)

13. Mirchandani, P., Francis, R. (eds.): Discrete Location Theory. Wiley-Intersience, New York (1990)
14. Charikar, M., Khuller, S., Mount, D.M., Narasimhan, G.: Algorithms for facility location problems with outliers. In: Proceedings of 12th ACM-SIAM Symposium on Discrete Algorithms, pp. 642–651 (2001)
15. Agarwal, P.K., Phillips, J.M.: An efficient algorithm for 2D Euclidean 2-center with outliers. In: Proceedings of 16th Annual European Symposium Algorithms, pp. 64–75 (2008)
16. McCutchen, R.M., Khuller, S.: Streaming algorithms for k-center clustering with outliers and with anonymity. In: Goel, A., Jansen, K., Rolim, J.D.P., Rubinfeld, R. (eds.) APPROX/RANDOM -2008. LNCS, vol. 5171, pp. 165–178. Springer, Heidelberg (2008). https://doi.org/10.1007/978-3-540-85363-3_14
17. Hatami, B., Zarrabi-Zade, H.: A streaming algorithm for 2-center with outliers in high dimensions. Comput. Geom. **60**, 26–36 (2017)
18. Jain, A.K.: Data clustering: 50 years beyond k-Means. Patt. Recogn. Lett. **31**(8), 651–666 (2010)
19. Shirkhorshidi, A.S., Aghabozorgi, S., Wah, T.Y., Herawan, T.: Big data clustering: a review. In: Murgante, B., et al. (eds.) ICCSA 2014, Part V. LNCS, vol. 8583, pp. 707–720. Springer, Cham (2014). https://doi.org/10.1007/978-3-319-09156-3_49
20. Bishop, C.M.: Pattern Recognition and Machine Learning. Springer, New York (2006)
21. James, G., Witten, D., Hastie, T., Tibshirani, R.: An Introduction to Statistical Learning. Springer, New York (2013)
22. Hastie, T., Tibshirani, R., Friedman, J.: The Elements of Statistical Learning, 2nd edn. Springer, New York (2009)
23. Aggarwal, C.C.: Data Mining: The Textbook. Springer, Switzerland (2015)
24. Goodfellow, I., Bengio, Y., Courville, A.: Deep Learning (Adaptive Computation and Machine Learning series). The MIT Press (2017)
25. Fu, T.-C.: A review on time series data mining. Eng. Appl. Artif. Intell. **24**(1), 164–181 (2011)
26. Kuenzer, C., Dech, S., Wagner, W.: Remote Sensing Time Series. Remote Sensing and Digital Image Processing, vol. 22. Springer, Switzerland (2015)
27. Liao, T.W.: Clustering of time series data – a survey. Patt. Recogn. **38**(11), 1857–1874 (2005)
28. de Waal, T., Pannekoek, J., Scholtus, S.: Handbook of Statistical Data Editing and Imputation. Wiley, New Jersey (2011)
29. Osborne, J.W.: Best Practices in Data Cleaning: A Complete Guide to Everything You Need to Do Before and After Collecting Your Data, 1st edn. SAGE Publication, Inc., Los Angeles (2013)
30. Farcomeni, A., Greco, L.: Robust Methods for Data Reduction. Chapman and Hall/CRC (2015)
31. Ageev, A.A., Kel'manov, A.V., Pyatkin, A.V., Khamidullin, S.A., Shenmaier, V.V.: Approximation polynomial algorithm for the data editing and data cleaning problem. Patt. Recogn. Image Anal. **27**(3), 365–370 (2017)
32. Ageev, A.A., Kel'manov, A.V., Pyatkin, A.V., Khamidullin, S.A., Shenmaier, V.V.: 1/2-Approximation polynomial-time algorithm for a problem of searching a subset. In: Proceedings of 2017 International MultiConference on Engineering, Computer and Information Sciences (SIBIRCON), 18–22 September 2017, Novosibirsk, Russia, pp. 8–12 (2017)

33. Kel'manov, A.V., Khamidullin, S.A., Khandeev, V.I., Pyatkin, A.V.: Exact algo-rithms for two quadratic Euclidean problems of searching for the largest subset and longest subsequence. In: Battiti, R., Brunato, M., Kotsireas, I., Pardalos, P.M. (eds.) LION 2018. LNCS, vol. 11353, pp. 308–318. Springer (2018). https://doi.org/10.1007/978-3-030-05348-2_28

34. Kel'manov, A.V., Khandeev, V.I., Panasenko A.V.: Exact algorithms for the special cases of two hard to solve problems of searching for the largest subset. In: van der Aalst, W.M.P., et al. (eds.) AIST 2018. LNCS, vol. 11179, pp. 294–304. Springer, Cham (2018). https://doi.org/10.1007/978-3-030-11027-7_28

35. Kel'manov, A.V., Pyatkin, A.V., Khamidullin, S.A., Khandeev, V.I., Shenmaier, V.V., Shamardin, Yu.V.: An approximation polynomial algorithm for a problem of searching for the longest subsequence in a finite sequence of points in Euclidean space. In: Eremeev, A., Khachay, M., Kochetov, Y., Pardalos, P. (eds.) OPTA 2018. CCIS, vol. 871, pp. 120–130. Springer, Cham (2018). https://doi.org/10.1007/978-3-319-93800-4_10

36. Garey, M.R., Johnson, D.S.: Computers and Intractability: A Guide to the Theory of NP-Completeness. Freeman, San Francisco (1979)

Improved Polynomial Time Approximation Scheme for Capacitated Vehicle Routing Problem with Time Windows

Michael Khachay[1,2,3]($\boxtimes$)(iD) and Yuri Ogorodnikov[1,2](iD)

[1] Krasovsky Institute of Mathematics and Mechanics, Ekaterinburg, Russia
{mkhachay,yogorodnikov}@imm.uran.ru
[2] Ural Federal University, Ekaterinburg, Russia
[3] Omsk State Technical University, Omsk, Russia

Abstract. The Capacitated Vehicle Routing Problem with Time Windows is the well-known combinatorial optimization problem having numerous valuable applications in operations research. In this paper, following the famous framework by M. Haimovich and A. Rinnooy Kan and technique by T. Asano et al., we propose a novel approximation scheme for the planar Euclidean CVRPTW. For any fixed $\varepsilon > 0$, the proposed scheme finds a $(1 + \varepsilon)$-approximate solution of CVRPTW in time

$$TIME(\mathrm{TSP}, \rho, n) + O(n^2) + O\left(e^{O\left(q\left(\frac{q}{\varepsilon}\right)^3 (p\rho)^2 \log(p\rho)\right)}\right),$$

where q is the given vehicle capacity bound, p is the number of time windows for servicing the customers, and $TIME(\mathrm{TSP}, \rho, n)$ is the time needed to find a ρ-approximate solution for an auxiliary instance of the metric TSP.

Keywords: Capacitated Vehicle Routing Problem · Time windows
Efficient Polynomial Time Approximation Scheme

1 Introduction

The Capacitated Vehicle Routing Problem (CVRP) is the well-known combinatorial optimization problem introduced by Dantzig and Ramser in 1959 in their seminal paper [4]. This problem is of great theoretic interest and has numerous practical applications in operational research (see, e.g. survey in [14]).

As it is known, the CVRP problem is strongly NP-hard even in the Euclidean plane. The problem is hardly approximable in general, but its geometric settings admit efficient approximation algorithms. The majority of known results back to the famous classical schemes proposed by Haimovich and Rinnooy Kan [6]

This research was supported by Russian Science Foundation, grant no. 14-11-00109.

Y. Evtushenko et al. (Eds.): OPTIMA 2018, CCIS 974, pp. 155–169, 2019.
https://doi.org/10.1007/978-3-030-10934-9_12

and by Arora [2]. Actually, the most recent results among them are Quasi-Polynomial Time Approximation Scheme (QPTAS) [5] for the Euclidean plane and the Efficient Polynomial Time Approximation Scheme (EPTAS) proposed in [7,9] for the Euclidean space of an arbitrary finite dimension $d > 1$.

Last decades, for the Capacitated Vehicle Routing Problem with Time Windows (CVPRTW) there is a significant progress in construction the branch-and-cut methods, local search heuristics, and metaheuristics that can be efficiently applied to some instances coming from the practice (see, e.g. [11,14]). But the results concerning approximation algorithms with performance guarantees for CVRPTW still remain quite rare. To the best of our knowledge, there are exhausted by the QPTAS scheme for the case of the finite number of mutually non-intersecting time windows introduced in [12,13] and the EPTAS for any fixed capacity and number of time windows proposed recently in [8]. Although the latter result appears to be the first PTAS for the CVRPTW, its time complexity bound $O(n^3 + \exp(\exp(1/\varepsilon)))$ remains huge even for a fixed capacity and number of time windows. In this paper, we try to bridge this gap and propose a new scheme, whose time complexity is better up to an exponent.

The rest of the paper is structured as follows. In Sect. 2, we describe a mathematical statement of CVRPTW. In Sect. 3, we provide the main idea and formal definition of our approximation scheme and announce its accuracy and time complexity bounds that are proved in Sect. 4. Finally, in Sect. 5, we summarize our results and discuss some questions that still remain open.

2   Problem Statement

We are given by a set of *customers* $X = \{x_1, \ldots, x_n\}$ and a set $T = \{t_1, \ldots, t_p\}$ of consecutive *time windows*. The set T is assumed to be ordered with respect to this precedence, i.e. $t_{j_1} \preceq t_{j_2}$ for any $1 \leq j_1 \leq j_2 \leq p$. Each customer x_i has a unit non-splittable demand that should be serviced in a given time window $t(x_i) \in T$. This service is performed by a fleet of *vehicles* initially arranged at a given point x_0 called *depot*. Vehicles have the same *capacity* q and visit customers by cyclic routes starting and finishing at the depot x_0. The goal is to provide a collection of capacitated vehicle routes visiting each customer once, in its time window, and minimizing the total transportation costs.

Mathematically an instance CVRPTW(X) of the CVRPTW problem is given by the weighted complete digraph $G = (X \cup \{x_0\}, E, w)$, partition

$$X_1 \cup \ldots \cup X_p = X, \tag{1}$$

and capacity bound $q \in \mathbb{N}$. Here,

(i) a route R is called *feasible*, if R is a simple directed cycle, has the form

$$R = x_0, x_{i_1}, \ldots, x_{i_s}, x_0 \tag{2}$$

and satisfies the following constraints

$$s \leq q \quad \text{(capacity)} \tag{3}$$

$$t(x_{i_j}) \preceq t(x_{i_{j+1}}) \quad \text{(time windows)} \tag{4}$$

(ii) non-negative weighting function w defines the transportation cost $w(x_i, x_j)$ for each ordered location pair $(x_i, x_j) \in E$, such that any route (2) has the cost

$$w(R) = w(x_0, x_{i_1}) + \sum_{j=1}^{s-1} w(x_{i_j}, x_{i_{j+1}}) + w(x_{i_s}, x_0)$$

(iii) any subset X_j of partition (1) consists of customers x, which should be serviced (visited) during the same time window $t(x) = t_j \in T$.

It is required to find a collection of feasible routes $S = \{R_1, \ldots, R_l\}$ that visits each customer once and has the minimum total transportation cost

$$w(S) = \sum_{i=1}^{l} w(R_i). \tag{5}$$

If the weighting function w is symmetric and satisfies the triangle inequality, then CVRPTW is called *metric* and transportation costs $w(x_i, x_j)$ are called distances between locations x_i and x_j. In this paper, we consider the Euclidean CVRPTW, where $X \cup \{x_0\} \subset \mathbb{R}^d$ and $w(x_i, x_j) = \|x_i - x_j\|_2$.

3   Approximation Scheme

The scheme proposed in Algorithm 1 improves our recent result [8] based on the famous problem decomposition approach and Iterated Tour Partition heuristic introduced by Haimovich and Rinnooy Kan in their seminal paper [6]. Applying the ideas from [3], we extend the parametric domain, where the algorithm remains PTAS, from $p = O(\log \log n)$ and $q = O(\log \log n)$ to $p^3 q^4 = O(\log n)$ and time complexity bound for any fixed p and q from $O(n^3 + \exp(\exp(O(1/\varepsilon))))$ to $O(n^3 + \exp(O(1/\varepsilon^3)))$.

To describe our scheme, we need the following auxiliary extremal problem, Partial Capacitated Vehicle Routing Problem with Time Windows (PCVRPTW) introduced in [3].

Settings of PCVRPTW and CVRPTW are very close to each other. Like for CVRPTW, an instance of PCVRPTW is given by a weighted complete digraph $G = (X \cup \{x_0\}, E, w)$ defining customer and depot locations augmented with corresponding transportation costs, partition $X_1 \cup \ldots \cup X_p = X$ of the customer set to subsets induced by the given time windows $\{t_1, \ldots, t_p\}$, and common vehicle capacity bound q.

The only difference is that, in PCVRPTW, feasible solutions are not required to visit all the customers. Instead, they should visit only *privileged* customers $x \in Q$ for some given subset $Q \subset X$. All other ones are optional for visiting but their omission is penalized in the objective function as follows. Suppose $S = \{R_1, \ldots, R_l\}$ is an arbitrary feasible solution. By $X[S]$ and $\bar{X}[S]$, we denote the subsets of customers visited[1] and omitted by S, respectively. Then, the cost of S is defined by the following equation

---

[1] By construction, $Q \subset X[S]$.

$$\text{cost}(S) = w(S) + \frac{2}{q} \sum_{x \in \bar{X}[S]} r(x), \tag{6}$$

where, as for CVRPTW, $w(S)$ represents total transportation expenses (5) and $r(x) = w(x_0, x)$ is a distance between the customer x and the depot x_0. In PCVRPTW, we are aimed to find a feasible solution U minimizing (6).

In the sequel, wherever this does not cause misunderstandings, we use the notation PCVRPTW(Y, Q) for an instance defined by the set X and the subset $Q \subset X$ of all and the privileged customers, respectively.

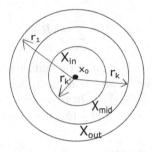

Fig. 1. Splitting the set X to subsets of *outer*, *intermediate*, and *inner* customers

The main idea of Algorithm 1 is as follows.

(i) Given by a relative error bound $\varepsilon > 0$, we reorder the customers in X by descending of their distances $r_i = r(x_i) = w(x_0, x_i)$ from the depot x_0. Then, for some numbers $k \leq k'$, which will be specified below, we split the set X to three disjoint subsets $X_{out} = \{x_1, \ldots, x_{k-1}\}$, $X_{mid} = \{x_k, \ldots, x_{k'-1}\}$, and $X_{in} = \{x_{k'}, \ldots, x_n\}$ of the *outer*, *intermediate*, and *inner* customers, respectively (Fig. 1). The main point here is that k and k' depend on ε, the capacity bound q, number of time windows p, but not on the number of customers n and their locations.

(ii) Then, applying Algorithm 2, we find an optimal solution U for the auxiliary instance of PCVRPTW$(X_{out} \cup X_{mid}, X_{mid})$. As for the classic Haimovich and Rinnooy Kan approach employed in [8], the time complexity of Algorithm 2 is independent on n, but its complexity bound is much more better.

(iii) By construction, the subset $\bar{X}[U]$ of customers omitted by U encloses X_{in} entirely and can also include some customers from X_{mid}. To find an approximate solution S_{ITP} for the subinstance CVRPTW$(\bar{X}[U])$ we apply our adaptation of the famous ITP heuristic to the case of CVRP with time windows (Algorithm 3).

(iv) Finally, we output $S_{APP} = U \cup S_{ITP}$.

Algorithm 1. Improved Approximation Scheme for the Euclidean CVRPTW

Input: an instance of the Euclidean CVRPTW defined by a complete graph $G(X \cup \{x_0\}, E, w)$, a capacity q, and partition $X_1 \cup \ldots \cup X_p = X$

Parameters: $\varepsilon > 0$ and $\rho \geq 1$

Output: an $(1 + \varepsilon)$-approximate solution S_{APP} of the given CVRPTW instance

1: relabel the customers in the order $r_1 \geq \ldots \geq r_n$
2: for the given $0 < \varepsilon < 1$, find the smallest integer $k \leq n$ such that

$$r_k \leq \left(\frac{\varepsilon}{q}\right)^2 \frac{1}{64\pi(p\rho)^2} \sum_{i=1}^{n} r_i \tag{7}$$

3: find the smallest integer k' such that $k \leq k' \leq n$ and

$$4(k' - 1) \cdot r_{k'} \leq \frac{\varepsilon}{q} \sum_{i=1}^{n} r_i \tag{8}$$

4: split the set X to the subsets $X_{out} = \{x_1, \ldots, x_{k-1}\}$, $X_{mid} = \{x_k, \ldots, x_{k'-1}\}$ and $X_{in} = X \setminus \{X_{out} \cup X_{mid}\}$ as shown at Fig. 1.
5: by Algorithm 2, find an optimal solution U for the $PCVRPTW(X_{out} \cup X_{mid}, X_{out})$ subinstance defined by the subgraph $G\langle X_{out} \cup X_{mid} \cup \{x_0\}\rangle$, partition

$$(X_1 \cup \ldots \cup X_p) \cap (X_{out} \cup X_{mid}),$$

and capacity q
6: applying Algorithm 3, find an approximate solution S_{ITP} for the subinstance $\text{CVRPTW}(\bar{X}[U])$
7: output the solution $S_{\text{APP}} = U \cup S_{\text{ITP}}$.

3.1 Exact Algorithm for PCVRPTW

Consider an instance $\text{PCVRPTW}(Y, Q)$ for some given $Q \subset Y$. The main idea of the proposed technique (Algorithm 2) is quite simple and based on the following observations.

(i) Since feasibility of any optimal solution $S = \{R_1, \ldots, R_l\}$ does not depend on the order of its routes, we can assume that the routes are ordered by decreasing of their individual capacities q_j, where q_j is equal to the number of customers visited by the route R_j.
(ii) Since any route of an optimal solution visits at least one[2] customer from Q and any $x \in Q$ should be visited only once, the number l of routes can not exceed $b = |Q|$.

As it follows from these observations, to each solution S, we can assign the *shape* $\lambda = \lambda(S) = q_1, \ldots, q_l$, where $q \geq q_1 \geq q_2 \geq \ldots \geq q_l \geq 1$ and $l \leq b$. Further, any two solutions S_1 and S_2 we call equivalent, if $\lambda(S_1) = \lambda(S_2)$. Thus,

---

[2] As it follows from Lemma 1.

the set of feasible solutions of the instance PCVRPTW(Y, Q) is partitioned into equivalence classes. It is convenient to represent each class with corresponding *Young diagram*[3] (see, e.g. [1]) as it is showed in Fig. 2. Each j-th row of such a

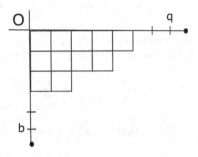

Fig. 2. Example: Young diagram for $q = 7$, $b = 5$ and the shape $\lambda = 5, 4, 2$

diagram corresponds to the j-th route excluding its starting and finishing node x_0. To obtain the one-to-one correspondence with a certain solution belonging to the considered equivalence class, we just fill each row of the diagram, left to right, cell by cell by customer locations in the order of visiting them by the corresponding route.

We start with an enumeration of all possible shapes for the given q and b. Further, having fixed a certain shape, we enumerate all the possible injections of the set Q into the cell set of the corresponding Young diagram (like in Fig. 3a) and filter out injections that induce infeasible solutions (violating time windows constraints).

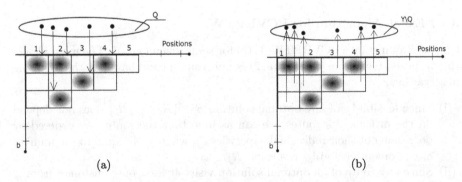

(a)   (b)

Fig. 3. Example: (a) injection of the set Q to a certain Young diagram; (b) mapping of free cells to $Y \setminus Q$

At the last stage, for any partially filled diagram (by the points of Q), we enumerate all injections of the subset of its free cells into the set $Y \setminus Q$.

---

[3] Also known as Ferrers board.

For any such an injection, we check out whether the candidate solution obtained is feasible and (if so) compute a value of objective function (6). Finally, we output the least cost feasible solution.

Algorithm 2. Exact algorithm for PCVRPTW(Y, Q)

Input: an instance PCVRPTW, defined by a complete graph $G(Y \cup \{y_0\}, E, w)$, a capacity q, partition $Y_1 \cup \ldots \cup Y_p = Y$, and subset $Q \subset Y$ of privileged customers
Output: an exact solution U of PCVRPTW(Y, Q)

1: let $b = |Q|$
2: **for each** shape $\lambda = q_1, \ldots, q_l$, s.t. $q \geq q_1 \geq \ldots q_l \geq 1$ and $1 \leq l \leq b$ **do**
3:     construct the corresponding Young diagram D_λ
4:     **for each** injection $\mu \colon Q \to D_\lambda$ **do**
5:         **if** the induced partial solution is feasible **then**
6:             let $F = D_\lambda \setminus \mu(Q)$
7:             **for each** injection $\nu : F \to Y \setminus Q$ **do**
8:                 obtain the corresponding candidate solution $S_{\lambda,\mu,\nu}$
9:                 **if** $S_{\lambda,\mu,\nu}$ is feasible **then**
10:                     compute $\mathrm{cost}(S_{\lambda,\mu,\nu})$ by formula (6)
11:                 **end if**
12:             **end for**
13:         **end if**
14:     **end for**
15: **end for**
16: output $U = \arg\min\{\mathrm{cost}(S_{\lambda,\mu,\nu}) \colon S_{\lambda,\mu,\nu} \text{ is feasible}\}$

3.2   Iterated Tour Partition (ITP) Heuristic for Metric CVRPTW

Algorithm 3 is our adaptation of the well known Haimovich and Rinnooy Kan ITP heuristic proposed in [6] for the metric CVRP to the case of additional time windows constraints. For the first time, Algorithm 3 was published in [8]. We present it in the current paper for the sake of clarity.

3.3   Result

The main result of our paper is claimed in Theorem 1. For the sake of simplicity, we present it for the simplest case of Euclidean plane and a single depot postponing more general result to the forthcoming paper.

Theorem 1. *For any $\varepsilon > 0$ an $(1 + \varepsilon)$-approximate solution for the Euclidean CVRPTW can be obtained in time*

$$TIME(\mathrm{TSP}, \rho, n) + O(n^2) + O\left(e^{O\left(q\left(\frac{q}{\varepsilon}\right)^3 (p\rho)^2 \log(p\rho)\right)}\right).$$

Algorithm 3. ITP heuristic for the metric CVRPTW

Input: an instance of the metric CVRPTW defined by a complete graph $G(X \cup \{x_0\}, E, w)$, a capacity q, and partition $X_1 \cup \ldots \cup X_p = X$

Parameter: ρ-approximation algorithm $\mathcal{A}_\rho$ for the metric TSP

Output: an approximate solution S_{ITP} of the given CVRPTW instance

1: using $\mathcal{A}_\rho$ obtain a ρ-approximate metric TSP solution H for the subgraph $G\langle X \rangle$
2: by shortcutting (Fig. 4), split the cycle H into smaller cycles $H_1, \ldots, H_p$, s.t. H_j spans customers from X_j
3: **for each** cycle H_j **do**
4:     **for each** $x \in X_j$ **do**
5:         starting from the vertex x, split the cycle H_j into $l_j = \lceil |X_j|/q \rceil$ chains, s.t. each of them, except maybe one, spans q vertices
6:         connecting endpoints of each chain with the depot x_0 directly, construct a set $S(x)$ of l_j routes
7:     **end for**
8:     put $S_j = \arg\min\{w(S(x)) \colon x \in X_j\}$
9: **end for**
10: output the solution $S_{\text{ITP}} = S_1 \cup \ldots \cup S_p$.

Remark 1. The scheme proposed is PTAS for $p^3 q^4 = O(\log n)$. Furthermore, for any fixed $p \geq 1$ and $q \geq 1$, the scheme is EPTAS with time complexity $O(n^3 + e^{O(1/\varepsilon^3)})$ provided the inner TSP instance in Algorithm 3 is solved by the Christofides algorithm.

4   Proof Sketch

It is convenient to present the proof of Theorem 1 in two subsections. In Subsect. 4.1, we show that, for any given $\varepsilon > 0$ Algorithm 1 do provide an $(1 + \varepsilon)$-approximate solution of CVRPTW. Further, in Subsect. 4.2, we provide its time complexity bound.

4.1   Accuracy Guarantee

To estimate the accuracy bound for our scheme, we need the following technical lemmas.

Lemma 1. *For any non-negative weight function w,*

$$\text{CVRPTW}^*(X) \geq \frac{2}{q} \sum_{x \in X}^{n} r(x).$$

Lemma 1 is a straightforward consequence of similar results proved in [6, 10] for the general CVRP. The proof of Lemma 2 is also follows from the literature [3].

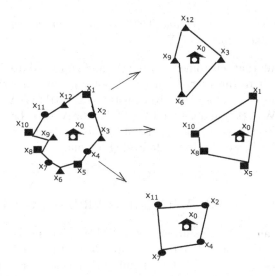

Fig. 4. Example: splitting of a Hamiltonian cycle, geometric shapes encode different time windows

Lemma 2. *For any non-negative weighting function w and $Q \subset X$, the following relation*

$$\text{PCVRPTW}^*(X, Q) \leq \text{CVRPTW}^*(X) \tag{9}$$

is valid.

Proof. Let U be an arbitrary optimum solution of $\text{CVRPTW}(X)$ and $U' \subset U$ be the minimal family of routes visiting all customers of the subset Q. Since U' is a feasible solution of $\text{PCVRPTW}(X, Q)$,

$$\text{PCVRPTW}^*(X, Q) \leq w(U') + \frac{2}{q} \sum_{x \in \bar{X}[U']} r(x) \leq w(U') + \text{CVRPTW}^*(\bar{X}[U']),$$

by Lemma 1. Therefore, all elements of the subset $\bar{X}[U']$ (and only them) should be visited by $U \setminus U'$, by the choice of U. Hence, $U \setminus U'$ is a feasible solution for $\text{CVRPTW}(\bar{X}[U'])$ and $\text{CVRPTW}^*(\bar{X}[U']) \leq w(U \setminus U')$. Finally, we have

$$\text{PCVRPTW}^*(X, Q) \leq w(U') + w(U \setminus U') = w(U) = \text{CVRPTW}^*(X).$$

Lemma 2 is proved.

Lemma 3 ([8]). *For any solution S_{ITP} found by Algorithm 3 for a given instance of the metric CVRPTW defined by a graph $G = (X \cup \{x_0\}, E, w)$, capacity q, and time windows induced partition $X_1 \cup \ldots \cup X_p = X$, the following bound is valid*

$$w(S_{\text{ITP}}) \leq p \left(1 - \frac{1}{q}\right) \rho \text{TSP}^*(X) + \frac{2}{q} \sum_{x \in X} r(x) + 2pr_{max}, \tag{10}$$

where $r_{max} = \max\{r_1, \ldots, r_{|X|}\}$ and TSP* is the optimum value of the metric TSP instance for the subgraph $G\langle X \rangle$.

Further, recall that customers $x_i \in X$ are supposed to be ordered by descending of their distances r_i from the depot x_0. Take an arbitrary integer $k \leq n = |X|$ and decompose the given instance CVRPTW(X) to two subinstances CVRPTW$(X(k))$ and CVRPTW$(X'(k))$ for $X(k) = \{x_1, \ldots, x_{k-1}\}$ and $X'(k) = X \setminus X(k)$. Following the argument of the similar result proposed in [6] for the metric CVRP, we obtain the following lemma.

Lemma 4

$$\text{CVRPTW}^*(X(k)) + \text{CVRPTW}^*(X'(k)) \leq \text{CVRPTW}^*(X) + 4(k-1)r_k \quad (11)$$

Proof. Let U be an optimal solution of CVRPTW$^*(X)$ and $R \in U$ be its arbitrary route. Split the route R into subroute R_0 visiting only customers from X'_k and subroutes $R_1, \ldots, R_s$ that visit only outer customers (from $X(k)$) (as in Fig. 5). Performing such a transformation for each route R of the initial solution U, we obtain feasible solutions S and S' for CVRPTW$(X(k))$ and CVRPTW$(X'(k))$, respectively, since time windows constraints evidently remain satisfied.

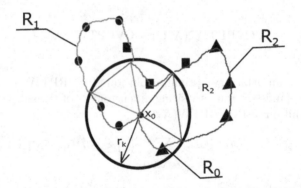

Fig. 5. Example: split the route

Obviously, this transformation can be done by adding at most $2(k-1)$ radii, each of length r_k, and at most $(k-1)$ chords, whose lengths again do not exceed $2r_k$, by triangle inequality. Therefore,

$$w(S) + w(S') \leq w(U) + 4(k-1)r_k,$$

and Lemma 4 is proved, since

$$\text{CVRPTW}^*(X(k)) + \text{CVRPTW}^*(X'(k)) \leq w(S) + w(S').$$

Lemma 5 ([8]). *For any instance $X = \{x_1, \ldots, x_n\}$ of the planar Euclidean TSP enclosed in the sphere with center x_0 and radius $\mathcal{R}$,*

$$\text{TSP}^*(X) \leq 2\mathcal{R} + 4\sqrt{\pi \mathcal{R} \sum_{i=1}^{n} r_i}, \tag{12}$$

where r_i is the distance between x_i and x_0.

At last, we can estimate the accuracy of the solution S_{APP} provided by Algorithm 1.

Lemma 6
$$w(S_{APP}) \leq (1 + \varepsilon)\,\text{CVRPTW}^*(X). \tag{13}$$

Proof. Since $w(S_{APP}) = w(U) + w(S_{ITP})$, then

$$w(S_{APP}) \leq w(U) + \frac{2}{q} \sum_{x \in \bar{X}[U]} r(x) + p\rho\text{TSP}^*(\bar{X}[U]) + 2pr_k,$$

by Lemma 3.

Since U is an optimal solution of $\text{PCVRPTW}(X_{out} \cup X_{mid}, X_{out})$, then

$$w(U) + \frac{2}{q} \sum_{x \in \bar{X}[U] \cap X_{mid}} r(x) = \text{PCVRPTW}^*(X_{out} \cup X_{mid}, X_{out})$$

$$\leq \text{CVRPTW}^*(X_{out} \cup X_{mid}),$$

by Lemma 2. Further, by Lemma 1,

$$\frac{2}{q} \sum_{x \in X_{in}} r(x) \leq \text{CVRPTW}^*(X_{in}).$$

Therefore,

$$w(S_{APP}) \leq w(U) + \frac{2}{q} \sum_{x \in \bar{X}[U] \cap X_{mid}} r(x) + \frac{2}{q} \sum_{x \in X_{in}} r(x) + p\rho\text{TSP}^*(\bar{X}[U]) + 2pr_k$$

$$\leq \text{CVRPTW}^*(X_{out} \cup X_{mid}) + \text{CVRPTW}^*(X_{in}) + p\rho\text{TSP}^*(\bar{X}[U]) + 2pr_k$$

$$\leq \text{CVRPTW}^*(X) + 4(k'-1)r_{k'} + p\rho\text{TSP}^*(\bar{X}[U]) + 2pr_k$$

$$\leq \text{CVRPTW}^*(X) + 4(k'-1)r_{k'} + 4p\rho\sqrt{\pi r_k \sum_{i=1}^{n} r_i} + 2p(\rho+1)r_k,$$

by Lemmas 4 and 5.

Applying Eqs. (7) and (8), for any $\varepsilon \in (0, 1)$ and $p, q, \rho \geq 1$, we obtain

$$w(S_{APP}) \leq \text{CVRPTW}^*(X) + \frac{\varepsilon}{2} \frac{2}{q} \sum_{i=1}^{n} r_i + \frac{\varepsilon}{4} \frac{2}{q} \sum_{i=1}^{n} r_i + \left(\frac{\varepsilon}{q}\right)^2 \frac{2p(\rho+1)}{64\pi(p\rho)^2} \sum_{i=1}^{n} r_i$$

$$\leq \text{CVRPTW}^*(X) + \frac{\varepsilon}{2} \frac{2}{q} \sum_{i=1}^{n} r_i + \frac{\varepsilon}{96} \frac{2}{q} \sum_{i=1}^{n} r_i + \frac{\varepsilon}{4} \frac{2}{q} \sum_{i=1}^{n} r_i.$$

Indeed, our assumptions imply that $(\varepsilon/q)^2 \leq \varepsilon/q$, $\rho + 1 \leq 2\rho$, and

$$\frac{p(\rho + 1)}{64\pi(p\rho)^2} \leq \frac{2p\rho}{64\pi(p\rho)^2} = \frac{1}{32\pi p\rho} \leq \frac{1}{32\pi} \leq \frac{1}{96}.$$

Again, applying Lemma 1, we obtain our final equation

$$w(S_{\text{APP}}) \leq (1 + \varepsilon)\,\text{CVRPTW}^*(X).$$

Lemma is proved.

4.2   Time Complexity Bound

Time complexity bound for Algorithm 1 can be obtained as a sum of the similar bounds for Algorithms 2 and 3. Whilest, the latter bound is known [6]

$$TIME(\text{TSP}, \rho, n) + O(n^2), \tag{14}$$

where $TIME(\text{TSP}, \rho, n)$ is complexity of finding a ρ-approximate solution for the planar Euclidean TSP, to estimate time complexity of Algorithm 2, we need some additional technical lemmas.

We start with estimation of $|X_{out}|$ and $|X_{mid}|$.

Lemma 7

$$|X_{out}| = k - 1 < \left(\frac{q}{\varepsilon}\right)^2 64\pi(p\rho)^2 \tag{15}$$

Proof. For $k = 1$, Lemma is evidently true. Consider the case $k > 1$. By choice of r_k in Algorithm 1,

$$r_1 \geq \ldots \geq r_{k-1} > \left(\frac{\varepsilon}{q}\right)^2 \frac{1}{64\pi(p\rho)^2} \sum_{i=1}^{n} r_i.$$

The assumption $k - 1 \geq \left(\frac{q}{\varepsilon}\right)^2 64\pi(p\rho)^2$ leads to the following evident contradiction

$$\sum_{i=1}^{n} r_i \geq \sum_{i=1}^{k-1} r_i \geq (k-1)r_{k-1} > \left(\frac{q}{\varepsilon}\right)^2 64\pi(p\rho)^2 \left(\frac{\varepsilon}{q}\right)^2 \frac{1}{64\pi(p\rho)^2} \sum_{i=1}^{n} r_i = \sum_{i=1}^{n} r_i.$$

Lemma is proved.

Hereinafter, we assume without loss of generality that $k > 1$.

Lemma 8

$$|X_{mid}| = k' - k, \text{ where } k \leq k' \leq O\left(ke^{4q/\varepsilon}\right) \tag{16}$$

Proof. Indeed, in the case $k' = k$, Lemma is true for any $\varepsilon \in (0,1)$ and $q \geq 1$. Consider the non-trivial case, when $k < k'$. By construction,

$$\frac{r_l}{\sum_{i=1}^{n} r_i} > \frac{\varepsilon}{4q(l-1)}$$

for any $k \leq l < k'$. Then,

$$1 \geq \sum_{l=k}^{k'-1} \frac{r_l}{\sum_{i=k}^{n} r_i} > \frac{\varepsilon}{4q} \sum_{l=k}^{k'-1} \frac{1}{l-1} = \frac{\varepsilon}{4q} \sum_{i=k-1}^{k'-2} \frac{1}{i} \geq \frac{\varepsilon}{4q} \int_{k-1}^{k'-2} \frac{dx}{x} \geq \frac{\varepsilon}{4q} \log \frac{k'-2}{k-1}.$$

Therefore,

$$k' < (k-1)\, e^{4q/\varepsilon} + 2 = O(k\, e^{4q/\varepsilon}).$$

Lemma is proved.

Lemma 9. *Algorithm 2 find an optimal solution of* PCVRPTW($X_{out} \cup X_{mid}, X_{out}$) *in time*

$$e^{O(q\,(q/\varepsilon)^3\,(p\rho)^2\,\log(p\rho))}.$$

Proof. Indeed, the number of shapes enumerated by Algorithm 2 at Step 2 is bounded from above by the number $\binom{b+q}{q} \leq O(b^q)$ of Young diagrams enclosed to $b \times q$-rectangle, where $b = |Q|$. Then, for any fixed shape, the number of injections of the set Q to cell set of the appropriate diagram (considered at Step 4) does not exceed $(bq)^{|Q|} = (bq)^b$. Each such an injection leaves some cells of the diagram unused (free). At Step 7, Algorithm 2 enumerates all injections of the set of unassigned cells into the set $Y \setminus Q$. The number of such maps is at most $|Y \setminus Q|^{bq}$. Finally, running time of Step 10 is $O(bq)$. Therefore, Algorithm 2 finds an optimal solution of PCVRPTW(Y, Q) in time

$$O(b^q)\,(bq)^{b+1}\,|Y \setminus Q|^{bq}. \tag{17}$$

For the instance PCVRPTW($X_{out} \cup X_{mid}, X_{out}$), $Q = X_{out}$, $Y \setminus Q = X_{mid}$. Therefore, $b = k - 1$, $|Y \setminus Q| = |X_{mid}| \leq O\left(ke^{4q/\varepsilon}\right)$, by Lemmas 7 and 6, respectively. Transforming (17) and applying Lemma 7 again, we obtain

$$O(k^q)\,(kq)^k(ke^{4q/\varepsilon})^{kq} = \exp(O(q\log k + k\log(qk) + qk\log k + 4kq\,q/\varepsilon))$$

$$\leq \exp(O(4qk(\log k + q/\varepsilon))) = \exp\left(O\left(q\left(\frac{q}{\varepsilon}\right)^3 (p\rho)^2 \log(p\rho)\right)\right).$$

Lemma is proved.

Proof of Theorem 1 can be obtained as a straightforward consequence of Lemmas 6 and 9.

5 Conclusion

In this paper, we introduced a new approximation scheme for the Capacitated Vehicle Routing Problem with Time Windows. Due to smart problem decomposition, we succeed in reducing its time complexity up to the exponent in contrast to our previous result based on classic Haimovich and Rinnooy Kan approach.

For any fixed capacity $q \geq 1$ and number of time windows $p \geq 1$, the proposed scheme is EPTAS finding a $(1+\varepsilon)$-approximate solution for the planar Euclidean CVRPTW in time $O(n^3 + \exp((1/\varepsilon)^3))$. In addition, the scheme remains PTAS, when

$$p^3 q^4 = O(\log n).$$

In the forthcoming paper, we extend our algorithm to the case of Euclidean space of an arbitrary fixed dimension $d > 1$ and multiple depots.

References

1. Andrews, G.E., Eriksson, K.: Integer Partitions, 2nd edn. Cambridge University Press, Cambridge (2004)
2. Arora, S.: Polynomial time approximation schemes for Euclidean traveling salesman and other geometric problems. J. ACM **45**, 753–782 (1998)
3. Asano, T., Katoh, N., Tamaki, H., Tokuyama, T.: Covering points in the plane by k-tours: towards a polynomial time approximation scheme for general k. In: Proceedings of the Twenty-Ninth Annual ACM Symposium on Theory of Computing, STOC 1997, pp. 275–283. ACM, New York (1997). https://doi.org/10.1145/258533.258602, http://doi.acm.org/10.1145/258533.258602
4. Dantzig, G.B., Ramser, J.H.: The truck dispatching problem. Manage. Sci. **6**(1), 80–91 (1959)
5. Das, A., Mathieu, C.: A quasipolynomial time approximation scheme for Euclidean capacitated vehicle routing. Algorithmica **73**, 115–142 (2015). https://doi.org/10.1007/s00453-014-9906-4
6. Haimovich, M., Rinnooy Kan, A.H.G.: Bounds and heuristics for capacitated routing problems. Math. Oper. Res. **10**(4), 527–542 (1985). https://doi.org/10.1287/moor.10.4.527
7. Khachai, M.Y., Dubinin, R.D.: Approximability of the vehicle routing problem in finite-dimensional Euclidean spaces. Proc. Steklov Inst. Math. **297**(1), 117–128 (2017). https://doi.org/10.1134/S0081543817050133
8. Khachay, M., Ogorodnikov, Y.: Efficient PTAS for the Euclidean CVRP with time windows. In: Analysis of Images, Social Networks and Texts - 7th International Conference (AIST 2018). LNCS, vol. 11179, pp. 296–306 (2018). https://doi.org/10.1007/978-3-030-11027-7_30
9. Khachay, M., Dubinin, R.: PTAS for the Euclidean capacitated vehicle routing problem in R^d. In: Kochetov, Y., Khachay, M., Beresnev, V., Nurminski, E., Pardalos, P. (eds.) DOOR 2016. LNCS, vol. 9869, pp. 193–205. Springer, Cham (2016). https://doi.org/10.1007/978-3-319-44914-2_16
10. Khachay, M., Zaytseva, H.: Polynomial time approximation scheme for single-depot Euclidean capacitated vehicle routing problem. In: Lu, Z., Kim, D., Wu, W., Li, W., Du, D.-Z. (eds.) COCOA 2015. LNCS, vol. 9486, pp. 178–190. Springer, Cham (2015). https://doi.org/10.1007/978-3-319-26626-8_14

11. Kumar, S., Panneerselvam, R.: A survey on the vehicle routing problem and its variants. Intell. Inf. Manage. **4**, 66–74 (2012). https://doi.org/10.4236/iim.2012. 43010

12. Song, L., Huang, H.: The Euclidean vehicle routing problem with multiple depots and time windows. In: Gao, X., Du, H., Han, M. (eds.) COCOA 2017. LNCS, vol. 10628, pp. 449–456. Springer, Cham (2017). https://doi.org/10.1007/978-3-319-71147-8_31

13. Song, L., Huang, H., Du, H.: Approximation schemes for Euclidean vehicle routing problems with time windows. J. Comb. Optim. **32**(4), 1217–1231 (2016). https://doi.org/10.1007/s10878-015-9931-5

14. Toth, P., Vigo, D.: Vehicle Routing: Problems, Methods, and Applications. MOS-SIAM Series on Optimization, 2nd edn. SIAM, Philadelphia (2014)

Piecewise Linear Bounding Functions
for Univariate Global Optimization

Oleg Khamisov[1] , Mikhail Posypkin[2,3,4] , and Alexander Usov[2]([✉])

[1] Melentiev Energy Systems Institute of Siberian Branch of the Russian
Academy of Sciences, Lermontov st.,130, Irkutsk 664033, Russia
globopt@mail.ru
[2] Dorodnicyn Computing Centre, FRC CSC RAS, Vavilov st. 40, Moscow, Russia
mposypkin@gmail.com, alusov@mail.ru
[3] Moscow Institute of Physics and Technology, Institutskiy Pereulok,
9 Dolgoprudny, Russia
[4] Institute for Information Transmission Problems RAS,
Bolshoy Karetny per. 19, build. 1, Moscow 127051, Russia

Abstract. The paper addresses the problem of constructing lower and
upper bounding functions for univariate functions. This problem is of a
crucial importance in global optimization where such bounds are used
by deterministic methods to reduce the search area. It should be noted
that bounding functions are expected to be relatively easy to construct
and manipulate with. We propose to use piecewise linear estimators for
bounding univariate functions. The rules proposed in the paper enable
an automated synthesis of lower and upper bounds from the function's
expression in an algebraic form. Numerical examples presented in the
paper demonstrate the high accuracy of the proposed bounds.

Keywords: Univariate global optimization
Piecewise linear functions · Estimators · Deterministic methods

1 Introduction

This paper is devoted to constructing lower and upper bounding functions for
univariate functions. A function $\phi(x)$ is called *lower (upper) bounding function*
(or simply *bound*) for a function $f(x)$ over an interval $[a,b]$ if $f(x) \geq \phi(x)$
$(f(x) \leq \phi(x))$ for all $x \in [a,b]$.

Lower and upper bounds for objective functions and constraints play an
important role in global optimization. Indeed, suppose we know a lower bounding
function $\phi(x)$ for an objective function $f(x)$. Than we can safely exclude from
the further search the set defined by the following inequality:

$$\phi(x) \geq f_r - \varepsilon, \tag{1}$$

The reported study was funded by RFBR according to the research project 17-07-00510.

where f_r is an incumbent value (best solution found along the search) and ε is a prescribed tolerance [6,7].

The inequality 1 can be solved efficiently only when the function $\phi(x)$ has a simple structure. In this work one of such function types: *piecewise linear function* (or PWL-function for brevity) is studied. We propose a method to obtain PWL bounds from the function's algebraic representation (formula). The evaluation of bounds is driven by rules that are applied iteratively from the bottom of the expression tree to its top similarly to computing function values, interval bounds or derivatives. We show that PWL bounds constructed with the help of the proposed approach are generally much tighter than bounds computed with the interval [10] or slope [20] arithmetic.

The paper is organized as follows. The Sect. 2 outlines related works and compares our approach with existing ones. The definition and properties of PWL bounds are discussed in Sect. 3. Numerical examples comparing the accuracy of PWL bounds and other approaches are presented in Sect. 4.

2   Related Work

Deterministic univariate global optimization stems from seminal works of Pijavskij [18] and Shubert [26]. In these papers authors proposed to use the Lipschitzian property of a function to determine the precision of found solutions. They used simple Lipschitzian underestimations:

$$\mu(x) = f(c) - L|x - c|,$$

where L is the Lipschitz constant.

These ideas were further developed in works of Strongin and Sergeyev [23,28] who established an elaborated theory ("information-statistical approach") for estimating function bounds over given intervals.

Second-order Lipschitzian bounds were studied in [1,3]. Authors proposed to use the following underestimation:

$$\mu(x) = f(c) + f'(c)(x - c) - L(x - c)^2, \tag{2}$$

where L is the Lipschitzian constant for the derivative. This underestimation was further improved by Sergeyev in [24]. Sergeyev introduced a smooth support function that is closer to the objective function than 2. In the paper [25], geometric and information frameworks are taken into consideration for constructing global optimization algorithms. Another paper [15] is devoted to the development of effective global optimization algorithms with Lipschitz functions and Lipschitz first derivatives.

The further progress in univariate global optimization was made by an important observation that interval bounds on the derivatives can replace a Lipschitz constant. In [4] authors combine ideas borrowed from the Pijavskij method and interval approaches. Besides new bounds the paper introduces powerful reduction rules that can significantly speed up the search process.

Another replacement of Lipschitz constant is provided by *slopes*. A slope is defined as an interval $S_f(c)$ that satisfies the following inclusion:

$$f(x) \subseteq f(c) + S_f(c) \cdot [\underline{x}, \overline{x}],$$

where c is a point within the interval $[\underline{x}, \overline{x}]$.

Clearly $S_f(c) \subseteq [\min_{x \in [\underline{x}, \overline{x}]} f'(x), \max_{x \in [\underline{x}, \overline{x}]} f'(x)]$. However, this inclusion is often strict: slopes can provide much tighter bounds than derivative estimations. In [20,21] efficient algorithms for evaluating slopes are proposed. Slopes are evaluated from an algebraic expression driving by rules similarly to automatic differentiation.

It worth to note powerful global optimization techniques [6,7,9,17,19,27] for a multivariate case that can serve as a source of good ideas for univariate optimization. See [11] for a good survey of such approaches.

Interval bounds, Lipschitzian bounds and slopes can be considered as a simple form of linear underestimations. More elaborate PWL lower bounds called "kites" are considered in [29]. Kites combine the centered forms and the linear boundary value forms thereby achieving better approximation w.r.t. both forms used separately.

Concave PWL lower and convex PWL upper bounds consisting of exactly two line segments were considered in [5,13]. Authors propose the rules to evaluate these bounds automatically from an algebraic representation of an expression.

We also should mention that the approach suggested in our paper differs from convex envelopes, convex underestimators and other convexification techniques developed in [2,8,12,14]. The main difference between the approaches outlined above and ours is that we consider generic PWL bounds not limiting to convex or concave cases with an arbitrary finite number of segments.

3   Piecewise Linear Bounds

3.1   Basic Properties of Piecewise Linear Functions

A *piecewise linear function* on an interval $[a, b]$ is defined as a sequence of segments z_i connecting points (x_i, y_i) and (x_{i+1}, y_{i+1}). More formally:

$$\psi(x) = y_i + \frac{y_{i+1} - y_i}{x_{i+1} - x_i}(x - x_i), \ x \in [x_i, x_{i+1}], i = 1, \dots, n-1.$$

where $n \geq 2$, $a = x_1 \leq \cdots \leq x_n = b$. In what follows we'll use the abbreviation *PWL* for "piecewise linear".

Where appropriate we will denote a PWL function by a sequence of its nodes enclosed in braces:

$$\{(x_1, y_1), \dots, (x_n, y_n)\}.$$

A set of PWL functions is closed under a set of basic algebraic operations including superposition. More formally this is stated in the two apparent propositions below.

Proposition 1. *Let $\psi(x)$ and $\phi(x)$ be PWL functions on an interval $[a, b]$. Then expressions*

$$\lambda\psi(x), \lambda \in \mathbb{R}$$
$$|\psi(x)|,$$
$$\psi(x) \pm \phi(x),$$
$$\max(\psi(x), \phi(x)),$$
$$\min(\psi(x), \phi(x))$$

are PWL functions on $[a, b]$.

Proposition 2. *Let $\phi(x)$ and $\psi(x)$ be PWL functions on intervals $[a, b]$, $[c, d]$, where $c = \min_{x \in [a,b]} \psi(x)$, $d = \max_{x \in [a,b]} \psi(x)$. Then $\omega(x) = \psi(\phi(x))$ is a PWL function on $[a, b]$.*

A *piecewise linear lower (upper) bound* for a function $f(x)$ on an interval $[a, b]$ is a piecewise linear function $\psi(x)$ such that $f(x) \geq \psi(x)$ $(f(x) \leq \psi(x))$ for all $x \in [a, b]$.

A desirable feature for practice is an ability to automatically construct PWL bounds from the function representation. Below we introduce a theory and rules for computing PWL bounds automatically assuming that the symbolic representation of a function is known.

3.2 PWL Bounds for Elementary Functions

PWL bounds are computed from the tree representation of an expression from the bottom (leaves) to the top (root). The rules to compute bounds rely on PWL bounds for elementary functions.

Under the elementary functions we understand the following univariate functions $\sin(x)$, $\cos(x)$, $\arcsin(x)$, $\arccos(x)$, $\tan(x)$, $\arctan(x)$, $\cot(x)$, e^x, $\ln(x)$, $\frac{1}{x}$ (for $x > 0$), x^α ($\alpha > 0$). Many of elementary functions (like e^x, $\ln(x)$, $\frac{1}{x}$) are convex or concave on a whole domain of definition. For remaining ones (like $\sin(x)$, $\cos(x)$, $\tan(x)$) the interval to compute bounds can be subdivided into smaller intervals where a function is concave or convex. Thus w.l.o.g. we can limit our consideration to a case when a function is convex (the concavity case is similar). It is obvious that the list of elementary functions can be easily enlarged.

Proposition 3. *Let $f(x)$ be a convex function over an interval $[a, b]$. Consider $n \geq 1$ points within this interval $a \leq x_1 < \cdots < x_n \leq b$. Define a function $\mu(x)$ as follows*

$$\mu(x) = \max_{1 \leq i \leq n} \psi_i(x),$$

where $\psi_i(x) = f(x_i) + d_i(x - x_i)$ and d_i is a subderivative of $f(x)$ at a point x_i on an interval $[a, b]$. Then $\mu(x)$ is a lower PWL bound for $f(x)$ (Fig. 1).

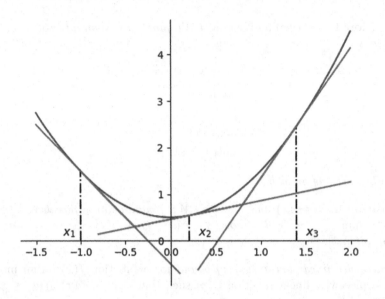

Fig. 1. The lower PWL bound for a convex function

Proposition 4. *Let $f(x)$ be a convex function over an interval $[a, b]$. Consider $n \geq 2$ points within this interval $a = x_1 < \cdots < x_n = b$. Define a function $\psi(x)$ as follows:*

$$\psi(x) = y_i + \frac{y_{i+1} - y_i}{x_{i+1} - x_i}(x - x_i), \ x \in [x_i, x_{i+1}], i = 1, \ldots, n-1.$$

where $y_i = f(x_i)$, $i = 1, \ldots, n$. Then $\psi(x)$ is an upper PWL bound for $f(x)$ (Fig. 2).

Propositions 3 and 4 provide grounds for constructing lower and upper bounds for elementary functions. Clearly the precision of bounds grows with the number of points in a set $x_1, \ldots, x_n$.

3.3   Automated Synthesis of PWL Bounds

In the previous section we proposed an approach to constructing linear bounds for elementary functions. However to compute bounds for a function defined by an expression we need rules to calculate bounds for a superposition of functions and basic operators used to construct formulas.

We start from simple rules for linear combination and min, max operators.

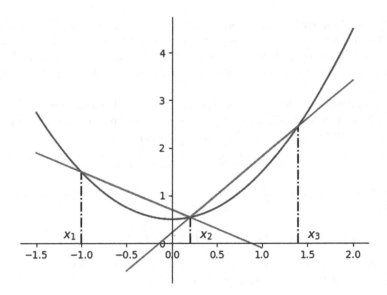

Fig. 2. The upper PWL bound for a convex function

Proposition 5. *Let* $\underline{\mu_f}(x)$, $\overline{\mu_f}(x)$ *be PWL lower and upper bounds for a function* $f(x)$ *on an interval* $[a, b]$. *Let* $\underline{\mu_g}(x)$, $\overline{\mu_g}(x)$ *be PWL lower and upper bounds for a function* $g(x)$ *on an interval* $[a, b]$. *Then the following properties hold:*

$$\lambda\underline{\mu_f}(x) \leq \lambda f(x) \leq \lambda\overline{\mu_f}(x), \lambda > 0, \tag{3}$$

$$\lambda\overline{\mu_f}(x) \leq \lambda f(x) \leq \lambda\underline{\mu_f}(x), \lambda < 0, \tag{4}$$

$$\underline{\mu_f}(x) + \underline{\mu_g}(x) \leq f(x) + g(x) \leq \overline{\mu_f}(x) + \overline{\mu_g}(x), \tag{5}$$

$$\underline{\mu_f}(x) - \overline{\mu_g}(x) \leq f(x) - g(x) \leq \overline{\mu_f}(x) - \underline{\mu_g}(x), \tag{6}$$

$$\min(\underline{\mu_f}(x), \underline{\mu_g}(x)) \leq \min(f(x), g(x)) \leq \min(\overline{\mu_f}(x), \overline{\mu_g}(x)), \tag{7}$$

$$\max(\underline{\mu_f}(x), \underline{\mu_g}(x)) \leq \max(f(x), g(x)) \leq \max(\overline{\mu_f}(x), \overline{\mu_g}(x)). \tag{8}$$

Obvious rules (3)–(5) allow to construct PWL bounds for linear combinations of elementary functions.

The situation with multiplication is more complex. If $x \in [a, b], y \in [c, d]$ then the following inequalities [10] hold for multiplication:

$$\min(\phi_1(x), \phi_2(x), \phi_3(x), \phi_4(x)) \leq f(x) g(x) \leq \max(\phi_1(x), \phi_2(x), \phi_3(x), \phi_4(x)), \tag{9}$$

where $\phi_1(x) = \underline{\mu_f}(x) \cdot \underline{\mu_g}(x)$, $\phi_2(x) = \underline{\mu_f}(x) \cdot \overline{\mu_g}(x)$, $\phi_3(x) = \overline{\mu_f}(x) \cdot \underline{\mu_g}(x)$, $\phi_4(x) = \overline{\mu_f}(x) \cdot \overline{\mu_g}(x)$. Observe that expressions $\phi_i(x)$, $i = 1, 2, 3, 4$ are piecewise quadratic. Since quadratic functions are either concave or convex the PWL lower and upper bounds for them are readily constructed according to Propositions 3 and 4. The lower bounds for $\min(\phi_1(x), \phi_2(x), \phi_3(x), \phi_4(x))$ and $\max(\phi_1(x), \phi_2(x), \phi_3(x), \phi_4(x))$ are obtained from (7), (8).

The PWL bounds for the reciprocal $1/x$ can be constructed following Propositions 3 and 4 for an interval $[a, b]$, when either $a > 0$ or $b < 0$. The remaining case $0 \in [a, b]$ is omitted in this paper and left for further studies. Having bounds for the reciprocal the division is reduced to the multiplication in a standard way $x/y = x\frac{1}{y}$.

To evaluate bounds for complex expressions we need rules to process function superposition. Consider a function $h(x) = f(g(x))$ on an interval $[a, b]$. Let $\underline{\mu_g}(x)$ and $\overline{\mu_g}(x)$ be PWL bounds for $g(x)$ on an interval $[a, b]$:

$$\underline{\mu_g}(x) \leq g(x) \leq \overline{\mu_g}(x), x \in [a, b]. \tag{10}$$

Denote $c = \min_{x \in [a,b]} \underline{\mu_g}(x)$, $d = \max_{x \in [a,b]} \overline{\mu_g}(x)$. Let $\underline{\mu_f}(x)$ and $\overline{\mu_f}(x)$ be PWL bounds for a function $f(x)$ on $[c, d]$:

$$\underline{\mu_f}(x) \leq f(x) \leq \overline{\mu_f}(x), x \in [c, d]. \tag{11}$$

Proposition 6. *If the function* $\underline{\mu_f}(x)$ *is non-decreasing monotonic on* $[c, d]$ *then* $\underline{\mu_f}(\underline{\mu_g}(x))$ *is a PWL lower bound for* $h(x)$ *on* $[a, b]$.

Proof. According to the Proposition 2 $\underline{\mu_f}(\underline{\mu_g}(x))$ is a PWL function. It remains to prove that $\underline{\mu_f}(\underline{\mu_g}(x)) \leq h(x)$ on $[a, b]$. Consider $x \in [a, b]$. From (10) we derive that $\underline{\mu_g}(x) \leq g(x)$. Since $\underline{\mu_g}(x), g(x) \in [c, d]$ and $\underline{\mu_f}(x)$ is non-decreasing monotonic on $[c, d]$ we obtain $\underline{\mu_f}(\underline{\mu_g}(x)) \leq \underline{\mu_f}(g(x))$. From (11) it follows that $\underline{\mu_f}(g(x)) \leq f(g(x)) = h(x)$. Thus $\underline{\mu_f}(\underline{\mu_g}(x)) \leq h(x)$ for $x \in [a, b]$.

Figure 3 shows the synthesis of a PWL lower bound for the composite function $h(x) = \sin^2(x)$ on an interval $[0, \pi]$. Notice that $\sin(x) \in [0, 1]$ when $x \in [0, \pi]$. The outer function x^2 as well as PWL lower bound $\underline{\mu_f}$ are non-decreasing monotonic on interval $[0, 1]$. That is why it suffices to consider a PWL lower bound $\underline{\mu_g}$ for the inner function $\sin(x)$. According to the Proposition 6 the composite function $\underline{\mu_f}(\underline{\mu_g})$ is a PWL lower bound for the function $\sin^2(x)$.

Proposition 7. *If the function* $\underline{\mu_f}(x)$ *is non-increasing monotonic on* $[c, d]$ *then* $\underline{\mu_f}(\overline{\mu_g}(x))$ *is a PWL lower bound for* $h(x)$ *on* $[a, b]$.

Proof. According to Proposition 2 $\underline{\mu_f}(\overline{\mu_g}(x))$ is a PWL function. It remains to prove that $\underline{\mu_f}(\overline{\mu_g}(x)) \leq h(x)$ on $[a, b]$. Consider $x \in [a, b]$. From (10) we derive that $g(x) \leq \overline{\mu_g}(x)$. Since $\overline{\mu_g}(x), g(x) \in [c, d]$ and $\underline{\mu_f}(x)$ is non-increasing monotonic on $[c, d]$ we obtain $\underline{\mu_f}(\overline{\mu_g}(x)) \leq \underline{\mu_f}(g(x))$. From (11) it follows that $\underline{\mu_f}(g(x)) \leq f(g(x)) = h(x)$. Thus $\underline{\mu_f}(\overline{\mu_g}(x)) \leq h(x)$ for $x \in [a, b]$.

Figure 4 shows the synthesis of a PWL lower bound for a composite function $h(x) = \sin^2(x)$ on $[\pi, 2\pi]$ interval. Notice that $\sin(x) \in [-1, 0]$ when $x \in [\pi, 2\pi]$. Unlike the previous example the outer function x^2 as well as the PWL lower bound $\underline{\mu_f}$ are non-increasing monotonic on the interval $[-1, 0]$. Therefore we take the PWL upper bound $\overline{\mu_g}$ for the inner function $\sin(x)$. According to the Proposition 7 the composite function $\underline{\mu_f}(\overline{\mu_g})$ is a PWL lower bound for the function $\sin^2(x)$.

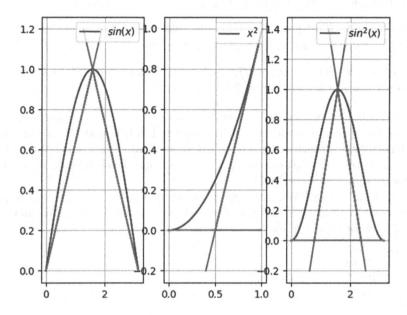

Fig. 3. Synthesis of PWL lower bound for $sin^2(x)$ on $[0, \pi]$

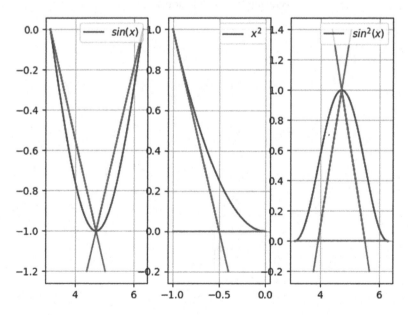

Fig. 4. Synthesis of PWL lower bound for $sin^2(x)$ on $[\pi, 2\pi]$

In the same way we can prove the following two propositions.

Proposition 8. *If the function $\overline{\mu_f}(x)$ is non-decreasing monotonic on $[c,d]$ then $\overline{\mu_f}(\overline{\mu_g}(x))$ is a PWL upper bound for $h(x)$ on $[a,b]$.*

Proposition 9. *If the function $\overline{\mu_f}(x)$ is non-increasing monotonic on $[c,d]$ then $\overline{\mu_f}(\underline{\mu_g}(x))$ is a PWL upper bound for $h(x)$ on $[a,b]$.*

If PWL bounds for an elementary function $f(x)$ on $[c,d]$ are monotonic we can directly apply Propositions 6–9 to compute PWL bounds for a composite function $f(g(x))$. Otherwise we need to somehow obtain monotonic PWL bounds for $f(x)$. Fortunately this can be easily done for any PWL bound. We explain this for an upper bound (Fig. 5).

Suppose $\mu(x)$ is a PWL function on an interval $[c,d]$. The Monotonize procedure constructs a PWL function $\tilde{\mu}(x)$ that is non-decreasing monotonic and $\mu(x) \leq \tilde{\mu}(x)$ for all $x \in [c,d]$.

$\tilde{\mu}(x_1) := \mu(x_1)$
for $i = 2$ **to** n **do**
$\quad$ **if** $\tilde{\mu}(x_{i-1}) > \mu(x_i)$ **then**
$\quad\quad \mid \tilde{\mu}(x_i) := \tilde{\mu}(x_{i-1})$
$\quad$ **else**
$\quad\quad \mid \tilde{\mu}(x_i) := \mu(x_i)$
$\quad$ **end**
end

Algorithm 1: Monotonize

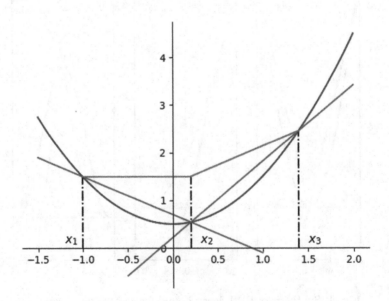

Fig. 5. "Lifting" PWL upper bounds

Starting from the leftmost node $x_1 = c$ the Monotonize algorithm compares successive nodes of a PWL function and if the monotonicity is violated fixes it by "lifting" one node until the link between successive nodes is horizontal. If $\mu(c) = \mu(x_1) > \mu(x_n) = \mu(d)$ it is usually better to construct a non-increasing PWL upper bound $\tilde{\mu}(x)$ going in the opposite direction (from d to c) in the same manner. The case of a lower bound is considered similarly.

4   Numerical Examples

Below we consider two examples and compare the proposed approach with the interval [10] and the slopes [20,21] arithmetic. For the sake of presentation quality we use approximate computations with a small precision. In practice the computations could be as accurate as needed.

Example 1. *Compute the PWL lower bound for the $h(x) = \sin(x) \cdot (-x^2 + x)$ function over the $[1, 3]$ interval.*

According to the rule (9) the lower bound $\underline{\mu}_h(x)$ is computed as follows:

$$\underline{\mu}_h(x) = \min\left(\underline{\mu}_f(x)\underline{\mu}_g(x), \underline{\mu}_f(x)\overline{\mu}_g(x), \overline{\mu}_f(x)\underline{\mu}_g(x), \overline{\mu}_f(x)\overline{\mu}_g(x)\right).$$

Observe that $\underline{\mu}_f(x) \geq 0$ while $\overline{\mu}_g(x) \leq 0$ for $x \in [1,3]$ (Fig. 6). Thus we can conclude that $\overline{\mu}_h(x) = \overline{\mu}_f(x)\underline{\mu}_g(x)$. Two tangents $0.54x + 0.3$ and $-0.99x + 3.11$ forming the upper bound for $\sin(x)$ intersect at $x = 1.83$. For simplicity we choose the same x for constructing chords $-1.83x + 1.83$ and $-3.83x + 5.50$ constituting the lower bound for $g(x)$.

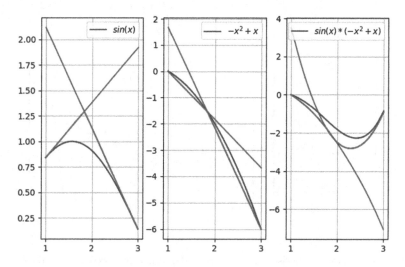

Fig. 6. Steps for synthesis lower PWL bound of function $sin(x)(-x^2 + x)$

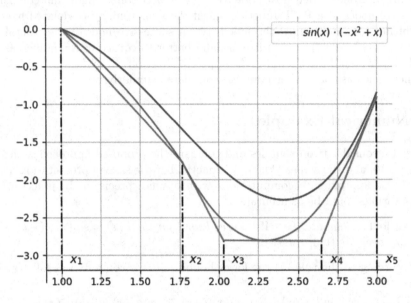

Fig. 7. Lower PWL bound for function $sin(x)(-x^2 + x)$

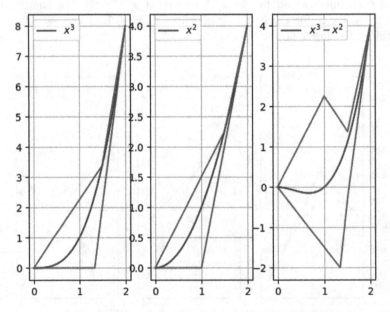

Fig. 8. Steps 1–3 for synthesis lower PWL bound of function $- \exp(x^3 - x^2)$ on $x \in [0, 2]$

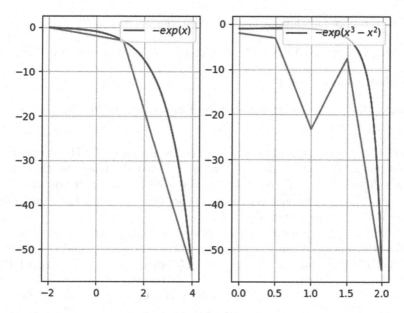

Fig. 9. Steps 4–5 for synthesis lower PWL bound of function $- \exp(x^3 - x^2)$ on $x \in [0, 2]$

Multiplying the obtained bounds we get

$$\overline{\mu_f}(x)\underline{\mu_g}(x) = \begin{cases} \phi(x) = (0.54x + 0.3)(-1.83x + 1.83), \ x \in [1, 1.83] \\ (-3.83x + 5.50)(-0.99x + 3.11), \ x \in [1.83, 3]. \end{cases}$$

However $\phi(x)$ is not piecewise linear. Thus we need to construct a PWL bound for $\phi(x)$. Figure 7 shows a PWL lower bound $\{(1.0, 0.0), (1.76, -1.76), (2.03, -2.8), (2.64, -2.8), (3, -0.91)\}$ for the $\phi(x)$ (and hence that for $f(x)$). This lower bound gives a lower estimate for the $f(x)$ equal to -2.8.

Interval analysis gives the following result:

$$h([1, 3]) \subseteq \sin([1, 3]) \cdot (-[1, 3]^2 + [1, 3]) = [0.14, 1] \cdot ([-9, -1] + [1, 3])$$
$$= [0.14, 1] \cdot [-8, 2] = [-8, 2],$$

yielding -8 as a lower bound.

To estimate $h(x)$ by means of slopes arithmetic we compute the enclosure Y_s of $s_h(c, A)$ for $A = [1; 3]$ and $c = 2$:

$$h((A, c, 1)) = \sin((A, c, 1)) \cdot (-(A, c, 1)^2 + (A, c, 1))$$
$$= \sin(([1, 3], 1, 1)) \cdot (-([1, 3], 1, 1)^2 + ([1, 3], 1, 1))$$
$$= ([0.14, 1], 0.84, [-0.99, 0.54]) \cdot (-([1, 9], 1, [2, 4]) + ([1, 3], 1, 1))$$
$$= ([0.14, 1], 0.84, [-0.99, 0.54]) \cdot ((([-8, 2], 0, [-3, -1]))$$
$$= ([-8, 2], 0, [-3, -0.14]) = (Y_x, Y_c, Y_s).$$

Thus the slopes arithmetic gives the following inclusion:

$$h([1,3]) \subseteq h(c) + Y_s(A - c) = 0 + [-3, -0.14] \cdot ([1,3] - 1) = [-6, 0],$$

yielding -6 as a lower bound.

Example 2. *Evaluate a lower bound for a composite function* $h(x) = -e^{x^3 - x^2}$, $x \in [0, 2]$.

Notice that $h(x)$ is neither convex nor concave on $[0, 2]$. The function is a composition of the outer function $f(x) = -e^x$ and the inner function $g(x) = x^3 - x^2$. Figures 8 and 9 demonstrate synthesis of a PWL lower bound.

Since $-e^x$ and its lower PWL bound $\underline{\mu_f}$ are non-increasing functions (Fig. 9) we can apply the Proposition 7.

According to the Proposition 7 we fist need to compute the image of the inner function which is a difference of two elementary functions x^3 and x^2. To obtain the upper PWL bound take a point $x = 1.5$ from $[0, 2]$. Then the upper PWL bound for x^3 is $\{(0,0), (1.5, 3.375), (2, 8)\}$. It consists of two chords $\overline{\mu_1}(x) = 2.25x$ and $\overline{\mu_2}(x) = 9.25x - 10.5$. Similarly the upper PWL bound for x^2 is $\{(0,0), (1.5, 2.25), (2, 4)\}$. It also consists of two chords $\overline{\nu_1}(x) = 1.5x$ and $\overline{\nu_2}(x) = 3.5x - 3$.

In order to build lower PWL bounds for x^3 and x^2 we draw tangents at points $x = 0$ and $x = 2$. The resulting lower PWL bound for x^3 is $\{(0,0), (\frac{4}{3}, 0), (2, 8)\}$. It consists of two tangents $\mu_1(x) = 0$ and $\mu_2(x) = 12x - 16$. The lower PWL bound for x^2 obtained in a similar way is $\{(0,0), (1, 0), (2, 2)\}$. It also consists of two tangents $\nu_1(x) = 0$ and $\nu_2(x) = 4x - 4$.

Now, we are ready to build upper and lower PWL bounds for the inner function $g(x) = x^3 - x^2$. According to the rule 6 the upper PWL bound is as follows:

$$\overline{\mu_g}(x) = \begin{cases} \overline{\mu_1}(x) - \nu_1(x) = 2.25x, & x \in [0, 1], \\ \overline{\mu_1}(x) - \nu_2(x) = -1.75x + 4, & x \in [1, 1.5], \\ \overline{\mu_2}(x) - \nu_2(x) = 5.25x - 6.5, & x \in [1.5, 2]. \end{cases}$$

The similarly computed lower PWL bound looks as follows:

$$\underline{\mu_g}(x) = \begin{cases} \mu_1(x) - \overline{\nu_1}(x) = -1.5x, & x \in [0, \frac{4}{3}], \\ \mu_2(x) - \overline{\nu_1}(x) = 10.5x - 16, & x \in [\frac{4}{3}, 1.5], \\ \mu_2(x) - \overline{\nu_2}(x) = 8.5x - 13, & x \in [1.5, 2]. \end{cases}$$

Using obtained bounds we get the following estimation of the inner function image: $g(x) \in [-2, 4]$. According to the Proposition 7 we should construct a lower PWL bound for the outer function $f(x) = -e^x$ over $[-2, 4]$. Choosing a point $x = 1.125$, $x \in [-2, 4]$ we obtain a PWL lower bound $\mu_f(x)$, with the following sequence of nodes $\{(-2, -0.13), (1.125, -3.08), (4, -54.\overline{59})\}$. It consists of the following two chords: $\phi_1(x) = -0.95x - 1.99$ for $x \in [-2, 1.125]$ and $\phi_2(x) = -17.91x + 17.06$ for $x \in [1.125, 4]$ depicted at the Fig. 9.

Finally we get the PWL lower bound $\underline{\mu_h}(x)$ for $h(x) = e^{x^3 - x^2}$ over $x \in [0, 2]$ as a composite function $\mu_f(\overline{\mu_g}(x))$:

$$\underline{\mu_h}(x) = \begin{cases} -0.95 \cdot (2.25x) - 1.99, & x \in [0, 0.5], \\ -17.91 \cdot (2.25x) + 17.06, & x \in [0.5, 1], \\ -17.91 \cdot (-1.75x + 4) + 17.06, & x \in [1, 1.5], \\ -17.91 \cdot (5.25x - 6.5) + 17.06, & x \in [1.5, 2.0]. \end{cases}$$

Finally we get the following lower PWL bound for $h(x) = -e^{x^3 - x^2}$ over $x \in [0, 2]$: $\{(0, -1.99), (0.5, -1.068), (1, -23.23), (1.5, -7.56), (2, -54.59)\}$. Observe that the lower estimation -54.59 of the function is precise: $f(2) = -e^{2^3 - 2^2} \approx -54.59$.

The interval analysis gives the following result:

$$h([0, 2]) \subseteq -\exp([0, 2]^3 - [0, 2]^2) = -\exp([0, 8] - [0, 4])$$
$$= -\exp([-4, 8]) = [-2980.95, -0.018].$$

yielding the lower estimate -2980.95.

The lower estimation of $h(x) = -e^{x^3 - x^2}$ by means of slopes arithmetic is computed as follows. First compute the enclosure Y_s of $s_h(c, A)$ for $A = [0; 2]$ and $c = 1$:

$$h((A, c, 1)) = -\exp(([0, 2], 1, 1)^3 - ([0, 2], 1, 1)^2)$$
$$= -\exp(([0, 8], 1, [1, 7]) - ([0, 4], 1, [1, 3]))$$
$$= -\exp(([-4, 8], 0, [-2, 6]))$$
$$= -([0.018, 2980.95], 1, [0.491, 496.65])$$
$$= ([-2980.95, 0.018], -1, [-496.65, -0.491])$$
$$= (Y_x, Y_c, Y_s).$$

The slope based enclosure is computed as follows:

$$h([A]) \subseteq h(c) + Y_s(A - c) = -1 + [-496.65, -0.491] \cdot ([0, 2] - 1)$$
$$= -1 + [-496.65, -0.491] \cdot [-1, 1] = [-497.65, 495.65].$$

Table 1 summaries the results obtained with the help of different approaches. The superiority of the proposed approach is clearly observed.

Table 1. Comparison of estimates obtained by different approaches

Function	PWL estimate	Interval estimate	Slope estimate
$\sin(x) \cdot (-x^2 + x))$	-2.8	-8	-6
$-e^{x^3 - x^2}$	-54.59	-2980.95	-497.65

5 Conclusions

We proposed an approach to an automated construction of bounding functions of one variable. The synthesis of bounds is driven by rules applied to an algebraic expression of a function. The proposed approach was experimentally compared with interval and slope arithmetic. Experiments demonstrated that for some functions the proposed method can significantly outperform standard approaches.

It should be noted that our approach can be helpful in separable programming problems [16,22]. We plan to study this topic in the future.

References

1. Baritompa, W.: Accelerations for a variety of global optimization methods. J. Global Optim. **4**(1), 37–45 (1994)
2. Bompadre, A., Mitsos, A.: Convergence rate of McCormick relaxations. J. Global Optim. **52**(1), 1–28 (2012)
3. Breiman, L., Cutler, A.: A deterministic algorithm for global optimization. Math. Program. **58**(1–3), 179–199 (1993)
4. Casado, L.G., Martínez, J.A., García, I., Sergeyev, Y.D.: New interval analysis support functions using gradient information in a global minimization algorithm. J. Global Optim. **25**(4), 345–362 (2003)
5. Ershov, A., Khamisov, O.V.: Automatic global optimization. Diskretnyi Analiz i Issledovanie Operatsii **11**(2), 45–68 (2004)
6. Evtushenko, Y.G.: A numerical method of search for the global extremum of functions (scan on a nonuniform net). Zhurnal Vychislitel'noi Matematiki i Matematicheskoi Fiziki **11**(6), 1390–1403 (1971)
7. Evtushenko, Y., Posypkin, M.: A deterministic approach to global box-constrained optimization. Optimization Letters **7**(4), 819–829 (2013)
8. Floudas, C., Gounaris, C.: A review of recent advances in global optimization. J. Global Optim. **45**(1), 3–38 (2009)
9. Gergel, V., Grishagin, V., Israfilov, R.: Local tuning in nested scheme of global optimization. Procedia Comput. Sci. **51**, 865–874 (2015)
10. Hansen, E., Walster, G.W.: Global Optimization Using Interval Analysis: Revised and Expanded. CRC Press, New York (2003)
11. Horst, R., Pardalos, P.M.: Handbook of Global Optimization, vol. 2. Springer Science & Business Media, Dordrecht (2013)
12. Khajavirad, A., Sahinidis, N.: Convex envelopes of products of convex and component-wise concave functions. J. Global Optim. **52**(3), 391–409 (2012)
13. Khamisov, O.: Explicit univariate global optimization with piecewise linear support functions, proc. DOOR 2016, CEUR-WS.org, vol. 1623, pp. 218–255. http://ceur-ws.org/Vol-1623/papermp19.pdf
14. Khamisov, O.: Optimization with quadratic support functions in nonconvex smooth optimization, aIP Conference Proceedings 1776, 050010 (2016). https://doi.org/10.1063/1.4965331
15. Lera, D., Sergeyev, Y.D.: Acceleration of univariate global optimization algorithms working with lipschitz functions and lipschitz first derivatives. SIAM J. Optim. **23**(1), 508–529 (2013)

16. Pardalos, P.M., Rosen, J.: Reduction of nonlinear integer separable programming problems. Int. J. Comput. Math. **24**(1), 55–64 (1988)
17. Paulavičius, R., Žilinskas, J.: Simplicial Global Optimization. Springer, New York (2014). 10.1007/978-1-4614-9093-7
18. Pijavskij, S.: An algorithm for finding the global extremum of function. Optimal Decisions **2**, 13–24 (1967)
19. Pintér, J.D.: Global Optimization in Action: Continuous and Lipschitz Optimization: Algorithms, Implementations and Applications, vol. 6. Springer Science & Business Media, New York (2013)
20. Ratz, D.: An optimized interval slope arithmetic and its application. Inst. für Angewandte Mathematik (1996)
21. Ratz, D.: A nonsmooth global optimization technique using slopes: the one-dimensional case. J. Global Optim. **14**(4), 365–393 (1999)
22. Rosen, J.B., Pardalos, P.M.: Global minimization of large-scale constrained concave quadratic problems by separable programming. Math. Program. **34**(2), 163–174 (1986)
23. Sergeyev, Y.D.: An information global optimization algorithm with local tuning. SIAM J. Optim. **5**(4), 858–870 (1995)
24. Sergeyev, Y.D.: Global one-dimensional optimization using smooth auxiliary functions. Math. Program. **81**(1), 127–146 (1998)
25. Sergeyev, Y.D., Mukhametzhanov, M.S., Kvasov, D.E., Lera, D.: Derivative-free local tuning and local improvement techniques embedded in the univariate global optimization. J. Optim. Theory Appl. **171**(1), 186–208 (2016)
26. Shubert, B.O.: A sequential method seeking the global maximum of a function. SIAM J. Numer. Anal. **9**(3), 379–388 (1972)
27. Strekalovsky, A.S.: Global optimality conditions for nonconvex optimization. J. Global Optim. **12**(4), 415–434 (1998)
28. Strongin, R.G., Sergeyev, Y.D.: Global Optimization with Non-convex Constraints: Sequential and Parallel Algorithms, vol. 45. Springer Science & Business Media, New York (2013)
29. Vinkó, T., Lagouanelle, J.L., Csendes, T.: A new inclusion function for optimization: Kite-the one dimensional case. J. Global Optim. **30**(4), 435–456 (2004)

The Scalability Analysis of a Parallel Tree Search Algorithm

Roman Kolpakov[1,2]([✉]) [iD] and Mikhail Posypkin[1,2,3,4] [iD]

[1] Lomonosov Moscow State University, GSP-1, Leninskie Gory, Moscow, Russia
foroman@mail.ru, mposypkin@gmail.com
[2] Dorodnicyn Computing Centre, FRC CSC RAS, Vavilov st. 40, Moscow, Russia
[3] Moscow Institute of Physics and Technology,
Institutskiy Pereulok, 9, Dolgoprudny, Russia
[4] Institute for Information Transmission Problems RAS,
Bolshoy Karetny per. 19, build. 1, Moscow 127051, Russia

Abstract. Increasing the number of computational cores is a primary way of achieving high performance of contemporary supercomputers. However, developing parallel applications capable to harness the enormous amount of cores is a challenging task. Thus, studying the scalability of parallel algorithms (the growth order of the number of processors required to accommodate the growing amount of work, below we give a clear definition of the scalability investigated in our paper) is very important. In this paper we propose a parallel tree search algorithm aimed at distributed parallel computers. For this parallel algorithm, we perform a theoretical analysis of its scalability and show that the achieved scalability is close to the theoretical maximum.

Keywords: Parallel scalability · Parallel efficiency
Complexity analysis · Parallel tree search · Global optimization

1 Introduction

The era of exascale computing introduced new challenges related to the emergent growth of the number of cores that can run in parallel. Therefore the design and study of parallel algorithms aimed at systems with 10^6–10^9 cores is a significant problem.

Many algorithms e.g. branch-and-bound, branch-and-cut and many others have a tree-like structure of computations. Different branches can be processed independently thereby opening ample opportunities for parallel execution. However, efficient parallelization is not an easy task. It is well-known that even load distribution among parallel processors is crucial for ensuring parallelization.

This work is supported by the program of RAS No. 26 "Fundamental basis of algorithms and software for perspective ultra-high-performance computing" and by the Russian Foundation for Basic Research, project 18-07-00566.

If processors are not loaded evenly then some of them will wait until others complete their tasks. This situation prevents an efficient utilization of the processor's power and decreases the performance. For tree-like computation, the even load distribution among processors is a difficult problem since the branch-and-bound tree is usually not well-balanced.

Various parallel tree search algorithms were proposed so far. However there still a lack of systematic theoretical study of such algorithms. Though the theoretical analysis is not a substitute for numerical experiments the former is crucial for understanding principal bounds on the parallel efficiency. It is worth noting that real experiments on exascale systems can't be used widely because they are rather expensive. This observation also increases the value of the theoretical analysis of parallel complexity.

In this paper, we study the parallel efficiency of tree search algorithms with the stable tree, i.e. when the number of nodes is not changing during the search. This desirable property doesn't usually hold for global optimization algorithms because the number of nodes in the search tree largely depends on the rate of incumbent improvement. However, for many important cases, the structure of the search tree doesn't depend on the order of the tree traversal. One notable example of such algorithms is proposed in [6] for solving systems of nonlinear inequalities.

The fundamental property of any parallel algorithm is its *scalability* which is defined as an ability to maintain a certain level of efficiency when the number of CPUs is growing [10,16]. Below a model of a distributed memory multiprocessor where processors communicate via message passing is considered. For this model, we propose a parallel algorithm that achieves the scalability close to the theoretical maximum. The paper is organized as follows. The definition and the formalization of a tree search algorithm are described in Sect. 2. Section 3 introduces a model of parallel computations considered in the paper. The proposed parallel algorithm ("balanced recursive scheme") is described in Sect. 4. Section 5 outlines the results of the scalability analysis of the proposed algorithm.

Parallelization of the tree search has been extensively studied last decades. Many algorithms for shared [4,9,17] and distributed memory multiprocessors [1,7] and GPU accelerators [2,22] were proposed. A number of software frameworks [5,7] that target several types of HPC platforms were developed. New approaches for solving parallelization of global optimization problems were proposed in [19–21].

Parallel tree search has been also studied from the theoretical point of view. In [8,18] authors study parallel convergence properties of global optimization algorithms.

In [14] the exact lower bound for the parallel complexity of so-called *frontal parallel branch-and-bound algorithm* is obtained. Paper [13] continues the study the efficiency of the frontal algorithm for a particular series of subset sum problems. It was shown that scalability depends on the ratio of the total weight of items to the weights bound.

A particular variant of the tree search problem: *backtrack search problem* (BSP) was considered in several papers. In [23] authors suggested an algorithm for parallel solving of BSP in a distributed memory model in time $O(T/m + md)$ where T is the time of solving of the problem by one processor, d is the depth of the search tree, and m is the number of processors. This approach was further developed in [3]. A randomized algorithm for parallel solving of BSP in a distributed memory model in optimal time $O(T/m + d)$ with a high probability was proposed in [12]. Some improvements in parallel solving of BSP were made for shared memory model. For example, a shared memory parallel algorithm for BSP proposed in [11] works in $O((T/m + md)(\log\log\log m)^2)$. In [15] a space-efficient algorithm for parallel solving of BSP in a shared memory model in time $O(T/m + d\log m)$ was proposed. However, the question of parallel solving of BSP in optimal time $O(T/m + d)$ is still open. Below we propose an algorithm that has $O(T/m + d)$ execution time thereby showing that this bound is tight. It is, however, worth noting that our algorithm requires an additional information about subproblems complexity. We would like to note that our algorithm uses new ideas which were not proposed before in the literature.

2   A Formalization of a Serial Tree-Like Algorithm

We consider *tree-like recursion algorithms*. Such algorithms can be applied to so-called *binary decomposable problems*, i.e. optimization problems P that can be either pruned, i.e. solved in a fixed time, or decomposed into two binary decomposable subproblems P_1 and P_2 such that the best of solutions of P_1 and P_2 is a solution of P. The core TRA procedure of a tree-like recursion algorithm for solving of a binary decomposable problem P works as follows.

1: **Procedure** TRA(P)
2: **if** *evaluate*(P) = **true then**
3:     $r := getSolution(P)$
4:     **return** r
5: **else**
6:     Decompose P into P_1, P_2
7:     $r_1 := $ TRA(P_1)
8:     $r_2 := $ TRA(P_2)
9:     $r := best(r_1, r_2)$
10:     **return** r
11: **end if**

The TRA procedure relies on some auxiliary functions listed in Table 1. At line 2 P is evaluated to decide whether P should be pruned or decomposed. In global discrete or continuous optimization, the evaluation normally means the feasibility and optimality checking. If the *evaluate* function returns **true** the procedure *getSolution* is invoked to yield a solution of P. Otherwise the problem P is decomposed into two new subproblems P_1 and P_2. The subproblems are recursively processed by TRA procedure at lines 7, 8 to obtain solutions r_1, r_2. The solution of P is obtained in this case at line 9 as the best of the solutions r_1 and r_2.

Table 1. Auxiliary procedures

evaluate(P)	Evaluates the subproblem P and returns **true** if it can be pruned, **false** otherwise
getSolution(P)	Returns a solution of a terminal sub-problem P
best(r_1, r_2)	Returns the best from solutions r_1, r_2

The TRA procedure described above is a generic implementation scheme for any tree-like recursive algorithm. For example, the simplest variant of Branch and Bound method is a tree-like recursive algorithm. Following the TRA procedure, we assume that the running time $T_A(P)$ of an algorithm A of solving a problem P is computed as

$$T_A(P) = \begin{cases} t_e + t_y, & \text{if } evaluate \text{ returns true} \\ t_e + t_d + T_A(P_1) + T_A(P_2) + t_c, & \text{otherwise,} \end{cases} \quad (1)$$

where t_e is the time for evaluating P, t_y is the time for yielding a solution, t_d is the time for decomposing P into P_1 and P_2, $T_A(P_1), T_A(P_2)$ are the times for solving subproblems P_1, P_2 respectively, and t_c is the time needed to compare two solutions and select the best solution. Thus, we have from (1) that $T_A(P)$ satisfies the following property.

Proposition 1. *Let a problem P be decomposed by a tree-like recursion algorithm A into two subprolems P_1 and P_2. Then $T_A(P) \geq T_A(P_1) + T_A(P_2)$.*

For a tree-like recursive algorithm A of solving a binary decomposable problem P denote by $S_A(P)$ the set of all subproblems of P processed by TRA procedure during the resolution of P by the algorithm A (including P itself).

For each subproblem P' from $S_A(P)$ define the *depth* $d_A(P')$ of P' in the following recursive way. Assume the depth $d_A(P)$ of the initial problem P equal to 0. Let P' be a subproblem from $S_A(P)$ which is decomposed into some subproblems P'_1 and P'_2 during the solving of P by A. Define $d_A(P'_1) = d_A(P'_2) = d_A(P') + 1$. Denote by $D_A(P)$ the maximum depth of subproblems from $S_A(P)$.

Note that $S_A(P)$ can be considered as a rooted binary tree with the root P where each subproblem P' decomposed during the solving of P by A is connected by edges with subproblems obtained by this decomposition. In this case for any subproblem, P' from $S_A(P)$ the depth $d_A(P')$ is the depth of P' in the tree $S_A(P)$, i.e. the length of the path[1] from the root P to the node P', and $D_A(P)$ is the depth of the tree $S_A(P)$, i.e. the maximum length of paths from the root P to leaves of $S_A(P)$. Note also that for any subproblem P' from $S_A(P)$ we can consider in the tree $S_A(P)$ the subtree for P' rooted in P'. Subproblems contained in the subtree for P' are called *descendents* of P'.

---

[1] By the length of a path we mean the number of edges contained in the path.

3   A Parallel Computing Model for Tree-Like Recursion Algorithms

We consider a parallel distributed memory computer having some number of processors. During the resolution of a problem P each processor can do three kinds of activity:

- perform computations;
- send data to another processor;
- receive data from another processor.

We assume that any two processors can communicate with each other.

We consider the synchronous model of communication when both the sender and the receiver are occupied during the whole data transmission, i.e. both the sender and the receiver are blocked during the data transmission. The communication time is computed as $t_l + t_b$ where t_b denotes the time needed for transmitting a unit of data, and t_l is the latency time (standby time for readiness of the communicated processor to execute the transmission). A unit of data is assumed to be either a subproblem of the initial problem sent to some processor for solving of this subproblem or a subproblem solution sent by the processor solved this subproblem.

For any parallel execution scheme we assume that at the beginning time moment T_{beg} of computations the solved problem P is stored on some processor which is called the *master* processor. The computations terminate at the time moment T_{end} when after finishing of all processors computations a solution of the problem P is stored on the same master processor. During the period between the time moments T_{beg} and T_{end} the computer processors operate with subproblems of the problem P (process the subproblems according to TRA procedure and exchange the subproblems or their solutions) according to the parallel execution scheme. The running time of the parallel execution scheme for solving of the problem P is defined as the duration $T_{end} - T_{beg}$ of computations (note that this time depends on the number of the computer processors).

4   Balanced Recursive Scheme

4.1   Algorithm Outline

We propose the following parallel execution scheme for tree-like recursion algorithms which we call *a balanced recursive scheme*.

By a group or subgroup of processors, we will mean a group of more than one processor. In the proposed scheme subproblems from $\mathcal{S}_\mathcal{A}(P)$ are processed by single processors or by groups of processors dynamically created during the resolution process (in particular, the initial problem P is processed by the group of all processors assigned for solving of P). In each group of processors assigned for solving of a subproblem from $\mathcal{S}_\mathcal{A}(P)$ a *master* processor which obtains the solution of the subproblem is selected. When a subproblem from $\mathcal{S}_\mathcal{A}(P)$ is decomposed into other subproblems from $\mathcal{S}_\mathcal{A}(P)$ the group of processors assigned for solving of this subproblems is divided into two new groups proportionally to the estimated solution times of the subproblems obtained by the decomposition, and the masters of the new created groups receive the obtained subproblems for solving. If a single processor is assigned for solving of a subproblem from $S_A(P)$ then the assigned processor solves completely this subproblem, using TRA procedure.

There is one exception from this general scheme: the proportion of running times of solving the obtained subproblems may not allow assigning a group of processors or even a single processor for solving the easiest of these subproblems. For example, if a group contains 4 processors and the obtained subproblems have respectively running times 2 and 20 of solving then this group can't be divided according to the ratio of the running times which is 0.1 in this case. In such situations, one of the processors in a group is assigned to a particular role of collecting "small" subproblems. In the sequel, such processors are called *collectors*.

Thus, in each of the groups of processors assigned to solve subproblems from $S_A(P)$ a collector processor can be additionally selected. Collectors can be either created by a group master or inherited from the parent group. A collector accumulates "small" subproblems until their total running time of solving reaches some threshold value which is called *minimum load* of the collector. Then the collector solves the whole bunch of accumulated subproblems and sends the best of the solutions of these subproblems back to the group master created this collector.

More detailed, the balanced recursive scheme is performed as follows. Let G' be a group of p' processors assigned for solving some subproblem P' from $S_A(P)$ (in particular, P' can be the initial problem P), and u be the master processor of the group G'. If P' can be pruned then, using TRA procedure, the processor u solves the subproblem P' and, if P' is received by u from another processor u', returns to the processor u' the obtained solution of P'. Let P' can not be pruned, i.e. P' be decomposed into some subproblems P_1' and P_2'. Without loss of generality we assume that $T_A[P_1'] \geq T_A[P_2']$. Consider separately two possible cases according to the existence of the collector processor in G'.

1. Let G' have no collector processor. Consider separately two possible subcases.
   a. Let $T_A[P_2'] \geq T_A[P']/p'$. Then the processor u assigns $\left\lfloor \frac{T_A[P_2']}{T_A[P']} p' \right\rfloor$ processors from G' for solving the subproblem P_2'. Other processors from G', including the master processor u, are assigned for solving of the subproblem P_1'. If $\left\lfloor \frac{T_A[P_2']}{T_A[P']} p' \right\rfloor = 1$, i.e. only one processor is assigned for solving of the subproblem P_2', then, using the algorithm A, this processor solves the subproblem P_2' and sends back to u the obtained solution of P_2'. If more than one processor is assigned for solving of P_2' then among the assigned processors a new master processor u' is selected. u' processes recursively the subproblem P_2' for solving this subproblem and returns to u the obtained solution of P_2'. In an analogous way, if only u is assigned for solving of P_1' then u solves the subproblem P_1', using the algorithm A, and if more than one processor is assigned for solving of P_1' then u

processes recursively the subproblem P_1' for solving this subproblem. After the solving of P_1' and P_2' the processor u finds a solution of P' as the best of the obtained solutions of P_1' and P_2'.

b. Let $T_A[P_2'] < T_A[P']/p'$. In this case, u selects in G' a new collector processor $\hat{u}$, defining the minimum load of $(\hat{u})$ as $T_A[P']/p'$, and sends to $\hat{u}$ the subproblem P_2' for solving. Further the processor $\hat{u}$ can receive additionally for solving some number of descendent subproblems of the subproblem P_1'. Using the algorithm $\mathcal{A}$, the processor $\hat{u}$ solves consequently all the received subproblems, chooses the best solution from the solutions of these subproblems, and returns this solution to the processor u. The group G' with the new collector processor $\hat{u}$ is assigned for solving of the subproblem P_1' (solving of a subproblem by a group of processors with a collector processor is described below). By a *quasisolution* of the subproblem P_1' we mean the best of solutions of all descendent subproblems of P_1' except the descendent subproblems received by $\hat{u}$ for solving. After processing of P_1' the processor u obtains a quasi-solution of P_1' and then chooses the best from this quasi-solution and the solution obtained from $\hat{u}$ as a solution of P'.

2. Let G' have a collector processor $\hat{u}$. Denote by $T_\Sigma(\hat{u})$ the total running time of solving of all subproblems received by $\hat{u}$ for solving at the moment. $T_\Sigma(\hat{u})$ is called the *current load* of the collector $\hat{u}$. In this case, we assume that $T_\Sigma(\hat{u})$ is less than the minimum load of $(\hat{u})$. Analogously to subcase 1.b, by a quasi-solution of the subproblem P' we mean the best of solutions of all descendent subproblems of P' except the descendent subproblems received by $\hat{u}$ for solving. Consider separately two possible subcases.

a. Let $T_A[P_2'] \geq T_A[P']/(p'-1)$. Then the processor u assigns $\left\lfloor \frac{T_A[P_2']}{T_A[P']}(p'-1) \right\rfloor$ processors from G' which are different from $\hat{u}$ for solving of the subproblem P_2'. Analogously to subcase 1.a, the subproblem P_2' is solved by these processors and a solution of P_2' is returned to u. The other processors from G', including the master processor u and the collector processor $\hat{u}$, are assigned for solving of the subproblem P_1'. Processing recursively the subproblem P_1', the processor u obtains a quasi-solution of P_1' and then chooses the best from this quasi-solution and the obtained solution of P_2' as a quasi-solution of P'.

b. Let $T_A[P_2'] < T_A[P']/(p'-1)$. In this case u sends the subproblem P_2' to the collector processor $\hat{u}$. If $T_\Sigma(\hat{u}) + T_A[P_2']$ is less than the minimum load of $(\hat{u})$ then the whole group G' with the collector processor $\hat{u}$ is assigned for solving of P_1', and the subproblem P_1' is processed recursively by the processor u for obtaining a quasisolution of P_1' which is also a quasisolution of P'. If $T_\Sigma(\hat{u}) + T_A[P_2']$ is greater or equal than the minimum load of $(\hat{u})$ then the collector processor $\hat{u}$ stops to receive subproblems and begins to solve consecutively all the received subproblems. In this case, all processors from G' except the processor $\hat{u}$ are assigned for solving P_1', and, similarly to case 1, P_1' is processed recursively for obtaining a solution of P_1' which is also a quasi-solution of P' (if only one processor u is assigned for solving of P_1' then u solves completely P_1', using the algorithm $\mathcal{A}$).

4.2   Formal Algorithm Description

To give a pseudocode description of the proposed scheme, we introduce some formal notations. We consider a distributed memory model where processors of a parallel computer communicate via message passing. We assume that a finite number m of processors is assigned for solving an initial problem. Processors are numbered from 1 to m and a processor can obtain its number through p_{id} variable. Two primitives (**send** and **recv**) are used for sending and receiving the data. The **send** command has the following syntax:

send τ **to** p_d,

where tuple τ is a message and p_d is the number of the destination processor. This command initiates the message transfer from the processor executing this command to the destination processor with the number p_d. To make the data transfer actually happen the processor p_d should eventually execute the peer **recv** command which has the following syntax:

recv τ **from** p_s,

where p_s is the number of the sending processor. The **recv** command blocks the processor with number p_d until the message τ has arrived from the processor with number p_s. If $p_s = $ undef the message τ is anticipated to be arrived from an arbitrary processor. The **recv** command returns the p_{id} of the processor sent the message. The message τ is actually a tuple $(\tau_0, \tau_1, \ldots, \tau_l)$ of data components which are stored respectively in the variables $\tau_0, \tau_1, \ldots, \tau_l$. To illustrate the processors communication, consider the following example where some data is transferred from the processor 1 to the processor 2.

Processor 1:

   send α, 1, 2 **to** 2

Processor 2:

   $p :=$ **recv** t, a, b **from** 1

In this example, the processor 1 sends the tuple $(\alpha, 1, 2)$ to the processor 2. The processor 2 receives the data. After the receiving of the data the variables p, t, a, b have values $1, \alpha, 1, 2$ respectively.

The message data components can be of four following types: special tags, subproblems, solutions of subproblems, and descriptions of groups of processors. Tags have special meanings used by the receiver to process other components of the message. All possible tags are listed in Table 2.

All processors execute the same code:

```
1: if p_id = 1 then /* master processor */
2:    /* read the initial problem data */
3:    r := GTRA(P, G, undef, 0, 0) /* P — initial problem; G — group of
      processors for solving P. */
4:    /* output the initial problem solution r */
5: else
6:    p := recv τ0, τ1, τ2 from undef
7:    /* τ0 — tag; τ1 — subproblem received for solving; τ2 — description of
      the attached group of processors. */
```

Table 2. Tags

Tag	Description
Single	Accompanies a single subproblem to solve
Stop	Causes a collector processor to exit
Collect	Accompanies a (single) subproblem sent to a collector processor
Group	Accompanies a (single) subproblem and an attached group description sent to the group master

```
 8:    P' := τ₁
 9:    if τ₀ = Single then /* single processor */
10:       r' := TRA(P')
11:    else if τ₀ = Collect then /* collector processor */
12:       r' := CTRA(P',p)
13:    else if τ₀ = Group then /* group master */
14:       G' := τ₂
15:       r' := GTRA(P', G', undef, 0, 0)
16:    end if
17:    send r' to p
18: end if
```

The 1-st processor (*master processor*) reads and stores the initial problem data at the beginning of computations and outputs the solution at the end. It initiates the resolution process by calling the GTRA procedure (line 3). Other processors are waiting for a message and, after receiving the message, perform TRA, GTRA or CTRA procedure depending on the tag contained in the message. The obtained result r' is sent back (line 16).

The simplest case of the procedure corresponds to the **Single** tag (line 9): the received subproblem P' is entirely solved with the TRA procedure. If the tag is equal to **Collect** then the following CTRA (Collector Tree Recursive Algorithm) procedure is executed:

Procedure CTRA(P',p)
Input:
  P' — the received subproblem
  p — the processor from which P' is received

```
 1: [auxiliary list of subproblems initiated by the inserted P'] L := (P')
 2: repeat
 3:    recv τ₀, τ₁ from p
 4:    if τ₀ = Collect then
 5:       P'' := τ₁
 6:       [append the received subproblem P'' to L] L := L, P''
 7:    else if τ₀ = Stop then /* stop signal */
 8:       r' := worst solution
 9:       while L is not empty do
```

10:        extract P'' from L
11:        $r'' := TRA(P'')$
12:        $r' := best(r', r'')$
13:     **end while**
14:     **return** r'
15: **end if**
16: **until** $\tau_0 = $ Stop

CTRA procedure performed by a collector accumulates subproblems received by the collector until the collector receives a message with the **Stop** tag. Then the received subproblems are solved in a batch and the best solution of these subproblems is sent back to the processor p from which these subproblems are received. The message with the **Stop** tag is sent by the processor p when the current load of the collector becomes no less than its minimum load (the processor p accounts i.e. the current load of the collector).

The GTRA (Group Tree Recursive Algorithm) procedure is aimed at solving of the subproblem P' by the group of processors G' and is performed by the group master processor that distributes computations among other processors of G'.

Procedure GTRA(P', G', c_{id}, T_Σ, T_{min})
Input:
    P' — the received subproblem
    G' — group of processors for solving of P'
    c_{id} — the collector processor id
    T_Σ — the current load of the collector processor
    T_{min} — the minimum load of the collector processor
1: $m' := |G'|$
2: **if** $evaluate(P') = $ **true then**
3:    **if** $c_{id} \neq$ undef **then**
4:       **send** Stop **to** c_{id}
5:    **end if**
6:    **return** $getSolution(P')$
7: **else if** $m' = 1$ **then**
8:    **return** TRA(P')
9: **else**
10:    Decompose P' into P'_1, P'_2
11:    **if** $c_{id} \neq$ undef **then**
12:       $m'_2 := \lfloor (m' - 1)T_{\mathcal{A}}[P'_2]/T_{\mathcal{A}}[P'] \rfloor$
13:       **if** $m'_2 = 0$ **then**
14:          **send** Collect, P'_2 **to** c_{id}
15:          **if** $T_\Sigma + T_{\mathcal{A}}[P'_2] \geq T_{min}$ **then**
16:             **send** Stop **to** c_{id}
17:             **return** GTRA(P'_1, $G' \setminus \{c_{id}\}$, undef, 0, 0)
18:          **else**
19:             **return** GTRA(P'_1, G', c_{id}, $T_\Sigma + T_{\mathcal{A}}[P'_2]$, T_{min})
20:          **end if**

```
21:      else if m'_2 = 1 then
22:          p := selectProcessor(G')
23:          send Single, P'_2 to p
24:          r_1 := GTRA(P'_1, G' \ {p}, c_id, T_Σ, T_min)
25:          recv r_2 from p
26:          return best(r_1, r_2)
27:      else /* m'_2 ≥ 2 */
28:          G'_2 := selectSubGroup(G', m'_2)
29:          send Group, P'_2, G'_2 to master(G'_2)
30:          r_1 := GTRA(P'_1, G' \ G'_2, c_id, T_Σ, T_min)
31:          recv r_2 from master(G'_2)
32:          return best(r_1, r_2)
33:      end if
34:  else /* no collector processor */
35:      m'_2 := ⌊m' T_A[P'_2]/T_A[P']⌋
36:      m'_1 := m' − m'_2
37:      if m'_2 = 0 then /* select collector processor */
38:          p := selectProcessor(G')
39:          send Collect, P'_2 to p
40:          r_1 := GTRA(P'_1, G', p, T_A[P'_2], T_A[P']/m')
41:          recv r_2 from p
42:      else if m'_2 = 1 then
43:          p := selectProcessor(G')
44:          send Single, P'_2 to p
45:          if m'_1 ≥ 2 then
46:              r_1 := GTRA(P'_1, G' \ {p}, undef, 0, 0)
47:          else
48:              r_1 := TRA(P'_1)
49:          end if
50:          recv r_2 from p
51:      else /* m'_2 ≥ 2 */
52:          G'_2 := selectSubGroup(G', m'_2)
53:          send Group, P'_2, G'_2 to master(G'_2)
54:          r_1 := GTRA(P'_1, G' \ G'_2, undef, 0, 0)
55:          recv r_2 from master(G'_2)
56:      end if
57:      return best(r_1, r_2)
58:  end if
59:  end if
```

The GTRA procedure processes the subproblem P' by the group G' of processors with an (optional) collector. If the subproblem P' is not pruned by the *evaluate* procedure it is decomposed into two new subproblems P'_1 and P'_2.

The processors in the group G' are separated into two new groups proportionally to the running times of solving of these subproblems. The numbers m'_1 and m'_2 of processors in the new groups are computed by the formulas:

$$m'_2 = \begin{cases} \lfloor (m' - 1)T_{\mathcal{A}}[P'_2]/T_{\mathcal{A}}[P'] \rfloor & \text{if there is a collector in } G', \\ \lfloor m' T_{\mathcal{A}}[P'_2]/T_{\mathcal{A}}[P'] \rfloor & \text{otherwise,} \end{cases}$$

$$m'_1 = m' - m'_2,$$

where without lost of generality we assume $T_{\mathcal{A}}[P'_2] \leq T_{\mathcal{A}}[P'_1]$. If $m'_2 = 1$, only one processor is assigned for solving of the subproblem P'_2 which is indicated by the **Single** tag in the message received by this processor. If $m'_2 = 0$, the subproblem P'_2 is sent to the collector processor which is indicated by the **Collect** tag in the message with P'_2.

5   The Scalability Analysis

We call the described above parallel execution scheme for a tree-like recursion algorithm $\mathcal{A}$ a *balanced recursive scheme* of $\mathcal{A}$. Let H be the balanced recursive scheme of a tree-like recursion algorithm $\mathcal{A}$ for solving of a binary decomposable problem P. We denote by $T^H[P; m]$ the running time of the scheme H for solving of the problem P by m processors. Denote $\alpha = \prod_{i=1}^{\infty}(1 + 2^{-i}) \approx 2.384231$. The main result of this work is the following theorem (we give a sketch of the proof due to the space restrictions).

Theorem 1. *Let $\mathcal{A}$ be a tree-like recursion algorithm for solving of a binary decomposable problem P. Then*

$$T^H[P; m] \leq 2\alpha^2 \frac{T_{\mathcal{A}}[P]}{m} + O(D_{\mathcal{A}}(P)). \tag{2}$$

For proving of Theorem 1 consider all decompositions of subproblems of P performed by processors in procedures GTRA during the process of the solving. Denote by $\mathcal{S}^H(P)$ the set consisting of P and all subproblems of P obtained by these decompositions. A subproblem from $\mathcal{S}^H(P)$ which is decomposed into other subproblems from $\mathcal{S}^H(P)$ is called *decomposed*. Note that for solving any decomposed subproblem from $\mathcal{S}^H(P)$ a group of processors is assigned. A subproblem from $\mathcal{S}^H(P)$ which is not decomposed is called *terminal*. Note that $\mathcal{S}^H(P)$ can be considered as a rooted binary tree with the root P. In this tree decomposed subproblems correspond to internal nodes and terminal subproblems correspond to leaves. Note that, the tree $\mathcal{S}^H(P)$ is a rooted subtree of the tree $\mathcal{S}_{\mathcal{A}}(P)$ with the same root P, so the depth $D^H(P)$ of the tree $\mathcal{S}^H(P)$ is not greater than $D_{\mathcal{A}}(P)$. Any decomposed subproblem P' from $\mathcal{S}^H(P)$ is called *the parent* for the subproblems from $\mathcal{S}^H(P)$ obtained by the decomposition of P'. Note that P' is decomposed into some subproblems P_1 and P_2 such that $T_A[P_1] \geq T_A[P_2]$. We call P_1 (P_2) *the senior child* (*the junior child*) of P'. If for solving of a terminal subproblem from $S^H(P)$ a group of processors is assigned then this subproblem is called *degenerate*. If a terminal subproblem is non-degenerate, only one processor is assigned for solving this subproblem. Note that this processor can be a collector processor. In this case, the given subproblem is called *collected*. We call the set of all terminal subproblems received by a collector processor for solving *collected multiproblem*. By *noncollected terminal subproblems* we will mean nondegenerate terminal subproblems such that the processors assigned for solving these subproblems are not collector processors. Note that any uncollected terminal subproblem is solved by only one processor which does not solve any other terminal subproblems.

We call the initial time of the solving of P by H *the start time*. Let P' be a decomposed or degenerate terminal subproblem from $S^H(P)$, and G' be the group of processors assigned for solving of P'. We denote by $T_{gen}[P']$ the period between the start time and the initial time of processing of P' by the master of G' (in particular, we assume $T_{gen}[P] = 0$). By $T_{get}[P']$ we denote the period between the start time and the time of obtaining the solution of P' by the master of G'. We also define $T_{sol}[P'] = T_{get}[P'] - T_{gen}[P']$ which is called *the solution time* of P' (in particular, $T^H[P; p]$ is the solution time $T_{sol}[P]$ of P). Note that if a degenerate terminal subproblem cannot be pruned then it has to be decomposed to other subproblems from $S^H(P)$. Hence any degenerate terminal subproblem can be pruned and so is resolved in constant time $t_e + t_y$ by the master processor of the group assigned for solving this subproblem. Thus, the solution time of any degenerate terminal subproblem is $O(1)$. Let $\tilde{P}'$ be a collected multiproblem consisting of s collected subproblems, and u be the collector processor assigned for solving of subproblems from $\tilde{P}'$. For convenience we denote $\sum_{P' \in \tilde{P}} T_A[P']$ by $T_A[\tilde{P}']$. We denote also by $T_{gen}[\tilde{P}']$ the period of time between the start time and the initial time of solving of subproblems from $\tilde{P}'$ and by $T_{get}[\tilde{P}']$ the period of time between the start time and the time of obtaining by u the best solution of subproblems from $\tilde{P}'$. We define the solution time $T_{sol}[\tilde{P}'] = T_{get}[\tilde{P}'] - T_{gen}[\tilde{P}']$ of $\tilde{P}'$. Note that for obtaining the best solution of subproblems from $\tilde{P}'$ the processor u has to solve all subproblems from $\tilde{P}'$ and to choose the best from the solutions of these subproblems. Thus, $T_{sol}[\tilde{P}']$ is defined as $T_{sol}[\tilde{P}'] = T_A[\tilde{P}'] + (s-1)t_c = T_A[\tilde{P}'] + O(s)$. Note that all subproblems from $\tilde{P}'$ have distinct depths, so $s \leq D^H(P)$. Thus, $T_{sol}[\tilde{P}'] = T_A[\tilde{P}'] + O(D^H(P))$. Now let P' be a noncollected terminal subproblem, and u be the processor assigned for solving of P'. We denote by $T_{gen}[P']$ the period of time between the start time and the initial time of solving of P' by u and by $T_{get}[P']$ the period of time between the start time and the time of obtaining the solution of P' by u. Note that in this case the solution time $T_{sol}[P'] = T_{get}[P'] - T_{gen}[P']$ of P' is equal to $T_A[P']$.

Let P' be a decomposed or degenerate terminal subproblem from $S^H(P)$ such that $P' \neq P$, P'' be the parent of P', and u be the master of the group assigned for solving of P''. Note that in the period of time between the initial times of processing of subproblems P'' and P' the processor u evaluates P'' in time t_e, decomposes P'' in time t_d, specifies the processor for processing of the junior child of P'' and sends the junior child of P'' to this processor in time t_b. Since all these operations are performed in constant time, we have

$$T_{gen}[P'] - T_{gen}[P''] = O(1). \tag{3}$$

From this relation we immediately obtain the following fact.

Proposition 2. $T_{gen}[P'] = O(d_{\mathcal{A}}(P'))$ *for any decomposed or degenerate terminal subproblem P' from $\mathcal{S}^H(P)$.*

If P' is a degenerate terminal subproblem, from the bound $T_{sol}[P'] = O(1)$ and Proposition 2 we obtain $T_{get}[P'] = T_{gen}[P'] + T_{sol}[P'] = O(d_{\mathcal{A}}(P')) = O(D^H(P))$.

Corollary 1. $T_{get}[P'] = O(D^H(P))$ *for any degenerate terminal subproblem P' from $\mathcal{S}^H(P)$.*

Now let P' be a noncollected terminal subproblem from $\mathcal{S}^H(P)$, and P'' be the parent of P'. Analogously to relation (3), in this case we also have $T_{gen}[P'] - T_{gen}[P''] = O(1)$ and, moreover, by Proposition 2 we have $T_{gen}[P''] = O(d_{\mathcal{A}}(P''))$. Thus, we obtain $O(d_{\mathcal{A}}(P'))$ bound on $T_{gen}[P']$.

Proposition 3. $T_{gen}[P'] = O(d_{\mathcal{A}}(P')) = O(D^H(P))$ *for any noncollected terminal subproblem P' from $\mathcal{S}^H(P)$.*

Let $\tilde{P}$ be a collected multiproblem, u be the collector assigned for solving of subproblems from $\tilde{P}$, P' be the last subproblem from $\tilde{P}$ received by u for solving, and P'' be the parent of P'. Note that $T_{gen}[\tilde{P}]$ is actually the time of receiving of the subproblem P' by u, so, analogously to relation (3), we have $T_{gen}[\tilde{P}] - T_{gen}[P''] = O(1)$. Thus, using Proposition 2 for the subproblem P'', for $T_{gen}[\tilde{P}]$ we also obtain $O(d_{\mathcal{A}}(P')) = O(D^H(P))$ bound.

Proposition 4. $T_{gen}[\tilde{P}] = O(D^H(P))$ *for any collected multiproblem $\tilde{P}$.*

For any subproblem P' define $\overline{T}_{\mathcal{A}}[P'] = T_{\mathcal{A}}[P']/m'$ where m' is the number of processors assigned for solving of P'.

Proposition 5. *For any noncollected terminal subproblem P'' the bound $T_{\mathcal{A}}[P''] < 2\alpha^2 \overline{T}_{\mathcal{A}}[P]$ is valid.*

Proposition 6. *For any collected multiproblem $\tilde{P}''$ the bound $T_{\mathcal{A}}[\tilde{P}''] < \frac{4}{3}\alpha^2 \overline{T}_{\mathcal{A}}[P]$ is valid.*

Denote by $\mathcal{F}'$ the set of all uncollected or degenerate terminal subproblems from $\mathcal{S}^H(P)$, and by $\mathcal{F}''$ the set of all collected multi-problems. Define $T'_{maxget} = \max_{P' \in \mathcal{F}'} T_{get}[P']$, $T''_{maxget} = \max_{\tilde{P}'' \in \mathcal{F}''} T_{get}[\tilde{P}'']$, and $T_{maxget} = \max(T'_{maxget}, T''_{maxget})$. From Corollary 1, Propositions 3 and 5 we obtain that

$T_{get}[P'] < 2\alpha^2 \overline{T}_{\mathcal{A}}[P] + O(D^H(P))$ for any subproblem P' from $\mathcal{F}'$, and from Propositions 4 and 6 we obtain that $T_{get}[\tilde{P}''] < \frac{4}{3}\alpha^2 \overline{T}_{\mathcal{A}}[P] + O(D^H(P))$ for any multiproblem $\tilde{P}''$ from $\mathcal{F}''$. Thus,

$$T_{maxget} < 2\alpha^2 \overline{T}_{\mathcal{A}}[P] + O(D^H(P)). \tag{4}$$

For any subproblem P' from $\mathcal{S}^H(P)$ by the height $h(P')$ of P' we mean the depth of the subtree of $\mathcal{S}^H(P)$ rooted in P'.

Lemma 1. $T_{get}[P'] \leq T_{maxget} + O(h(P'))$ *for any basic decomposed subproblem P' from $\mathcal{S}^H(P)$.*

Note that $h(P) = D^H(P)$, so, applying Lemma 1 to the problem P, we obtain that

$$T_{get}[P] \leq T_{maxget} + O(h(P)) = T_{maxget} + O(D^H(P)).$$

Thus, using relation (4) and taking into account that $T^H[P;m] = T_{get}[P]$, we conclude the bound

$$T^H[P;m] \leq 2\alpha^2 \overline{T}_{\mathcal{A}}[P] + O(D^H(P)) = 2\alpha^2 \overline{T}_{\mathcal{A}}[P] + O(D_{\mathcal{A}}(P))$$

which is stated in Theorem 1, so Theorem 1 is proved.

It is proved in [12] that the time parallel solving of BSP is $\Omega(T/m + d)$ where T is the time of solving the problem by one processor, d is the depth of the search tree, and m is the number of processors. This proof can be directly applied to the case of our problem, i.e. by the same arguments, it can be shown that the running time of any parallel execution scheme for the algorithm $\mathcal{A}$ for solving of the problem P is $\Omega(T_{\mathcal{A}}[P]/m + D_{\mathcal{A}}(P))$. From Theorem 1 it follows that $T^H[P;m] = O\left(\frac{T_{\mathcal{A}}[P]}{m} + D_{\mathcal{A}}(P)\right)$ and thus the proposed balanced recursive scheme is time optimal.

6   Conclusion

For a distributed memory parallel computer we proposed a time optimal parallel tree search algorithm. The main drawback of the proposed parallel execution scheme is that it requires an a priori knowledge of the running times of subproblems processed during the solving of the initial problem. In practice, at best, we can have some estimations for these running times. So one of the directions for further research is an adaptation of the proposed parallel execution scheme for the case when only estimations for these running times are known.

References

1. Baldwin, A., Asaithambi, A.: An efficient method for parallel interval global optimization. In: 2011 International Conference on High Performance Computing and Simulation (HPCS), pp. 317–321. IEEE (2011)
2. Barkalov, K., Gergel, V.: Parallel global optimization on GPU. J. Global Optim. **66**(1), 3–20 (2016)

3. Bhatt, S., Greenberg, D., Leighton, T., Liu, P.: Tight bounds for on-line tree embeddings. SIAM J. Comput. **29**(2), 474–491 (1999)
4. Casado, L.G., Martinez, J., García, I., Hendrix, E.M.: Branch-and-bound interval global optimization on shared memory multiprocessors. Optim. Methods Softw. **23**(5), 689–701 (2008)
5. Eckstein, J., Phillips, C.A., Hart, W.E.: Pico: an object-oriented framework for parallel branch and bound. Stud. Comput. Math. **8**, 219–265 (2001)
6. Evtushenko, Y., Posypkin, M., Rybak, L., Turkin, A.: Approximating a solution set of nonlinear inequalities. J. Global Optim. **71**(1), 129–145 (2018)
7. Evtushenko, Y., Posypkin, M., Sigal, I.: A framework for parallel large-scale global optimization. Comput. Sci. Res. Dev. **23**(3–4), 211–215 (2009)
8. Gergel, V., Sergeyev, Y.D.: Sequential and parallel algorithms for global minimizing functions with Lipschitzian derivatives. Comput. Math. Appl. **37**(4–5), 163–179 (1999)
9. Gmys, J., Leroy, R., Mezmaz, M., Melab, N., Tuyttens, D.: Work stealing with private integer-vector-matrix data structure for multi-core branch-and-bound algorithms. Concurr. Comput. Pract. Exp. **28**(18), 4463–4484 (2016)
10. Grama, A., Kumar, V., Gupta, A., Karypis, G.: Introduction to Parallel Computing. Pearson Education, Upper Saddle River (2003)
11. Herley, K.T., Pietracaprina, A., Pucci, G.: Deterministic parallel backtrack search. Theor. Comput. Sci. **270**(1–2), 309–324 (2002)
12. Karp, R.M., Zhang, Y.: Randomized parallel algorithms for backtrack search and branch-and-bound computation. J. ACM (JACM) **40**(3), 765–789 (1993)
13. Kolpakov, R., Posypkin, M.: Estimating the computational complexity of one variant of parallel realization of the branch-and-bound method for the knapsack problem. J. Comput. Syst. Sci. Int. **50**(5), 756 (2011)
14. Kolpakov, R.M., Posypkin, M.A., Sigal, I.K.: On a lower bound on the computational complexity of a parallel implementation of the branch-and-bound method. Autom. Remote Control **71**(10), 2152–2161 (2010)
15. Pietracaprina, A., Pucci, G., Silvestri, F., Vandin, F.: Space-efficient parallel algorithms for combinatorial search problems. J. Parallel Distrib. Comput. **76**, 58–65 (2015)
16. Rauber, T., Rünger, G.: Parallel Programming: For Multicore and Cluster Systems. Springer, Heidelberg (2013). https://doi.org/10.1007/978-3-642-37801-0
17. Roucairol, C.: A parallel branch and bound algorithm for the quadratic assignment problem. Discret. Appl. Math. **18**(2), 211–225 (1987)
18. Sergeyev, Y.D., Grishagin, V.: Sequential and parallel algorithms for global optimization. Optim. Methods Softw. **3**(1–3), 111–124 (1994)
19. Sergeyev, Y., Grishagin, V.: Parallel asynchronous global search and the nested optimization scheme. J. Comput. Anal. Appl. **3**(2), 123–145 (2001)
20. Strongin, R., Sergeyev, Y.: Global multidimensional optimization on parallel computer. Parallel Comput. **18**(11), 1259–1273 (1992)
21. Strongin, R., Sergeyev, Y.: Global optimization: fractal approach and non-redundant parallelism. J. Glob. Optim. **27**(1), 25–50 (2003)
22. Vu, T.T., Derbel, B.: Parallel branch-and-bound in multi-core multi-CPU multi-GPU heterogeneous environments. Futur. Gener. Comput. Syst. **56**, 95–109 (2016)
23. Wu, I.C., Kung, H.T.: Communication complexity for parallel divide-and-conquer. In: Proceedings of 32nd Annual Symposium on Foundations of Computer Science, pp. 151–162. IEEE (1991)

Minimization of the Weighted Total Sparsity of Cosmonaut Training Courses

Alexander Lazarev[1,2,3], Nail Khusnullin[1], Elena Musatova[1(✉)],
Denis Yadrentsev[4], Maxim Kharlamov[4], and Konstantin Ponomarev[4]

[1] V.A. Trapeznikov Institute of Control Science of Russian Academy of Sciences,
Moscow, Russia
jobmath@mail.ru, nhusnullin@gmail.com, nekolyap@mail.ru
[2] Lomonosov Moscow State University, Moscow, Russia
[3] International Laboratory of Decision Choice and Analysis,
National Research University Higher School of Economics, Moscow, Russia
[4] Yu.A. Gagarin Research & Test Cosmonaut Training Center, Star City, Russia
{d.yadrentsev,m.kharlamov,k.ponomarev}@gctc.ru

Abstract. The paper is devoted to a cosmonaut training planning problem, which is some kind of resource-constrained project scheduling problem (RCPSP) with a new goal function. Training of each cosmonaut is divided into special courses. To avoid too sparse courses, we introduce a special objective function—the weighted total sparsity of training courses. This non-regular objective function requires the development of new methods that differ from methods for solving the thoroughly studied RCPSP with the makespan criterion. New heuristic algorithms for solving this problem are proposed. Their efficiency is verified on real-life data. In a reasonable time, the algorithms let us find a solution that is better than the solution found with the help of the solver CPLEX CP Optimizer.

Keywords: Resource-constrained project scheduling problem
Heuristic algorithms · Planning · Priority rule

1 Introduction

Cosmonauts training is a long process of preparing cosmonauts for their space missions. This process includes physical training, medical tests, extra-vehicular activity training, procedure training, as well as training on experiments they will perform during their stay on the space station. For the training process in the Yu.A. Gagarin Research & Test Cosmonaut Training Center (GCTC), high-tech training facilities are used. These include integrated simulators for space crafts and space stations; a water tank for spacewalking training; centrifuges for simulating g-loads during launch and so on. Because of their high cost, the number

This work was supported by the Russian Science Foundation (grant 17-19-01665).

of these facilities is limited. As several crews are trained at the same time, it is important to share the equipment among cosmonauts so that each of them has enough time to complete the training plan before launching. In this paper, we consider the final stage of cosmonaut training—training of approved crews for a specific space flight on a manned spacecraft. Simultaneously there are several crews training in the GCTC, but this article presents results for the major crew, i.e. the crew whose flight is planned for the nearest future. This crew has priority in the use of simulators and other equipment. Each member of the crew has his own training plan which should be performed with respect to resources and time constraints. Training of each cosmonaut is divided into thematic courses. Each training course is a set of special tasks. A cosmonaut can study several courses simultaneously. The study of the courses should take place in accordance with a curriculum which specifies precedence relations for courses. Resource constraints and precedence relations are supplemented by some constraints that take into account the specifics of training of cosmonauts. In other words, the cosmonauts training problem has the same constraints as the resource-constrained project scheduling problem (RCPSP) has, but it also has some additional restrictions. Note that RCPSP is NP-hard in the strong sense [1]. Different heuristic methods are proposed for solving RCPSP: priority-rule based scheduling methods truncated branch-and-bound, integer programming based heuristics, disjunctive arc concepts, local constraint-based analysis, sampling techniques, evolutionary algorithms and local search techniques (see, for example, [3,4,6,8,11]). In [8] a large number of modern heuristics that have been proposed for RCPSP solving are summarized and categorized. Note that most of these solving methods are designed for objective functions that are different from the one given in this article. As a rule, the objective function in such problems is the project duration (makespan). Another popular criteria are the maximum lateness, the total tardiness, the total flow time [3]. Most of them are regular, i.e. nondecreasing with respect to all finishing times of tasks. In our problem, we use unusual and non-regular objective function. To increase the efficiency of training, it is necessary to establish the correct duration of each course. Duration of a course is a length of a time period between the start of the first task of this course and the end of the last task of this course. If the duration of a course is either too large or too short, then the efficiency of training is reduced. Too fast study of a course is avoided by imposing constraints on the number of tasks of this course per week. To avoid too long duration of a course, we introduce a special objective function—the total weighted sparsity of training courses. It requires the development of new methods that differ from methods for solving the thoroughly studied RCPSP with the makespan criterion.

Since the problem has a high dimension, the use of exact methods does not allow to find an optimal solution in an acceptable time. Previously, for finding a feasible solution of this problem an approach based on methods of integer linear programming was proposed [10]. However, this approach turned out to be ineffective for high-dimensional problems. In [9] comparison of two approaches for finding a feasible solution to this problem for a medium dimension was presented.

The first approach was based on integer linear programming and the second one was on the basis of constraint programming (CP) [5]. A significant advantage of CP has been shown in that paper. Therefore in this paper, we use a mathematical model in terms of CP and run real-life test problems on the solver CPLEX CP Optimizer [12]. The article has the following structure. Sections 2 and 3 presents a description and mathematical formulation of the problem in terms of CP. Results of computational experiments based on CP Optimizer are given. In Sect. 4 we present a heuristic algorithm, its modifications, and comparison of different approaches on some test problems.

2   Problem Description

A crew consists of three cosmonauts. Each of them has his own individual training plan: a set of courses and a set of tasks in each course. Training schedule for cosmonauts of the crew is subject to the following constraints:

(c1) All tasks must be completed before launching. There exist special time boundaries for some tasks. These boundaries can be both precise and rather extended. In the first case a task has an exact date of its execution and in the second case—some period of time (for example, a winter forest landing training has to be in winter). We will call the tasks with time boundaries by fixed tasks.

(c2) There are precedence relations between some courses. A cosmonaut cannot start some course before finishing the previous course. There are also precedence relations between tasks into each course. For some tasks, there are strict precedence relations, i.e. strict time intervals between tasks. The peculiarity of this problem lies in the fact that fixed tasks are not included in the precedence relations graph.

(c3) At each moment of time the resources must be sufficient to execute all current tasks.

(c4) For some sets of tasks there are constraints on the total volume of tasks per day, or per week for one cosmonaut.

(c5) Each task has to be completed before the end of a working day. Some tasks can be executed only during a definite part of a day.

(c6) Some tasks must be performed by all crew members or by two of them simultaneously.

As it was mentioned in the Introduction, we describe an unusual objective function. Each course into a schedule has two characteristics: a volume and a duration (see Fig. 1). The volume of a course is a sum of durations of all tasks in this course. This value does not depend on a schedule and is known in advance. The duration of a course in some schedule is its volume plus all time intervals between tasks of this course. The volume of a course is a lower bound of its duration. Due to constraints (c4) this bound is not achievable. We define *sparsity* of a course as a ratio of its duration to its volume. Our goal is to minimize the total weighted sparsity which is a weighted sum of sparsities of all courses.

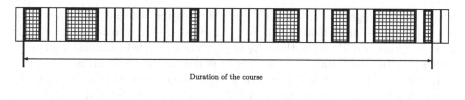

Duration of the course

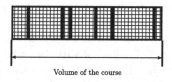

Volume of the course

Fig. 1. Characteristics of a course.

To formalize the objective function, input the following notations. Let $I = \{1, 2, 3\}$ be the set of cosmonauts, and B_i be the set of courses of cosmonaut $i \in I$. Each course b has a weight (significance) w_b, which can take one of two values ω_1 or ω_2 ($\omega_1 < \omega_2$) established by an expert. Denote as $J_{i,b}$ the set of tasks of cosmonaut $i \in I$ from course $b \in B_i$. Each task $j \in J_{i,b}$ has duration p_j. $S_{i,b}^{first}(\pi)$ and $C_{i,b}^{last}(\pi)$ are the start time of the first task and the finish time of the last task of course $b \in B_i$ for cosmonaut $i \in I$ respectively in a schedule π. If for some course an order of tasks is not defined or several tasks can be the first (last) then we input dummy start and end tasks. Then the objective function has the following form:

$$F(\pi) = \sum_{i \in I} \sum_{b \in B_i} w_b \frac{C_{i,b}^{last}(\pi) - S_{i,b}^{first}(\pi)}{\sum_{j \in J_{i,b}} p_j}. \tag{1}$$

It is worth noting that such objective function can arise in other areas of planning. For example, instead of an educational process, the following production planning can be considered. There is a set of complex jobs. Each job consists of a number of operations that require the use of different resources (machines). It is necessary to minimize the stay of each job in the production process. Such an objective function can be explained by additional storage costs or other expenses.

Since in the considered problem there is a constraint on makespan (all tasks must be completed before launching), the proof of NP-hardness can be obtained trivially by reduction from the classical RCPSP. The decision version of RCPSP can be transformed into the decision version of the considered problem with zero number of courses, the sparsity of which must be minimized.

3   Problem Statement in Terms of Constraint Programming

The first stage of our investigation of the problem is using an exact method. As previously we have demonstrated the advantage of constraint programming over integer linear programming for the search of a feasible solution [9], now we try to use CP for minimization of (1). So, formalization of constraints (c1)–(c6) is made in terms of CP and with using of Optimization Programming Language (OPL) [13]. Integer linear programming formulation can be found in [9].

In CP formulation of an optimization problem constraints are not necessarily linear. In addition, the concept of global variables is widely used in CP. This concept allows us to reformulate the group of constraints into "global" one and treat the values of variables as a single entity on the whole planning horizon (or on its part). Another powerful tool of CP modeling is cumulative functions. They allow us to define the "behavior" of variables with help of piecewise linear functions. Due to the fact that IBM ILOG Optimizer and OPL were used for implementation of the model and for solving the problem, we use in the article names of built-in functions and the syntax of OPL.

We will assume that the planning of the cosmonaut training takes place at a certain given time interval, which specifies a planning horizon. We take as a unit of time a half-hour interval. Denote as W a set of all weeks and as D a set of all days on the planning horizon. Set H is a set of all half-hour intervals in a day. Possible moments of the beginning of a task are in a set $T = \{0, 1, \ldots, \mathcal{T}\}$, where $\mathcal{T}$ is the number of half-hour intervals in the planning horizon. Let J be a set of all tasks. For each task $j \in J$ input interval variable s_j with duration p_j:

$$interval \ \ s_j \in [t_1^j..t_2^j] \ \ size \ \ p_j, \tag{2}$$

where $[t_1^j..t_2^j]$ is a subset of T which is defined by some preliminary estimations or by constraints (c1).

We also need auxiliary integer variables $day_j \in \{1, 2, \ldots, |W| \cdot |D|\}$ and $week_j \in \{1, 2, \ldots, |W|\}$, $j \in J$, which are the numbers of the day and of the week respectively on the planning horizon for task j. The relations of variables in terms of CP can be written as follows:

$$day_j == element(days, startOf(s_j)),$$

$$week_j == element(weeks, startOf(s_j)),$$

where $days = (d_1, d_2, \ldots, d_{\mathcal{T}})$ and $weeks = (w_1, w_2, \ldots, w_{\mathcal{T}})$ are vectors whose components are associated with the numbers of days (weeks), corresponding to the respective time intervals from $T = \{0, 1, \ldots, \mathcal{T}\}$. Function $element(a, i)$ returns i-th element of vector a and function $startOf(s_j)$ is used to access the start time of interval variable s_j.

Let $G_i = (J_i, \Gamma_i)$ be a graph of precedence relations between tasks of cosmonaut $i \in I$. If $(j_1, j_2) \in \Gamma_i$, then the task j_1 has to be completed before the beginning of the task j_2. Then constraint (c2) can be written with help of built-in function $endBeforeStart$ as

$$endBeforeStart(s_{j_1}, s_{j_2}), \quad \forall(j_1, j_2) \in \Gamma_i, \quad \forall i \in I. \tag{3}$$

Constraints on resource usage over time can be modeled with constraints on cumulative function expressions. A cumulative function expression is a step

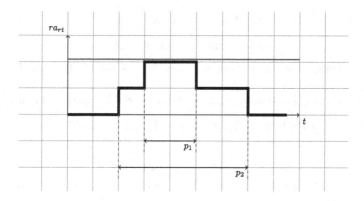

Fig. 2. Cumulative resource usage function.

function that can be incremented or decremented in relation to a fixed time or an interval. It is convenient to use cumulative function expression $pulse(a, b)$, which represents the contribution b to the cumulative function of an individual interval variable or fixed interval a. Figure 2 illustrates cumulative resource usage function for two intervals of length p_1 and p_2.

Let R be a set of renewable resources (instructors, simulators, classrooms, special equipment). Each cosmonaut is also a resource, available in a quantity of 1 during a working day. Denote by ra_{rt} the amount of resource $r \in R$ which is available at time $t \in T$, and by rc_{jr}—the amount of resource $r \in R$ required for the task $j \in J$. Let JR_r be the set of tasks which use resource $r \in R$. Using *step* function for constraint (c3) we get

$$\sum_{j \in JR_r} pulse(s_j, rc_{jr}) \leq ra_{rt}, \ \forall r \in R. \tag{4}$$

For realization of constraints (c4) we use built-in function $count(a, b)$. This function counts how many of the elements in the array given as a are equal to the value given as b. Then, if the number of tasks from some set J_i^A of cosmonaut i should be no more then a_i per week,

$$\sum_{j \in J_i^A} count(week_j, w) \leq a_i, \forall i \in I, \forall w \in W, \tag{5}$$

and, if number of tasks from some set J_i^C of cosmonaut i should be no more then c_i per day,

$$\sum_{j \in J_i^C} count(day_j, d) \leq c_i, \forall i \in I, \forall d \in D. \tag{6}$$

In order to take into account constraints of kind (c5), we use OPL expression $forbidStart(s_j, f)$, which constrains an interval variable s_j to not start where the associated step function f has a zero value:

$$forbidStart(s_j, f), \ \forall j \in J. \tag{7}$$

Similar constraints can be written not only for the set of all tasks J, but also for arbitrary subsets of it.

The constraint *synchronize* makes interval variables start and end together:

$$synchronize(s_{j_1}, s_{j_2}, s_{j_3}), \ \forall (j_1, j_2, j_3) \in J^{123}. \tag{8}$$

or

$$synchronize(s_{j_1}, s_{j_2}), \ \forall (j_1, j_2) \in J^{12}, \tag{9}$$

where J^{123} and J^{12} are sets of tasks which cosmonauts have to execute together (constraints (c6)).

So, constraints (c1)–(c6) can be formulated with help of (2)–(9). Goal function (1) can be rewritten using functions $StartOf$ and $EndOf$. We used for the computational experiment 5 problems with real-life data provided by the GCTC. Characteristics of the problems (the number of courses, the total number of tasks and the planning horizon in weeks) are presented in Table 1.

Table 1. Characteristics of test problems.

N	Courses	Tasks	Planning horizon, weeks
1	10	202	5
2	18	383	15
3	37	838	40
4	48	1214	60
5	11	193	6

The graphs of precedence relations between all possible courses for test problems are presented on Fig. 3. Each cosmonaut has a set of courses which is a subset of all possible courses. Each graph G_i is obtained from the graph of precedence relations between courses and from graphs of precedence relations between tasks in the courses.

As we use the exact method, its runtime can be too much for a high-dimension problem. Therefore the runtime was restricted by some value. In Table 2 numerical results for solving problem (1)–(9) with help of ILOG CP Optimizer are presented.

We use the following notations: t is a time restriction (the solver stopped its work after this time), F is the found value of the objective function, *Solutions* is a number of solutions which the solver found out for time t, *Branches* is a number of branches, $Av.Sparsity$ is an average sparsity of courses, which is defined as

$$\Phi(\pi) = \frac{1}{\sum\limits_{i \in I} |B_i|} \sum_{i \in I} \Big(\sum_{b \in B_i} \frac{C_{i,b}^{last}(\pi) - S_{i,b}^{first}(\pi)}{\sum\limits_{j \in J_{i,b}} p_j} \Big),$$

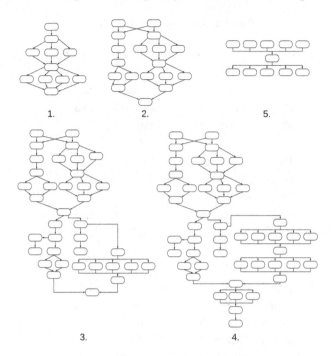

Fig. 3. Graphs of precedence relations between courses for test problems.

where $|B_i|$ is the number of courses of cosmonaut i. The average sparsity is a handy characteristic of a solution, which shows how many times the average length of the course is greater than its minimum possible length (volume). The calculations were performed on a workstation with an Intel Xeon processor E5-2673, 2.4 GHz, and 15 Gb of RAM, solver IBM ILOG CPLEX CP 12.6.2 was used. In these problems we have $\omega_1 = 1$ and $\omega_2 = 5$.

The solver could not find an optimal solution for 3 h. In the problem of the highest dimension (problem 4) the value of the objective function is not improved after 30 min. It is worth noting that problem 5, despite its low dimension, has also proved to be very difficult: the average sparsity is very high and the value of the objective function almost is not improved. This can be explained by the structure of the precedence relations graph (see Fig. 3).

In the next section heuristics algorithms will be presented for solving the problem.

4   Heuristic Algorithms

Despite a wide variety of heuristic methods for solving RCPSP, the priority rule-based scheduling is still one of the most important solution techniques. This can be explained by the simplicity of its implementation, its speed, and efficiency. A priority rule-based scheduling heuristic is made up of two parts, a schedule

Table 2. Testing the CP model.

N	t	F	Solutions	Branches	Av.Sparsity
1	1 min	58.89	1	21 205	2.05
	5 min	57.51	5	78 030	2.03
	15 min	54.36	7	202 587	2.03
	30 min	53.99	10	389 910	2.01
	1 h	53.27	13	465 358	1.99
	3 h	51.27	17	533 425	1.87
2	1 min	106.98	2	36 579	2.41
	5 min	106.92	4	95 403	2.41
	15 min	97.97	8	155 712	2.22
	30 min	97.97	9	258 495	2.22
	1 h	96.87	13	507 082	2.03
	3 h	96.71	14	653 121	2.01
3	1 min	488.86	2	14 785	4.88
	5 min	487.86	4	45 673	4.76
	15 min	487.74	5	130 437	4.75
	30 min	487.21	7	243 897	4.7
	1 h	482.26	13	612 342	4.34
	3 h	302.21	23	756 934	3.09
4	1 min	692.95	1	13 635	5.07
	5 min	691.29	3	43 801	5.06
	15 min	690.60	7	66 075	5.01
	30 min	690.07	8	93 085	5.01
	1 h	690.07	8	142 191	5.01
	3 h	690.07	8	304 989	5.01
5	1 min	197.51	3	25 011	4.99
	5 min	197.25	16	105 672	4.98
	15 min	196.41	24	238 235	4.98
	30 min	195.87	25	439 793	4.93
	1 h	195.83	32	828 890	4.91
	3 h	195.83	39	1 906 453	4.91

generation scheme and a priority rule. Two best known and the oldest schedule generation and the parallel scheduling schemes (see [7]). These schemes generate a feasible schedule by extending a partial schedule step by step. In each step, a generation scheme defines the decision set, i.e. set of all schedulable tasks. Then a used to choose one (or more in parallel case) task from the decision set is scheduled.

In this article, some kind of scheduling heuristics with a serial generation scheme is presented. Firstly we define a priority rule which takes into account objective function (1). Then a schedule generation scheme is presented.

4.1   The Priority Rule

Each task $j \in J_{i,b}$ has a weight $w_b \in \{\omega_1, \omega_2\}$ which is defined by the weight of the course $b \in B_i$. Let $A = \{1, 2, 3\}$ be a set of possible priorities of a task. Assume that we already have a partial schedule, i.e. a schedule where only a task has been assigned a finish time. This subset of tasks will be called a *scheduled* set. We say that a task $j \in J_{i,b}$ from a decision set has priority $\alpha_j = 1$, if there are no any tasks from course b in the scheduled set. In other words, $\alpha_j = 1$, if j is the first task of a course. Otherwise task j has priority $\alpha_j = 2$ in case of $w_b = \omega_1$ and $\alpha_j = 3$ in case of $w_b = \omega_2$. Thus, the main priority is given to tasks from the already started courses with the high weight. So, for each task $j \in J_{i,b}$, $b \in B_i, i \in I$, we have

$$\alpha_j = \begin{cases} 1, \text{ if } j \text{ is the first task of course } b, \\ 2, \text{ if } j \text{ is not the first task of course } b \text{ and } w_b = \omega_1, \\ 3, \text{ otherwise.} \end{cases} \quad (10)$$

4.2   Schedule Generation Scheme

To present the proposed algorithm, define the following functions:

- $GetTask(w, d, i, start)$ returns a schedulable task of the maximal possible priority and its start moment for cosmonaut i. The depth-first search is used. This function checks whether it is possible to schedule a task in the day d of the week w at moment $start$ within the constraints (c1)–(c5). If there are no any schedulable tasks for this moment, the function returns the next possible time moment of this day or the empty set (if there are no any schedulable tasks in this day). As constraint (c6) is connected with several cosmonauts, this function does not check, whether the respective synchronous task of another cosmonaut can be scheduled at the same time.
- $task.IsShared$ returns $TRUE$, if the $task$ must be performed simultaneously with some task of another cosmonaut.
- $Release(start, task)$ adds the $task$ to a partial schedule at time moment $start$.
- $DealContract(task.SharedTasks, start, w, d)$ is a function, which defines the nearest possible common moment in day d of week w (starting from moment $start$) for the $task$ and for all its synchronous tasks from the set $task.SharedTasks$.

Let enumerate cosmonauts of the crew in the decreasing order of their training loads. We take the first unplanned time moment from the planning horizon and try to schedule for cosmonaut 1 a task of the highest possible priority. Then we turn to schedule for cosmonaut 2 and so on. For details see Algorithm 1.

Algorithm 1.

1: **for** w in W, d in D **do**
2:     **repeat**
3:        **for** i in I **do**
4:           $(start, task) = GetTask(w, d, i, start)$
5:           **if** $task == null$ **then**
6:              continue
7:           **end if**
8:           **if** not $task.IsShared$ **then**
9:              $Release(start, task)$
10:             continue
11:          **end if**
12:          $DealContract(task.SharedTasks, start, w, d)$
13:        **end for**
14:     **until** not $isAnyonePlanned()$
15: **end for**

We used two variants of function *DealContract* for Algorithm 1. In the first variant, this function only checks whether it is possible to establish all tasks from *task.SharedTasks* at a given time moment *start*. If it is possible, it adds all shared tasks to a partial schedule. In the second variant, the function tries to find out the possible time moment if the shared tasks cannot be performed at a time moment *start* (see Algorithm 2). In both variants of *DealContract* and in *GetTask* a function $TryToDeal(task, w, d, start)$ is used, which check whether it is possible to schedule the *task* in day d of week w at time moment *start* within the constraints (c1)–(c5). So, this function is the main brick of the algorithm and in the future we will measure the complexity of the algorithm as the number of calls to this function.

4.3 Computational Results

At the first stage of the computational experiment, we solved 5 practical previously declared problems. Results of using Algorithm 1 with two variants of function *DealContract* described above are presented in Table 3. The following notations were used: F is the value of the objective function, *Av.Sparsity* is the average sparsity of courses, *TryToDeal count* is the number of calls to the function *TryToDeal*.

It is easy to see that for small problems 1 and 2 the heuristic algorithm with both variants of the function *DealContract* received identical results. The problem of the highest dimension is the only problem in which the second variant of the algorithm turned out to be better than the first one. For all problems using the heuristic algorithms significantly improved the value of the objective function in comparison with the value obtained with the help of the CP solver (see Table 2).

At the second stage of the experiment, we generated a series of 500 different problems for each graph of precedence relations between courses from Fig. 3.

Algorithm 2. $DealContract(task.SharedTasks, start, w, d)$

```
 1: for task in SharedTasks do
 2:     if not task.AllParents.IsSchedulled then
 3:         return
 4:     end if
 5: end for
 6: repeat
 7:     for task in SharedTasks do
 8:         (isDealed, start) = TryToDeal(task, w, d, start)
 9:         if not isDealed then
10:             return
11:         end if
12:         task.Start = start
13:     end for
14: until SharedTasks.AllStartsAreEqual()
15: for task in SharedTasks do
16:     Release(start, task)
17: end for
```

Table 3. Testing the heuristic algorithms on real-life problems.

N	The first variant			The second variant		
	F	$Av.Sparsity$	$TryToDeal$ count	F	$Av.Sparsity$	$TryToDeal$ count
1	36.204	1.248	466	36.204	1.248	466
2	79.225	1.650	1048	79.225	1.650	1048
3	251.394	2.646	2353	254.020	2.673	2241
4	346.805	2.774	3629	270.258	2.162	3766
5	77.475	2.671	534	83.612	2.883	567

Duration p_j for each task of a problem was chosen as an arbitrary integer from the interval $[1; 8]$. Results of this experiment are presented in Table 4. Here N is a number of graph from Fig. 3, which is used for generation of a series of problems; $F_1 < F_2$ ($F_1 > F_2$) is a number of problems from a series, in which the function value is strictly better for variant 1 (variant 2) of the algorithm; $T_1 < T_2$ ($T_1 > T_2$) is a number of problems, in which the makespan is strictly better for variant 1 (variant 2) of the algorithm; Avg, Max and Min are the average, the minimal and the maximal sparsities respectively for a series of problems.

The experiment shows that for the first two series of problems two approaches give the same results. For the problems of high dimensions (series 3 and 4) almost in half of the cases, the best results were given by one approach, in half - by another. The second variant of the function $DealContract$ has a slight advantage in the number of problems with the best value of the objective function for large-scale problems. So, both variants can be used for such problems together. The runtime of algorithms did not exceed 2 min. If to say about the makespan, the

Table 4. Testing the heuristic algorithms on series of test problems.

N	$F_1 < F_2$	$F_1 > F_2$	$T_1 < T_2$	$T_1 > T_2$	The first variant			The second variant		
					Avg	Max	Min	Avg	Max	Min
1	3	1	4	0	1.51	1.87	1.30	1.51	1.87	1.30
2	0	3	0	3	2.07	3.20	1.67	2.07	3.20	1.67
3	230	264	227	267	2.72	3.88	1.97	2.72	4.53	1.97
4	245	253	272	226	2.74	4.45	1.96	2.74	5.06	1.92
5	350	149	384	115	2.92	4.38	2.18	2.92	4.65	2.02

trend is the same, but in some cases, the better value of the objective function is achieved due to the worse value of the makespan. This is clearly seen from Table 5, which contains summary data for all series of problems. This tendency is a distinctive feature of this problem. Indeed, if we did not have a constraint on the planning horizon and other constraints, the optimal value of the objective function would be achieved in the schedule where courses pass one after another without intersections. Therefore, it is necessary to maintain a balance between the value of the objective function and the makespan.

Table 5. The relations between the values of the objective function and the planning horizons in the series of problems.

	$T_1 < T_2$	$T_1 > T_2$	$T_1 = T_2$
$F_1 < F_2$	761	67	0
$F_1 > F_2$	126	544	0
$F_1 = F_2$	0	0	1002

5 Conclusion

A new real-life problem statement for cosmonaut training planning with an unusual objective function (the weighted total sparsity) is presented. A heuristic algorithm with two variants of realization is proposed for solving this problem. Comparison of the algorithm with an exact method is carried out. Numerical testing of algorithms on real data, provided by the GCTC, demonstrated the efficiency of the approach. For all problems, the heuristic algorithm found schedules in which the value of the objective function is significantly better than the value obtained with a help of the CPLEX CP Optimizer in 3 h. Comparison of the two variants of the heuristic algorithm showed that it is sensible to apply both variants to high-dimensional problems since they give different solutions. In this case, the best solution can be obtained by both the first and the second variant.

The directions of further research may include obtaining more flexible priority rules, lower bounds for the given objective function and developing approximate algorithms.

References

1. Artigues, C., Demassey, S., Neron, E. (eds.): Resource-Constrained Project Scheduling: Models, Algorithms, Extensions and Applications. Wiley-ISTE, Hoboken-London (2008)
2. Bartusch, M., Mohring, R.H., Radermache, F.J.: Scheduling project networks with resource constraints and time windows. Ann. Oper. Res. **16**, 201–240 (1988)
3. Brucker, P., Drexl, A., Mohring, R., Neumann, K., Pesch, E.: Resource-constrained project scheduling: notation, classification, models, and methods. Eur. J. Oper. Res. **112**, 3–41 (1999)
4. Debels, D., Vanhoucke, M.: A decomposition-based genetic algorithm for the resource-constrained project-scheduling problem. Oper. Res. **55**(3), 457–469 (2007)
5. Dechter, R.: Constraint Processing. Morgan Kaufmann Publishers, San Francisco (2003)
6. Homberger, J.: A multi-agent system for the decentralized resource-constrained multi-project scheduling problem. Int. Trans. Oper. Res. **14**, 565–589 (2007)
7. Kolisch, R.: Serial and project scheduling methods revisited: theory and computation. Eur. J. Oper. Res. **90**, 320–333 (1996)
8. Kolisch, R., Hartmann, S.: Experimental investigation of heuristics for resource-constrained project scheduling: an update. Eur. J. Oper. Res. **174**(1), 23–37 (2006)
9. Lazarev, A.A., et al.: Mathematical modeling of the astronaut training scheduling. UBS **63**, 129–154 (2016)
10. Musatova, E., Lazarev, A., Ponomarev, K., Yadrentsev, D., Bronnikov, S., Khusnullin, N.: A mathematical model for the astronaut training scheduling problem. IFAC PapersOnLine **49**(12), 221–225 (2016)
11. Valls, V., Quintanilla, M.S., Ballestin, F.: Resource-constrained project scheduling: a critical activity reordering heuristic. Eur. J. Oper. Res. **149**(2), 282–301 (2003)
12. IBM Homepage. https://www.ibm.com/analytics/data-science/prescriptive-analytics/cplex-cp-optimizer. Accessed 1 July 2018
13. IBM Homepage. https://www.ibm.com/analytics/data-science/prescriptive-analytics/optimization-modeling. Accessed 1 July 2018

Genetic Local Search for Conflict-Free Minimum-Latency Aggregation Scheduling in Wireless Sensor Networks

Roman Plotnikov[1]($\boxtimes$) (iD), Adil Erzin[1,2] (iD), and Vyacheslav Zalyubovskiy[1] (iD)

[1] Sobolev Institute of Mathematics, Novosibirsk, Russia
{prv,adilerzin,slava}@math.nsc.ru
[2] Novosibirsk State University, Novosibirsk, Russia

Abstract. We consider a Minimum-Latency Aggregation Scheduling problem in wireless sensor networks when aggregated data from all sensors are required to be transferred to the sink. During one time slot (time is discrete) each sensor can either send or receive one message or be idle. Moreover, only one message should be sent by each sensor during the aggregation session, and the conflicts caused by interference of radio waves must be excluded. It is required to find a min-length conflict-free schedule for transmitting messages along the arcs of the desired spanning aggregation tree (AT) with the root in the sink. This problem is NP-hard in a general case, and also remains NP-hard in a case when AT is given. In this paper, we present a new heuristic algorithm that uses a genetic algorithm and contains the local search procedures and the randomized mutation procedure. The extensive simulation demonstrates a superiority of our algorithm over the best of the previous approaches.

Keywords: Wireless sensor networks · Aggregation
Minimum latency · Genetic local search · Simulation

1 Introduction

One of the most common applications of wireless sensor networks (WSNs) is the collection of sensing information to a designated node called a sink [1]. This kind of all-to-one communication pattern is also known as *convergecast*. Since sensor nodes are equipped with radio transmitters with limited transmission ranges, hop-by-hop communications are usually used to deliver the data from the sensor nodes to the sink. WSN is commonly modeled as a graph, where vertices represent sensor nodes, and any two nodes are connected if the distance between them is within their transmission range. The convergecast process, in this case, is modeled by building a logical tree on top of the physical topology with the sink located at the root, assuming that packets are routed along arcs of the tree.

Since energy limitation is the most critical issue of WSNs application, energy efficiency becomes one of the primary design goals for a convergecast protocol.

© Springer Nature Switzerland AG 2019
Y. Evtushenko et al. (Eds.): OPTIMA 2018, CCIS 974, pp. 216–231, 2019.
https://doi.org/10.1007/978-3-030-10934-9_16

Obviously, the convergecast of all raw data will cause a burst traffic load. To reduce the traffic load, in-network data aggregation can be applied. Here aggregation refers to the process when the relay nodes merge their received data with their own data by means of data compression, data fusion or aggregation function [2]. In this case, each sensor node has to transmit only once and the transmission links form a tree, which is called the *aggregation tree*.

When a node sends its data to its receiver, a *collision* or *interference* can occur at the receiver if the transmission interferes with signals concurrently sent by other nodes, and thus the data should be retransmitted. Since the retransmissions cause both extra energy consumption and an increase of convergence time, protocols able to eliminate the collisions are necessary. A common approach is to assign a sending timeslot to each node in such a way that all data can be aggregated without any collision on their way to the sink node, known in the literature as *time division multiple access* (TDMA)-based scheduling. Most of the scheduling algorithms adopt the *protocol interference model* [3], which enables the use of simple graph-based scheduling schemes. The *physical model* is based on the signal-to-interference-plus-noise-ratio (SINR) and provides a better solution in terms of realistic capturing of interference from multiple transmissions [4].

In terms of common objectives of TDMA-based scheduling algorithms, the following two are the most fundamental with respect to data aggregation in WSNs: minimizing schedule length or latency and minimizing energy consumption. In terms of design assumptions, the algorithms differ mainly in the following categories: use of communication and interference models, centralized or distributed implementation, topology assumption, and types of data collection [5]. In this paper, we consider the problem of minimization aggregation latency assuming collision-free transmission under the protocol model and uniform transmission range. This problem is known as Minimum-Latency Aggregation Scheduling (MLAS) [6] or Aggregation Convergecast Scheduling [7].

In terms of computational complexity, the MLAS problem is NP-hard even in a special case when the communication graph is unit disk graph [8], so most of the existing results in the literature are heuristic algorithms, that are usually comprised of two independent phases: construction of an aggregation tree and link transmission scheduling. It is worth mentioning that both these problems are very hard to solve. On one hand, there is no result describing the structure of an optimal aggregation tree for a given graph. On the other hand, even for a given aggregation tree, optimal time slot allocation is still NP-hard [9].

To overcome the above-mentioned problems, we propose a metaheuristic algorithm to reduce aggregation delay by constructing a proper suboptimal aggregation tree and make the following contributions.

- We suggest a new approach to generate and evaluate different aggregation trees within a genetic local search (GLS) metaheuristic. To the best of our knowledge, it is the first use of evolutionary algorithms for the MLAS problem.
- We conduct extensive simulation experiments to demonstrate the quality of the solutions provided by the proposed method vs. Integer Programming-based optimal algorithm and the current state-of-the-art heuristics.

The rest of this paper is organized as follows: Sect. 2 reviews the related work. In Sect. 3, we provide assumptions and formulation of the problem. A new heuristic algorithm based on genetic algorithm and local search metaheuristics is presented in Sect. 4. Section 5 contains the results and analysis of an experimental study, and Sect. 6 concludes the paper.

2   Related Work

In one of the early works on MLAS, Chen et al. [8] consider a slightly general-ized version of the MLAS, Minimum Data Aggregation Time (MDAT) problem, where only a subset of nodes generates data. They proved that MDAT is NP-hard and designed an approximation algorithm with a guaranteed performance ratio of $\Delta - 1$, where Δ is the maximal number of sensor nodes within the transmission range of any sensor. In [10] the authors propose an approximation algorithm with guaranteed performance ratio $\frac{7\Delta}{\log_2 |S|} + c$, where S is the set of sensors containing source data, and c is a constant.

As mentioned before, most MLAS algorithms solve the problem in two con-secutive phases: aggregation tree construction and link scheduling. Shortest Path Tree (SPT) and Connected Dominating Set (CDS) are the usual patterns for the aggregation tree. In SPT based algorithms [2,5,8], a sensor transmits data through a path with minimum length, which reduces aggregation delay, but they do not take potential collisions into consideration. As for CDS based algorithms, due to the topological properties of CDS, it is often possible to prove their upper bounds of data aggregation delay, which usually depend on network radius R and maximum node degree Δ. Huang et al. [11] proposed an aggregation scheduling method based on CDS with the latency bound $23R + \Delta - 18$. Based on the deeper study of the properties of neighboring dominators in CDS, Nguyen et al. [12] provided a proof of an upper bound $12R + \Delta - 12$ for their algorithm.

Despite the ability to have a delay upper bound for CDS based algorithms, the upper bounds are much greater than their real performance shows. Moreover, a dominating node is likely to be a node of large degree, which may have a negative effect on aggregation delay. SPT has similar problems for the sink node, which takes all its neighbors as its children. It was shown in [13] that an optimal solution could be neither SPT nor CDS based. De Souza et al. [14] constructed an aggregation tree by combining an SPT and a minimum interference tree built by Edmond's algorithm [15]. In [7], the authors proposed a Minimum Lower bound Spanning Tree (MLST) algorithm for aggregation tree construction. To achieve a small delay lower bound, they use the sum of the receivers' depth and child number as the cost of the transmission link. However, the problem of finding the optimal aggregation tree for the MLAS problem remains unsolved.

Genetic algorithm [16] (GA) is one of the most common metaheuristics that is used for the approximate solution of NP-hard discrete optimization prob-lems including those in the WSN domain. In particular, in [17–20], different GA based approaches were used to solve problems associated with the minimiza-tion of energy consumption. In addition, researchers in [21] proposed GAs for the multiple QoS (quality of service) parameter multicast routing problem in a Mobile ad hoc network (MANET). In [22] a GA based framework was presented for the custom performance metric optimization of WSN. In [23] a GA based algorithm was proposed to solve the convergecast scheduling problem with an unbounded number of channels, where only conflicts between the transmitters to the same addressee were taken into account.

3   Problem Formulation

We consider a WSN consisting of stationary sensor nodes with one sink. All sensors are homogeneous. We use a protocol interference model [3], which is a graph-theoretic approach that assumes the correct reception of a message if and only if there are no simultaneous transmissions within proximity of the receiver. For simplicity, we assume that the interference range is equal to the transmission range, which is the same for each sensor. Then the WSN with sink node s can be represented as a graph $G = (V, E)$, where V denotes all the sensor nodes and $s \in V$. An edge (u, v) belongs to E if the distance between the nodes u and v is within the transmission range. We also assume that time is divided into equal-length slots under the assumption that each slot is long enough to send or receive one packet. The problem considered in this paper is defined as follows: Given a connected undirected graph $G = (V, E)$, $|V| = n$ and a sink node $s \in V$, find the minimum length schedule of data aggregation from all the vertices of $V \setminus \{s\}$ to s (i.e., assign a sending time slot and a recipient to each vertex) under the following conditions:

- each vertex sends a message only once during the aggregation session (except the sink which always can only receive messages);
- once a vertex sends a message, it can no longer be a destination of any transmission;
- if some vertex sends a message, then during the same time slot none of the other vertices within a receiver's interference range can send a message;
- a vertex cannot receive and transmit at the same time slot.

As it follows from this formulation, the data aggregation have to be performed along the directed edges (arcs) of a spanning tree rooted in s. Since it is convenient to consider the arcs when constructing an aggregation tree, we also introduce a directed graph $G_{dir} = (V, A)$ constructed from G by replacing each edge with two oppositely directed arcs and excluding the arcs starting from s.

4   Genetic Local Search

As was mentioned before, there are a lot of known heuristic approaches for the approximate solution to the considered problem. As a rule, each of these methods consists of two stages. At the first stage, an aggregation tree is built, and at the second stage, a conflict-free schedule is constructed. We suggest a new approach where different aggregation trees are examined within an algorithm, based on a GA metaheuristic, that applies a local search procedure to the offspring as well as a fully randomized mutation procedure. Such an algorithm is often called *memetic algorithm* or *genetic local search* (GLS).

Since the considered problem remains NP-hard even for a given aggregation tree [9], the problem of defining the difference between the values of minimum length aggregation schedules of two aggregation trees is intractable, too. Therefore, any type of local search is not applicable, within an acceptable time, to this problem. Instead, we propose performing a local search for the reduced problem, where only the primary conflicts are taken into account. In this case, the difference of the objective values between two neighboring aggregation trees may be calculated efficiently. Additionally, after applying a local search procedure, an approximate scheduling algorithm for the initial problem can be applied to the aggregation tree. As an approximate algorithm of calculation of a conflict-free schedule for a given aggregation tree, we use the heuristic algorithm Neighbor Degree Ranking (NDR) [7]. Note that this method can change the aggregation tree because of the Supplementary Scheduling subroutine.

A brief description of the main steps of the proposed algorithm is presented in Algorithm 1. Like other GA based methods, this algorithm imitates an evolutionary process. Once created by the Initialization procedure *population*, a set of feasible solutions to the considered problem is iteratively updated within the **while** loop. In each iteration of this loop, the current population generates an offspring after applying Selection and Crossover procedures, and then the elements of the offspring are modified by Mutation and LocalSearch procedures. Then, the fitness of each offspring element is calculated, and inside the Join procedure, the fittest elements are included in the next generation.

Algorithm 1. Genetic local search

1: *Input*: $G_{dir} = (V, A)$ is a communication graph, $PopSize, OffspSize, FPItCount,$ $SPProportion, PM, PLS, K_{max}$—algorithm parameters;
2: *Output*: T—spanning tree in G rooted in s;
3: Initialization;
4: FitnessCalculation(population);
5: **while** stop condition is not met **do**
6:     Selection;
7:     Crossover;
8:     Mutation;
9:     LocalSearch;
10:     FitnessCalculation(offspring);
11:     Join;
12:     Let T be the best tree among the current population;
13: **end while**

The algorithm takes as an input a communication graph G_{dir} and the following parameters:

- *PopSize*—the size of population;
- *OffspSize*—the size of offspring;
- *FPItCount*—the number of iterations in the first population construction procedure;
- *SPProportion*—the ratio of shortest-path trees in the starting population;
- *PM*—the probability of mutation;
- *PLS*—the probability of a local search.
- K_{max}—the maximum possible number of iterations in the mutation procedure.

The main steps are described in detail in the next subsections.

4.1   Initialization

The first population is generated within the Initialization procedure. At first, we generate three trees with the most efficient known heuristics: a shortest-path tree (e.g., constructed by the Dijkstra algorithm), a tree constructed by the Round Heuristic (RH) [24], that appears to be very effective for the simplified problem with only the primary conflicts, and Minimum Latency Spanning Tree (MLST) introduced in [7]. After the shortest-path tree is constructed, the length of the shortest path from each vertex to the sink is known. Let $l(v)$ be the length (number of edges) of the shortest path from vertex $v \in V$ to s. The next trees added to the population are generated by two procedures: *RandomShortestPath* and *RandomMinDegree*. The procedure *RandomShortestPath* starts with a tree $T = (\{s\}, \emptyset)$, and then for each vertex $v \in V \setminus \{s\}$ an arc from a set $A_v = \{(u, v) | (u, v) \in A, l(u) = l(v) - 1\}$ is chosen at random and added to the current tree. The procedure *RandomMinDegree* starts with a tree $T = (\{s\}, \emptyset)$ as well, and then an arc from A that connects a vertex from the current tree with a vertex from V that does not belong to the current tree is sequentially chosen at random and added to the current tree, and the probability of an arc choice is inversely proportional to the degree of a corresponding vertex in the current tree. A new tree is added to the population only if it is not a copy of an existing one. The Initialization step requires three parameters: *PopSize*—the maximum size of the population, *SPProportion*—an approximate part of the trees generated by the procedure *RandomShortestPath*, and *FPItCount*—the maximum number of successive attempts to generate a tree. The pseudocode of the Initialization procedure can be found in the Algorithm 2.

4.2   Fitness Calculation

In order to estimate the quality of every tree in the population, its fitness should be calculated. Fitness is a positive value which is higher when the value of the objective function is closer to optimal. Let $L(T)$ be the length of an aggregation

---

Algorithm 2. Initialization

1: *Input:* $G_{dir} = (V, A)$ — a communication graph, *PopSize*, *FPItCount*,
   SPProportion — algorithm parameters;
2: *Output:* P — population (a set of spanning trees in G rooted in s);
3: Set $P \leftarrow \emptyset$, $i \leftarrow 0$;
4: Add the trees constructed by Dijkstra algorithm, Round Heuristic, and Minimum
   Latency Spanning Tree to P ;
5: **while** $i < FPItCount$ and $|P| < PopSize$ **do**
6:     $x \leftarrow$ random real value between 0 and 1
7:     **if** $x < SPProportion$ **then**
8:         Set $T \leftarrow RandomShortestPath()$;
9:     **else**
10:        Set $T \leftarrow RandomMinDegree()$;
11:    **end if**
12:    **if** P does not contain T **then**
13:        Add T to P;
14:    **end if**
15:    Set $i \leftarrow i + 1$;
16: **end while**

---

schedule for a spanning tree T. Then the fitness is $1/L(T)$. As was mentioned before, conflict-free scheduling on a given tree is an NP-hard problem. Therefore, instead of searching for an optimal schedule, the approximate solution to the subproblem of finding a minimum length schedule for a given aggregation tree is constructed using the NDR heuristic.

4.3   Selection

Within the selection procedure, a set of prospective parents of the next offspring is filled with solutions from the current population in the following way. Sequentially, two trees are taken from the current population in proportion to their fitness probability: the first tree of each pair is chosen randomly from the entire population, and the second tree is chosen from the remaining part of the population. Each pair should contain different trees, but the same tree may be included in many pairs. In such manner, $OffspSize$ pairs are selected.

4.4   Crossover

At the crossover procedure each pair of previously selected parents reproduces a child. Namely, a pair of parents $T_p^1 = (V, A_p^1)$ and $T_p^2 = (V, A_p^2)$ generates a child tree T_c in the following way. Let us consider a vertex $v \in V \setminus \{s\}$ and two vertices $v_1, v_2 \in V : a_1 = (v, v_1) \in A_p^1$, $a_2 = (v, v_2) \in A_p^2$. Choose an arc from $\{a_1, a_2\}$ and add it to T_c. If $v_1 = v_2$ then the arc a_1 is chosen. If adding one arc from $\{a_1, a_2\}$ to T_c leads to the appearance of cycles, then another arc is chosen. In the remaining case let us introduce the weight $w_i = 1/\delta(v_i) + 1/|l(v) - l(v_i) - 2|$, where $\delta(v_i)$ is a degree of the vertex v_i in the tree $T_p^i, i \in \{1, 2\}$. Then the arc is chosen randomly from $\{a_1, a_2\}$ with probability $P(a_i) = w_i/(w_1 + w_2), i \in \{1, 2\}$.

4.5   Mutation

Mutation is a randomized procedure which is applied to the tree in the current offspring. The mutation procedure is applied with probability PM (a parameter of GLS) to each offspring. The mutation procedure takes as an argument (an integer parameter) K—the maximum difference (number of different arcs in the initial tree and in the modified one). This parameter is taken randomly from the interval $[1, ..., K_{max}]$, where K_{max} is another parameter, inverse to its value probability (i.e., smaller modifications are more possible). The pseudocode of the mutation procedure is given in Algorithm 3.

Algorithm 3. Mutation

1: *Input*: $G_{dir} = (V, A)$—a communication graph, $T = (V, A(T))$ — a spanning tree on G rooted in s, K — an integer parameter;
2: *Output*: T — spanning tree in G rooted in s;
3: **for all** $k \in \{1, ..., K\}$ **do**
4:     Set $(v, u) \leftarrow$ random arc from $A \setminus A(T)$;
5:     **if** v is not descendant of u **then**
6:         Remove the arc $(v, Parent(v))$ from T and add the arc (v, u) to T;
7:     **end if**
8: **end for**

4.6   Local Search

As well as mutation, the local search procedure is applied to a subset of offspring defined by the probability PLS—another algorithm parameter. We suggest two different local search procedures. The first one, *BranchReattaching* algorithm is already proposed in [23]. The pseudocode of this local search procedure is presented in Algorithm 4. At each iteration, the procedure performs a search of such arc $a = (v_1, v_2) \in A \setminus A(T)$ whose addition to T (together with detaching of v_1 from its parent in T) leads to the maximum decrease of the objective function. The method *ReattachingEffect*(T, v, u) (see Algorithm 5) calculates the change of the schedule length after detaching of v from its parent in T and adding an arc (v, u). The whole procedure continues while the solution is improved.

Let us describe another local search procedure *ArcInversion* that we also apply within GLS. As an elementary movement the sequence of the following operations can be executed within *ArcInversion* for any vertex $v \in V$ of a current tree T except its root's children: at first, the arc from v to its parent p is inverted; then the arc from p to its parent is removed from T, and after that the most efficient arc that starts from v and joins two obtained connected components is added to T. In the first steps of the algorithm and after each change of a tree, a schedule (i.e., time slots assignment) is recalculated taking into account only the primary conflicts. The pseudo code of this method is shown in Algorithm 6.

Algorithm 4. *BranchReattaching* local search

1: *Input*: $G_{dir} = (V, A)$ — a communication graph, $T = (V, A(T))$ — a spanning tree on G rooted in s;
2: *Output*: T — spanning tree in G rooted in s;
3: Calculate a schedule on T with only the primary conflicts;
4: Set *improved* $\leftarrow$ *true*;
5: **while** *improved* **do**
6:    Set *improved* $\leftarrow$ *false*;
7:    Set $u^* \leftarrow \emptyset$, $v^* \leftarrow \emptyset$, *bestImpr* $\leftarrow 0$;
8:    **for all** arcs $(u, v) \in A \setminus A(T)$ where u is not a descendant of v **do**
9:       Set *effect* $\leftarrow$ *ReattachingEffect*(T, v, u);
10:      **if** *effect* $<$ *bestEffect* **then**
11:         Set $u^* \leftarrow u$, $v^* \leftarrow v$, *bestEffect* $\leftarrow$ *effect*, *improved* $\leftarrow$ *true*;
12:      **end if**
13:   **end for**
14:   **if** *improved* **then**
15:      Remove the arc $(v^*, Parent(v^*))$ from T and add the arc (v^*, u^*) to T;
16:      Calculate a schedule on T with only the primary conflicts;
17:      **break**;
18:   **end if**
19: **end while**

Algorithm 5. *ReattachingEffect*

1: *Input*: an initial tree $T = (A, V)$ rooted in v_0 with time slot assigned to each vertex; a starting vertex v of an arc (v, p) that is considered for inversion; a vertex u that is considered to be a new parent of v in T.
2: *Output*: $L(T) - L(T_1)$ — the difference of schedule length after removing the arc (v, p) and adding the arc (v, u).
3: Find the first common predecessor s between v and u and two paths: *path$_1$* that starts at v and ends at q and *path$_2$* that starts at u and ends at r (q and r are both children of s);
4: Find the new minimum sending time of q after removing the arc (v, p) from T.
5: **if** sending time of q was not decreased **then**
6:    return 0.
7: **end if**
8: Find the new minimum sending time of u after adding of v to its list of children.
9: Given new minimum sending time of u find the new minimum sending time of r.
10: Given new minimum sending times of q and r find the new minimum sending time of s.
11: Given new minimum sending time of s find and **return** the difference of a schedule length traversing the vertices from s to v_0.

Similar to how it is done in *BranchReattaching* local search procedure, in order to speed up the *ArcInversion* local search procedure, instead of performing the modification of a tree and recalculating the schedule at each step, we calculate only the value of the schedule length change by the method *ArcInversionEffect*. This method is presented in Algorithm 7. It uses the following idea. If all minimum possible time slot values for a list of children of some vertex $v \in V$ are known, then one can easily calculate the minimum possible time slot of v, such that its children do not conflict with each other. If this value differs from the initial time slot of v, then the minimum time slot of a parent of v can be updated in the same manner, and so on to the root.

Both methods $ArcInversionEffect$ and $ReattachingEffect$ in the worst case have linear complexity $O(|V|)$, but they are often performed in constant time because they consider only 2 vertices, their neighborhood, and paths in the root direction. Note, that if detachment of an arc does not decrease the length of a schedule, then the schedule length cannot be decreased by the entire local tree modification, and since only those cases when the tree is improved are taken into account inside the local search procedure, then $ReattachingEffect$ stops in line 6 (and, for the same reason, the $ArcInversionEffect$ stops in line 6).

Algorithm 6. $ArcInversion$ local search

1: *Input*: an initial tree $T = (A, V)$ rooted in v_0;
2: Calculate a schedule on T with only the primary conflicts;
3: $improved \leftarrow true$;
4: **while** $improved$ **do**
5:     $improved \leftarrow false$;
6:     **for all** $v \in V \setminus \{\{v_0\} \cup \{$children of $v_0\}\}$ **do**
7:         Set $p^* \leftarrow Parent(v)$, $bestEffect \leftarrow 0$;
8:         **for all** $u \in N(v)$ where u is not a descendant of v in T **do**
9:             Set $effect \leftarrow CalcArcInversionEffect(T, v, u)$;
10:            **if** $effect < bestEffect$ **then**
11:                Set $p^* \leftarrow u$, $bestEffect \leftarrow effect$;
12:            **end if**
13:        **end for**
14:        **if** $bestEffect < 0$ **then**
15:            Inverse the arc (v, p), remove the arc $(p, Parent(p))$ and add the arc (v, p^*) to the tree T;
16:            Calculate a schedule on T with only the primary conflicts;
17:            Set $improved \leftarrow true$;
18:        **end if**
19:    **end for**
20: **end while**

4.7 Join

At the join procedure $PopSize$ solutions from the current population and the current offspring, which have the largest fitness values, are chosen to fill the population of the next generation.

Algorithm 7. *ArcInversionEffect*

1: *Input*: an initial tree $T = (A, V)$ rooted in v_0 with time slot assigned to each vertex; a starting vertex v of an arc (v, p) that is considered for inversion; a vertex u that is considered to be a new parent of v in T.

2: *Output*: $L(T) - L(T_1)$ — the difference of schedule length after arc inversion and adding the arc (v, u).

3: Find the first common predecessor s between v and u and two paths: $path_1$ that starts at v and ends at q and $path_2$ that starts at u and ends at r (q and r are both children of s);

4: Find the new minimum sending time of q after removing the arc (v, p) from T.

5: **if** sending time of q was not decreased **then**

6:    return 0.

7: **end if**

8: Find the new minimum sending time of p after exclusion of v from its list of children.

9: Find the new minimum sending time of v after adding of p to its list of children.

10: Find the new minimum sending time of u after adding of v to its list of children.

11: Given new minimum sending time of u find the new minimum sending time of r.

12: Given new minimum sending times of q and r find the new minimum sending time of s.

13: Given new minimum sending time of s find find and **return** the difference of a schedule length traversing the vertices from s to v_0.

---

5   Simulation

In order to verify the efficiency of the proposed algorithm, we have implemented it in C++ programming language. We have also implemented some of the best state-of-the-art methods MLST and RH to generate an aggregation tree and NDR to construct a schedule on the aggregation tree.

The ILP formulation from [13] was used to obtain optimal solutions in cases of small dimension using the CPLEX package.

Each instance has been generated in the following way. A set of n sensors have been randomly uniformly spread inside a planar square with a side of unit length. To simulate the communication network we generated a unit disk graph (UDG) based topology on the given set of nodes. We use the following definition of UDG taken from [25]. Given n points in the plane and some specified bound $d \in \mathbb{R}^+$, the *unit disk graph* is an undirected graph with n vertices corresponding to the given n points, where an edge connects two vertices if and only if the Euclidean distance between the two corresponding points does not exceed the *critical distance* d. This is one of the most commonly used models for communication networks, where each element has the same transmission range equal to the critical distance of the corresponding UDG.

We have tested the different pairs of (n, d). Obviously, in cases of small density (when d is too small for a given n) the UDG may be disconnected. Therefore, we always validate connectivity and regenerate a new network in the case of failure.

During the experiment, we tested the following heuristics for generating an aggregation tree: Dijkstra algorithm (Dij), since the shortest-path trees often appear to be rather efficient in low arc density cases; RH, that appears to be very efficient for the simplified problem with only primary conflicts, and MLST. Then, for each generated tree NDR was applied to get a near-optimal schedule. Our novel GA-based approach was implemented in two variants: GA_AI, where *ArcInversion* local search is used, and GA_BR that includes *BranchReattaching* local search. Overall, the following heuristic algorithms were compared:

- Generation of aggregation tree and scheduling:
  - Dij_NDR;
  - RH_NDR;
  - MLST_NDR;
- Genetic local search variants:
  - GA_AI;
  - GA_BR;

For the GA-based algorithms the following parameters allow us to get the best results: $PopSize = 50$, $OffspSize = 20$, $FPItCount = 150$, $SPProportion = 0.6$, $PM = 0.5$, $PLS = 0.5$, $K_{max} = \lfloor n/3 \rfloor$. As a stopping criterion we used the following rule: the maximum fitness value among all solutions does not change during last 5 iterations.

In Table 1 the results for small size samples are presented. In these cases, the optimal solution was obtained by CPLEX. The last column (called *OptCount*) shows the number of cases when the solution is optimal. Note that GA-based algorithms in 85% of cases yield the optimal solution, whereas the best of other algorithms—only in 35% of cases.

Table 1. Schedule lengths obtained by different algorithms in small dimension cases.

n	15	15	15	15	15	15	15	15	15	15	20	20	20	20	20	20	20	20	20	20	*OptCount*
d	.4	.4	.4	.4	.4	.4	.4	.4	.4	.4	.3	.3	.3	.3	.3	.3	.3	.3	.3	.3	
CPLEX	6	6	6	6	6	5	6	6	6	7	6	6	6	7	7	7	6	8	7	6	—
Dij_NDR	7	7	7	7	7	8	6	6	9	9	7	7	6	7	8	8	8	8	8	7	4
RH_NDR	7	8	8	6	8	6	6	6	6	8	7	7	6	7	7	8	7	8	7	7	7
MLST_NDR	7	7	8	7	7	7	6	6	8	8	9	7	6	7	9	9	7	8	8	7	6
GA_AI	6	6	6	6	6	6	6	6	6	7	6	7	6	7	7	8	6	8	7	7	17
GA_RB	6	6	7	6	6	6	6	6	6	7	6	6	6	7	7	8	6	8	7	6	17

For a larger problem size, we have tested 20 instances for the same pair (n, d). The average values of schedule length are shown in Fig. 1. Vertical segments correspond to the standard deviation for the confidence interval of 90%. Again, GA-based heuristics construct better solutions than other methods. In all cases, the graphics of all GA-based heuristics are positioned very close to each other,

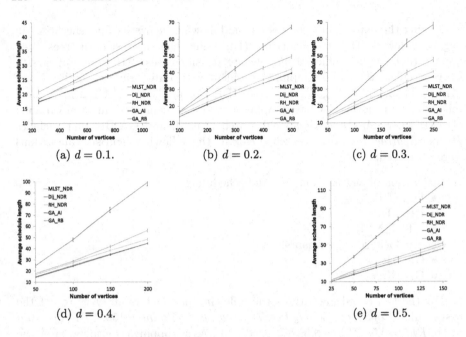

Fig. 1. Average schedule lengths.

and that means that these methods almost always construct a schedule of the same length. Note that, on average, RH_NDR appeared to be more efficient than MLST_NDR in all tested cases, and Dij_NDR appeared to be more efficient than MLST_NDR in the case when $d = 0.1$ (see Fig. 1a). The quality of a solution constructed by Dij_NDR becomes significantly worse with an increase of d. This is a predictable effect caused by the fact that the shortest-path tree becomes inefficient when the communication graph arc density is high (for example, in the case of the complete graph the shortest-path spanning tree is a star). The difference between schedule lengths of any of GA-based methods and MLST_NDR grows with the increase of n. In the case $d = 0.1$ it is about 3.5 time slots when $n = 250$ and grows to 11 for $n = 1000$. When $d = 0.2$, again, this value grows from 3.5 ($n = 100$) to 11 ($n = 500$), it grows from 2 ($n = 50$) to 10 ($n = 250$) in the case $d = 0.3$, from 3 ($n = 50$) to 10 ($n = 200$) in the case $d = 0.4$, and finally from 1.5 ($n = 25$) to 6 ($n = 150$) in the case $d = 0.5$. Herein, RH_NDR always yields a solution with a schedule length 2–4 time slots greater than both GA-based heuristics.

The average running times of different algorithms in the same test cases are presented in Table 2. Of course, GA-based heuristics require more computational time. In the case of a large dimension GA_AI becomes more time consuming than GA_BR when $d = 0.1$ and it runs faster than GA_BR when $d \geq 0.2$. If 50–100 s is not an acceptable running time, then one can choose RH_NDR, which spends less time than any of GA-based methods and outperforms MLST_NDR.

Table 2. Average running time of different algorithms in seconds.

n	d	Dij_NDR	MLST_NDR	RH_NDR	GA_AI	GA_RB
250	0.1	0.001	0.004	0.11	3.57	3.17
500	0.1	0.003	0.04	0.48	18.5	14.4
750	0.1	0.007	0.16	1.81	52.8	39.7
1000	0.1	0.01	0.42	4.65	114	96.5
100	0.2	0.0005	0.002	0.05	1.97	1.63
200	0.2	0.001	0.02	0.16	8.12	6.8
300	0.2	0.002	0.07	0.36	19.5	21.9
400	0.2	0.003	0.17	0.67	42.3	51.8
500	0.2	0.005	0.39	1	74.8	99.5
50	0.3	0.0001	0.0008	0.03	0.9	0.83
100	0.3	0.0004	0.005	0.08	3.66	3.79
150	0.3	0.001	0.02	0.177	9.07	11.7
200	0.3	0.002	0.06	0.29	16.4	30.2
250	0.3	0.002	0.1	0.42	33	59
50	0.4	0.0005	0.0006	0.03	1.11	1.23
100	0.4	5E-05	0.01	0.09	5.98	7.51
150	0.4	0.001	0.04	0.22	17.2	26.9
200	0.4	0.002	0.09	0.35	33.6	66.6
25	0.5	0.0002	0.0004	0.02	0.43	0.42
50	0.5	0.0003	0.002	0.04	1.68	2.03
75	0.5	0.0007	0.007	0.08	4.76	6.6
100	0.5	0.001	0.02	0.13	11.9	17.4
125	0.5	0.001	0.04	0.19	20.6	33.6
150	0.5	0.002	0.07	0.31	33.2	69.6

6 Conclusion

In this paper, we considered an NP-hard problem of conflict-free minimum length aggregation scheduling in a wireless sensor network. We proposed a new heuristic algorithm based on the Genetic Algorithm metaheuristic that uses two different local search procedures. Extensive simulation has shown that our algorithms outperform the best of the known approaches.

Acknowledgments. The research is partly supported by the Russian Science Foundation (project 18-71-00084) (Sect. 4–5), by the Russian Foundation for Basic Research (project 16-07-00552) (Sect. 2–3), and by the program of fundamental scientific researches of the SB RAS No. I.5.1 (project 0314-2016-0014) (Sect. 1).

References

1. Bagaa, M., Challal, Y., Ksentini, A., et al.: Data aggregation scheduling algorithms in wireless sensor networks: solutions and challenges. IEEE Commun. Surv. Tutorials **16**, 1339–1367 (2014)
2. Malhotra, B., Nikolaidis, I., Nascimento, M.A.: Aggregation convergecast scheduling in wireless sensor networks. Wireless Netw. **17**, 319–335 (2011)
3. Hromkovic, J., Klasing, R., Monien, B., Peine, R.: Dissemination of information in interconnection networks (broadcasting & gossiping). Combinatorial Network Theory. Applied Optimization, vol. 1, pp. 125–212. Springer, Boston (1996). https://doi.org/10.1007/978-1-4757-2491-2_5
4. Gupta, P., Kumar, P.R.: The capacity of wireless networks. IEEE Trans. Inf. Theory **46**, 388–404 (2000)
5. Incel, O.D., Ghosh, A., Krishnamachari, B., et al.: Fast data collection in tree-based wireless sensor networks. IEEE Trans. Mob. Comput. **11**, 86–99 (2011)
6. Xu, X., Li, X.Y., Mao, X., et al.: A delay-efficient algorithm for data aggregation in multihop wireless sensor networks. IEEE Trans. Parallel Distrib. Syst. **22**, 163–175 (2011)
7. Pan, C., Zhang, H.: A time efficient aggregation convergecast scheduling algorithm for wireless sensor networks. Wireless Netw. **22**, 2469–2483 (2016). https://doi.org/10.1007/s11276-016-1337-5
8. Chen, X., Hu, X., Zhu, J.: Minimum data aggregation time problem in wireless sensor networks. In: Jia, X., Wu, J., He, Y. (eds.) MSN 2005. LNCS, vol. 3794, pp. 133–142. Springer, Heidelberg (2005). https://doi.org/10.1007/11599463_14
9. Erzin, A., Pyatkin, A.: Convergecast scheduling problem in case of given aggregation tree: the complexity status and some special cases. In: 10th International Symposium on Communication Systems Networks and Digital Signal Processing (CSNDSP), pp. 1–6. IEEE (2016) https://doi.org/10.1109/CSNDSP.2016.7574007
10. Zhu, J., Hu, X.: Improved algorithm for minimum data aggregation time problem in wireless sensor networks. J. Syst. Sci. Complex. **21**, 626–636 (2008)
11. Huang, S.C.-H., Wan, P.J., Vu, C.T., et al.: Nearly constant approximation for data aggregation scheduling in wireless sensor networks. In: IEEE Conf. on Computer Communications (INFOCOM 2007), pp. 366–472 (2007)
12. Nguyen, T.D., Zalyubovskiy, V., Choo, H.: Efficient time latency of data aggregation based on neighboring dominators in WSNs. In: IEEE Globecom, 6133827 (2011)
13. Tian, C., Jiang, H., Wang, C., et al.: Neither shortest path nor dominating set: aggregation scheduling by greedy growing tree in multihop wireless sensor networks. IEEE Trans. Veh. Technol. **60**, 3462–3472 (2011)
14. de Souza, E., Nikolaidis, I.: An exploration of aggregation convergecast scheduling. Ad Hoc Netw. **11**, 2391–2407 (2013)
15. Edmonds, J.: Optimum branchings. J. Res. Natl. Bur. Stan. B **71**, 233–240 (1967)
16. Sivanandam, S., Deepa, S.: Introduction to Genetic Algorithms. Springer, Heidelberg (2008). https://doi.org/10.1007/978-3-540-73190-0
17. Gruber, M., van Hemert, J., Raidl, G.R.: Neighbourhood searches for the bounded diameter minimum spanning tree problem embedded in a VNS, EA, and ACO. In: Proceedings of the 8th Annual Conference on Genetic and Evolutionary Computation, July 08–12, Seattle, Washington, USA (2006). https://doi.org/10.1145/1143997.1144185

18. Hussain, S., Matin, A.W., Islam, O.: Genetic algorithm for hierarchical wireless sensor networks. J. Netw. **2**, 87–97 (2007)
19. Sudha, N., Valarmathi, M.L., Neyandar, T.C.: Optimizing energy in WSN using evolutionary algorithm. In: Proceeding of IJCA on International Conference on VLSI, Communications and Instrumentation (ICVCI 2011), Kerala, India, 7–9 April 2011, pp. 26–29 (2011)
20. Liu, J., Ravishankar, C.V.: LEACH-GA: genetic algorithm-based energy-efficient adaptive clustering protocol for wireless sensor networks. Int. J. Mach. Learn. Comput. **1**, 79–85 (2011)
21. Yen, Y.S., Chao, H.C., Chang, R.S., Vasilakos, A.: Flooding-limited and multi-constrained QoS multicast routing based on the genetic algorithm for MANETs. Math. Comput. Model. **53**, 2238–2250 (2011)
22. Zhu, N., O'Connor, I.: iMASKO: a genetic algorithm based optimization framework for wireless sensor networks. J. Sens. Actuator Netw. **2**, 675–699 (2013). https://doi.org/10.3390/jsan2040675
23. Plotnikov, R., Erzin, A., Zalyubovsky, V.: Convergecast with unbounded number of channels. In: MATEC Web of Conferences 125, 03001 (2017). https://doi.org/10.1051/matecconf/201712503001
24. Beier, R., Sibeyn, J.F.: A powerful heuristic for telephone gossiping. In: 17th International Colloquium on Structural Information & Communication Complexity (SIROCCO 2000), pp. 17–36 (2000)
25. Clark, B.N., Colbourn, C.J., Johnson, D.S.: Unit disk graphs. Discrete Math. **86**(1–3), 165–177 (1990)

Optimal Control

Numerical Implementation of the Contact of Optimal Trajectory with Singular Regime in the Optimal Control Problem with Quadratic Criteria and Scalar Control

Alexander P. Afanas'ev[1,2,3]([✉]), Sergei M. Dzyuba[4], Irina I. Emelyanova[1,2,3,4], and Elena V. Putilina[1]

[1] Institute for Information Transmission Problems RAS,
19, build. 1, Grand Carriage line, Moscow 127994, Russia
`apa@iitp.ru`
[2] National Research University Higher School of Economics,
20, Myasnitskaya str., Moscow 101000, Russia
[3] Lomonosov Moscow State University,
GSP-1, Vorobievy Gory, Moscow 119991, Russia
[4] Tver State Technical University, 22, Af. Nikitin emb., Tver 170026, Russia

Abstract. Previous works by these authors offer the numerical method of successive approximations for developing the solutions of the problem of stabilization of nonlinear systems with standard functional. This paper considers applying this method for studying the problem with singular control. It is achieved by introducing an auxiliary problem. The solution for the auxiliary problem provides a smooth approximation to the solution of the initial problem. The paper presents the algorithms for constructing an approximate solution for the initial problem. It is demonstrated that unlike direct algorithms of optimal control, these algorithms allow registering the saturation point, thus enabling one to register and study singular regimes.

Keywords: Method of successive approximations
Problem of minimizing a quadratic functional for a class of nonlinear systems with scalar control · Singular control

1 Introduction

Problems of optimal control with a quadratic criterion are of great importance for automatic control theory, mathematical economics, and many other branches of science.

The statement of the present problem is as follows.

Let us consider a nonlinear dynamical system which is characterized by the differential equation

$$\dot{x} = g(x, u), \tag{1}$$

© Springer Nature Switzerland AG 2019
Y. Evtushenko et al. (Eds.): OPTIMA 2018, CCIS 974, pp. 235–246, 2019.
https://doi.org/10.1007/978-3-030-10934-9_17

where $x = (x^1, \ldots, x^n)$ is an n-dimensional state variable vector, $u = (u^1, \ldots, u^m)$ is an m-dimensional control vector, and $g = (g^1, \ldots, g^n)$ is a vector function, defined and continuous together with its partial derivatives

$$\frac{\partial g^i}{\partial x^j}, \quad i, j = 1, \ldots, n,$$

and

$$\frac{\partial g^i}{\partial u^j}, \quad i = 1, \ldots, n, \quad j = 1, \ldots, m,$$

in Euclidean vector space $\mathbb{R}^{n+m}$.

Let us assume that the initial state

$$x(0) = c \tag{2}$$

is fixed and the objective of system control (1) is to minimize the functional

$$J(u) = \frac{1}{2} \int_0^T [\langle x(t), Qx(t) \rangle + \langle u(t), Ru(t) \rangle] \, dt + \frac{1}{2} \langle x(T), Px(T) \rangle, \tag{3}$$

in which T is a fixed finite time, Q and P are positive semi-definite $(n \times n)$ matrices and R is a positive definite $(m \times m)$ matrix.

The Pontryagin function for our problem is

$$H(x, \varphi, u) = -\frac{1}{2} [\langle x, Qx \rangle + \langle u, Ru \rangle] + \langle \varphi, g(x, u) \rangle.$$

If on interval $[t_1, t_2]$

$$\frac{\partial}{\partial u^j} H(x, \varphi, u) \equiv 0, \quad j = 1, \ldots, m,$$

the pair $(x(t), u(t))$ on $t \in [t_1, t_2]$ is called singular regime (e.g., see [1]). The landing point on singular regime is called saturation point.

Presence of the singular regime is a typical situation. Actually, if matrices Q and R are positive definite ones in the problem (1)–(3), then the problem reaches the absolute minimum at the solution

$$x(t) \equiv 0, \quad u(t) \equiv 0. \tag{4}$$

Why, if the time value T is great enough, generally speaking, it is worth keeping to the regime (4) for a while. This is a singular regime.

Initially, the problem (1)–(3) was considered in the case of a linear system (1), that is, to a system of the form:

$$\dot{x} = Ax + Bu, \tag{5}$$

where A and B are $(n \times n)$ and $(n \times m)$ matrices. In this case, the solution of the problem is well-known and provided by the linear feedback control law (e.g., see [1]).

In general, getting a solution (or solutions ratings) of the problem (1)–(3) requires various methods (e.g., see [2–14]). One of the principal methods herein is a method of successive approximations, which is described in [2]. Seemingly simple and intuitive, it (the method) reduces the initial problem to a certain sequence of linear-quadratic problems. However, this method is not widely used because of its excessive bulkiness.

Let us now note that the modification of the method [2], performed in the paper [11], allowed, in some cases, to establish the existence of a solution of the problem (1)–(3) and provided such a solution in the form of a nonlinear feedback control law when the system (1) has the form

$$\dot{x} = Ax + Bu + f(x, u). \tag{6}$$

Separating the linear part in the system (1) and putting the said system into the form (6) allowed constructing the method of successive approximations in the papers [11,14]. In many cases, this method helped to establish the existence of a solution for the problem (6), (2), (3) and to acquire the form of optimal control. Within the framework of this paper, we will apply a modification of the said method in order to construct an approximate solution for the following problem.

Let us consider the problem of minimization of the functional

$$J(u) = \frac{1}{2} \int_0^T \langle x(t), Qx(t) \rangle \, dt + \frac{1}{2} \langle x(T), Px(T) \rangle, \tag{7}$$

with the relations

$$\dot{x} = Ax + bu + f(x, u), \quad x(0) = c \tag{8}$$

and

$$|u(t)| \le 1, \tag{9}$$

where $x(t)$ is a real n-dimensional state vector and $u(t)$ is a scalar control function, A and B are real $(n \times 1)$ and $(n \times m)$ matrices, and $f = (f^1, \dots, f^n)$ is a vector function, defined and continuous together with its partial derivatives

$$\frac{\partial f^i}{\partial x^j}, \quad i, j = 1, \dots, n,$$

and

$$\frac{\partial f^i}{\partial u}, \quad i = 1, \dots, n,$$

in a Euclidean vector space $\mathbb{R}^{n+1}$; the other assumptions remain the same as in the problem (1)–(3).

For a long time, researchers thought that landing into the singular regime happens either with a finite number of switches or smoothly. However, in 1961 A.T. Fuller drew a simple example [15], when landing on the singular regime happens with a discontinuity of the second kind. The modern interpretation of that example looks as follows:

$$J(u) = \int_0^T x^2(t)\, dt \to \min, \quad \ddot{x} = u, \quad |u(t)| \leq 1.$$

The optimal control takes on the values -1 and $+1$ in turn; besides, the time interval between subsequent switches becomes shorter, the number of switches approaches infinity by the moment t_1, and at the moment t_1 the control takes on the value $u(t) = 0$ (see Fig. 1).

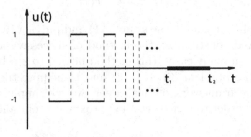

Fig. 1. Landing on singular control

Singular regimes and their coming in contact with a regular trajectory are studied in the works [16–19]. The phenomenon of a regular and a singular trajectories coming in contact is studied most profoundly in the work [20]. The paper considers the system (1) and the problem comprises minimization of the functional

$$J(u) = \int_0^T [\langle x(t), Cx(t) \rangle + \langle x(t), Du(t) \rangle]\, dt$$

where $n = m$, U is a convex solid compact set and $0 \in \mathrm{int}\, U$. For the case when $U = \{u : \langle u, u \rangle \leq 1\}$ it is demonstrated that at the moment t_1 the regular trajectory comes in contact with the singular one in such a manner that at the moment t_1 there appears a discontinuity of the second kind. At the same time, within $[0, t_1)$ the regular trajectory is a smooth function. The stated fact, that the presence of a discontinuity of the second kind in the case of a contact with the singular regime is a typical behavior, comprises an important result of the current paper.

The even more complex behavior of the regular part of the optimal trajectory occurring before the contact with the singular regime is discovered in the work [21]. A linear-quadratic problem is studied in the following form

$$J(u) = \frac{1}{2} \int_0^{+\infty} \langle x(t), x(t) \rangle\, dt \to \min, \quad \ddot{x} = u, \quad u(t) \in \Omega,$$

where $n = m = 2$, Ω is a triangle and $0 \in \text{int } \Omega$. It is determined that before the contact with the singular regime occurs, the vertices of the triangle are circumvented an infinite number of times and in a chaotic way.

Obviously, it is impossible to conduct numerical modeling of a discontinuity if the second kind. However, it is crucial to develop an approximate algorithm in such a way that it would "sense" the presence of the point where the regular solution and the singular regime come in contact and, wherever possible, would identify this point with certain precision. It is difficult to apply direct methods based on difference approximation. These methods miss the point of contact with the singular regime (e.g., see [22]). Applying these methods, one must know in advance of this point and the singular regime being present. Besides, in order to catch this situation, one should conduct several runs with different approximation steps.

The approach suggested in the present paper provides a partial workaround for these issues and helps to model the landing into the singular regime with controlled precision.

2   Method and Main Theorem

We will perform separating of the linear part (in order to acquire the system (8)) in such a way that the linear system

$$\dot{x} = Ax + bu, \tag{10}$$

would be Kalman-controlled, that is, so that the rank of the matrix

$$G = [b \ Ab \ A^2b \ \dots \ A^{n-1}b]$$

was equals to n. Further on, such separating will play a great role in constructing an approximate solution of the problem (7)–(9).

In order to construct an approximate solution for the problem (7)–(9), let us introduce an auxiliary problem and study the method of constructing a solution for it. Namely, let us consider the problem of minimization of the functional

$$J_\varepsilon(u) = \frac{1}{2} \int\limits_0^T [\langle x(t), Qx(t)\rangle + \varepsilon u^2(t)] \, dt + \frac{1}{2}\langle x(T), Px(T)\rangle, \tag{11}$$

with the relations (8) and (9), where ε is a small positive parameter. To construct a solution of this problem, we shall apply the method of successive approximations described in the paper [11]. To achieve this, we shall study a succession of problems in reducing the functional

$$J_{\varepsilon,N+1}(u) =$$

$$= \frac{1}{2} \int\limits_0^T [\langle x_{N+1}(t), Qx_{N+1}(t)\rangle + \varepsilon u_{N+1}^2(t)] \, dt + \frac{1}{2}\langle x_{N+1}(T), Px_{N+1}(T)\rangle \tag{12}$$

to a minimum under the relations

$$\dot{x}_{N+1} = Ax_{N+1} + bu_{N+1} + f(x_N, u_N), \quad x_{N+1}(0) = c \tag{13}$$

and

$$|u_{N+1}(t)| \leq 1. \tag{14}$$

Then it is easy to demonstrate that, if x_N and u_N are fixed functions, the solution of the problems (12)–(14) is provided by the feedback control law

$$u_{N+1}(t) =$$

$$= \begin{cases} -1, & \varepsilon^{-1}b'[h_{N+1}(t) - K(t)x_{N+1}(t)] < -1, \\ \varepsilon^{-1}b'[h_{N+1}(t) - K(t)x_{N+1}(t)], & |\varepsilon^{-1}b'[h_{N+1}(t) - K(t)x_{N+1}(t)]| \leq 1, \\ 1, & \varepsilon^{-1}b'[h_{N+1}(t) - K(t)x_{N+1}(t)] > 1, \end{cases} \tag{15}$$

where $K(t)$ is a solution of Riccati matrix equation

$$\dot{K}(t) = -K(t)A - A'K(t) + \varepsilon^{-1}K(t)bb'K(t) - Q \tag{16}$$

with boundary condition

$$K(T) = P, \tag{17}$$

while $h_{N+1}(t)$ is a solution of the linear equation

$$\dot{h}_{N+1}(t) = -[A - \varepsilon^{-1}bb'K(t)]'h_{N+1}(t) + K(t)f(x_N(t), u_N(t)) \tag{18}$$

with boundary condition

$$h_{N+1}(T) = 0. \tag{19}$$

We also note that, for simplicity, the initial approximation is to be determined by the relations

$$x_0(t) \equiv c \tag{20}$$

and

$$u_0(t) \equiv -\varepsilon^{-1}b'K(t)c. \tag{21}$$

Remark 1. The autonomy of the system (8) and constancy of matrix Q in this paper are not used anywhere else and are taken to simplify the notation.

As demonstrated by many computational experiments, if the system (10) is controlled according to Kalman, while the function f is a Lipschitzian one, then, in most cases, the sequence $(x_N, h_N)_{N \in \mathbb{N}}$ is defined for all $N \in \mathbb{N}$ and meets the following conditions (see [14]):

(i) the sequence (x_N, h_N) is uniformly bounded;
(ii) the set of points $\mathfrak{T} \in [0, T]$, in which the sequence converges, contains interior points, i.e.

$$\mathfrak{T} \cap (0, T) \neq \varnothing.$$

From now on, we will consider the conditions (i) and (ii) fulfilled. Then the above-described method of successive approximations allows to establish the existence of a solution of the problem (11), (8), (9) and to acquire a solution for it in the form of a nonlinear feedback control law. The foregoing is established by the following

Theorem 1. *Let ε be an arbitrary positive number and the set $\mathfrak{T}$ dense in the interval $[0, T]$. Then the optimal control $u_\varepsilon(t)$ of the problem (11), (8), (9) does exist. Moreover, for all $t \in [0, T]$ we have*

$$u_\varepsilon(t) =$$

$$= \begin{cases} -1, & \varepsilon^{-1}b'[h_\varepsilon(t) - K(t)x_\varepsilon(t)] < -1, \\ \varepsilon^{-1}b'[h_\varepsilon(t) - K(t)x_\varepsilon(t)], & |\varepsilon^{-1}b'[h_\varepsilon(t)^*(t) - K(t)x_\varepsilon(t)]| \leq 1, \\ 1, & \varepsilon^{-1}b'[h_\varepsilon(t) - K(t)x_\varepsilon(t)] > 1, \end{cases} \quad (22)$$

where $x_\varepsilon(t)$ is a solution of the differential equation

$$\dot{x}_\varepsilon = Ax_\varepsilon + bu_\varepsilon + f(x_\varepsilon, u_\varepsilon), \quad x_\varepsilon(0) = c \quad (23)$$

and $h_\varepsilon(t)$ is a solution of the differential equation

$$\dot{h}_\varepsilon(t) = -[A - \varepsilon^{-1}bb'K(t)]'h_\varepsilon(t) + K(t)f(x_\varepsilon(t), u_\varepsilon(t)), \quad h_\varepsilon(T) = 0. \quad (24)$$

Proof. For the sake of simplicity of the notation, let us assume that

$$\varphi_N(t) = (x_N(t), h_N(t)).$$

As the sequence $(\varphi_N)_{N \in \mathbb{N}}$ is uniformly bounded, then, by virtue of the conditions (13)–(21), the sequence $(\dot{\varphi}_N)_{N \in \mathbb{N}}$ is uniformly bounded. Therefore, the sequence $(\varphi_N)_{N \in \mathbb{N}}$ is not only uniformly bounded but also equicontinuous.

Let us now note that the set $\mathfrak{T}$ is dense in the interval $[0, T]$. So, according to the second Ascoli theorem (e.g., see [23]), by virtue of uniform boundedness and equicontinuity of the sequence $(\varphi_N)_{N \in \mathbb{N}}$, there exists the function

$$\varphi^*(t) = (x_\varepsilon(t), h_\varepsilon(t))$$

which is bounded and continuous in the interval $[0, T]$ and meets the following conditions

$$\lim_{N \to +\infty} \max_{t \in [0,T]} \|x_N(t) - x_\varepsilon(t)\| = 0 \quad (25)$$

and

$$\lim_{N \to +\infty} \max_{t \in [0,T]} \|h_N(t) - h_\varepsilon(t)\| = 0. \quad (26)$$

By virtue of the condition (15), there exists function u_ε is defined and continuous in the interval $[0, T]$ and meets the condition

$$\lim_{N \to +\infty} \max_{t \in [0,T]} |u_N(t) - u_\varepsilon(t)| = 0. \quad (27)$$

Taking into account the Eqs. (25)–(27), we can see that the functions x_ε and h_ε comply with the Eqs. (23) and (24) accordingly. Moreover, the function u_ε meets the condition (22).

Thus the Theorem 1 is proved.

Theorem 1 has two standalone significations. First, it allows, for all the values of $\varepsilon > 0$, to speak about the existence and structure of a solution of the problems (11), (8), (9). Secondly, according to Theorem 1, the reducibility of the sequence (x_N, h_N) provides a method (13)–(21) for developing the solution of this problem.

We denote by u^* the optimal control in problem (7)–(9). One more crucial signification of Theorem 1 is the fact that when $\varepsilon \to 0$, the solutions of the problem (11), (8), (9) can be transformed into the solutions of the problem (7)–(9). This issue requires additional investigation. Here we just note that, if the following relation

$$\lim_{\varepsilon \to 0} J_\varepsilon(u_\varepsilon) = J(u^*) \tag{28}$$

is true, then, for each sufficiently small value of $\varepsilon > 0$, the solution of the problem (11), (8), (9) may be considered as an approximate solution of the problem (7)–(9). This approximate solution may be acquired through the method of successive approximations (13)–(21).

3   Computational Aspects

The said specific feature of the procedure for the transition from relay control to singular control makes the procedure of constructing the optimal control by numerical methods complicated, as the singular regime becomes difficult to achieve (e.g., see [22]).

Before we start speaking of developing a solution for the problem (7)–(9), let us discuss the issue of approximate development of a solution for the problem (11), (8), (9).

Theorem 1 provides a simple enough algorithm of developing an approximate solution for the problem (11), (8), (9) (algorithm $\mathcal{A}$).

Algorithm $\mathcal{A}$

1. We develop a solution of Riccati Eq. (16). This procedure is performed once. As Riccati equation is an equation with a polynomial right-hand side, a solution for it may be developed as a multidimensional polynomial of the matrix P and time t (see [24]). Further on the value of this polynomial can be calculated efficiently with the help of Horner scheme (see [25]).
2. At each step, the method of successive approximations of developing the functions x_{N+1}, h_{N+1} and u_{N+1} comes down to solving a standard linear two-point boundary value problem. A solution of this problem can be obtained, for instance, using the sweep method. A sign of the end of the computational process can be the fulfillment of the conditions

$$\max_{i=1,\dots,l} \|x_N(\tau_i) - x_{N+1}(\tau_i)\| < \mu,$$

$$\max_{i=1,\dots,l} \|h_N(\tau_i) - h_{N+1}(\tau_i)\| < \mu$$

and
$$\max_{i=1,\ldots,l} |u_N(\tau_i) - u_{N+1}(\tau_i)| < \mu,$$

where $\tau_1, \ldots, \tau_l$ is a computational grid and μ is a computational accuracy value.

According to Theorem 1, the approximate optimal control for all $\varepsilon > 0$ is constructed as a continuous function with some intervals, in which the controlling action remains constant, i.e., saturation occurs. In this case, the saturation points are determined by the relation (22)–(24). Moreover, according to (22)–(24), the occurrence of singular optimal control areas can be identified and registered.

Obviously, the decrease of the ε value in the functional (11) makes the optimal control, though continuous, still ever-closer to a relay control, with registered singular areas. In other words, at small enough values of ε, the optimal control, set by the relations (22)–(24), can be considered as an approximate landing procedure from regular control to a singular. Thus, the method (shown in Sect. 2) of constructing an approximate solution of the problem (7)–(9) as a continuous function of the optimal control gives a qualitative characteristic of the landing of the relay control on a singular control. Unlike standard numerical methods of optimal control, the set method allows not only defining the control saturation points, but also checking the moments when the relay control landing on the singular regime (see Fig. 2).

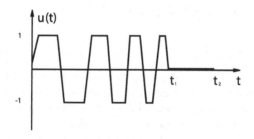

Fig. 2. Approximate landing on singular control

The algorithm of developing an approximate solution for the problem (7)–(9) (algorithm $\mathcal{B}$) implies the condition (28) fulfilled. Let us study this algorithm.

Algorithm $\mathcal{B}$

1. Using some number generator, we create a sequence $(\varepsilon_i)_{i\in\mathbb{N}}$, which satisfies the condition

$$\varepsilon_1 > \varepsilon_2 > \ldots > \varepsilon_i > \ldots,$$

and memorize this sequence.

2. We assume

$$\varepsilon = \varepsilon_j$$

and for $j = 1, 2, \ldots$ we develop an approximate solution for the problem (11), (8), (9), using algorithm $\mathcal{A}$. If singular control appears, we register the saturation point. The completion of the computation process may be marked by the following inequality being true:

$$|J_{\varepsilon_j}(u_{\varepsilon_j}) - J_{\varepsilon_{j+1}}(u_{\varepsilon_{j+1}})| < \nu, \tag{29}$$

where ν is a computational accuracy (see relation (28)) and u_{ε_j} is an approximate optimal control in problem (11), (8), (9), obtained by means of algorithm $\mathcal{A}$.

3. If the relation (29) cannot be obtained for a reasonable number of steps in the algorithm $\mathcal{B}$, we generate the sequence $(\varepsilon_i)_{i\in\mathbb{N}}$ a new and go back to step 2 of algorithm $\mathcal{B}$.

To sum up all the above, let us note that the efficiency of the algorithm $\mathcal{B}$ is determined by the relation (28) being fulfilled and depends significantly on the method, we use to generate the sequence $(\varepsilon_i)_{i\in\mathbb{N}}$.

4   Conclusion

The key method applied in the present paper is the method of successive approximations described in Sect. 2. This method developed from the results of the classical monograph [2] and the previous works by the authors [11,14].

Convergence of this method under the conditions of Theorem 1 allows establishing the existence of a solution for the auxiliary problem (10), (7), (8) in the form of a nonlinear feedback control law. As referred to in Sect. 2, the solution of the auxiliary problem provides a method for developing an approximate solution for the main problem (7)–(9). The given proof of Theorem 1 in simple enough, but it is based on the little-known second Ascoli theorem.

The numerical implementation of the above-stated methods is considered in Sect. 3 (algorithms $\mathcal{A}$ and $\mathcal{B}$). These algorithms have a distinctive feature: unlike standard algorithms of optimal control, they allow to register the singular control and the saturation point, that is, the point of contact between the regular control and the singular one. This allows to evaluate very accurately and to research the character of the optimal control in the main problem (7)–(9).

Acknowledgments. This research was supported by the Russian Science Foundation, grant no. 16-11-10352.

References

1. Athans, M., Falb, P.: Optimal Control. McGraw-Hill, New York (1966)
2. Bellman, R.: Adaptive Control Processes. Princeton University Press, Princeton (1961)
3. Lukes, D.L.: Optimal regulation of nonlinear systems. SIAM J. Control Optim. **7**, 75–100 (1969)
4. Yamamoto, Y.: Optimal control of nonlinear systems with quadratic performance. J. Math. Anal. Appl. **64**, 348–353 (1978)
5. Dacka, C.: On the controllability of a class of nonlinear systems. IEEE Trans. Automat. Control. **25**, 263–266 (1980)
6. Balachandran, K., Somasundaram, D.: Existence of optimal control for nonlinear systems with quadratic performance. J. Austral. Math. Soc. Ser. B. **29**, 249–255 (1987)
7. Afanas'ev, A.P., Dzyuba, S.M., Lobanov, S.M., Tyutyunnik, A.V.: Successive approximation and suboptimal control of systems with separated linear part. Appl. Comput. Math. **2**(1), 48–56 (2003)
8. Afanas'ev, A.P., Dzyuba, S.M., Lobanov, S.M., Tyutyunnik, A.V.: On a suboptimal control of nonlinear systems via quadratic criteria. Appl. Comput. Math. **3**(2), 158–169 (2004)
9. Tang, G.-Y., Gao, D.-X.: Approximation design of optimal controllers for nonlinear systems with sinusoidal disturbances. Nonlinear Anal. Theory Methods Appl. **66**(2), 403–414 (2007)
10. Zhang, S.-M., Tang, G.-Y., Sun, H.-Y.: Approximate optimal rejection to sinusoidal disturbance for nonlinear systems. In: IEEE International Conference on Systems, Man and Cybernetics, pp. 2816–2821 (2008)
11. Afanas'ev, A.P., Dzyuba, S.M.: Optimal control of nonlinear systems via quadratic criteria. Trans. ISA RAS **32**, 49–62 (2008). (in Russian)
12. Gao, De-Xin: Disturbance attenuation and rejection for systems with nonlinearity via successive approximation approach. In: IEEE Proceedings of the 30th Chinese Control Conference, pp. 250–255 (2011)
13. Ma, S.Y.: A successive approximation approach of nonlinear optimal control with R-rank persistent disturbances. Appl. Mech. Mater. **130**, 1862–1866 (2012)
14. Afanas'ev, A.P., Dzyuba, S.M., Emelyanova, I.I., Pchelintsev, A.N., Putilina, E.V.: Optimal control of nonlinear systems with separated linear part via quadratic criteria. Optim. Lett. (2018). https://doi.org/10.1007/s11590-018-1309-z
15. Fuller, A. T.: Optimization of relay control systems with respect to different qualitative criteria. In: Proceedings of the First IFAC Conference, AN SSSR, pp. 584–605 (1961)
16. Kelley, H.J., Kopp, R.E., Moyer, H.G.: Singular extremals. In: Leitmann, G. (ed.) Topics of Optimization, pp. 63–101. Academic Press, New York and London (1967)
17. Neustadt, L.W.: Optimization, A Theory of Necessary Conditions. Princeton University Press, Princeton (1976)
18. Krener, A.: The high order maximal principle and its application to singular extremals. Siam J. Control Optim. **15**(2), 256–293 (1977)
19. Bershchanskii, Y.M.: Fusing of singular and nonsingular parts of optimal control. Autom. Remote Control **40**(3), 325–330 (1979)
20. Chukanov, S.V., Milutin, A.A.: Qualitative study of singularities for extremals of quadratic optimal control problem. Russ. J. Math. Phys. **2**(1), 31–48 (1994)

21. Zelikin, M.I., Lokutsievskii, L.V., Hildebrand, R.: Typicality of chaotic fractal behavior of integral vortices in Hamiltonian systems with discontinuous right hand side. In: Optimal Control, Contemporary Mathematics. Fundamental directions(SMFN), the Peoples' Friendship University of Russia (RUDN), Moscow, vol. 56, pp. 5–128 (2015)
22. Fedorenko, R.P.: Approximate Solution of Optimal Control Problems. Nauka, Moscow (1978). (in Russian)
23. Schwartz, L.: Analyse Mathématique, vol. II. Hermann, Paris (1967). (in French)
24. Afanas'ev, A.P., Dzyuba, S.M.: Method for constructing approximate analytic solutions of differential equations with a polynomial right-hand side. Comput. Math. Math. Phys. **55**(10), 1665–1673 (2015)
25. Afanas'ev, A.P., Dzyuba S.M., Emelyanova, I.I., Putilina, E.V.: Decomposition algorithm of generalized horner scheme for multidimensional polynomial evaluation. In: Proceedings of the OPTIMA-2017 Conference, Petrovac, Montenegro, 02 October 2017, pp. 7–11 (2017)

On the Stability of the Algorithm of Identification of the Thermal Conductivity Coefficient

Alla Albu[1] and Vladimir Zubov[1,2]($\boxtimes$)

[1] Dorodnicyn Computing Centre, Federal Research Center
"Computer Science and Control" of Russian Academy of Sciences, Moscow, Russia
alla.albu@yandex.ru, vladimir.zubov@mail.ru
[2] Moscow Institute of Physics and Technology, Moscow, Russia

Abstract. The paper is devoted to the inverse problem of determining the thermal conductivity coefficient of substance depending on the temperature. The consideration is based on the first boundary value problem for the non-stationary heat equation. The mean-root-square deviation of the temperature distribution field and the heat flux on the boundary of the domain from the experimental data is used as the cost functional. The algorithm for the numerical solution of the problem based on the modern approach of Fast Automatic Differentiation was proposed by the authors in previous works. In the present paper, a numerical stability analysis of the obtained solutions is carried out. It is shown that the perturbation of the restored thermal conductivity coefficient is of the same order as the perturbation of the experimental data that caused it. Many illustrative examples are presented.

Keywords: Inverse coefficient problems · Heat equation
Numerical algorithm

1 Introduction

The growing need for the creation of new materials in recent years has led to an increase in interest in inverse problems. This is due to the fact that some of the characteristics of these materials turn out to be unknown in advance and they can be determined by solving inverse problems. One of the important inverse problems is the problem of determining the dependence of the thermal conductivity coefficient of the substance on the temperature using the results of experimental observations of the dynamics of the temperature field in the substance under study (see, for example, [1]). In [2] the genetic algorithm to identify the temperature dependent thermal conductivity of a solid material is presented. In [3] a modified Levenberg-Marquardt algorithm for simultaneous estimation

This work was partially supported by the Russian Foundation for Basic Research (project no. 17-07-00493 a).

© Springer Nature Switzerland AG 2019
Y. Evtushenko et al. (Eds.): OPTIMA 2018, CCIS 974, pp. 247–263, 2019.
https://doi.org/10.1007/978-3-030-10934-9_18

of multi-parameters of boundary heat flux by solving transient nonlinear inverse heat conduction problems is considered.

In [4] one of the possible statements of the inverse coefficient problem is studied. The research is based on the first boundary value problem for the non-stationary heat equation. The inverse coefficient problem was reduced to the variational problem: it was required to find such dependence of the thermal conductivity coefficient on the temperature at which the temperature field and heat flux at the boundary of the object obtained as a result of solving the direct problem differed little from the data obtained experimentally. The mean-root-square deviation of the temperature field and the heat flux on the left boundary of the domain from the experimental data is used as the objective functional. An algorithm for the numerical solution of the inverse coefficient problem was proposed. It was based on the modern approach of Fast Automatic Differentiation technique (FAD-technique, see [5,6]), which allowed to successfully solve a number of complex optimal control problems for dynamic systems. In [4] the case of continuous thermal conductivity coefficient was considered (there are no jumps of the coefficient in the neighborhood of any point). The inverse problem in the case when the thermal conductivity coefficient had jumps of the first kind was examined in [7].

In [8] also was investigated the inverse problem, of the one that was examined in [4]. But, in contrast to [4], the consideration in [8] was carried out for the two-dimensional nonstationary heat equation. The mean-root-square deviation of the temperature field obtained as a result of solving the direct problem from the experimental data is used as the objective functional. The implicit scheme with weights was used for approximation of the direct problem. In this case, the FAD-technique also builds an implicit scheme with weights to approximate the conjugate problem and consequently to determine the conjugate variables that are necessary to organize an iterative process.

Taking into account this fact, a scheme of alternating directions was used in [9] to approximate a direct problem. In this case, no iterative process is required to solve the conjugate equations and the conjugate variables are determined as the solution of the system of linear algebraic equations with the aid of the tridiagonal Gaussian elimination method. In [9] the special features of the solution related to the two-dimensional spatial character of the problem are indicated.

The work [10] is a continuation of the investigations carried out in [9]. In contrast to [9], the mean-root-square deviation of the calculated heat flux at the boundary of the two-dimensional object from the experimentally determined heat flux was used as the cost functional. An algorithm for the numerical solution of the inverse coefficient problem, in this case, was proposed. The performed numerical computations demonstrate the efficiency of the proposed algorithm.

The investigations carried out in [4,8–10] are made under the assumption that all the initial data is precise. In the present paper, a numerical analysis of the stability of the obtained solutions is carried out. It is shown that the perturbation of the restored thermal conductivity coefficient is of the same order as the perturbation of the experimental data that caused it.

2   Formulation of the Problem

We consider a domain $Q \subset R^n$ with piecewise-smooth boundary S. This domain is filled with the substance being investigated. The distribution of the temperature field at each moment is described by the following initial boundary value (mixed) problem:

$$C(x)\frac{\partial T}{\partial t} = div_x(K(T)\nabla_x T), \qquad x \in Q, \quad 0 < t \leq \Theta, \tag{1}$$

$$T(x,0) = w_0(x), \qquad x \in Q, \tag{2}$$

$$T(x,t) = w_s(x,t), \qquad x \in S, \quad 0 \leq t \leq \Theta. \tag{3}$$

Here $x = (x_1, ..., x_n)$ are the Cartesian coordinates; t is time; $T(x,t)$ is the temperature of the material at the point with the coordinates x at time t; $C(x)$ is the volumetric heat capacity of the material; $K(T)$ is the thermal conductivity coefficient; $w_0(x)$ is the given temperature at the initial time $t = 0$; $w_s(x,t)$ is the given temperature on the boundary of the object. The volumetric heat capacity of a substance $C(x)$ is considered a known function of the coordinates.

If the dependence of the thermal conductivity coefficient $K(T)$ on the temperature T is known, then we can solve the mixed problem (1)–(3) to find the temperature distribution $T(x,t)$ in $Q \times (0, \Theta]$. If the dependence of the thermal conductivity coefficient of the material on the temperature is not known, it is of interest to determine this dependence. A possible statement of this problem is as follows: find the dependence $K(T)$ on T under which the temperature field $T(x,t)$, obtained by solving the mixed problem (1)–(3), is close to the field $Y(x,t)$ obtained experimentally, and the heat flux $-K(T(x,t))\frac{\partial T(x,t)}{\partial n}$ on the boundary of the domain is close to the experimentally flux $P(x,t)$. The quantity

$$\Phi(K(T)) = \int\limits_0^\Theta \int\limits_Q [T(x,t) - Y(x,t)]^2 \cdot \mu(x,t)dx\,dt +$$

$$+ \int\limits_0^\Theta \int\limits_S \beta(x,t)\left[-K(T(x,t))\frac{\partial T(x,t)}{\partial n} - P(x,t)\right]^2 ds\,dt + \varepsilon \int\limits_a^b (K'(T))^2 dT \tag{4}$$

can be used as the measure of difference between these functions. Here, $\varepsilon \geq 0$, $\beta(x,t) \geq 0$, $\mu(x,t) \geq 0$ are given weight parameters; $Y(x,t)$ is a given temperature field; $P(x,t)$ is a given heat flux at the boundary S of the domain Q, $\frac{\partial T}{\partial n}$ is the derivative of the temperature along the direction of the outer normal to the boundary of the domain; $[a;b]$ is the interval on which the function $K(T)$ will be restored. Thus, the optimal control problem is to find the optimal control $K(T)$ and the corresponding solution $T(x,t)$ of problem (1)–(3) that minimize functional (4).

The optimal control problem formulated above was solved numerically. One of the main elements of the proposed numerical method for solving the inverse coefficient problem is the solution of the mixed problem (1)–(3). Spatial and time grids (generally nonuniform) have been introduced to solve the problem numerically. At each node of the computational domain $\overline{Q} \times [0, \Theta]$, all the functions are determined by their point values. To approximate the heat equation we used two finite difference schemes (a two-layer implicit scheme with weights, a scheme of alternating directions).

The temperature interval $[a, b]$ (the interval of interest) on which the function $K(T)$ will be restored is defined as the set of values of the given functions $w_0(x)$ and $w_s(x,t)$. This interval is partitioned by the points $\widetilde{T}_0 = a, \widetilde{T}_1, \widetilde{T}_2, \ldots, \widetilde{T}_M = b$ into M parts (they can be equal or of different lengths). Each point $\widetilde{T}_m$ ($m = 0, \ldots, M$) is connected with a number $k_m = K(\widetilde{T}_m)$. The function $K(T)$ to be found is approximated by a continuous piecewise linear function with the nodes at the points $\left\{ (\widetilde{T}_m, k_m) \right\}_{m=0}^{M}$ (see [4]). If the temperature at the point fell outside the boundaries of the interval $[a; b]$, then the linear extrapolation was used to determine the function $K(T)$.

The objective functional (4) was approximated by a function $F(k_0, k_1, \ldots, k_M)$ of the finite number of variables using the method of the rectangles. The heat fluxes at the boundary of the domain were approximated according to the formulas given in [4] and [10]. In [4,8–10], it was assumed that in formula (4) there is no last term, i.e. $\varepsilon = 0$. In the present paper this term is approximated by means of formula

$$\int_a^b (K'(T))^2 dT = \sum_{m=1}^{M} \frac{(k_m - k_{m-1})^2}{(\widetilde{T}_m - \widetilde{T}_{m-1})}.$$

Minimization of the function $F(k_0, k_1, \ldots, k_M)$ was carried out numerically using the gradient method. It is well known that it is very important for the gradient methods to determine accurate values of the gradients.

The analytic expressions of the gradient are given in [4] (for the case $n = 1$) and in [10] (for the case $n = 2$). It is easy to see that it is difficult to determine the gradient using analytic expressions. For this reason, we used the efficient approach of Fast Automatic Differentiation to calculate the components of gradient ([5,6]). The effectiveness of this methodology is ensured by using the solution of the conjugate problem for calculating the gradient of the function. The FAD-methodology allows us to formulate an adjoint problem that is coordinated with the chosen approximation of the direct problem. As a result, the FAD-methodology delivers canonical formulas by means of which the calculated value of the gradient of the cost functional is precise for the chosen approximation of the optimal control problem. For the one-dimensional inverse problem, the conjugate problem and formulas for calculating the gradient of the functional are given in [4], and for the two-dimensional problem - in [8–10]. If in (4) $\varepsilon > 0$, then the last term in this formula gives the following contribution to the formula for determining the gradient components $\frac{\partial F}{\partial k_m}$:

$$2\varepsilon A \frac{(k_m - k_{m-1})}{(\widetilde{T}_m - \widetilde{T}_{m-1})} - 2\varepsilon B \frac{(k_{m+1} - k_m)}{(\widetilde{T}_{m+1} - \widetilde{T}_m)},$$

$$A = 1, B = 1, \quad if \quad m = 1, ..., M - 1, \quad A = 0, B = 1, \quad if \quad m = 0,$$
$$A = 1, B = 0, \quad if \quad m = M.$$

3 The Numerical Analysis of the Stability of Solutions (One-Dimensional Case)

To numerically solve the mixed problem

$$C(x)\frac{\partial T(x,t)}{\partial t} = \frac{\partial}{\partial x}\left(K(T)\frac{\partial T(x,t)}{\partial x}\right), \qquad 0 < x < L, \quad 0 < t \le \Theta,$$

$$T(x,0) = w_0(x), \qquad 0 \le x \le L,$$

$$T(0,t) = w_1(t), \qquad T(L,t) = w_2(t), \qquad 0 \le t \le \Theta,$$

the domain $G = \{(x,t) : 0 < x < L, \ 0 < t < \Theta\}$ is decomposed by the grid lines $\{\widetilde{x}_i\}_{i=0}^{I}$ and $\{\widetilde{t}^j\}_{j=0}^{J}$ into rectangles. At each node $(\widetilde{x}_i, \widetilde{t}^j)$ of G characterized by the pair of indices (i,j), all the functions are determined by their values at the point $(\widetilde{x}_i, \widetilde{t}^j)$ (e.g., $T(\widetilde{x}_i, \widetilde{t}^j) = T_i^j$).

The cost functional in this case has the form

$$\Phi(K(T)) = \int_0^\Theta \int_0^L [T(x,t) - Y(x,t)]^2 \cdot \mu(x,t)dx \, dt + \varepsilon \int_a^b (K'(T))^2 dT +$$

$$+\int_0^\Theta \beta(t)\left\{\left[K(T(0,t))\frac{\partial T}{\partial x}(0,t) - P(0,t)\right]^2 + \left[-K(T(L,t))\frac{\partial T}{\partial x}(L,t) - P(L,t)\right]^2\right\} dt.$$

To approximate the heat equation in the one-dimensional case, we used a two-layer implicit scheme with weights ([4]). The weight coefficient was chosen $\sigma = 0.55$.

For the solution of the direct and inverse problems, we used the uniform grid with the parameters $I = 300$ (the number of intervals along the axis x) and $J = 6000$ (the number of intervals along the axis t), which ensures the sufficient accuracy of computation of the temperature field and of the field of conjugate variables. The interval $[a,b]$ was partitioned into 64 intervals ($M = 64$). In all the examples below the experimental data was perturbed by a random number generator.

3.1. In the first series we wanted to find the thermal conductivity coefficient for the following input data: $L = 1, \quad \Theta = 1, \quad C(x) \equiv 1,$

$$w_0(x) = sin(x), \qquad\qquad\qquad 0 \le x \le L,$$
$$w_1(t) = 0, \qquad w_2(t) = sin1 \cdot exp(-4t), \qquad 0 \le t \le \Theta,$$
$$Y(x,t) = sinx \cdot exp(-4t), \qquad 0 \le x \le L, \qquad 0 \le t \le \Theta,$$
$$a = 0, \qquad\qquad b = sin1.$$

Note that the inverse problem with this input data has an analytical solution, indeed, the function $Y(x,t)$ is the solution of the mixed problem (1)–(3) with the input data indicated above for $K(T) \equiv 4$.

In the **first example**, we assumed that in the cost functional (4) the weighting function $\mu(x,t) \equiv 1$ and $\beta(t) \equiv 0$ (the thermal conductivity coefficient is restored by the temperature field). The function $K(T) \equiv T$ is selected as the initial control.

The perturbations intensity Δ was determined as the maximum deviation of the perturbed data from unperturbed data divided by the average value of the unperturbed data. Further, the symbol δ designated the appropriate maximum variation of thermal conductivity divided by its average value.

At first we attempted to solve the inverse problem with $\varepsilon = 0$. For low perturbations intensities of the temperature field proposed algorithm gives small deviations of the inverse problem's solutions. So, for $\Delta = 1.4 \cdot 10^{-5}$ the value $\delta = 1.003 \cdot 10^{-5}$ and for $\Delta = 2.1 \cdot 10^{-3}$ the value $\delta = 1.308 \cdot 10^{-3}$.

Increase in the perturbations intensities Δ with $\varepsilon = 0$ leads to the substantial growth of the deviations of the inverse problem's solutions. It should also be noted that the decrease in the number M (the number of partitions of temperature interval on which the conductivity coefficient was restored) allows one to visibly reduce the deviation δ. So, Fig. 1 shows the functions $K(T)$ obtained by the proposed algorithm for $\Delta = 2.1 \cdot 10^{-1}$ and $M = 4, 8, 16$.

To obtain stable solutions to the inverse problem for larger values Δ we use $\varepsilon > 0$ in the cost functional (4). Table 1 presents dependence of δ upon ε for $\Delta = 2.1 \cdot 10^{-1}$ and $M = 64$, from which it follows that when $\varepsilon \cong 10^{-3}$ the deviation δ is minimal. On the Fig. 2 the dependence of thermal conductivity coefficient upon temperature for different ε is presented ($M = 64$). Here a line marked by negative number p corresponds to $K(T)$ obtained for $\varepsilon = 10^p$. Line obtained for $\varepsilon \cong 10^{-3}$ is not represented, since it practically does not differ from lines obtained for $\varepsilon \cong 10^{-7}$. The results presented in Fig. 2, also led to the conclusion that the best solution is achieved when $\varepsilon = 10^{-3}$.

Table 1. Series 3.1, first example, dependence of δ upon ε for $\Delta = 2.1 \cdot 10^{-1}$

ε	10^{-0}	10^{-3}	10^{-5}	10^{-6}	10^{-7}	10^{-8}	10^{-9}
$\delta \cdot 10^3$	7.791569	7.791563	7.791967	7.796929	7.822980	8.135212	10.74287

In the **second example** of the first series, the weighting function $\mu(x,t)$ was zero and $\beta(t) \equiv 1$ (the thermal conductivity coefficient is restored by the heat flux on the boundary).

In this example, as in previous one, for low perturbations intensities of the heat flux proposed algorithm gives small deviations of the inverse problem's solutions. So, for $\Delta = 2.1 \cdot 10^{-5}$ the value $\delta = 5.411 \cdot 10^{-6}$ and for $\Delta = 2.1 \cdot 10^{-3}$ the value $\delta = 5.113 \cdot 10^{-4}$.

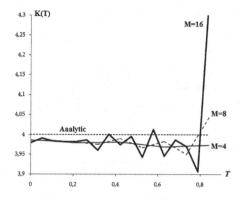

Fig. 1. The functions $K(T)$, series 3.1, first example, $\Delta = 2.1 \cdot 10^{-1}$, $\varepsilon = 0$

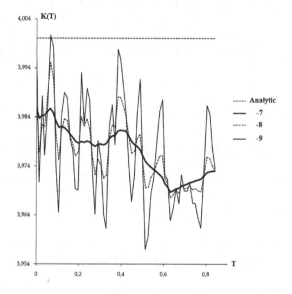

Fig. 2. The functions $K(T)$, series 3.1, first example, $\Delta = 2.1 \cdot 10^{-1}$, $M = 64$, different ε

Here again, the increase in the intensity Δ of the boundary heat flux perturbations ($\varepsilon = 0$) leads to a marked increase in deviations, and reducing the M number permits to reduce the value δ. The Fig. 3 shows the functions $K(T)$ obtained for $\Delta = 2.1 \cdot 10^{-1}$ and $M = 4, 8, 16$.

If we use $\varepsilon > 0$ in the cost functional (4) than solutions of inverse problem will be more smooth. Table 2 presents dependence of δ upon ε for $\Delta = 2.1 \cdot 10^{-1}$ and $M = 64$, from which it follows that when $\varepsilon \cong 10^{-5}$ the deviation δ is minimal. On the Fig. 4 the dependence of thermal conductivity coefficient upon temperature for different ε is presented. Here the notations are the same as in Fig. 2.

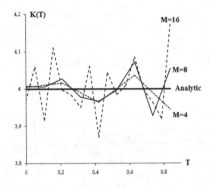

Fig. 3. The functions $K(T)$, series 3.1, second example, $\Delta = 2.1 \cdot 10^{-1}$, $\varepsilon = 0$

Table 2. Series 3.1, second example, dependence of δ upon ε for $\Delta = 2.1 \cdot 10^{-1}$

ε	10^{-1}	10^{-3}	10^{-4}	10^{-5}	10^{-6}	10^{-7}
$\delta \cdot 10^3$	3.499455	3.499455	3.425485	3.347866	11.55226	19.31322

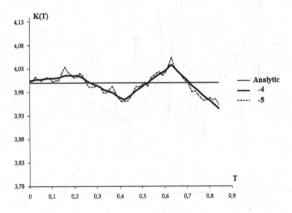

Fig. 4. The functions $K(T)$, series 3.1, second example, $\Delta = 2.1 \cdot 10^{-1}$, $M = 64$, different ε

3.2. In the second series, we wanted to find the coefficient of the thermal conductivity for the following input data: $L = 1$, $\Theta = 1$, $C(x) \equiv 1$,

$$
\begin{aligned}
w_0(x) &= \sqrt{2(1.5 - x)}, & & & & 0 \le x \le L, \\
w_1(t) &= \sqrt{2(1.5 + t)}, & w_2(t) &= \sqrt{2(0.5 + t)}, & & 0 \le t \le \Theta, \\
Y(x,t) &= \sqrt{2(1.5 + t - x)}, & & 0 \le x \le L, & & 0 \le t \le \Theta, \\
a &= 1, & b &= \sqrt{5}.
\end{aligned}
$$

Note that the inverse problem with these input data has an analytical solution, indeed, the function $Y(x,t)$ is the solution of the mixed problem (1)–(3) with the input data indicated above for $K(T) \equiv T^2$.

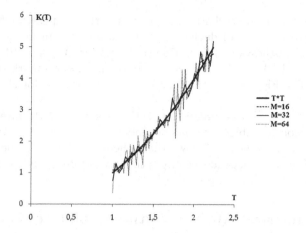

Fig. 5. The functions $K(T)$, series 3.2, $\Delta = 2.0 \cdot 10^{-1}$, $\varepsilon = 0$

Table 3. Series 3.2, dependence of δ upon ε for $\Delta = 2.1 \cdot 10^{-1}$

ε	10^{-3}	10^{-4}	10^{-5}	10^{-6}
$\delta \cdot 10^2$	5.261507	5.243166	5.155661	9.762721

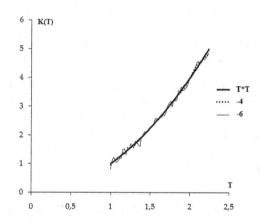

Fig. 6. The functions $K(T)$, series 3.2, $\Delta = 2.0 \cdot 10^{-1}$, $M = 64$, different ε

In the studies given below, it was assumed that the weighting function $\mu(x, t)$ was zero and $\beta(t) \equiv 1$ (the thermal conductivity coefficient is restored by the heat flux on the boundary). The function $K(T) \equiv 7$ is selected as the initial control. The main objective of this example is to show that the identification algorithm works when the solution of the inverse problem is not constant.

Here again, for low perturbations intensities of the temperature flux the proposed algorithm gives small deviations of the inverse problem's solutions. So, for $\Delta = 2.1 \cdot 10^{-5}$ the value $\delta = 2.056 \cdot 10^{-5}$ and for $\Delta = 2.0 \cdot 10^{-3}$ the value $\delta = 4.121 \cdot 10^{-3}$. If Δ is not small and $\varepsilon = 0$ we encounter a marked increase in deviations δ. The Fig. 5 shows the functions $K(T)$ obtained for $\Delta = 2.0 \cdot 10^{-1}$ and $M = 16, 32, 64$.

If we use $\varepsilon > 0$ in the cost functional (4) than solutions of the inverse problem will be more smooth. Table 3 presents dependence of δ upon ε for $\Delta = 2.0 \cdot 10^{-1}$ and $M = 64$, from which it follows that when $\varepsilon \cong 10^{-5}$ the deviation δ is minimal. On Fig. 6 the dependence of the thermal conductivity coefficient upon temperature for different ε is presented.

4   The Numerical Analysis of the Stability of Solutions (Two-Dimensional Case)

A plate of material of length L and width R is considered. The points $x = (x_1, x_2)$ of the plate form a domain $Q = \{(0, L) \times (0, R)\}$ with a boundary $S = \partial Q$. The distribution of the temperature field in the plate at each moment of time is described by the following initial-boundary (mixed) problem:

$$C(x)\frac{\partial T(x,t)}{\partial t} = div_x(K(T(x,t))\nabla_x T(x,t)), \quad (x,t) \in G = \{Q \times (0, \Theta)\}, \quad (5)$$

$$T(x, 0) = w_0(x), \qquad\qquad\qquad x \in \overline{Q}, \qquad\qquad (6)$$

$$T(x, t) = w_S(x, t), \qquad\qquad\qquad x \in S, \quad 0 \le t \le \Theta. \qquad (7)$$

Let l be the arc length of the boundary contour S, calculated from the point $x = (0, 0)$ in the positive direction. Then the contour S can be parametrically given in the form $x(l) = (x_1(l), x_2(l))$, and any function $f(x) = f(x_1, x_2)$ given at the points of this contour is reduced to a complex function $\widetilde{f}(l) = f(x(l)) = f(x_1(l), x_2(l))$.

The cost functional in this case has the form

$$\Phi(K(T)) = \int_0^\Theta \int_Q [T(x,t) - Y(x,t)]^2 \cdot \mu(x,t)dx\,dt + \varepsilon \int_a^b (K'(T))^2 dT +$$

$$+ \int_0^\Theta \oint_S \beta(x,t) \left[-K(T(x,t))\frac{\partial T(x,t)}{\partial n}\bigg|_{x=x(l)\in S} - P(x(l),t) \right]^2 dl\,dt. \quad (8)$$

Here, $Y(x, t)$ is a given temperature field, and $P(x(l), t)$ is a given heat flux at the boundary points of the domain.

To numerically solve the mixed problem (5)–(7) the domain $[0, L] \times [0, R] \times [0, \Theta]$ is decomposed by the grid lines $\{x_1^n\}_{n=0}^N$, $\{x_2^i\}_{i=0}^I$, and $\{t^j\}_{j=0}^J$ into parallelepipeds as it is shown in [8]. At each node (x_1^n, x_2^i, t^j) characterized by the indices (n, i, j), all the functions are determined by their values at the point

(x_1^n, x_2^i, t^j) (e.g., $T(x_1^n, x_2^i, t^j) = T_{ni}^j$). To approximate the heat equation, an implicit scheme of alternating directions was used (see [9]) (splitting in the x_1 and x_2 directions and formulation of two subproblems). In the examples considered in this section the uniform grid with the parameters $N = 60$, $I = 60$ (the number of intervals along the axis x_1 and x_2) and $J = 60$ (the number of intervals along the axis t), which ensures the sufficient accuracy of computation of the temperature field. The interval $[a, b]$ was partitioned into 64 intervals ($M = 64$).

4.1. In the first series, we wanted to find the thermal conductivity coefficient for the following input data: $L = 2$, $R = 1$, $\Theta = 1$, $C(x) \equiv 1$,

$$w_0(x_1, x_2) = x_1^2 + 7x_1x_2 - x_2^2 + \sin\left(\frac{\pi}{2}x_1\right) \cdot \sin(2\pi x_2),$$

$$(0 \le x_1 \le 2, \quad 0 \le x_2 \le 1),$$

$$w_1(x_2, t) = -x_2^2, \qquad w_2(x_2, t) = 14x_2 - x_2^2 + 4, \quad 0 \le x_2 \le 1, \quad 0 \le t \le 1,$$

$$w_3(x_1, t) = x_1^2, \qquad w_4(x_1, t) = x_1^2 + 7x_1 - 1, \quad 0 \le x_1 \le 2, \quad 0 \le t \le 1$$

$$a = -1, \qquad b = 17.$$

The formulated problem has an analytical solution, since the function $Y(x_1, x_2, t) = x_1^2 + 7x_1x_2 - x_2^2 + \exp(-t) \cdot \sin\left(\frac{\pi}{2}x_1\right) \cdot \sin(2\pi x_2)$ is a solution of the mixed problem (5)–(7) with the above parameters for $K(T) \equiv \frac{4}{17\pi^2}$.

In the studies given below it was assumed the weighting function $\mu(x_1, x_2, t)$ was zero and $\beta(x_1, x_2, t) = 1$ (the thermal conductivity coefficient is restored by the heat flux on the boundary). The function $K(T) = 1$ was selected as the initial control.

Here, as in the one-dimensional case, small changes in the heat flux on the boundary of the plate lead to small changes in the solution of the inverse problem. So, for $\Delta = 10^{-4}$ the value $\delta = 1.141 \cdot 10^{-4}$ and for $\Delta = 10^{-3}$ the value $\delta = 1.150 \cdot 10^{-3}$. Let us note that all these results were obtained without a smoothing function (i.e. with $\varepsilon = 0$). But if we continue to increase the deviation of Δ, then, as in the one-dimensional case, an increase in the deviation of the solution of the inverse problem is observed. Therefore, for large values of Δ, it is necessary to use $\varepsilon > 0$ in the functional (8). Table 4 presents dependence of δ upon ε for $\Delta = 2.7 \cdot 10^{-1}$. The best δ was achieved for $\varepsilon = 10^2$.

Table 4. Series 4.1, dependence of δ upon ε for $\Delta = 2.7 \cdot 10^{-1}$

ε	10^{+2}	10^{+1}	10^0	10^{-1}
$\delta \cdot 10^2$	3.152464	3.420390	6.034271	12.51151

Figures 7 and 8 illustrate the thermal conductivity coefficients, built with the aid of the proposed algorithm for different values of ε for $M = 64$. It should be noted that the decrease in the number M, as in the one-dimensional case, leads to the smoothing of the obtained solution of the inverse problem even for $\varepsilon = 0$.

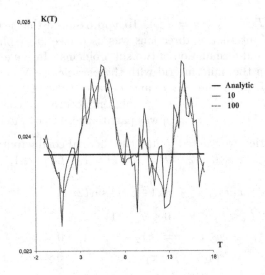

Fig. 7. The functions $K(T)$, series 4.1, $\Delta = 2.7 \cdot 10^{-1}$, $M = 64$, different ε

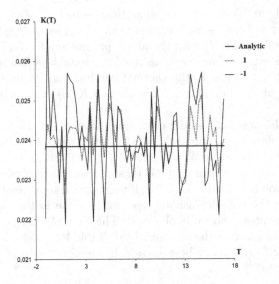

Fig. 8. The functions $K(T)$, series 4.1, $\Delta = 2.7 \cdot 10^{-1}$, $M = 64$, different ε

Figures 9 and 10 show the thermal conductivity coefficients obtained with $\varepsilon = 0$ for different M.

4.2. In the second series of the computation experiments, the inverse coefficient problem was considered for the following input data:

$$L = 1, \quad R = 1, \quad \Theta = 1, \quad C(x) \equiv 1,$$
$$w_0(x_1, x_2) = 9/(x_1 + 2x_2 + 1), \qquad (0 \leq x_1 \leq 1, \quad 0 \leq x_2 \leq 1),$$
$$w_1(x_2, t) = 9/(2x_2 + 5t + 1), \qquad w_2(x_2, t) = 9/(2x_2 + 5t + 2),$$
$$0 \leq x_2 \leq 1, \quad 0 \leq t \leq 1,$$
$$w_3(x_1, t) = 9/(x_1 + 5t + 1), \qquad w_4(x_1, t) = 9/(x_1 + 5t + 3),$$
$$0 \leq x_1 \leq 1, \quad 0 \leq t \leq 1$$
$$a = 1, \qquad b = 9.$$

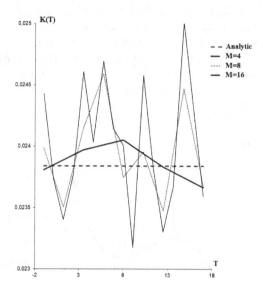

Fig. 9. The functions $K(T)$, series 4.1, $\Delta = 2.7 \cdot 10^{-1}$, $\varepsilon = 0$

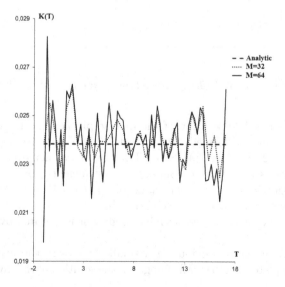

Fig. 10. The functions $K(T)$, series 4.1, $\Delta = 2.7 \cdot 10^{-1}$, $\varepsilon = 0$

In the **first example**, we assumed that in the cost functional (8) the weighting function $\mu(x_1, x_2, t) = 1$ and $\beta(x_1, x_2, t) = 0$ (the thermal conductivity coefficient is restored by the temperature field). As the experimental field $Y(x_1, x_2, t)$ has been chosen the temperature field that was obtained as a result of solving the direct problem (5)–(7) with the given thermal conductivity coefficient:

$$K_*(T) = \begin{cases} 0.1 \cdot (T-3)(T-6)(T-7) + 3.4, & T \geq 3, \\ 1.2 \cdot (T-3) + 3.4, & T < 3. \end{cases}$$

The function $K(T) = 4.5$ was selected as the initial control.

For such curious example, the conclusions about the stability of the algorithm remain the same. So, for $\Delta = 10^{-4}$ the value $\delta = 3.842 \cdot 10^{-4}$ and for $\Delta = 2.2 \cdot 10^{-3}$ the value $\delta = 7.780 \cdot 10^{-3}$. These results were obtained without a smoothing function (i.e. with $\varepsilon = 0$) and for $M = 64$. If we continue to increase the deviation in the input data from the exact values, then, if the number M grows, an increase in the deviation of the solution of the inverse problem is observed (see Fig. 11).

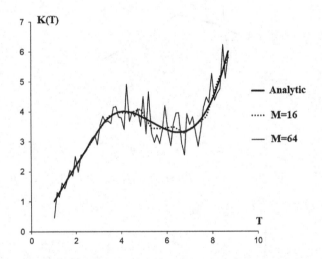

Fig. 11. The functions $K(T)$, series 4.2, first example, $\Delta = 2.2 \cdot 10^{-1}$, $\varepsilon = 0$

To obtain a smoothed solution of the inverse problem for large values of Δ, the calculations were performed with $\varepsilon > 0$. Table 5 presents dependence of δ upon ε for $\Delta = 2.2 \cdot 10^{-1}$ and $M = 64$. The best δ was achieved for $\varepsilon = 10^{-6}$. Figure 12 illustrates the thermal conductivity coefficients built with the aid of the proposed algorithm for different values of ε ($M = 64$).

In the **second example**, the weighting function $\mu(x_1, x_2, t)$ was zero and $\beta(x_1, x_2, t) \equiv 1$ (the thermal conductivity coefficient is restored by the heat flux on the boundary).

In this example, as in previous one, small changes in the heat flux on the boundary of the plate lead to small changes in the solution of the inverse problem.

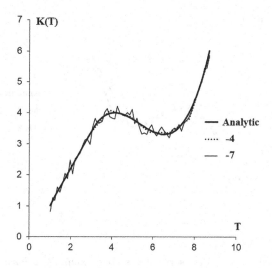

Fig. 12. The functions $K(T)$, series 4.2, first example, $\varDelta = 2.2 \cdot 10^{-1}$, $M{=}64$, different ε

Table 5. Series 4.2, first example, dependence of δ upon ε for $\varDelta = 2.2 \cdot 10^{-1}$

ε	10^{-4}	10^{-5}	10^{-6}	10^{-7}
$\delta \cdot 10^2$	5.909976	5.894398	5.756982	10.06103

Table 6. Series 4.2, second example, the error of the restored thermal conductivity coefficient with $\varepsilon = 0$ for different M and $\varDelta = 2.6 \cdot 10 - 1$

M	2	4	8	16	32	64
$\delta \cdot 10^2$	26.29114	6.126572	3.217278	1.307354	1.327451	1.356869

For example, for $\varDelta = 10^{-4}$ the value $\delta = 1.520 \cdot 10^{-4}$ and for $\varDelta = 10^{-2}$ the value $\delta = 1.079 \cdot 10^{-2}$.

This example is characterized by the fact that during the large perturbations in the experimental data (26%), the error of the restored thermal conductivity coefficient with $\varepsilon = 0$ and $M > 8$ is about 1%. This fact is confirmed by the data from the following Table 6 ($\varepsilon = 0$). Figure 13 shows that there are no oscillations in the solution for $M = 8$. It is interesting to note that oscillations do not appear for $M > 8$ and $\varepsilon = 0$. The restored thermal conductivity coefficient even for $M = 128$ and $\varepsilon = 0$ graphically does not differ from the analytical one.

Using the cost functional with $\varepsilon > 0$ does not improve the quality of the obtained solution with $\varepsilon = 0$.

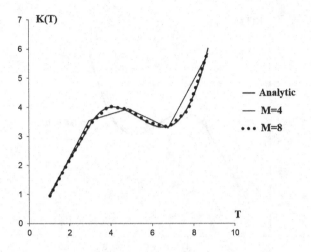

Fig. 13. The functions $K(T)$, series 4.2, second example, $\Delta = 2.6 \cdot 10^{-1}$, $\varepsilon = 0$

5   Conclusion

Based on the studies performed in this work, we conclude:

(1) For small changes in the input data, the proposed algorithm leads to small changes in the solution of the inverse problem with $\varepsilon = 0$.
(2) To reduce oscillations in the solution of the inverse problem in the case when the perturbations in the input data are not small ($\cong 20 - 30\%$), it is advisable to use the smoothing term in the cost functional ($\varepsilon > 0$).
(3) If the perturbations in the input data are not small, then one can achieve a smoothed solution by choosing a small number M of partitions of the interval $[a, b]$.
(4) For the efficient work of the proposed algorithm, it is advisable to solve the inverse problem beginning with small ($\cong 8$) values of M. The obtained solution is used as the initial approximation for larger values of M.

References

1. Alifanov, O.M., Cherepanov, V.V.: Mathematical simulation of high-porosity fibrous materials and determination of their physical properties. High Temp. **47**, 438–447 (2009). https://doi.org/10.1134/S0018151X09030183
2. Czel, B., Grof, G.: Simultaneous identification of temperature-dependent thermal properties via enhanced genetic algorithm. Int. J. Thermophys. **33**, 1023–1041 (2012). https://doi.org/10.1007/s10765-012-1226-9
3. Cui, M., Yang, K., Xu, X.-L., Wang, S.-D., Gao, X.-W.: A modified Levenberg-Marquardt algorithm for simultaneous estimation of multi-parameters of boundary heat flux by solving transient nonlinear inverse heat conduction problems. Int. J. Heat Mass Transf. **97**, 908–916 (2016). https://doi.org/10.1016/j.ijheatmasstransfer.2016.02.085

4. Zubov, V.I.: Application of fast automatic differentiation for solving the inverse coefficient problem for the heat equation. Comp. Math. and Math. Phys. **56**(10), 1743–1757 (2016). https://doi.org/10.7868/S0044466916100148

5. Evtushenko, Y.G., Zubov, V.I.: Generalized fast automatic differentiation technique. Comp. Math. Math. Phys. **56**(11), 1819–1833 (2016). https://doi.org/10.1134/S0965542516110075

6. Evtushenko, Y.G.: Computation of exact gradients in distributed dynamic systems. Optim. Methods Softw. **9**, 45–75 (1998). https://doi.org/10.1080/10556789808805686

7. Albu, A.F., Evtushenko, Y.G., Zubov, V.I.: Identification of discontinuous thermal conductivity coefficient using fast automatic differentiation. In: Battiti, R., Kvasov, D.E., Sergeyev, Y.D. (eds.) LION 2017. LNCS, vol. 10556, pp. 295–300. Springer, Cham (2017). https://doi.org/10.1007/978-3-319-69404-7_21

8. Zubov, V.I., Albu, A.F.: The FAD-methodology and recovery the thermal conductivity coefficient in two dimension case. In: Proceedings of the VIII International Conference on Optimization Methods and Applications "Optimization and applications", pp. 39–44 (2017). http://ceur-ws.org/Vol-1987/

9. Albu, A.F., Zubov, V.I.: Identification of thermal conductivity coefficient of substance using the temperature field. Comp. Math. Math. Phys. 58(10) (2018)

10. Albu, A.F., Zubov, V.I.: Identification of the coefficient of thermal conductivity of the substance using the surface heat flux. Comp. Math. Math. Phys. 58(12) (2018)

On the Effectiveness of the Fast Automatic Differentiation Methodology

Alla Albu[1], Andrei Gorchakov[1,2], and Vladimir Zubov[1,3]

[1] Dorodnicyn Computing Centre, Federal Research Center "Computer Science and Control" of Russian Academy of Sciences, Moscow, Russia
alla.albu@yandex.ru, andrgor12@gmail.com, vladimir.zubov@mail.ru
[2] Nuclear Safety Institute of the Russian Academy of Sciences, Moscow, Russia
[3] Moscow Institute of Physics and Technology, Moscow, Russia

Abstract. In this paper, we compare the three approaches for calculating the gradient of a complex function of many variables. The compared approaches are: the use of precise, analytically derived formulas; the usage of formulas derived with the aid of the Fast Automatic Differentiation methodology; the use of standard software packages that implement the ideas of Fast Automatic Differentiation methodology. Comparison of approaches is carried out with the help of a complex function that represents the energy of atoms system whose interaction potential is the Tersoff potential. As a comparison criterion, the computer time required to calculate the gradient of the function is used. The results show the superiority of the Fast Automatic Differentiation methodology in comparison with the approach using analytical formulas. Standard packages compute the function gradient around the same time as using the formula of the Fast Automatic Differentiation methodology.

Keywords: Fast Automatic Differentiation
Standard software packages · Tersoff potential

1 Introduction

Let us assume that the differentiable scalar function $f(u)$ of vector argument $u \in R^r$ is specified explicitly. The problem consists in calculating the partial derivatives of the function $f(u)$ with respect to u. In computational practice, we often meet elementary functions for which the exact calculation of these derivatives is a very difficult problem.

In such a case, you can use one of several ways of finding the partial derivatives of the function $f(u)$:

(1) **Finite difference approximation.** If $f(u)$ is a function of r-dimensional vector u, that using the simplest numerical differentiation formulas to determine all derivatives must be at least $(r + 1)$ times to calculate the value of

This work was partially supported by the Russian Foundation for Basic Research (project no. 17-07-00493 a).

Y. Evtushenko et al. (Eds.): OPTIMA 2018, CCIS 974, pp. 264–276, 2019.
https://doi.org/10.1007/978-3-030-10934-9_19

a function $f(u)$ and perform r subtractions and divisions. Results of such calculations will be accurate only for linear functions; in all other cases, the gradient is defined with some error and sometimes with some difficulty. For example, in [1] the energy of atoms group is calculated. All atoms interact with the add of Tersoff potential. When calculating the partial derivatives of the energy on the specific parameters of the Tersoff potential using this method, it is necessary to choose the appropriate increment of each parameter and do it every time for new energy value.

(2) **Symbolic differentiation.** When applying this technique we meet with great difficulties, if a vector u has a large dimension.

(3) **The use of Fast Automatic Differentiation (FAD)** [2]. This approach is often referred to as the inverse method with a conjugate problem.

(4) **Analytical differentiation.** The calculation of $f(u)$ is represented as some multistep process (superposition of any functions) and the derivatives of $f(u)$ are calculated analytically as the derivatives of functions superpositions. This approach may be referred to as the modified direct method.

(5) **Program packages.** The use of a common package where the direct and inverse Fast Automatic Differentiation technique are realized.

In [2] a general approach to the differentiation of complex functions that occur in multistep processes is proposed. It is also shown that from the obtained results follow FAD formula for elementary functions. In the present paper mentioned methods for calculating the gradient of complex functions are compared.

2   Comparison of FAD-Technique and Analytical Differentiation

In order to use the above-indicated methods of calculating the derivatives, it is necessary to represent the calculation of the elementary function $f(u)$ as a multistep process. To this end, we introduce a vector $z \in R^k$ of phase variables (a vector of auxiliary, intermediate values of the function):

$$z_i = F(i, Z_i, U_i), \qquad\qquad 1 \le i \le k \qquad (1)$$

where Z_i is the set of vectors z_j, that appear at the right part of equality (1), and U_i is the set of vectors u_j, that appear at the right part of the same equality (1). Usually, the vectors $z \in R^k$ and the vectors $u \in R^r$ are called dependent (phase) and independent (control) variables respectively. Sequence (1) is constructed in such a way that the process would be explicit, and the last component z_k coincides with the value of the function $f(u)$, i.e. $z_k = f(u)$.

We assume that all functions $F(i, Z_i, U_i)$ are differentiable with respect to all components of the vectors z and u. We introduce the vector of conjugate variables $p \in R^k$, which is determined from the following system of linear algebraic equations:

$$p_l = \sum_{q \in \overline{Q}_l} F_{z_l}(q, Z_q, U_q) p_q, \qquad \overline{Q}_l = \{i : 1 \le i \le k, \quad z_l \in Z_i\}. \qquad (2)$$

Then, according to the FAD-technique formulas, the gradient of the function $f(u)$ with respect to the independent variables u_m is given by the formula

$$\frac{\partial f}{\partial u_m} = \sum_{q \in \overline{K}_m} F_{u_m}(q, Z_q, U_q) p_q, \qquad \overline{K}_m = \{i : 1 \leq i \leq k, \quad u_m \in U_i\}. \quad (3)$$

In [2] it is shown that the use of FAD-technique for the differentiation of elementary functions leads to the remarkable result: the time, necessary for calculating the gradient, referred to the time, necessary for calculating the value of the function, does not exceed 3.

Using the multistep process (1), it is possible to calculate the derivatives $\frac{\partial f}{\partial u_m}$, $m = \overline{1, r}$ analytically (without using conjugate variables) as follows. Namely, the following derivatives are calculated consecutively:

$$\frac{\partial z_1}{\partial u_m}; \qquad \frac{\partial z_2}{\partial u_m} = \frac{\partial z_2}{\partial z_1} \frac{\partial z_1}{\partial u_m}; \qquad \frac{\partial z_3}{\partial u_m} = \frac{\partial z_3}{\partial z_2} \frac{\partial z_2}{\partial u_m}; \dots;$$
$$\frac{\partial z_k}{\partial u_m} = \frac{\partial z_k}{\partial z_{k-1}} \frac{\partial z_{k-1}}{\partial u_m}; \qquad m = \overline{1, r}. \quad (4)$$

In order to illustrate the application of formulas (2), (3) and (4), we slightly changed the elementary function, which was given in [2] in such a way that its partial derivatives were not equal to each other. Assume that the following elementary function is given:

$$f(u) = f(u_1, u_2) = \exp(u_1 + (u_2)^2) + \sin((u_1)^3 + u_2)^2.$$

We build the multistep process of its calculation as follows:

$$z_1 = u_1 + (u_2)^2; \qquad z_2 = \exp(z_1); \qquad z_3 = (u_1)^3 + u_2;$$
$$z_4 = (z_3)^2; \qquad z_5 = \sin(z_4); \qquad z_6 = z_2 + z_5. \quad (5)$$

Using formulas (2) and (3), we have

$$p_1 = \exp(z_1) p_2; \qquad p_2 = p_6; \qquad p_3 = 2 z_3 p_4;$$
$$p_4 = \cos(z_4) p_5; \qquad p_5 = p_6; \qquad p_6 = 1; \quad (6)$$

$$\frac{\partial f}{\partial u_1} = p_1 + 3(u_1)^2 p_3; \qquad \frac{\partial f}{\partial u_2} = 2 u_2 p_1 + p_3. \quad (7)$$

Conjugate variables are calculated in the reverse order: $p_6, p_5, \dots, p_1$.

In order to obtain partial derivatives of function $f(u)$ without using the adjoint problem (using formulas (4)), we introduce the notation: $\tilde{z}_i = \frac{\partial z_i}{\partial u_1}$, $\tilde{\tilde{z}}_i = \frac{\partial z_i}{\partial u_2}$. Values $\tilde{z}_i$ and $\tilde{\tilde{z}}_i$ are calculated by the formulas:

$$\tilde{z}_1 = 1; \qquad \tilde{z}_2 = \tilde{z}_1 \exp(z_1); \qquad \tilde{z}_3 = 3(u_1)^2;$$
$$\tilde{z}_4 = 2 \tilde{z}_3 z_3; \qquad \tilde{z}_5 = \tilde{z}_4 \cos(z_4); \qquad \tilde{z}_6 = \tilde{z}_2 + \tilde{z}_5. \quad (8)$$

$$\tilde{z}_1 = 2u_2; \qquad \tilde{z}_2 = \tilde{z}_1 \exp(z_1); \qquad \tilde{z}_3 = 1;$$
$$\tilde{z}_4 = 2\tilde{z}_3 z_3; \qquad \tilde{z}_5 = \tilde{z}_4 \cos(z_4); \qquad \tilde{z}_6 = \tilde{z}_2 + \tilde{z}_5. \tag{9}$$

Then

$$\frac{\partial f}{\partial u_1} = \tilde{z}_6; \qquad \frac{\partial f}{\partial u_2} = \tilde{z}_6.$$

In this case, instead one multistep process (6), which is required for calculation of conjugate variables, it is necessary to use $r = 2$ (number of independent variables) multistep processes: (8) and (9). Consequently, the time required to calculate the gradient with the FAD-technique formulas is about r times less than the time required to calculate this gradient using formulas (4). The effectiveness of using FAD-technique increases if the number of independent variables increases.

All computational experiments were carried out for a function E that calculates the total interatomic energy of the atoms' system whose interaction potential is the Tersoff potential. Let $\bar{r}_i = (x_{1i}, x_{2i}, x_{3i})$ are the coordinates of some lattice atom. As to total energy $E(\bar{r}_1, \bar{r}_2, \ldots, \bar{r}_I)$ it is calculated with the help of formulae $E(\bar{r}_1, \bar{r}_2, \ldots, \bar{r}_I) = \sum_{1=1}^{I} \sum_{j=1; j \neq i}^{I} V_{ij}$, where V_{ij} is the interaction potential between atoms marked i and j (i-atom and j-atom):

$$V_{ij} = f_c(r_{ij}) \left(V_R(r_{ij}) - b_{ij} V_A(r_{ij}) \right),$$

$$f_c(r) = \begin{cases} 1, & r < R - R_{cut}, \\ \frac{1}{2}\left(1 - \sin\left(\frac{\pi(r-R)}{2R_{cut}}\right)\right), & R - R_{cut} < r < R + R_{cut}, \\ 0, & r > R + R_{cut}, \end{cases}$$

$$V_{ij}^R = V_R(r_{ij}) = \frac{D_e}{S-1} \exp\left(-\beta\sqrt{2S}(r_{ij} - r_e)\right),$$

$$V_{ij}^A = V_A(r_{ij}) = \frac{SD_e}{S-1} \exp\left(-\beta\sqrt{\frac{2}{S}}(r_{ij} - r_e)\right),$$

$$b_{ij} = (1 + (\gamma\zeta_{ij})^\eta)^{-\frac{1}{2\eta}}, \quad \zeta_{ij} = \sum_{k=1; k \neq i,j}^{I} f_c(r_{ik})g_{ijk}\omega_{ijk}, \quad \omega_{ijk} = \exp(\lambda^3\tau_{ijk}),$$

$$\tau_{ijk} = (r_{ij} - r_{ik})^3, \qquad g_{ijk} = 1 + \left(\frac{c}{d}\right)^2 - \frac{c^2}{d^2 + (h - \cos\Theta_{ijk})^2}.$$

Here I is the number of atoms in considered system; r_{ij} is the distance between i-atom and j-atom:

$$r_{ij} = \sqrt{(x_{1i} - x_{1j})^2 + (x_{2i} - x_{2j})^2 + (x_{3i} - x_{3j})^2};$$

Θ_{ijk} is the angle between two vectors, first vector begins at i-atom and finishes at j-atom, second vector begins at i-atom and finishes at k-atom and

$$\cos \Theta_{ijk} = q_{ijk} = \frac{r_{ij}^2 + r_{ik}^2 - r_{jk}^2}{2r_{ij}r_{ik}};$$

R and R_{cut} are known parameters, identified from experimental geometric properties of substance. Tersoff Potential depends on ten parameters $(m = 10)$, specific to modeled substance: $D_e, r_e, \beta, S, \eta, \gamma, \lambda, c, d, h$.

To tackle the problems of computer modeling the crystal structures gradient optimization methods are often used. This raises the need to determine the exact gradient of the total energy E of atoms' system with respect to specific parameters of Tersoff Potential.

When using different optimization methods, there is also a need for smoothing the function $f_c(r)$. It is proposed to replace the function $f_c(r)$ as follows:

$$f_c(r) = \begin{cases} 0, & r \geq R + R_{cut}, \\ 1, & r \leq R - R_{cut}, \\ C \cdot (f_*)^{\varphi(r)}, & R \leq r < R + R_{cut}, \\ C \cdot \left(2f_* - (f_*)^{\psi(r)}\right), & R - R_{cut} < r \leq R, \end{cases}$$

where

$$C = \frac{1}{2f_*}, \quad f_* = \exp\left(-\frac{3}{2}\right), \quad \varphi(r) = \frac{R_{cut}^2}{(r - R - R_{cut})^2}, \quad \psi(r) = \frac{R_{cut}^2}{(r - R + R_{cut})^2}.$$

Derivative of function $f_c(r)$ with respect to r is calculated by the formulae:

$$\frac{\partial f_c(r)}{\partial r} = \begin{cases} 0, & r \geq R + R_{cut}, \\ 0, & r \leq R - R_{cut}, \\ C \cdot (f_*)^{\varphi(r)} \ln(f_*) \cdot \widetilde{\varphi}(r), & R \leq r < R + R_{cut}, \\ -C \cdot (f_*)^{\psi(r)} \ln(f_*) \cdot \widetilde{\psi}(r), & R - R_{cut} < r \leq R, \end{cases}$$

where

$$\widetilde{\varphi}(r) = \frac{-2R_{cut}^2}{(r - R - R_{cut})^3}, \qquad \widetilde{\psi}(r) = \frac{-2R_{cut}^2}{(r - R + R_{cut})^3}.$$

The optimization problems are solved with determined, fixed position of basic atoms of the considered crystal structure. After solving the parameters identification problem in such a statement, there is no certainty that the positions of basic atoms will correspond to the minimum of the system's energy. Therefore, the following step of studies is to find the particles coordinates, that minimizing the summary potential energy of the considered system of atoms. At this stage, there is a need to determine the gradient of the energy of atoms'system with respect to the coordinates of the atoms.

The considered function E is the obvious case of the fact that without the use of a multistep process it is practically impossible to calculate its partial derivatives with respect to the chosen arguments. Formulas for calculating the exact gradient of the function E with respect to 10 parameters specific to modeled substance, as well as formulas for calculating its partial derivatives with respect to the coordinates of atoms using FAD-methodology are given in [3].

To obtain the derivatives $\frac{\partial E}{\partial x_{lm}}$ ($l = 1, 2, 3;\ m = \overline{1, I}$) of the function E with respect to the coordinates of atoms using formulas (4), we use the same multistep process of calculating this function, which was built in [3]. We introduce vectors $\overline{u}$ and $\overline{z}$ having the following coordinates: $\overline{u}^T = [u_1, u_2, ..., u_{10}]^T$, $\overline{z}^T = [z_1, z_2, ..., z_{10}]^T$, where $u_1 = D_e$, $u_2 = r_e$, $u_3 = \beta$, $u_4 = S$, $u_5 = \eta$, $u_6 = \gamma$, $u_7 = \lambda$, $u_8 = c$, $u_9 = d$, $u_{10} = h$;

$$z_1 = \left\{ z_1^{ijk} = \sqrt{(x_{1i} - x_{1k})^2 + (x_{2i} - x_{2k})^2 + (x_{3i} - x_{3k})^2} \right\} \equiv F(1, Z_1, U_1),$$

$$z_2 = \left\{ z_2^{ijk} = \sqrt{(x_{1j} - x_{1k})^2 + (x_{2j} - x_{2k})^2 + (x_{3j} - x_{3k})^2} \right\} \equiv F(2, Z_2, U_2),$$

$$z_3 = \left\{ z_3^{ijk} = q_{ijk} = \frac{(z_{13}^{ij})^2 + (z_1^{ijk})^2 - (z_2^{ijk})^2}{2 z_1^{ijk} z_{13}^{ij}} \right\} \equiv F(3, Z_3, U_3),$$

$$z_4 = \left\{ z_4^{ijk} = f_c(z_1^{ijk}) \right\} \equiv F(4, Z_4, U_4),$$

$$z_5 = \left\{ z_5^{ijk} = g_{ijk} = 1 + \left(\frac{u_8}{u_9}\right)^2 - \frac{(u_8)^2}{(u_9)^2 + (u_{10} - z_3^{ijk})^2} \right\} \equiv F(5, Z_5, U_5),$$

$$z_6 = \left\{ z_6^{ijk} = \tau_{ijk} = (z_{13}^{ij} - z_1^{ijk})^3 \right\} \equiv F(6, Z_6, U_6),$$

$$z_7 = \left\{ z_7^{ijk} = \omega_{ijk} = \exp((u_7)^3 z_6^{ijk}) \right\} \equiv F(7, Z_7, U_7),$$

$$z_8 = \left\{ z_8^{ijk} = f_c(r_{ik}) g_{ijk} \omega_{ijk} = z_4^{ijk} z_5^{ijk} z_7^{ijk} \right\} \equiv F(8, Z_8, U_8),$$

$$z_9 = \left\{ z_9^{ij} = \zeta_{ij} = \sum_{k=1; k \neq i, j}^{I} z_8^{ijk} \right\} \equiv F(9, Z_9, U_9),$$

$$z_{10} = \left\{ z_{10}^{ij} = \gamma \zeta_{ij} = u_6 z_9^{ij} \right\} \equiv F(10, Z_{10}, U_{10}),$$

$$z_{11} = \left\{ z_{11}^{ij} = (\gamma \zeta_{ij})^\eta = (z_{10})^{u_5} \right\} \equiv F(11, Z_{11}, U_{11}),$$

$$z_{12} = \left\{ z_{12}^{ij} = b_{ij} = (1 + z_{11}^{ij})^{-\frac{1}{2 u_5}} \right\} \equiv F(12, Z_{12}, U_{12}),$$

$$z_{13} = \left\{ z_{13}^{ij} = \sqrt{(x_{1i} - x_{1j})^2 + (x_{2i} - x_{2j})^2 + (x_{3i} - x_{3j})^2} \right\} \equiv F(13, Z_{13}, U_{13}),$$

$$z_{14} = \left\{ z_{14}^{ij} = V_{ij}^R = \frac{u_1}{u_4 - 1} \exp\left(-u_3 \sqrt{2 u_4}(z_{13}^{ij} - u_2)\right) \right\} \equiv F(14, Z_{14}, U_{14}),$$

$$z_{15} = \left\{ z_{15}^{ij} = V_{ij}^A = \frac{u_1 u_4}{u_4 - 1} \exp\left(-u_3 \sqrt{\frac{2}{u_4}}(z_{13}^{ij} - u_2)\right) \right\} \equiv F(15, Z_{15}, U_{15}),$$

$$z_{16} = \left\{ z_{16}^{ij} = f_c(z_{13}^{ij}) \right\} \equiv F(16, Z_{16}, U_{16}),$$

$$z_{17} = \left\{ z_{17}^{ij} = V_{ij} = z_{16}^{ij}(z_{14}^{ij} - z_{12}^{ij}z_{15}^{ij}) \right\} \equiv F(17, Z_{17}, U_{17}),$$

$$(i = \overline{1,I}, \quad j = \overline{1,I}, \quad j \neq i, \quad k = \overline{1,I}, \quad k \neq i,j).$$

The energy E of the atoms in the system with the help of new variables may be rewritten as follows:

$$E = E(x_{11}, x_{21}, x_{31}, ..., x_{1I}, x_{2I}, x_{3I}) = \sum_{i=1}^{I} \sum_{j=1; j\neq i}^{I} z_{17}^{ij}.$$

Variables $z_1, z_2, ..., z_{17}$ (the phase variables) are determined by the specified above multistep algorithm $z_l = F(l, Z_l, U_l)$, $(l = 17)$, where Z_l is the set of elements z_n in the right part of the equation $z_l = F(l, Z_l, U_l)$, and U_l is the set of elements u_n that appear in the right side of this equation. Note that each component z_l depends on a number of other components (z_l^{ij} or z_l^{ijk}).

Let us introduce also the following designations: $\tilde{z}_1, \tilde{z}_2, ..., \tilde{z}_{17}$, where

$$\tilde{z}_s = \left\{ \tilde{z}_s^{ijk} : \tilde{z}_s^{ijk} = \frac{\partial z_s^{ijk}}{\partial x_{lm}} \right\}, \quad s = \overline{1,8}, \qquad \tilde{z}_s = \left\{ \tilde{z}_s^{ij} : \tilde{z}_s^{ij} = \frac{\partial z_s^{ij}}{\partial x_{lm}} \right\}, \quad s = \overline{9,17}.$$

These values are calculated by the formulas:
For all $l = 1, 2, 3$ and $i, j, k, m = \overline{1,I}$ we have

$$\tilde{z}_1^{ijk} = \frac{\partial z_1^{ijk}}{\partial x_{lm}} = \begin{cases} (x_{lm} - x_{lk})/z_1^{mjk}, & m = i, m \neq k, \\ (x_{lm} - x_{li})/z_1^{ijm}, & m = k, m \neq i, \\ 0, & \text{in other cases;} \end{cases}$$

$$\tilde{z}_2^{ijk} = \frac{\partial z_2^{ijk}}{\partial x_{lm}} = \begin{cases} (x_{lm} - x_{lk})/z_2^{imk}, & m = j, m \neq k, \\ (x_{lm} - x_{lj})/z_2^{ijm}, & m = k, m \neq j, \\ 0, & \text{in other cases;} \end{cases}$$

$$\tilde{z}_3^{ijk} = \frac{z_1^{ijk}(z_{13}^{ij})^2 \tilde{z}_{13}^{ij} + z_{13}^{ij}(z_1^{ijk})^2 \tilde{z}_1^{ij} - 2z_1^{ijk} z_{13}^{ij} z_2^{ijk} \tilde{z}_2^{ijk}}{2(z_{13}^{ij})^2(z_1^{ijk})^2} +$$

$$+ \frac{z_1^{ijk}(z_2^{ijk})^2 \tilde{z}_{13}^{ij} - (z_1^{ijk})^3 \tilde{z}_{13}^{ij} - (z_{13}^{ij})^3 \tilde{z}_1^{ij} + z_{13}^{ij} \tilde{z}_1^{ijk}(z_2^{ijk})^2}{2(z_{13}^{ij})^2(z_1^{ijk})^2};$$

$$\tilde{z}_4^{ijk} = \begin{cases} 0, & z_1^{ijk} \geq R + R_{cut}, \\ 0, & z_1^{ijk} \leq R - R_{cut}, \\ -2C \cdot R_{cut}^2 \ln(f_*)(f_*)^{\varphi(z_1^{ijk})} \cdot \dfrac{\tilde{z}_1^{ijk}}{(z_1^{ijk} - R - R_{cut})^3}, & R \leq z_1^{ijk} < R + R_{cut}, \\ 2C \cdot R_{cut}^2 \ln(f_*)(f_*)^{\psi(z_1^{ijk})} \cdot \dfrac{\tilde{z}_1^{ijk}}{(z_1^{ijk} - R + R_{cut})^3}, & R - R_{cut} < z_1^{ijk} \leq R; \end{cases}$$

$$\tilde{z}_5^{ijk} = \frac{-2(u_8)^2(u_{10} - z_3^{ijk})\tilde{z}_3^{ijk}}{\left((u_9)^2 + (u_{10} - z_3^{ijk})^2\right)^2}; \qquad \tilde{z}_6^{ijk} = 3(z_{13}^{ij} - z_1^{ijk})^2(\tilde{z}_{13}^{ij} - \tilde{z}_1^{ijk});$$

$$\tilde{z}_7^{ijk} = (u_7)^3 \exp\left((u_7)^3 z_6^{ijk}\right) \tilde{z}_6^{ijk};$$

$$\tilde{z}_8^{ijk} = \tilde{z}_4^{ijk} z_5^{ijk} z_7^{ijk} + \tilde{z}_5^{ijk} z_4^{ijk} z_7^{ijk} + \tilde{z}_7^{ijk} z_5^{ijk} z_4^{ijk};$$

$$\tilde{z}_9^{ij} = \sum_{k=1;\ k\neq i,j}^{I} \tilde{z}_8^{ijk}; \qquad \tilde{z}_{10}^{ij} = \tilde{z}_9^{ij} u_6; \qquad \tilde{z}_{11}^{ij} = \tilde{z}_{10}^{ij} u_5 (z_{10}^{ij})^{u_5-1};$$

$$\tilde{z}_{12}^{ij} = -\frac{\tilde{z}_{11}^{ij}(1 + z_{11}^{ij})^{-1/2u_5-1}}{2u_5};$$

$$\tilde{z}_{13}^{ij} = \frac{\partial z_{13}^{ij}}{\partial x_{lm}} = \begin{cases} (x_{lm} - x_{lj})/z_{13}^{mj}, & m = i, m \neq j, \\ (x_{lm} - x_{li})/z_{13}^{im}, & m = j, m \neq i, \\ 0, & \text{in other cases}; \end{cases}$$

$$\tilde{z}_{14}^{ij} = -u_3\sqrt{2u_4} z_{14}^{ij} \tilde{z}_{13}^{ij}; \qquad\qquad \tilde{z}_{15}^{ij} = -u_3\sqrt{2/u_4} z_{15}^{ij} \tilde{z}_{13}^{ij};$$

$$\tilde{z}_{16}^{ij} = \begin{cases} 0, & z_{13}^{ij} \geq R + R_{cut}, \\ 0, & z_{13}^{ij} \leq R - R_{cut}, \\ -2C \cdot R_{cut}^2 \ln(f_*)(f_*)^{\varphi(z_{13}^{ij})} \cdot \frac{\tilde{z}_{13}^{ij}}{(z_{13}^{ij}-R-R_{cut})^3}, & R \leq z_{13}^{ij} < R + R_{cut}, \\ 2C \cdot R_{cut}^2 \ln(f_*)(f_*)^{\psi(z_{13}^{ij})} \cdot \frac{\tilde{z}_{13}^{ij}}{(z_{13}^{ij}-R+R_{cut})^3}, & R - R_{cut} < z_{13}^{ij} \leq R; \end{cases}$$

$$\tilde{z}_{17}^{ij} = \tilde{z}_{16}^{ij} z_{14}^{ij} + \tilde{z}_{16}^{ij} z_{12}^{ij} z_{15}^{ij} + \tilde{z}_{14}^{ij} z_{16}^{ij} - \tilde{z}_{12}^{ij} z_{16}^{ij} z_{15}^{ij} - \tilde{z}_{15}^{ij} z_{12}^{ij} z_{16}^{ij}.$$

Then the derivatives of the function E with respect to the coordinates of the atoms will be calculated by formulas:

$$\frac{\partial E}{\partial x_{lm}} = \sum_{i=1}^{I} \sum_{j=1;\ j\neq i}^{I} \tilde{z}_{17}^{ij}, \qquad l = 1,2,3; \quad m = \overline{1,I}.$$

A large series of computational experiments was carried out, in which the dimensionality of the material (the three-dimensional and two-dimensional model was considered) and the number of atoms in the material fragment varied. If a three-dimensional model is considered ($l = 1,2,3$), then for each m-atom, computation by a multistep process $\tilde{z}_1, \tilde{z}_2, ..., \tilde{z}_{17}$ must be done 3 times, and if a two-dimensional model is considered ($l = 1,2$), then - 2 times. All numerical experiments showed that in the first case (the three-dimensional model) the gradient of the energy function with respect to the coordinates of the atom, determined with the help of the FAD-technique formulas, which are given in [2], is calculated three times faster than with the help of a multistep process $\tilde{z}_1, \tilde{z}_2, ..., \tilde{z}_{17}$. When a two-dimensional model of a material is used, the effectiveness of formulas grows in 2 times.

3   Comparison of FAD-Technique and Software Packages

As a computational experiment for comparing the determination of derivatives using the above methods implemented manually and with the aid of the software packages, the two-dimensional model of a multilayer piecewise-homogeneous material proposed in [4,5] is considered. In this model, the material is represented as a periodic piecewise homogeneous multilayer structure in which the types of atoms in different layers may be different. This model imposes the following constraints on the structure of the layers:

1. Each layer consists of identical atoms, but different layers may consist of different atoms.
2. The distances between adjacent atoms in the same level are identical, but they may be different in different layers.
3. There is a group of K parallel layers that are periodically repeated in the direction of the axis y.
4. The number of atoms in each layer and the total number of layers is potentially unbounded.

Figure 1 gives an example of the model in which a group of three layers is repeated. Each layer consists of atoms of a specific type.

In this model, the position of the atoms is determined by the following parameters:

h_i, $i = 1, \ldots, K$ is the distance between the layer number i and the preceding layer;

d_i, $i = 1, \ldots, K$ is the offset of the first atom in layer i with the positive abscissa relative to the zero point;

s_i, $i = 1, \ldots, K$ is the distance between the atoms in layer i.

The set of values of these parameters is called a configuration. It is required to determine the configuration corresponding to the minimum interaction energy of the atoms which enter into the simulated material fragment. The optimization problem consists in minimizing the energy $E(u)$ of the atoms' group (the potential of the interaction of atoms - the Tersoff potential) located on the K adjacent layers. The parameters of the optimization problem are the variables $u = (h_1, d_1, s_1, \ldots, h_K, d_K, s_K)$. The size of the vector u is $3K$. When we use gradient methods of function minimization, it is necessary to calculate the derivatives: $\frac{\partial E}{\partial h_k}, \frac{\partial E}{\partial d_k}, \frac{\partial E}{\partial s_k}$.

If x_{1i} is the first coordinate of the i-th atom of a considered structure; x_{2i} is its second coordinate, and the $i - th$ atom is an atom with an ordinal number j on the $k - th$ layer, then: $x_{1i} = d_k + (j-1)k$;

$$x_{2i} = 0, \quad if \quad k = 0, \qquad x_{2i} = \sum_{m=2}^{k} h_m, \quad if \quad k = 2, 3, \ldots, K.$$

The corresponding derivatives are calculated by the formulas:

$$\frac{\partial E}{\partial h_k} = \sum_{i=1}^{I} \frac{\partial E}{\partial x_{2i}} \frac{\partial x_{2i}}{\partial h_k}, \quad \frac{\partial E}{\partial d_k} = \sum_{i=1}^{I} \frac{\partial E}{\partial x_{1i}} \frac{\partial x_{1i}}{\partial d_k}, \quad \frac{\partial E}{\partial s_k} = \sum_{i=1}^{I} \frac{\partial E}{\partial x_{1i}} \frac{\partial x_{1i}}{\partial s_k}.$$

In the case when the determination of derivatives $\frac{\partial E}{\partial x_{1i}}$ and $\frac{\partial E}{\partial x_{2i}}$ was implemented manually, the FAD-technique formulas (see [3]) or the multistep process $\tilde{z}_1, \tilde{z}_2, ..., \tilde{z}_{17}$ (previous section) were used.

The test stand consisted of the following computer facilities:

1. Intel Core i7 4770 3.4 Hz, 8 Gb RAM – 4-core processor Haswell architecture, with support for Hyper-Threading technologies (two compute circuits per core, 8 logical cores), AVX2 -Advanced Vector Extensions (in the context of the problem being solved, processing data in floating point format in groups of 256 bits in length - that is, for a double-precision number, a vector of 4 numbers is simultaneously processed), and FMA-FusedMultiply-Add - performing a combined operation of multiplication-addition of the form a = a + b * c. The GNUC ++ compiler was used 6.4.0.
2. 2 × Intel XeonE5-2683V4 2.1 Hz, 512 Gb RAM – two-processor servers, each processor contains 16 cores supporting Hyper-Threading, AVX2 and FMA (64 logical cores). The GNUC ++ 7.2.0 compiler was used.

In the work we used well-proven earlier [6] packages CoDiPack 1.6.0 [7] and Adept 1.1 [8].

The methods were compared:

- Manual implementation of a direct method FAD, with some optimization of the obtained formulas, we will call it – **Hand1**;
- Manual implementation of the reverse method FAD – **Hand2**;
- Direct method implemented in the package CoDiPack [6] – **CoDiPack**;
- The reverse method implemented in the package Adept [7] – **Adept**;

For Hand1 and Hand2 methods, the optimizer option -Ofast, for Adept and CoDiPack -O2, was used. In addition, two parallelization methods are shown for the CoDiPack package, the first one using AVX2 technology (one-core processing of a vector of four numbers with double precision) and the second using OpenMP directives.

Calculations were made for a different number of layers and atoms:
- 4 layers and 94 atoms, the dimension of the problem r = 12
- 8 layers and 188 atoms, the dimension of the problem r = 24
- 12 layers and 204 atoms, the dimension of the problem r = 36

When calculating the gradient with the CoDiPack, the RealForward data type was used, consisting of two double precision numbers, the value and the partial derivative of the given value, using one of the independent variables. When using this data type to obtain a gradient, it is necessary to start the process of calculating the function and its partial derivative r times. In the parallel version of calculations, these calculations were parallelized using the OpenMP directive - *#pragma omp for*. In the second parallelization method, the RealForwardVecGen data type was used, consisting of r + 1 double precision numbers - the value of the variable and the vector of its partial derivatives. Using this type of data allowed to use AVX2 and accelerate calculations from 1.5 to 4 times without any labor, it was enough to specify the optimization option -Ofast for the compiler. The time for calculating the gradient and the function is shown in Table 1. When investigating the performance of methods of fast automatic differentiation, an important indicator is the ratio of the time of calculation of the gradient and the function to the time of calculation of the function itself. These data are given in Table 2.

Table 1. The calculation time (seconds) of the function and the gradient (simultaneously with the function).

Computer	IntelCore i7			IntelXeon		
Number of layers × Number of atoms	4×94	8×188	12×204	4×94	8×188	12×204
Function	0.25	0.92	1.16	0.27	1.11	1.38
Hand1	25.89	218.89	325.18	22.96	154.43	210.42
Hand2	2.58	22.16	28.27	3.95	49.53	60.75
CoDiPack	5.63	44.08	83.24	4.96	39.06	74.25
Adept	1.25	5.88	8.03	2.43	10.51	13.60
Parallel versions CoDiPack						
CoDiPack + AVX2	1.66	14.30	70.29	3.48	33.66	135.07
CoDiPack + OpenMP	1.64	10.00	20.55	0.72	5.10	4.75

Table 2. The ratio of the time for calculating the gradient (simultaneously with the function) to the time of calculating the function.

Computer	IntelCore i7			IntelXeon		
Number of layers × Number of atoms	4×94	8×188	12×204	4×94	8×188	12×204
Hand1	103.56	237.42	281.22	84.44	139.47	152.00
Hand2	10.31	24.03	24.45	14.52	44.73	43.89
CoDiPack	22.52	47.91	71.76	18.37	35.19	53.80
Adept	5.00	6.39	6.92	9.00	9.47	9.86

As you can see from Table 1, the Hand1 method has the lowest performance. The direct method from the CoDiPack package outperforms it 3–5 times in calculations using the one processor core and without using the SIMD technology. Expectedly the best performance is shown by the methods Hand2 and Adept. In this case, the reverse method of the Adept package is 2–4 times higher than the manual implementation. Table 2 shows that the calculation time of the gradient (and function) using the Hand1 and CoDiPack methods depends linearly on the

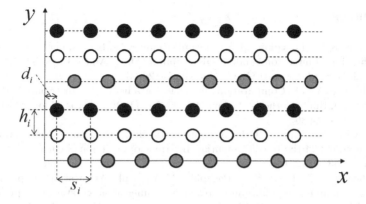

Fig. 1. Two-dimensional model of substance

dimension and time of the function calculation. Also, the computation time by the inverse method of the Adept package does not depend on the dimension of the problem. This is consistent with theoretical estimates.

For information purposes, Table 1 shows two ways of parallelizing the direct method of the CoDiPack package. For Intel Core i7, with small dimensions of the task, the use of technology AVX2 allows you to speed up the calculation in 3 times. Approximately the same acceleration is provided by the use of OpenMP (3–4 times, regardless of the size of the problem). For Intel Xeon, it is advisable to use OpenMP (acceleration is 6–15 times).

To perform the calculations, the computing resources of Federal Research Center Computer Science and Control of Russian Academy of Sciences were used (http://frccsc.ru).

4   Conclusion

The carried out investigations showed that FAD-methodology is an effective instrument to the calculation of the gradients of complex functions. During the solution of specific problems, it is possible to derive FAD-formulas by hand and to use them for calculation of the gradient or to use the existing program packages. How to work in a specific case depends on the problem. Both approaches make it possible to calculate gradients with the machine precision. The deficiencies of "handle" approach are next: deriving the FAD-formulas can be attributed by high intellectual expenditures. However, the use of program packages can collide with large limitations to the dimensionality of the problem and to the resources of computer.

References

1. Abgaryan, K.K., Posypkin, M.A.: Optimization methods as applied to parametric identification of interatomic potentials. Comput. Math. Math. Phys. **54**, 1929–1935 (2014). https://doi.org/10.1134/S0965542514120021
2. Evtushenko, Y.G.: Computation of exact gradients in distributed dynamic systems. Optim. Methods Softw. **9**, 45–75 (1998). https://doi.org/10.1080/10556789808805686
3. Albu, A.F.: Application of the fast automatic differentiation to the computation of the gradient of the tersoff potential. Informacionnye tekhnologii i vychislitel'nye sistemy **1**, 43–49 (2016)
4. Evtushenko, Y.G., Lurie, S.A., Posypkin, M.A., et al.: Application of optimization methods for finding equilibrium states of two-dimensional crystals. Comput. Math. Math. Phys. **56**, 2001–2010 (2016). https://doi.org/10.1134/S0965542516120083
5. Evtushenko, Y., Lurie, S., Posypkin, M.: New optimization problems arising in modelling of 2D-crystal lattices. In: AIP Conference Proceedings, vol. 1776 (2016). https://doi.org/10.1063/1.4965341
6. Albring, T., et al.: An aerodynamic design framework based on algorithmic differentiation. ERCOFTAC Bull. **102**, 10–16 (2015)
7. Hogan, R.J.: Fast reverse-mode automatic differentiation using expression templates in C++. ACM Trans. Math. Softw. (TOMS) **40**(4), 26–42 (2014). https://doi.org/10.1145/2560359
8. Gorchakov, A.Y.: On software packages of fast automatic differentiation. Informacionnye tekhnologii i vychislitel'nye sistemy **1**, 30–36 (2018)

Numerical Damping of Forced Oscillations of an Elastic Beams

Andrey Atamuratov[1], Igor Mikhailov[1,2], and Nikolay Taran[1(✉)]

[1] Moscow Aviation Institute (National Research University), Moscow, Russia
gooffydog@mail.ru, mikh_igor@mail.ru, n1ckolaytaran@gmail.com
[2] Federal Research Center "Informatics and Control" of RAS, Moscow, Russia

Abstract. The beam oscillations are modeled by the fourth-order hyperbolic partial differential equation. The minimized functional is the energy integral of an oscillating beam. Control is implemented via certain function appearing in the right side of the equation. It was shown that the solution of the problem exists for any given damping time, but with decreasing this time, finding the optimal control becomes more complicated. In this work, numerical damping of beam oscillations is implemented via several fixed point actuators. Computational algorithms have been developed on the basis of the matrix sweep method and the second order Marquardt minimization method. To find a good initial approximation empirical functions with a smaller number of variables are used. Examples of damping the oscillations via a different number of actuators are given. It is shown that the amplitude of the oscillations of any control functions increases with the reduction of the given damping time. Examples of damping the oscillations in the presence of constraints on control functions are given; in this case, the minimum damping time exists. The damping of oscillations is considered also in the case when different combinations of actuators are switched on at different time intervals of oscillation damping.

Keywords: Marquardt minimization method · Oscillations damping
Fixed point actuators · Matrix sweep method

1 Introduction

Methods of damping of oscillations of elements of complex mechanical systems began to develop intensively in the 70s of the XX century. The most significant were the works of Lagness [1], Russell [2], Butkovskiy [3,4], in which the problem of damping of string oscillations was considered and conditions for the existence of a solution to the problem were obtained. In particular, Butkovskiy proposed to use a point actuator for damping the oscillations of the string. However, later it was shown that in the case if a solution appears in the form of standing waves, if the actuator is in a node of standing waves, then the solution of the problem

Supported by Russian Science Foundation, Project 17-19-01247.

Y. Evtushenko et al. (Eds.): OPTIMA 2018, CCIS 974, pp. 277–290, 2019.
https://doi.org/10.1007/978-3-030-10934-9_20

may not exist. To avoid such a situation, Muravey [5,6] suggested using a point actuator moving along a small section of a string, but the practical implementation of such actuator is very difficult. In [7,8] it was shown that the solution of the problem exists for any positive time T, however, as T decreases, finding the optimal control becomes more complicated. In this paper, we consider the possibility of damping the beam oscillations using several fixed point actuators.

2 Problem Statement

2.1 Oscillations Damping

The purpose of this work is to develop numerical method for damping forced transverse oscillations of a beam via multiple fixed point actuators. The transverse oscillations of a beam are described by the Petrovsky-hyperbolic equation

$$u_{tt} = -a^2 u_{xxxx} + g(x,t), (x,t) \in \prod = \{0 \leq x \leq l, 0 \leq t \leq T\}. \tag{1}$$

Here, the time t and the linear dimension x are related to the characteristic values t^* and x^*. We will consider the initial displacement and the velocity of the beam movement

$$u|_{t=0} = h_0(x), \ u_t|_{t=0} = h_1(x), 0 \leq x \leq l \tag{2}$$

as initial perturbations. At the ends of the beam, the conditions of articulation are superimposed.

$$u|_{x=0} = u_{xx}|_{x=0} = 0, u|_{x=l} = u_{xx}|_{x=l} = 0, 0 \leq t \leq T. \tag{3}$$

The energy of the oscillating beam is

$$E(t) = \int_0^l \left[u_t^2(x,t) + a^4 u_{xx}^2(x,t) \right] dx. \tag{4}$$

The problem of damping is to find the control function $g(x,t)$, which transfers the beam from the initial state (2) to the state

$$u|_{t=T} = 0, \ u_t|_{t=T} = 0, 0 \leq x \leq l \tag{5}$$

in time T. According to Lions [9], this property of the system is called strict controllability.

Thus, the problem of damping of oscillations consists in finding the optimal control function $g(x,t) \in L_2((0,T) \times (0,l))$ such, that for any initial perturbations $h_0(x), h_1(x)$

$$E(T) = 0. \tag{6}$$

As a control function, we consider p fixed point actuators

$$g(x,t) = \sum_{i=1}^p w_i(t) \delta(x - x_i), \tag{7}$$

where $w_i(t), i = 1, \ldots, p$ - control functions, δ - Dirac delta-function, x_i - points in which the actuators are placed. We will assume that $w_i(t) \in L_2(0,T), i = 1, \ldots, p$.

2.2   Numerical Solution

Equation (1) can be reduced to a system of two equations of the second order

$$\begin{cases} u_t = av_{xx}, \\ v_t = -au_{xx} + \sum_{i=1}^{p} f_i(x,t); \end{cases} \tag{8}$$

where

$$f_i(x,t) = \begin{cases} w_i(t)\left(-\frac{x}{al}(l-x_i)\right), x < x_i, \\ w_i(t)\left(\frac{1}{a}(x-x_i) - \frac{x}{al}(l-x_i)\right), x \geq x_i. \end{cases} \tag{9}$$

We solve it by the matrix sweep method [10]. We approximate the control functions $w_i(t)$, $i = 1,\ldots,p$ with piecewise-constant functions: $\forall t \in [t_i, t_{i+1})$ assume $w_i(t)$ where w_j^i - const, $i = 1,\ldots,p, j = 0,\ldots,N_{T-1}$. Then the integral of the beam energy will be a function of the variables w_j^i

$$E(T) = L\left(w_0^1,\ldots,w_{N_T}^p\right)$$
$$= \int_0^l \left[u_t^2\left(w_0^1,\ldots,w_{N_T}^p,x,T\right) + a^4 u_{xx}^2\left(w_0^1,\ldots,w_{N_T}^p,x,T\right)\right] dx \tag{10}$$

For the numerical computation of the energy integral (10) we use the Simpson method.

2.3   Minimization

The optimal values $w_0^1,\ldots,w_{N_T}^p$, which minimize (10) with a specified accuracy ε, are the required solution of the problem. To solve the oscillation damping problem, we use the Marquardt method [11].

For large sizes of a finite-difference grid or when using a sufficiently large number of actuators, the numerical computation of control functions by using second-order minimization methods can be a computationally complex problem that requires a lot of computation time. However, it is possible to significantly reduce the computation time by finding a good initial approximation from minimizing some empirical function that depends on a small number of parameters.

The basic idea of using empirical functions is to replace the initial minimizable function with another continuous function $w(t)_{\text{EMP}}$ which depends on a small number of parameters. Suppose that each of the control functions has the following form

$$w(e_1,\ldots,e_7,t)_{\text{EMP}} = e_1 \sin(e_2 t + e_3) + e_4 \sin(e_5 t + e_6)\sin(e_7 t), \tag{11}$$

where constant values $e_1,\ldots,e_7$ are not yet known. We introduce a special transformation function

$$L_{\text{EMP}}(e) = L\left(w(e)_{\text{EMP}},\ldots,w(e)_{\text{EMP}}\right). \tag{12}$$

To find the empirical coefficients $e_1^1,\ldots,e_7^1,\ldots,e_1^p,\ldots,e_7^p$, we will solve the problem of finding the minimum of the function (12) using the Marquardt minimization method. The resulting control functions are used as the initial approximation for minimizing (10) with a specified accuracy ε.

Empirical formulas are best used as an initial approximation or in tasks where precision is not a high priority.

3   Use of Multiple Actuators

3.1   Solution Existence Problem

Consider the following example. The initial conditions are $h_0(x) = 0.1 \sin(2\pi x)$, $h_1(x) = 0$. The input parameters are $a = 1, l = 1$, and we set the required damping time equal to $T = 0.5$, the size of the finite-difference grid will be $N \times K = 20 \times 250$, so $h_x = 0.05, h_t = 0.002$. We will assume that the oscillation damping task is solved if $L(w(t)) \leq \varepsilon$, where $\varepsilon = 10^{-4}$.

The actuator placed at the point $x_0 = 0.5$ can not dampen the beam oscillations, since the point $x_0 = 0.5$ is a node of standing waves. This is clearly seen in the Fig. 1.

The model suggested by Butkovsky has the following drawback: in the case of the appearance of a solution (1) in the form of standing waves, if x_0 falls into a node of standing waves, then the solution of the problem may not exist. Let us consider the same conditions of the example, but for damping the oscillations we use two fixed point actuators at the points $x_1 = 0.25, x_2 = 0.75$, respectively. We rewrite the condition for solving the task in the form $L(w_1(t), w_2(t)) \leq 10^{-4}$. The size and steps of the finite-difference grid are the same.

We solve the task for two cases: using the minimization of the function (12) with initial control of the form $w_1(t) \equiv 0, w_2(t) \equiv 0$ and using empirical functions (12) to obtain the initial approximation. In the first case, the task was solved with the error $L(w_1(t), w_2(t)) = 5.0701 \cdot 10^{-13}$.

In the second case, using the Marquardt minimization method, we find the following empirical coefficients e:

$$e = \begin{pmatrix} 1000 & -0.1196 & 0.0317 & 1000 & -6.6844 & 2.9977 & -6.5086 \\ 1000 & -2.7326 & 3.2003 & 1000 & -2.5081 & -3.1785 & 9.9958 \end{pmatrix} \tag{13}$$

Substituting them into (12), we obtain the control functions $w_1(t)_{\text{EMP}}$, $w_2(t)_{\text{EMP}}$, allowing to solve the system with an error of $L_{\text{EMP}}(e_1^1, \ldots, e_7^2) = 0.32495$. Next, we take them as the initial approximation and use the Marquardt method again for the final determination of the control functions $w_1(t), w_2(t)$. As a result, we minimized the value of the beam energy integral (12) with the error $L(w_1(t), w_2(t)) = 4.8804 \cdot 10^{-13}$.

The graphs of the values of the function $u(x, t)$, illustrating the process of damping the beam oscillations, and the final form of the control functions $w_1(t)$ and $w_2(t)$ for both cases are shown on Figs. 2 and 3 respectively.

Thus, the task is solved for a time $T = 0.5$. It is noticeable that, despite the same initial conditions and grid parameters, the form of the control functions differs depending on the initial approximation, and the damping proceeds in different ways.

3.2   Alternation of Actuators

To dampen oscillations in the case of certain initial conditions, it may be necessary to use different groups of actuators at different time intervals. Let the initial

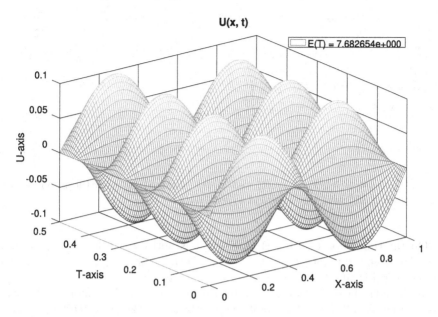

Fig. 1. The process of damping of oscillations via an actuator at the point $x_0 = 0.5$ (damping does not occur)

conditions be $h_0(x) = 0.25 \exp(x) \sin(2\pi x), h_1(x) = 0$. The input parameters are $a = 1, l = 1$, the required damping time is $T = 0.2$, the size of the finite-difference grid will be $N \times K = 20 \times 250$, so $h_x = 0.05, h_t = 0.0008$. We assume, that the task of damping the oscillations is solved if

$$L(w_1(t), \ldots, w_p(t))) \le \varepsilon, \qquad (14)$$

where $\varepsilon = 10^{-4}$.

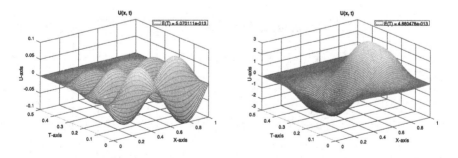

Fig. 2. The oscillation damping process via two actuators at the points $x_1 = 0.25, x_2 = 0.75$ (a) with the initial values $w_0^1, \ldots, w_{N_T}^2 = 0$, (b) with the empirical approximation (3.2)

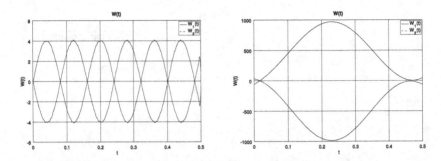

Fig. 3. The control functions $w_1(t)$ and $w_2(t)$, obtained (a) with the initial values $w_1, \ldots, w_2 = 0$, (b) with the empirical approximation (3.2)

On the time slice $T = 0.2$ (Fig. 4(b)), it is noticeable that the two actuators placed at the points $x_1 = 0.25$ and $x_2 = 0.75$ can not dampen the initial oscillations (Fig. 4(a)).

To dampen the oscillations, we divide the task into two time intervals. We will use 4 actuators placed at the points $x_1 = 0.15, x_2 = 0.25, x_3 = 0.65, x_4 = 0.75$, but in the first interval $T \in [0; 0.1]$ we will use only two of them at the points x_2 and x_4. The other two actuators at the points x_1 and x_3 on this interval will be left inactive.

Minimizing the function (12), we obtain the empirical coefficients e:

$$e = \begin{pmatrix} 1000 & -3.2196 & 3.0779 & 1000 & -5.9374 & -0.0239 & 7.5418 \\ 1000 & 30.536 & 3.0844 & 1000 & 25.405 & 0.1781 & 29.208 \end{pmatrix} \tag{15}$$

Further we get $w_2(t)_{\text{EMP}}, w_4(t)_{\text{EMP}}$ and use them as the initial approximation for $w_2(t), w_4(t)$. We solve the task with the error $L(w_2(t), w_4(t)) = 1.1797$. In Fig. 5 the process of partial damping of oscillations in the time interval $T = [0; 0.1]$ and a time slice of $T = 0.1$ is shown.

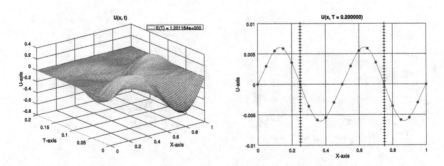

Fig. 4. The process of oscillation damping via two actuators at the points $x_1 = 0.25, x_2 = 0.75$ (partial damping), (b) the cut of the values of the function $u(x, t), t = T = 0.2$ (further damping is not possible)

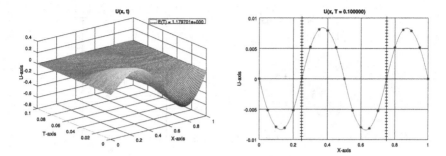

Fig. 5. The oscillation damping process via two actuators at the points $x_2 = 0.25, x_4 = 0.75$ on the interval T $in[0; 0.1]$ (partial damping), (b) the cut of the values of the function $u(x, t), t = T = 0.1$ (further damping is not possible)

In the second time interval $T \in [0.1; 0.2]$ to dampen the oscillations we use two remaining actuators at the points $x_1 = 0.15, x_3 = 0.65$. The previous two actuators at the points x_2 and x_4 are left inactive. We will use the solution of the previous subtask as a new initial perturbation. For the initial velocity, in this case we put $v(l, t)$ of the previous subtask as $v(0, t)$ of the second subtask.

By minimizing (12), we obtain e:

$$e = \begin{pmatrix} 1000 & 2.8741 & 3.1415 & 1000 & 10.572 & 0.9895 & 3.0815 \\ 1000 & 6.8311 & 3.1421 & 1000 & 13.532 & -2.3983 & -7.1904 \end{pmatrix} \quad (16)$$

By using $w_1(t)_{\text{EMP}}$, $w_3(t)_{\text{EMP}}$ as the initial approximation for $w_1(t)$ and $w_3(t)$ we get the error $L(w_1(t), w_3(t)) = 2.2163 \cdot 10^{-13}$ for the second subtask. Figure 6 shows the process of damping of oscillations in the time interval $T \in [0.1; 0.2]$ (a scale is used for $u(x, t)$, which is 20 times smaller than in Fig. 5) and a time slice of $T = 0.2$.

Thus, the task is solved in time $T = 0.2$ via two actuators $x_2 = 0.25$ and $x_4 = 0.75$ on $T \in [0; 0.1]$ and two actuators $x_1 = 0.15$ and $x_3 = 0.65$ on $T \in [0.1; 0.2]$ with the resulting error $L(w_1(t), \ldots, w_p(t)) = 2.2163 \cdot 10^{-13}$. Combining $u_1(x, t)$ and $u_2(x, t)$ into $u(x, t)$, we illustrate on Fig. 7 a complete process of damping the oscillations in this task.

Thus, a numerical method for damping the beam oscillations is developed via several fixed point actuators. It makes it possible to research the process of damping the oscillations for a different time T.

3.3   Dependence of the Damping Process on Time

Suppose that the initial conditions have the following form $h_0 = 0.1 \sin(2\pi x), h_1(x) = 0$. The input parameters are $a = 1, l = 1$, the size of the finite-difference grid is $N \times K = 20 \times 50$, so $h_x = 0.05, h_t = 0.002$. To dampen the oscillations, we use 4 actuators placed at the points $x_1 = 0.2, x_2 = 0.4, x_3 = 0.6, x_4 = 0.8$, respectively. The condition for damping the oscillations, as before, will be assumed

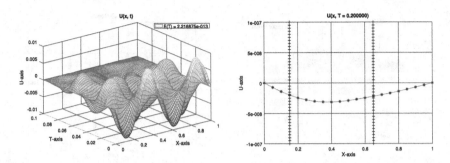

Fig. 6. (a) The oscillations damping process via two actuators at the points $x_1 = 0.15, x_3 = 0.65$, on the interval $T \in [0.1; 0.2]$, (b) the cut of the values of the function $u(x, t), t = T = 0.2$

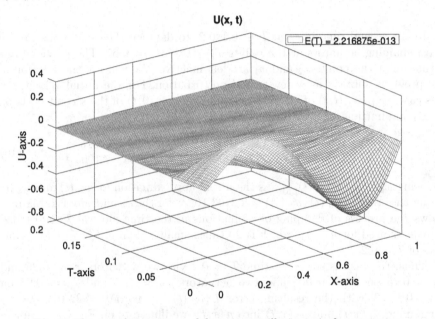

Fig. 7. The complete process of damping oscillations via four actuators

$$L(w_1(t), w_2(t), w_3(t), w_4(t)) \leq \varepsilon, \tag{17}$$

where $\varepsilon = 10^{-4}$. We set the damping time to $T = 0.1$. By default, the initial approximation for all control functions will be assumed to be zero.

On Fig. 8 the process of damping of the oscillations $u(x, t)$ and the control functions $w_i(t), i = 1, \ldots, 4$ are shown.

Consider the same conditions of the example, but put $T = 0.01$. On Fig. 9 the damping process and control functions are shown.

Drawing attention to Figs. 8(a) and 9(a), it is possible to clearly notice the difference in the process of damping of the oscillations depending on the prescribed damping time T. Thus, if we set a sufficiently long time, the damping

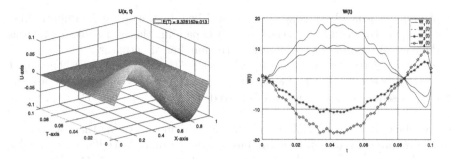

Fig. 8. (a) The oscillations damping process via four actuators at the points $x_1 = 0.2, x_2 = 0.4, x_3 = 0.6, x_4 = 0.8$ for the time $T = 0.1$, (b) the control functions $w_1(t), w_2(t), w_3(t)$, and $w_4(t)$

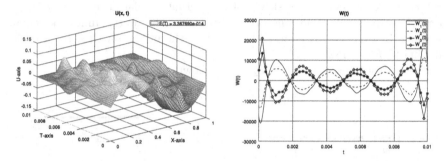

Fig. 9. (a) The oscillations damping process via four actuators at the points $x_1 = 0.2, x_2 = 0.4, x_3 = 0.6, x_4 = 0.8$ for the time $T = 0.01$, (b) the control functions $w_1(t), w_2(t), w_3(t)$, and $w_4(t)$

proceeds more smoothly. Conversely, if a small time is set, a multitude of micro-oscillations on the beam arise, which are then smoothed by the control functions. The amplitude of the oscillations of any control functions increases almost exponentially along with a decrease in the prescribed damping time.

3.4   Constrained Control Functions

Let us consider the case when constraints are imposed on control functions. This case is more approximate to the practical implementation since in the design of actuator mechanisms it is necessary to lay the maximum permissible power of the drives. To find constrained control functions, it is necessary to use methods of finding a conditional minimum. In this work, the penalty minimization method is used using the Marquardt method to solve the corresponding problem of finding an unconditional minimum.

Since the maximum amplitude of each of the control functions begins to increase as the damping time approaches zero, it is necessary to select a damping time for which $L(w_1(t), \dots, w_p(t)) \le \varepsilon$ and $w_i(t) \in [a; b]$. We call the minimal time T, under which both conditions are satisfied, by the optimal damping time. The damping time can be reduced either by expanding the admissible boundaries of the control function or by increasing the number of actuators.

Let the initial conditions be $h_0(x) = 0.2x(1-x), h_1(x) = 0$. The input parameters are $a = 1, l = 1$, the size of the grid is $N \times K = 40 \times 120$. We assume that the oscillation damping task is solved if $L(w_1(t), \ldots, w_p(t)) \le \varepsilon$, where $\varepsilon = 10^{-4}$.

We show that it is possible to reduce the minimum damping time by increasing the number of actuators with constant constraints. Initially, we will solve the task using a single fixed point actuator, placed in $x_1 = 0.5$. On the control function, we impose the constraints $w(t) \in [-2; 2]$. The minimum time required for damping is $T = 0.2265$.

The control function $w(t)$ is shown on Fig. 10(a).

Now we solve the same task via of two actuators placed in $x_1 = 0.25$ and $x_2 = 0.75$ respectively. The constraints imposed on the control functions remain the same. In this case, it is possible to reduce the minimum time required for damping to $T = 0.1825$.

The control functions $w_1(t), w_2(t)$ are shown on Fig. 10(b).

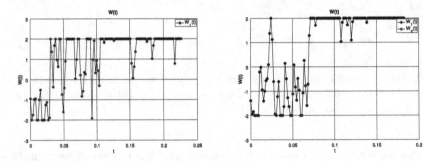

Fig. 10. Control functions w_i $left(t$ $right)$ with constraints $w_i(t) \in [-2; 2], i = 1, \ldots, p$ for (a) $p = 1, T = 0.2265$, (b) $p = 2, T = 0.1825$

And, finally, we solve the task with the use of 4 actuators at the points $x_1 = 0.25, x_2 = 0.4, x_3 = 0.6, x_4 = 0.75$. We impose the previous constraints on all control functions. In this case, the minimum time was reduced to $T = 0.1237$.

The oscillation damping process and control functions are shown on Fig. 11.

3.5   Realtime Oscillations Damping

In the previous examples, the problem of searching for some "ideal" control is considered, which is assumed a priori known before the direct beginning of oscillations damping. However, in fact, at the moment of the beginning of the damping of the oscillations, we do not know anything about how the actuators should behave, and the task of finding the control arises. Of course, we could

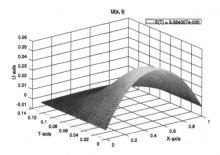 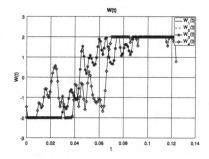

Fig. 11. (a) Control functions $w_1(t), w_2(t), w_3(t)$ and $w_4(t)$ with constraints $w_1(t), w_2(t), w_3(t), w_4(t) \in [-2; 2]$ for $T = 0.1237$, (b) the oscillations damping process via four actuators at the points $x_1 = 0.25, x_2 = 0.4, x_3 = 0.6, x_4 = 0.75$ for the time $T = 0.1237$ with constraints $w_1(t), w_2(t), w_3(t), w_4(t) \in [-2; 2]$

set the required damping time as a constant and start the process of finding control functions using the minimization methods right when registering the displacement, but it is obvious that this should take some time (which can be very, very impressive, up to several hours for some conditions). Moreover, even if we spend some time after the oscillations begin to look for control functions and try to apply the resulting control, it is likely that the displacement of the beam at this point in time will be completely different, and our control will, at best, not have the required smoothing effect on the beam, or even worse, will intensify the oscillations. We show that using the method of numerical damping described above it is possible to obtain the required control directly during the damping process. In this task, we made the transition from the dimensionless T to the dimensional damping time.

Let the parameters of the beam be $a = 1, l = 1$. On the beam are set $N + 1 = 21$ sensors at the same distance $h_x = l/N = 0.05$ from each other, which register the oscillations. Consider the case when sensors has detected the oscillations of the beam shown in the Fig. 12.

Now we will search the damping time T in seconds.

Numerous computations have shown that the time in which one iteration of the Marquardt method is calculated is on the average 90 ms. Under the iteration calculation time, here is meant the total execution time of the OpenCL kernels of calculating the integral of the energy, the gradient and the Hessian matrix of the function (10), obtained using the runtime profiler. Suppose that the response time of the actuators is 10 ms. Thus, the total response time of the system will be $h_t = 90$ ms $+ 10$ ms $= 0.1$ s.

By the condition, the time spent by the actuators for damping the beam oscillations on each individual time interval $t_i \in [0, T], t_i - t_{i-1} = h_t$ is exactly equal the real program execution time on calculation the iteration of the numerical damping method corresponding to this time interval. In other words, the time of damping T, which was considered in all previous tasks, will be equal to the time of the program execution in seconds.

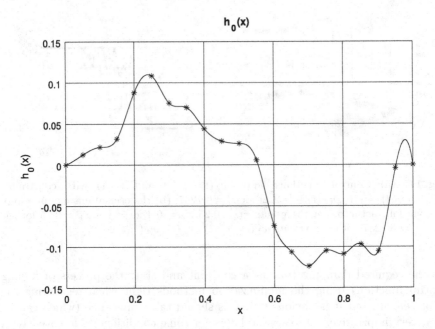

Fig. 12. The initial displacement of $h_0(x)$

The calculation process ends when $L(w_1(t), w_2(t)) \leq \varepsilon$, where $\varepsilon = 10^{-4}$. In this case, oscillations damping was performed with an allowable error of $L(w_1(t), w_2(t)) = 7.7879 \cdot 10^{-5}$ for 30 iterations of minimization method. Multiplying the number of iterations by the response time, we get that the damping time of the oscillations was $T = 3\,\text{s}$.

The process of damping of the oscillations $u(x, t)$ and the form of the control functions $w_1(t)$, $w_2(t)$ are shown on Fig. 13.

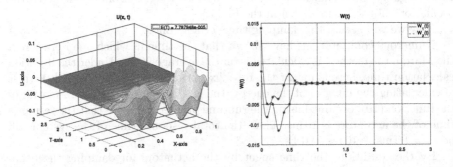

Fig. 13. (a) The process of damping the oscillations in real time via two actuators at the points $x_1 = 0.25, x_2 = 0.75$ for the time $T = 3\,\text{s}$, (b) the control functions $w_1(t)$ and $w_2(t)$

4   Conclusion

The problem of damping of forced transverse oscillations of an elastic beam for an arbitrary predetermined time T is considered. A computational algorithm with second-order convergence is developed for approximating the calculation of the oscillations of a freely supported beam with a given control. Control is considered via several fixed point actuators. Finding the control that performs oscillations damping is accomplished via minimization of a certain function of multiple variables by the Marquardt method.

For sufficiently large sizes of a finite-difference grid, the process of finding a numerical solution can be computationally complex and time-consuming. Empirical formulas are proposed, which make it possible to significantly reduce the calculation time. Empirical formulas are best used as an initial approximation or in tasks where precision is not a high priority.

The case is considered when different groups of actuators are used for different periods of time to dampen oscillations.

The oscillations damping is considered in the presence of constraints imposed on the control functions. In this case, the minimum time for damping the oscillations exists. It is shown that this time can be reduced by increasing the number of actuators.

In conclusion, an estimate was made of the real-time of finding the control of oscillations damping and, as a consequence, the possibility of practical implementation of the proposed algorithm.

References

1. Lagness, J.: Control of wave process with distributed controls supported on a subregion. SIAM J. Control Optim. **1**(1), 68–85 (1983)
2. Russel, D.: Controllability and stabilization theory for linear partial differential equations. SIAM Rev. **20**(5), 639–739 (1978)
3. Butkovsky, A.G.: Metody upravleniia sistemami s raspredelennymi parametrami [Methods of controlling systems with distributed parameters]. Nauka, Moscow (1975)
4. Butkovsky, A.G.: Prilozhenie nekotorykh rezultatov teorii chisel k probleme finitnogo upravleniia i upravliaemosti v raspredelennykh sistemakh [Application of some results of number theory to the problem of finite control and controllability in distributed systems]. Proc. USSR Acad. Sci. **227**(2), 309–311 (1976)
5. Muravey, L.: On the suppression on membrane oscillations. In: Summaries of IUTAM Symposium "Dynamical Problems of Rigid-elastic System", Moscow, pp. 50–51 (1990)
6. Muravey, L.: Mathematical problems on the damp of vibration. In: Preprint of IFAC Conference "Identification and system parameter estimations", Budapest, vol. 1, pp. 746–747 (1991)
7. Atamuratov, F., Mikhailov, I., Muravey, L.: The moment problem in control problems of elastic dynamic systems. Mechatron. Autom. Control **17**(9), 587–598 (2016)

8. Atamuratov, A., Mikhailov, I., Taran, N.: Numerical damping of oscillations of beams by using multiple point actuators. Mathematical Modeling and Computational Physics: Book of Abstracts of the International Conference, p. 158. JINR, Dubna (2017)
9. Lions, J.L.: Optimal Control of Systems Governed by Partial Differential Equations. Springer, New York (1971)
10. Samarsky, A.A., Gulin, A.V.: Chislennye metody: Ucheb. posobie dlia vuzov [Numerical methods. Textbook for Universities]. Nauka, Moscow (1989)
11. Panteleev, A.V., Letova, T.A.: Metody optimizatsii v primerakh i zadachakh: Ucheb. Posobie [Optimization methods in examples and tasks. Textbook for Universities]. Vysshaia shkola, Moscow (2005)

Solutions of Traveling Wave Type for Korteweg-de Vries-Type System with Polynomial Potential

Levon A. Beklaryan[1] , Armen L. Beklaryan[1,2](✉) ,
and Alexander Yu. Gornov[3]

[1] Central Economics and Mathematics Institute RAS,
Nachimovky prospect 47, 117418 Moscow, Russia
`beklar@cemi.rssi.ru`
[2] National Research University Higher School of Economics,
Kirpichnaya Ulitsa 33, 105187 Moscow, Russia
`abeklaryan@hse.ru`
[3] Institute for System Dynamics and Control Theory of SB RAS,
Lermontova street 134, 664033 Irkutsk, Russia
`gornov@icc.ru`
`http://www.hse.ru/en/staff/beklaryan`

Abstract. This paper deals with the implementation of numerical methods for searching for traveling waves for Korteweg-de Vries-type equations with time delay. Based upon the group approach, the existence of traveling wave solution and its boundedness are shown for some values of parameters. Meanwhile, solutions constructed with the help of the proposed constructive method essentially extend the class of systems, possessing solutions of this type, guaranteed by theory. The proposed method for finding solutions is based on solving a multiparameter extremal problem. Several numerical solutions are demonstrated.

Keywords: Korteweg-de Vries equation
Functional differential equations · Traveling waves

1 Introduction

The theory of differential equations with delay, which has developed rapidly in recent decades [9,15,23], has in many ways acquired a complete form and is now actively used in modeling various objects. Numerical algorithms for solving equations with a delay of various types were also developed [8,21]. At the same time, numerical methods for equations with advanced arguments (and, moreover, with mixed type deviations) are practically not studied, although references to them have been encountered for a long time, mainly in connection with the classification of equations with deviating argument [10]. As a rule, there is a parsing of equations of a particular kind with further obtaining the existence and uniqueness theorems based on the use of the properties of the right-hand side [2]

© Springer Nature Switzerland AG 2019
Y. Evtushenko et al. (Eds.): OPTIMA 2018, CCIS 974, pp. 291–305, 2019.
https://doi.org/10.1007/978-3-030-10934-9_21

and the application of methods such as the study of the roots of the characteristic quasi-polynomial [11], collocation methods and finite element scheme (expansion of the solution in terms of basis functions of some finite-dimensional space) [1, 19,20] or through the construction of a Hilbert space of the reproducing kernels on the basis of boundary conditions [18].

In spite of the fact that within the framework of this article we consider a functional-differential equation with only a delay in time, nevertheless, both the theoretical approach and the methods of numerical solution construction, which are proposed, are successfully applied to functional differential equations of pointwise type (FDEPT, also known as mixed type functional differential equations) [7,30].

The remaining part of this paper is organized as follows. In Sect. 2, the description of the developed program complex for construction of numerical solutions, and also the description of used methods are given. Section 3 is devoted to the fundamentals of the theoretical approach to investigating traveling wave-type solutions. In Sect. 4, we describe examples of several numerical solutions for Korteweg-de Vries-type equation's initial-boundary value problem with time delay.

2   Optimization Problem and Numerical Methods

Let's give a formal statement of the optimization problem for the search for numerical solutions of FDEPT. We consider the system

$$F_i(t, \dot{x}(t), x(t + n_1), \ldots, x(t + n_s)) = 0, \quad i = \overline{1, k}, \quad t \in [t_l, t_r],$$

where $F : \mathbb{R} \times \mathbb{R}^n \times \mathbb{R}^{ns} \longrightarrow \mathbb{R}^n$ – mapping of $C^{(0)}$ class; $n_j \in \mathbb{Z}, j = \overline{1, s}$; $t_l, t_r \in \mathbb{R}$; the values of the derivatives of the phase variables are defined on the extended interval $t \in [t_{ll}, t_{rr}]$, $t_{ll} = t_l + \min\{0, n_1, \ldots, n_s\}$, $t_{rr} = t_r + \max\{0, n_1, \ldots, n_s\}$

$$\dot{x}(t) = h_l(t), \quad t \in [t_{ll}, t_l],$$
$$\dot{x}(t) = h_r(t), \quad t \in [t_r, t_{rr}],$$

where $h_l, h_r : \mathbb{R}^n \longrightarrow \mathbb{R}^n$ – mapping of $C^{(0)}$ class. The initial-boundary conditions are given by the functionals

$$K_m(\dot{x}(\tau), x(\tau_1), \ldots, x(\tau_p)) = 0, \quad m = \overline{1, q}, \quad \tau, \tau_i \in [t_l, t_r], \quad i = \overline{1, p}.$$

The optimization problem consists in finding a trajectory $\hat{x}(t)$ that delivers the minimum of the residual functional

$$I(\hat{x}(t)) = v^{(N)} \left(\sum_{i=1}^{k} \int_{t_l}^{t_r} F_i^2(t, \dot{\hat{x}}(t), \hat{x}(t + n_1), \ldots, \hat{x}(t + n_s)) dt \right.$$

$$\left. + \int_{t_{ll}}^{t_l} [\dot{\hat{x}}(t) - h_l(t)]^2 dt + \int_{t_r}^{t_{rr}} [\dot{\hat{x}}(t) - h_r(t)]^2 dt \right)$$

$$+ v^{(K)} \sum_{m=1}^{q} K_m^2(\dot{\hat{x}}(\tau), \hat{x}(\tau_1), \ldots, \hat{x}(\tau_p)),$$

where $v^{(N)}, v^{(K)} \in \mathbb{R}_+$ – weighting coefficients.

The proposed approach to the investigation of boundary value problems is based on the Ritz method and spline-collocation constructions and was implemented in [7,29,30]. To solve the problems of the class under consideration, the trajectories of the system are discretized on a grid with a constant step, and a generalized residual functional is formulated that includes both the weighted residual of the original differential equation and the residual of the boundary conditions. To evaluate the derivatives of the desired trajectories of the system, a spline differentiation technique is used, based on two spline approximation designs: using cubic natural splines and using a special type of spline whose second derivatives at the edges are also controlled using optimized parameters.

For solving the stated finite-dimensional problems, a set of algorithms for local optimization (quasi-Newtonian method BFGS; two versions of the Powell's method; Barzilai-Borwein method; version of the method of confidence domains; stochastic search methods in random subspaces of dimension 3, 4, and 5, and others) and global optimization (method of random multistart; method of curvilinear searching; tunnel method; parabolic method, and others) was implemented. The used technology includes: an algorithm for sequentially increasing the accuracy of approximation by multiplying the number of nodes in the grid of the discretization; the algorithms for the difference evaluation of the derivatives of the functional – from the first to the sixth degree of accuracy inclusive; the method of successively increasing the precision of spline differentiation.

The corresponding software complex (SC) *OPTCON-F* was implemented in the language *C* under the control of operating systems *OS Windows*, *OS Linux* and *Mac OS* using compilers *BCC 5.5* and *GCC*. SC was designed to obtain a numerical solution of boundary value problems, parametric identification problems and optimal control for dynamical systems described by FDEPT [14].

In the SC *OPTCON-F* there is the possibility of sequential application of various algorithms within the framework of constructing a solution for one task. Thus, the constructed intermediate solution in the previous step becomes the starting solution ("baseline") for the following algorithm. In this case, such an implementation does not prevent the global search algorithms from "popping out" of the local solution. Separately, we note the presence of a programming module that allows predetermining the order of application of algorithms, as well as the construction of complex chain of steps (conditional statements, cycles, etc.) depending on the current or historical values of a number parameters (for example, error estimation or number of iterations). For the examples presented below, the following scheme was used in cycle: the generalized quasi-Newtonian and Powell-Brent's methods (with bi-directional line search along each dimension) were used alternately, and after the error changed by less than 10^{-h} the adaptive modification of the Hooke-Jeeves method was used l times (h and l are computable functions on the basis of the loop iteration number, as well as the Lipschitz constants of the equation itself and a number of other technical characteristics). The stopping criterion depended on the number of iterations in the first part of the cycle, as well as the current error estimate and its dynamics. In view of the above, as well as stochastic elements in the applied algorithms, the presented value of the residual functional (RF, i.e. error of a solution) can not be used to estimate the theoretical rate of convergence.

The heuristic search algorithm for the solution $\hat{x}(t)$ can be justified on the basis of the existence and uniqueness theorems for initial-boundary value problems for the investigated FDEPT, as well as theorems on approximating solutions of such equations on the whole line by solutions of the initial-boundary value problem on a sequence of expanding intervals. A description of such equations and the corresponding results are presented in the following sections.

3   Functional Differential Equations

3.1   Solutions of the Traveling Wave Type

For equations of mathematical physics, which are the Euler-Lagrange equation of the corresponding variational problem, an important class of solutions is traveling wave solutions (soliton solutions) [22,25]. In a number of models, such solutions are well approximated by traveling wave type solutions for finite-difference analogs of the original equations, which, in place of a continuous environment, describe the interaction of clumps of an environment placed at lattice sites [12,25]. Emerging systems belong to the class of infinite-dimensional dynamical systems.

Further the following infinite-dimensional dynamical system is studied

$$\ddot{y}_i = \psi(i, \dot{y}_i, y_{i-k_1}, \dots, y_{i-k_p}), \quad i, k_j \in \mathbb{Z}, \quad y_i \in \mathbb{R}, \quad t \in \mathbb{R}, \tag{1}$$

where $\psi(\cdot)$ is a smooth function. The Eq. (1) could be a system with a different potentials, for example, a system with the Frenkel-Kontorova potential [7,12]. In this case, such a system is a finite difference analog of the nonlinear wave equation. The study of such systems in general case is one of the intensively developing directions in the theory of dynamical systems. For these systems, the central task is to study solutions of the traveling wave type as one of the observed wave classes.

Definition 1. *We say that the solution $\{y_i(\cdot)\}_{-\infty}^{+\infty}$ of the system (1), defined for all $t \in \mathbb{R}$, has a traveling wave type, if there is $\tau > 0$, independent of t and i, that for all $i \in \mathbb{Z}$ and $t \in \mathbb{R}$ the following equality holds*

$$y_i(t + \tau) = y_{i+1}(t).$$

The constant τ will be called a characteristic of a traveling wave.

The proposed approach is based on the existence of a one-to-one correspondence of solutions of traveling wave type for infinite-dimensional dynamical systems with solutions of induced FDEPT [4]. To study the existence and uniqueness of solutions of traveling wave type, it is proposed to localize solutions of induced FDEPT in spaces of functions, majorized by functions of a given exponential growth, where the exponent is the parameter of the selected family of functions, which is defined as follows

$$\mathcal{L}_\mu^n C^{(k)}(\mathbb{R}) = \left\{ x(\cdot) : x(\cdot) \in C^{(k)}\left(\mathbb{R}, \mathbb{R}^n\right), \ \max_{0 \le r \le k} \sup_{t \in \mathbb{R}} \|x^{(r)}(t)\mu^{|t|}\|_{\mathbb{R}^n} < +\infty \right\}$$

with the norm

$$\|x(\cdot)\|_\mu^k = \max_{0 \le r \le k} \sup_{t \in \mathbb{R}} \|x^{(r)}(t)\mu^{|t|}\|_{\mathbb{R}^n}.$$

This approach is particularly successful for systems with Frenkel-Kontorova potentials and polynomial potentials. In this way, it is possible to obtain a "correct" extension of the concept of a traveling wave in the form of solutions of the quasi-traveling wave type, which is related to the description of processes in inhomogeneous environments for which the set of traveling wave solutions is trivial [5, 6].

In fact, the described connection between solutions of the traveling wave type of the infinite-dimensional dynamical system and solutions of the induced functional-differential equation is a fragment of a more general scheme that goes beyond the scope of this article.

3.2   Initial-Boundary Value Problem

In Sect. 3.1 we noted that the study of solutions of the traveling wave type is equivalent to the study of solutions of an induced FDEPT. We now turn to the study of such general equations. The most important goal in the study of FDEPT is the study of the *basic initial-boundary value problem*

$$\dot{x}(t) = f(t, x(q_1(t)), \dots, x(q_s(t))), \quad t \in B_R, \tag{2}$$

$$\dot{x}(t) = \varphi(t), \quad t \in \mathbb{R} \backslash B_R, \quad \varphi(\cdot) \in L_\infty(\mathbb{R}, \mathbb{R}^n), \tag{3}$$

$$x(\bar{t}) = \bar{x}, \quad \bar{t} \in \mathbb{R}, \quad \bar{x} \in \mathbb{R}^n, \tag{4}$$

where: $f : \mathbb{R} \times \mathbb{R}^{ns} \longrightarrow \mathbb{R}^n$ – mapping of the $C^{(0)}$ class; $q_j(\cdot)$, $j = 1, \dots, s$ – diffeomorphisms of the line preserving orientation; B_R is either closed interval $[t_0, t_1]$ or closed half-line $[t_0, +\infty)$ or line $\mathbb{R}$. $L_\infty(\mathbb{R}, \mathbb{R}^n)$ – space of measurable functions that are essentially bounded on each finite interval.

The solution of the Eq. (2) is any absolutely continuous function $x(t)$, $t \in \mathbb{R}$ that satisfies this equation almost everywhere. *The solution of the basic initial-boundary value problem* is any solution of the Eq. (2) that satisfies the boundary condition (3) and the initial condition (4).

If the mapping $f : \mathbb{R} \times \mathbb{R}^{n \cdot s} \longmapsto \mathbb{R}^n$, in the right-hand side of the equation (2), is a mapping of the class $C^{(0)}(\mathbb{R} \times \mathbb{R}^{n \cdot s}, \mathbb{R}^n)$ and satisfies the conditions of a quasilinear growth

$$\|f(t, z_1, \dots, z_s)\|_{\mathbb{R}^n} \le M_0(t) + M_1 \sum_{j=1}^s \|z_j\|_{\mathbb{R}^n}, \quad M_0(\cdot) \in C^{(0)}(\mathbb{R}, \mathbb{R}),$$

and the Lipschitz condition

$$\|f(t, z_1, \ldots, z_s) - f(t, \bar{z}_1, \ldots, \bar{z}_s)\|_{\mathbb{R}^n} \leq M_2 \sum_{j=1}^{s} \|z_j - \bar{z}_j\|_{\mathbb{R}^n},$$

and satisfies some additional conditions then a number of key results are true. For example, we have a theorem on the existence and uniqueness of solution [3, p. 570]. Moreover, in this case the solution depends continuously both on the initial-boundary conditions $\bar{x} \in \mathbb{R}^n$, $\varphi(\cdot) \in L_\infty(\mathbb{R}, \mathbb{R}^n)$ and on the right-hand side $f(\cdot)$ (as an element of the space with Lipschitz norm). It is important to note that the continuity condition, the growth conditions with respect to the phase variables and time variable, as well as the Lipschitz condition, are standard conditions in the theory of ordinary differential equations.

Along with the initial-boundary value problem (2)–(4), we consider some of its reformulation. To this end, we define a normed function space

$$\mathcal{L}_\mu^n L_\infty(\mathbb{R}) = \left\{ x(\cdot) : x(\cdot) \in L_\infty(\mathbb{R}, \mathbb{R}^n), \sup_{t \in \mathbb{R}} vrai \|x(t)\mu^{|t|}\|_{\mathbb{R}^n} < +\infty \right\}, \ \mu \in (0, 1]$$

with a norm

$$\|x(\cdot)\|_\mu = \sup_{t \in \mathbb{R}} vrai \|x(t)\mu^{|t|}\|_{\mathbb{R}^n}.$$

Reformulation of the initial-boundary value problem is as follows

$$\dot{x}(t) = f(t, x(q_1(t)), \ldots, x(q_s(t))), \quad t \in B_R,$$
$$\dot{x}(t) = \varphi(t), \quad t \in \mathbb{R} \backslash B_R, \quad \varphi(\cdot) \in \mathcal{L}_\mu^n L_\infty(\mathbb{R}),$$
$$x(\bar{t}) = \bar{x}, \quad \bar{t} \in \mathbb{R}, \quad \bar{x} \in \mathbb{R}^n.$$

It is obvious that the solvability of the original initial-boundary value problem and its reformulation are equivalent. Theorem of the existence and uniqueness of solution is also valid for the reformulation of the initial-boundary value problem, only with the replacement of the boundary condition $\varphi(\cdot) \in L_\infty(\mathbb{R}, \mathbb{R}^n)$ by the boundary condition $\varphi(\cdot) \in \mathcal{L}_\mu^n L_\infty(\mathbb{R})$ [4].

In what follows, we confine ourselves to considering the case of constant commensurable deviations. Without loss of generality, in this case we can assume that the deviations are integer, i.e. $q_j(t) = t + n_j$, $n_j \in \mathbb{Z}$, $j = 1, \ldots, s$. Similarly, we can assume that the initial moment $\bar{t}$ and the ends t_0, t_1 of the interval of the definition of the equation are also an integer.

Under the same conditions on the mapping $f(\cdot)$, a result was obtained regarding the approximation of solutions of an initial-value problem defined on the whole line by solutions of the initial-boundary value problem defined on the interval $[-k, k]$ as $k \to +\infty$. We considered the initial-value problem on the whole line $B_R = \mathbb{R}$

$$\dot{x}(t) = f(t, x(t + n_1), \ldots, x(t + n_s)), \quad t \in \mathbb{R}, \tag{5}$$
$$x(\bar{t}) = \bar{x}, \quad \bar{t} \in \mathbb{R}, \quad \bar{x} \in \mathbb{R}^n \tag{6}$$

and for each $k \in \mathbb{Z}$ the initial-boundary value problem on a finite interval $B_R = [-k, k]$

$$\dot{x}(t) = f(t, x(t + n_1), \ldots, x(t + n_s)), \quad t \in [-k, k], \tag{7}$$
$$\dot{x}(t) = \varphi(t), \quad t \in \mathbb{R}\backslash[-k, k], \quad \varphi(\cdot) \in \mathcal{L}_1^n L_\infty(\mathbb{R}), \tag{8}$$
$$x(\bar{t}) = \bar{x}, \quad \bar{t} \in \mathbb{R}, \quad \bar{x} \in \mathbb{R}^n. \tag{9}$$

Under these conditions, the solution $\hat{x}(\cdot)$ of the initial-value problem (5)–(6), as an element of the space $\mathcal{L}_\mu^n C^{(0)}(\mathbb{R})$, is approximated by solutions $\hat{x}_k(\cdot)$ of the initial-boundary value problem (7)–(9) as $k \to +\infty$ [7]. It is important to note that in the sequence of initial-boundary value problems on intervals $[-k, k]$ the boundary function can be an arbitrary fixed essentially bounded function $\varphi(\cdot) \in \mathcal{L}_1^n L_\infty(\mathbb{R})$.

The full text of the proofs of mentioned theorems, as well as a detailed description of the proposed approach, are given in the papers [4–7].

3.3 Korteweg-de Vries Equation

There has been a particular interest in the theory of Korteweg-de Vries (KdV) equation due to its significance in nonlinear dispersive wave theory. Many different real-world nonlinear physical problems are modeled by this well-known equation [13, 17]. For example, this equation has many direct physical applications to solids, liquids, gases, pedestrian flow models [27], car-following models [24, 28] and so on.

It has widely been argued and accepted [16, 26] that for various reasons, time delay should be taken into consideration in modeling. Zhao and Xu [32] have considered solitary wave solutions of the KdV equation with delays, also Zhao dealt with the initial-value problem of the delay KdV equation [31]. Therefore, we want to incorporate a single discrete time delay $\tau > 0$ into KdV equation and consider the delay KdV-type equation's initial-boundary value problem.

KdV-type class of equations has different versions of the representation, but further we will concentrate on finding numerical solutions for the equation of the following form

$$\ddot{x}(t) = c[\alpha x(t) + (1 - \alpha)x(t - \tau)] + \frac{1}{2}x^2(t) + \tau\dot{x}(t - \tau), \tag{10}$$
$$t \in \mathbb{R}, \quad \alpha \in [0, 1], \quad c, \tau \in \mathbb{R}_+.$$

It should also be noted that following [32], the second-order FDEPT (10) can be obtained by integrating an equation after substitution a solitary wave solution $U(x, t) = \phi(x + ct)$ to the KdV equation with time delay

$$[\alpha U_t(x, t) + (1 - \alpha)U_t(x, t - \tau)] + U(x, t)U_x(x, t) + \tau U_{xx}(x, t - \tau) - U_{xxx}(x, t) = 0.$$

In the previous sections, there was noted a one-to-one correspondence between solutions of traveling wave type for infinite-dimensional systems of ordinary differential equations and solutions of induced FDEPT. In particular, such infinite-dimensional differential equations can arise as finite-difference analogs of continuous systems. Here we consider traveling waves for the KdV equation as a continuous system. Moreover, we consider the KdV equation with a delay. The arising equations, satisfied by traveling waves, are also FDEPT. For such equations with a quasilinear right-hand side, conditions for the existence of bounded solutions were obtained, as well as an estimate of the radius of the ball in which such solutions change. For equations of traveling waves with a polynomial right-hand side, redefining the right-hand side outside a certain sphere, one can achieve that it becomes quasilinear. Applying the corresponding result on the existence of a bounded solution for equations with a quasilinear right-hand side, one can obtain bounded solutions for the initial equation of traveling waves, and also a description of the range of parameters under which this boundedness takes place. Below, we demonstrate how the algorithm for constructing a numerical solution, following the logic of the correspondences noted above, allows efficient construction of traveling wave solutions. At the same time, it can be shown that for large initial values there are no bounded traveling waves.

4   Numerical Experiments

Next, the results of the computational experiments on the study of initial-boundary value problems for systems of FDEPT using OPTCON-F software will be presented. Let's consider the FDEPT of the following form

$$\ddot{x}(t) = c[\alpha x(t) + (1 - \alpha)x(t - c\tau)] + \frac{1}{2}x^2(t) + \tau\dot{x}(t - c\tau), \quad t \in \mathbb{R}, \qquad (11)$$

where $\alpha \in [0, 1], c, \tau \in \mathbb{R}_+$. Using a time-variable transformation the Eq. (11) can be rewritten in the form of the following system of equations of the first order

$$\begin{cases} \dot{z}_1(t) = c\tau z_2(t), \\ \dot{z}_2(t) = c\tau \left(c[\alpha z_1(t) + (1 - \alpha)z_1(t - 1)] + \frac{1}{2}z_1^2(t) + \tau z_2(t - 1) \right). \end{cases} \qquad (12)$$

Under this system, we have the following real parameters: α, c, τ. In the following examples, for a given system, we consider such parameter values that conditions of the modified existence theorem for bounded solutions can be violated, so we get an extension of the solution space guaranteed by theory.

4.1 Example 1

We consider dynamical system in the following form:

$$\begin{cases} \dot{z}_1(t) = 0.01 z_2(t), \\ \dot{z}_2(t) = 0.01 \left(0.1 z_1(t) + 0.9 z_1(t-1) + \dfrac{1}{2} z_1^2(t) + 0.01 z_2(t-1) \right), \end{cases} \quad t \in \mathbb{R},$$

initial conditions

$$\begin{cases} z_1(0) = -5, \\ z_2(0) = 0. \end{cases}$$

$$\text{(13)}$$

Here, with respect to the system (12), we have $\alpha = 0.1, \tau = 0.01, c = 1$.

Taking into account the impossibility of considering the numerical solution of the system on an infinite interval, we introduce the parameter k and the corresponding family of expanding initial-boundary value problems

$$\begin{cases} \dot{z}_1(t) = 0.01 z_2(t), \\ \dot{z}_2(t) = 0.01 \left(0.1 z_1(t) + 0.9 z_1(t-1) + \dfrac{1}{2} z_1^2(t) + 0.01 z_2(t-1) \right), \end{cases} \quad t \in [-k, k],$$

boundary conditions

$$\begin{cases} \dot{z}_1(t) = 0, \\ \dot{z}_2(t) = 0, \end{cases} \quad t \in (-\infty, -k] \cup [k, +\infty),$$

initial conditions

$$\begin{cases} z_1(0) = -5, \\ z_2(0) = 0. \end{cases}$$

$$\text{(14)}$$

According to the mentioned theoretical results, the solution of the system (14) converges to the solution of the system (13) as $k \to \infty$. The graphs of the solution of the system (14) at different values of k are shown in Fig. 1.

Since the Eq. (11) is autonomous, the solution space of such equation is invariant with respect to time-variable shifts. Therefore, it suffices to consider a family of solutions of the initial problem (13) with different values of $z_1(0)$. Figure 2 shows the integral curves for different values of the parameter $r = z_1(0)$ for the system (14). Note that for values of $|r|$ greater than about 20, "destruction" of traveling waves occurs and the residual functional of the numerical solution begins to grow rapidly, although for smaller values of $|r|$ it does not exceed 3×10^{-4}.

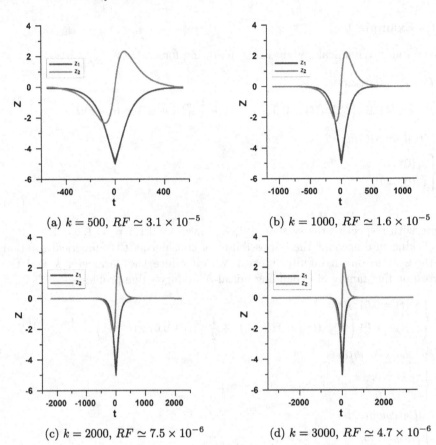

Fig. 1. Trajectories of the system (14) at different k.

4.2   Example 2

Next we consider dynamical system in the following form:

$$\begin{cases} \dot{z}_1(t) = 0.015 z_2(t), \\ \dot{z}_2(t) = 0.015 \left(0.45 z_1(t) + 1.05 z_1(t-1) + \frac{1}{2} z_1^2(t) + 0.01 z_2(t-1) \right), \end{cases} \quad t \in \mathbb{R},$$

initial conditions

$$\begin{cases} z_1(0) = -5, \\ z_2(0) = 0. \end{cases}$$

(15)

Here, with respect to the system (12), we have $\alpha = 0.3, \tau = 0.01, c = 1.5$.

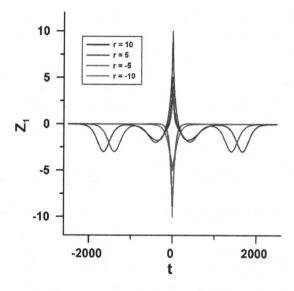

Fig. 2. Trajectories of the system (14) at different r

Again taking into account the impossibility of considering the numerical solution of the system on an infinite interval, we introduce the parameter k and the corresponding family of expanding initial-boundary value problems

$$\begin{cases} \dot{z}_1(t) = 0.015z_2(t), \\ \dot{z}_2(t) = 0.015\left(0.45z_1(t) + 1.05z_1(t-1) + \frac{1}{2}z_1^2(t) + 0.01z_2(t-1)\right), \end{cases} \quad t \in [-k, k],$$

boundary conditions

$$\begin{cases} \dot{z}_1(t) = 0, \\ \dot{z}_2(t) = 0, \end{cases} \quad t \in (-\infty, -k] \cup [k, +\infty),$$

initial conditions

$$\begin{cases} z_1(0) = -5, \\ z_2(0) = 0. \end{cases}$$

$$(16)$$

According to the mentioned theoretical results, the solution of the system (16) converges to the solution of the system (15) as $k \to \infty$. The graphs of the solution of the system (16) at different values of k are shown in Fig. 3.

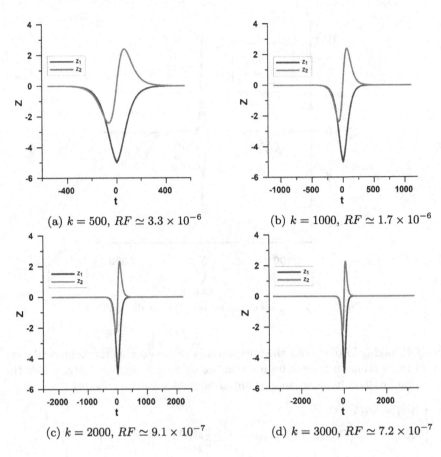

(a) $k = 500$, $RF \simeq 3.3 \times 10^{-6}$

(b) $k = 1000$, $RF \simeq 1.7 \times 10^{-6}$

(c) $k = 2000$, $RF \simeq 9.1 \times 10^{-7}$

(d) $k = 3000$, $RF \simeq 7.2 \times 10^{-7}$

Fig. 3. Trajectories of the system (16) at different k.

Since the Eq. (11) is autonomous, the solution space of such equation is invariant with respect to time-variable shifts. Therefore, it suffices to consider a family of solutions of the initial problem (15) with different values of $z_1(0)$. Figure 4 shows the integral curves for different values of the parameter $r = z_1(0)$ for the system (16). Note that for values of $|r|$ greater than about 10, "destruction" of traveling waves also occurs and the residual functional of the numerical solution begins to grow rapidly, although for smaller values of $|r|$ it does not exceed 4×10^{-5}. Thus, with increasing delay, the solution oscillates on the support around zero, and the threshold of the initial value for the destruction of traveling waves decreases.

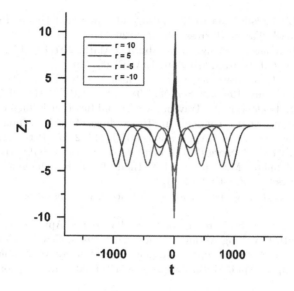

Fig. 4. Trajectories of the system (16) at different r

5 Conclusion

The construction of numerical solutions of the traveling wave type for Korteweg-de Vries-type equations with time delay using the developed software package was demonstrated. The obtained numerical results make it possible to hope for the possibility of obtaining a theorem on the existence of a bounded solution for systems with a polynomial potential on the basis of a redefinition of the right-hand side of the equation outside some ball. This redefinition of the right-hand side makes it quasilinear, and the resulting bounded solution, for the redefined system, must lie in the mentioned ball. This analytical result is the subject of further research.

Acknowledgments. This work was partially supported by Russian Science Foundation, Project 17-71-10116. Also, the reported study was partially funded by RFBR according to the research project 16-01-00110 A.

References

1. Abell, K., Elmer, C., Humphries, A., Van Vleck, E.: Computation of mixed type functional differential boundary value problems. SIAM J. Appl. Dyn. Syst. **4**(3), 755–781 (2005). https://doi.org/10.1137/040603425
2. Baotong, C.: Functional differential equations mixed type in banach spaces. Rendiconti del Seminario Matematico della Università di Padova, **94**, 47–54 (1995). http://www.numdam.org/item?id=RSMUP_1995_94_47_0
3. Beklaryan, L.: A method for the regularization of boundary value problems for differential equations with deviating argument. Soviet Math. Dokl. **43**, 567–571 (1991)

4. Beklaryan, L.: Introduction to the Theory of Functional Differential Equations. Group Approach. Factorial Press, Moscow (2007)
5. Beklaryan, L.: Quasitravelling waves. Sbornik: Math. **201**(12), 1731–1775 (2010). https://doi.org/10.1070/SM2010v201n12ABEH004129
6. Beklaryan, L.: Quasi-travelling waves as natural extension of class of traveling waves. Tambov Univ. Reports. Ser. Nat. Tech. Sci. **19**(2), 331–340 (2014)
7. Beklaryan, L., Beklaryan, A.: Traveling waves and functional differential equations of pointwise type. what is common? In: Proceedings of the VIII International Conference on Optimization and Applications (OPTIMA-2017), Petrovac, Montenegro, October 2–7. CEUR-WS.org (2017). http://ceur-ws.org/Vol-1987/paper13.pdf
8. Bellen, A., Zennaro, M.: Numerical Methods for Delay Differential Equations. Oxford University Press, USA (2003)
9. Bellman, R., Cooke, K.: Differential-Difference Equations. Academic Press, New York (1963)
10. El'sgol'ts, L., Norkin, S.: Introduction to the Theory and Application of Differential Equations with Deviating Arguments. Academic Press, New York (1973)
11. Ford, N., Lumb, P.: Mixed-type functional differential equations: a numerical approach. J. Comput. Appl. Math. **229**(2), 471–479 (2009). https://doi.org/10.1016/j.cam.2008.04.016
12. Frenkel, Y., Contorova, T.: On the theory of plastic deformation and twinning. J. Exp. Theor. Phys. **8**(1), 89–95 (1938)
13. Gardner, C., Greene, J., Kruskal, M.: Method for solving the korteweg-de vries equation. Phys. Rev. Lett. **19**(19), 1095–1097 (1967). https://doi.org/10.1103/PhysRevLett.19.1095
14. Gornov, A., Zarodnyuk, T., Madzhara, T., Daneeva, A., Veyalko, I.: A Collection of Test Multiextremal Optimal Control Problems, pp. 257–274. Springer, New York (2013). https://doi.org/10.1007/978-1-4614-5131-0_16
15. Hale, J.: Theory of Functional Differential Equations. Springer, New York (1977)
16. Hale, J., Lunel, S.V.: Introduction to Functional Differential Equations. Applied Mathematical Sciences, vol. 99. Springer, New York (1993). https://doi.org/10.1007/978-1-4612-4342-7
17. Kortweg, D., De Vries, G.: On the change of form of long waves advancing in a rectangular canal, and on a new type of long stationary waves. Philos. Mag. **5**(39), 422–443 (1895)
18. Li, X., Wu, B.: A continuous method for nonlocal functional differential equations with delayed or advanced arguments. J. Math. Anal. Appl. **409**(1), 485–493 (2014). https://doi.org/10.1016/j.jmaa.2013.07.039
19. Lima, P., Teodoro, M., Ford, N., Lumb, P.: Analytical and numerical investigation of mixed-type functional differential equations. J. Comput. Appl. Math. **234**(9), 2826–2837 (2010). https://doi.org/10.1016/j.cam.2010.01.028
20. Lima, P., Teodoro, M., Ford, N., Lumb, P.: Finite element solution of a linear mixed-type functional differential equation. Numer. Algorithms **55**(2–3), 301–320 (2010). https://doi.org/10.1007/s11075-010-9412-y
21. Maset, S.: Numerical solution of retarded functional differential equations as abstract cauchy problems. J. Comput. Appl. Math. **161**(2), 259–282 (2003). https://doi.org/10.1016/j.cam.2003.03.001
22. Miwa, T., Jimbo, M., Date, E.: Solitons: Differential Equations, Symmetries and Infinite Dimensional Algebras. Cambridge University Press, Cambridge (2000)
23. Myshkis, A.: Linear Differential Equations with Retarded Arguments. Nauka, Moscow (1972)

24. Sun, D., Chen, D., Zhao, M., Liu, W., Zheng, L.: Linear stability and nonlinear analyses of traffic waves for the general nonlinear car-following model with multi-time delays. Phys. A: Stat. Mech. Appl. **501**, 293–307 (2018). https://doi.org/10.1016/j.physa.2018.02.179

25. Toda, M.: Theory of Nonlinear Lattices, vol. 20. Springer, Berlin, Heidelberg (1989). https://doi.org/10.1007/978-3-642-83219-2

26. Wu, J.: Theory and Applications of Partial Functional Differential Equations, Applied Mathematical Sciences, vol. 119. Springer, New York (1996). https://doi.org/10.1007/978-1-4612-4050-1

27. Xu, L., Lo, S., Ge, H.: The korteweg-de vires equation for bidirectional pedestrian flow model. Procedia Eng. **52**, 495–499 (2013). https://doi.org/10.1016/j.proeng.2013.02.174

28. Yu, L., Shi, Z., Li, T.: A new car-following model with two delays. Phys. Lett. A **378**(4), 348–357 (2014). https://doi.org/10.1016/j.physleta.2013.11.030

29. Zarodnyuk, T., Anikin, A., Finkelshtein, E., Beklaryan, A., Belousov, F.: The technology for solving the boundary value problems for nonlinear systems of functional differential equations of pointwise type. Mod. Technol. Syst. Anal. Model. **49**(1), 19–26 (2016)

30. Zarodnyuk, T., Gornov, A., Anikin, A., Finkelstein, E.: Computational technique for investigating boundary value problems for functional-differential equations of pointwise type. In: Proceedings of the VIII International Conference on Optimization and Applications (OPTIMA-2017), Petrovac, Montenegro, October 2–7. CEUR-WS.org (2017). http://ceur-ws.org/Vol-1987/paper82.pdf

31. Zhao, Z., Rong, E., Zhao, X.: Existence for korteweg-de vries-type equation with delay. Adv. Differ. Equ. **2012**(1), 64 (2012). https://doi.org/10.1186/1687-1847-2012-64

32. Zhao, Z., Xu, Y.: Solitary waves for korteweg-de vries equation with small delay. J. Math. Anal. Appl. **368**(1), 43–53 (2010). https://doi.org/10.1016/j.jmaa.2010.02.014

The Synthesis of the Switching Systems Optimal Parameters Search Algorithms

Olga Druzhinina[1]([📧])[iD], Olga Masina[2][iD], and Alexey Petrov[2][iD]

[1] Federal Research Center "Computer Science and Control" of RAS, Moscow, Russia
ovdruzh@mail.ru
[2] Bunin Yelets State University, Yelets, Russia
olga121@inbox.ru, xeal91@yandex.ru

Abstract. The problems of the optimal motion parameters search for generalized models of the dynamical systems are considered. The switching dynamic models taking into account action of non-stationery forces and optimality conditions are studied. The method for designing the dynamical models using polynomial regression is proposed. The optimal analytical solutions for some types of parametric curves are found. The algorithms of the optimal motion parameters search by means of the intelligent control methods are elaborated. The indicated algorithms and the software package allowed to execute a series of computational experiments and to carry out the stability analysis. The prospects of the results development in terms of generalization and modification of the models and the methods are presented. The results and the algorithms can be applied to the problems of automated transport design, robotics, and aircrafts motion control.

Keywords: Dynamical switching model
Optimal parameters of motion · Optimal control
Algorithms · Computational experiment · Artificial neural networks

1 Introduction

The development of analytical and numerical methods of mathematical modeling of complex dynamic systems is an actual scientific and technical direction. In particular, modeling based on new qualitative and numerical methods aimed at finding optimal trajectories of technical systems is widely used in designing automated transport, aircraft and in robotic [1,2]. Due to the uncertainty of the data on the process or object under investigation, and also because of the variety of operating physical effects, it is necessary to develop stochastic and intelligent methods of optimization. Different aspects of modeling based on the mentioned methods are presented in [3,4].

The purpose of this work is the synthesis of mathematical models of the dynamics of technical systems in conditions of switching modes of operation and the development of algorithms for searching of optimal motion parameters with subsequent implementation in the form of a software package.

© Springer Nature Switzerland AG 2019
Y. Evtushenko et al. (Eds.): OPTIMA 2018, CCIS 974, pp. 306–320, 2019.
https://doi.org/10.1007/978-3-030-10934-9_22

The effect of applying the algorithms proposed in the work is to reduce the time and energy costs for creating a vector thrust, as well as in high-precision positioning. The developed software package allows formalizing a wide class of dynamic systems using artificial neural networks in a graphical interactive mode.

The formulation of the problem of finding optimal parameters can include various types of criteria [5]. These include problems with minimization of the functional, depending on the control and trajectory, problems with a constraint on the trajectory also called optimal speed problems and problems with control constraints. As a rule, similar problems reduce to solving boundary value problems with mixed boundary conditions under certain additional constraints, which allows us to find a range of parameters that satisfies the optimality criterion. However, in most cases, the analytical or numerical solution of this problem is difficult, and modeling by means of ordinary differential equations does not give a complete picture of the process under consideration.

The problems of modeling technical systems using differential inclusions are considered in [6–8] and in other works. In some cases, the assumptions on the dynamical system can conveniently be written using a multivalued function $f(x, p)$, where x is the trajectory of the system, p is the parameter vector. With the help of differential inclusion, it is possible to describe a fairly wide class of objects, including managed systems with the variable control area, as well as controlled systems with phase constraints [6,8]. In constructing the model, it is convenient to write the assumption on the system under study in the form of the differential inclusion of the type

$$\dot{x} \in f(x, t, P), \tag{1}$$

where P is the vector of system parameters, x is the state vector [8]. The device of differential inclusions is used for modeling of technical systems in conditions of switching of operating modes. Inclusions of the type (1) also generalize differential equations with discontinuous right-hand sides [9]. The theory of differential inclusions has applications in modern works in the field of intelligent technologies, which include [10–12] and other works.

Systems whose right-hand side represents a polynomial form are of interest in applied control problems. To solve applied problems for systems with polynomial regression, various optimization methods are often used. The study of systems with polynomial regression, used to simplify the solution of the boundary value problem, is devoted to many modern works, in particular, [13–15].

In this work, we consider the models of variable-thrust technical switching systems under the assumption of the replacement of unknown forces by a set of parametric polynomial curves. In addition, a basic two-dimensional switched model is described and a transition is made to generalized and finite-dimensional switched models. The algorithms of the motion optimal parameters search are proposed. The software package is developed for modeling technical systems in conditions of switching operation modes. The instrumental software for modeling these systems on the basis of open platforms is created. The algorithms for calculating the trajectories and optimal parameters search developed for the software package are realized taking into account the modular structure of the complex. The modular structure and the combination of formal and heuristic methods make it possible to use a universal approach to the research of the studied classes of models. One of the basic algorithms is based on the use of artificial neural networks. The software complex can find application in the problems of the unmanned aerial vehicles motion modeling, automated transport systems, robotic systems and in other problems.

2   Search for Optimal Parameters of Switching Models

2.1   Initial Model

A linear two-dimensional switched model is studied in [7]. This model is considered in Cartesian coordinates x, y and the initial point $(0,0)$ under the influence of constant positive forces f_x, f_y from the vector space F. Let the object move in two stages: in time intervals $(0, t_1)$ and (t_1, t_2) for the first and the second stages, respectively. At the first stage, the object of mass m moves in the plane (x, y) under the influence of the constant vector thrust (f_{x_1}, f_{y_1}), until the maximum height $H(x(t_1), h)$. At the second stage, the object moves under the influence of the constant vector thrust (f_{x_2}, f_{y_2}), until the point $L(l, 0)$. The considered model takes into account the effect of the gravitational force with the magnitude of the acceleration of gravity g. Taking into account

$$x(0) = 0, y(0) = 0, y(t_1) = h, y(t_2) = 0, x(t_2) = l, \tag{2}$$

we have $\forall x(t), t \in (0, t_2) : \dot{x} > 0; \forall y(t), t \in (0, t_1] : \dot{y} \geq 0; \forall y(t), t \in (t_1, t_2] : \dot{y} < 0$.

The differential inclusions describing the system have the form:

$$\begin{aligned} m\ddot{x} &\in f_x, \\ m\ddot{y} &\in f_y - mg, \end{aligned} \tag{3}$$

where $f_x > 0, f_y > 0$.

Single-valued realizations of the model (3) can be represented in the form of the equation systems

$$\begin{aligned} m\ddot{x} &= f_{x_1}, \\ m\ddot{y} &= f_{y_1} - mg, \end{aligned} \quad t \in [0, t_1], \tag{4}$$

$$\begin{aligned} m\ddot{x} &= f_{x_2}, \\ m\ddot{y} &= -f_{y_2} - mg, \end{aligned} \quad t \in (t_1, t_2]. \tag{5}$$

The criterion of optimality taking into account (4) and (5) has the form

$$\int_0^{t_1} (f_{x_1} + f_{y_1})dt + \int_{t_1}^{t_2} (f_{x_2} + f_{y_2})dt \rightarrow \min. \tag{6}$$

The physical meaning of the criterion is the minimum fuel consumption in the case of jet propulsion.

The parameters $f_{x_1}, f_{x_2}, f_{y_1}, f_{y_2}, t_1, t_2$ in the case of single-valued implementation have the form:

$$f_{x_1} = \frac{2l\tau(\tau-1)^2 gm}{4l(\tau-1)^2 + 2h\tau^2}, \quad f_{y_1} = \frac{2h\tau^2(\tau-1)^2 gm}{4l(\tau-1)^2 + 2h\tau^2} + mg,$$

$$f_{x_2} = \frac{2l\tau(\tau-1)gm}{4l(\tau-1) + 2h\tau^2}, \quad f_{y_2} = \frac{2h\tau^2(2\tau-1)gm}{4l(\tau-1)^2 2h\tau^2} + mg, \tag{7}$$

$$\tau = 1 + \frac{1}{\sqrt{3}}, \quad t_1 = \sqrt{\frac{4l(\tau-1)^2 + 2h\tau^2}{\tau^2(\tau-1)^2 g}}, \quad t_2 = \tau t_1.$$

The formulas (7) are obtained by means of the Mathematics package. We solve differential equations of the form (4) and (5) at the different intervals, taking into account the initial conditions (2) and the optimality conditions (6). Runge-Kutta method is used for the numerical solution of the differential equations. The set of trajectories of the system (4) and (5) taking into account (7) is represented in Fig. 1, where $h \in [1, 15]$. The switch is made from (4) to (5) when the point $(0, h)$ is reached.

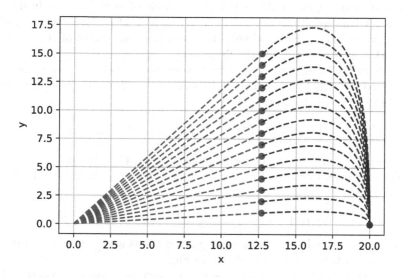

Fig. 1. The graphical representation of trajectories at $h \in [1, 15]$

The multivaluence that can be caused by the resistance factor of a sparse environment in the described model is taken into account in constructing a set of single-valued realizations of form (4) and (5) in the case when these realizations describe the motion of an object within the circle. The model (4) and (5) is greatly simplified for describing the motion of the controlled object since it does not take into account the possibility of considering a mobile final point (variability condition). In addition, the model (4) and (5) does not take into account air resistance and needs to be generalized and clarified [16–18].

2.2   Three-dimensional Switching Model

Next, the modification of the model (4) and (5) or the case of the final point motion in the complementary plane xoz. Next, we present an algorithm for constructing trajectories of motion, which is a generalization and modification of the algorithm proposed in [7]. The algorithm (algorithm 1) consists of the following steps.

Step 1. The object moves in the plane xoy under the influence of a constant thrust vector before reaching the point $(x(t_1), 0, h)$. During the time t_1, the final point reaches of the positions circle.

Step 2. Out of the all possible trajectories of the object motion we select the one that leads to an intermediate destination point. For the remaining time $t = t_2 - t_1$ the object will reach the plane xoz taking into account the direction to the center of the circle of positions. Next, we consider the motion for the time $\frac{t}{2}$ and establish a new circle of positions and select the trajectory leading to its center.

Step 3. If the radius of the positions circle is larger than the radius of the conformity, we repeat step 2. Wherein the time of the motion, when the object passes half of the route in the center of the circle of positions is halved. As a result of several repetitions of step 2, we get a circle of positions with a radius of less than σ that allows continuing the motion in the center of the positions circle with a guaranteed hit at the final point.

The described algorithm is used for generalized model, when the motion of the final point in the plane is taken into account xoz. Namely, from the system (4) and (5) we go to the system

$$m\ddot{x} = A + mG, \quad x(0) = (a_1, a_2, a_3), \quad \dot{x}(0) = (b_1, b_2, b_3), \tag{8}$$

where $x \in R^n$, A are the vector forces in the system, $G = (0, -g, 0)$.

Under the conditions $x(t_2) = (0, 0, \delta)$, $a_1 = \frac{h}{2}$, $t_2 = \tau t_1$ the values A take the form:

$$A = \left(-\frac{mb_1}{t_1(1-\tau)}, \frac{mb_2}{t_1(1-\tau)} - g, -\frac{mb_3}{t_1(\tau-1)} + m\delta \right). \tag{9}$$

The algorithm is implemented as a program in the language Octave. Graphical visualization of the trajectory is shown in Fig. 2.

The set of single-valued realizations of the model (8), which is a generalization of the model (4) and (5), describes the motion of the system taking into account the optimality and variability of the final point position.

2.3   Nonstationary Models

Next, we consider the generalization of the model (4) and (5) to the multidimensional case, taking into account time-dependent perturbation. For this generalization, we propose a numerical interpretation of the problem of the optimal trajectories search. The model (4) and (5) moves to the model in which the thrust vector is given by a function of time. Using the function from time allows

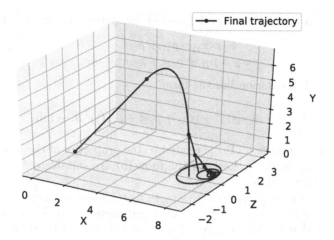

Fig. 2. Graphical representation of the trajectory of the model (8)

you to take into account the additional constraint on zero speed at the final point and move from two-stage motion to one-step motion corresponding to the interval $t \in (0, t_2)$. We introduce the restriction $\|\dot{x}(t_2)\| = 0$ on the zero velocity at the final point. This constraint is impossible for (4) and (5) because of the absence of non-stationary forces. In the case of nonstationary perturbations, the multidimensional generalization of the model (4) and (5) takes the form

$$m\ddot{x} = Q + mG, \tag{10}$$

where $x \in R^n$, G is the vector-column of the gravity acceleration, Q is the vector of polynomial functions on the interval $(0, t_2)$.

We give (10) to the matrix form:

$$m\ddot{x} = PT + mG, \tag{11}$$

where

$$P = \begin{pmatrix} p_{11} & p_{12} & p_{13} & \cdots & p_{1k} \\ p_{21} & p_{22} & p_{23} & \cdots & p_{2k} \\ \multicolumn{5}{c}{\cdots\cdots\cdots} \\ p_{n1} & p_{n2} & p_{n3} & \cdots & p_{nk} \end{pmatrix}, \quad T = \begin{pmatrix} t^0 \\ t^1 \\ t^2 \\ \vdots \\ t^{k-1} \end{pmatrix}. \tag{12}$$

With the help of the Eq. (11) it is possible to simulate with undetermined accuracy unallocated mechanical systems without feedback. The search for the right-hand sides reduces to the matrix of coefficients search P. In the simplest case, the matrices (12) have the form:

$$P = \begin{pmatrix} p_{11} & p_{12} & 0 \\ p_{21} & p_{22} & p_{23} \end{pmatrix}, \quad T = \begin{pmatrix} t^0 \\ t^1 \\ t^2 \end{pmatrix}. \tag{13}$$

Taking into account (13) the special case of the system (11) has the form:

$$m\ddot{x}_1 = p_{11} + p_{12}t,$$
$$m\ddot{x}_2 = p_{21} + p_{22}t + p_{23}t^2 - mg, \tag{14}$$

The optimality criterion has the form

$$\int_0^{t_2} (p_{11} + p_{12}t)^2 + (p_{21} + p_{22}t + p_{23}t^2)^2 dt \to \min. \tag{15}$$

The choice of the criterion (15) is related to the properties of the matrix P. We note that the optimality criteria of the form (6) are suitable only for systems with nonnegative matrices P.

Let $f_1(t) = p_{11} + p_{12}t$, $f_2(t) = p_{21} + p_{22}t + p_{23}t^2$. Based on the problem setting, one can determine the region of the required parameters. To achieve the height h, it is necessary that the following conditions be fulfilled $f_1(0) > 0$, $f_2(0) > g$. To achieve zero speed at the final point, it is necessary that the following conditions be fulfilled $f_1(t_2) < 0$, $f_2(t_2) > g$, $\dot{f}_2(0) < 0$, $\dot{f}_2(t_2) > 0$. Taking into account (14) and (15), we obtain the following system of inequalities from the above conditions

$$p_{11} > 0, p_{11} + p_{12}t_2^2 < 0,$$
$$p_{21} > g, \ p_{21} + p_{22}t_2 + p_{23}t_2^2 > g, \tag{16}$$
$$p_{22} < 0, \ p_{22} + 2p_{23}t_2 > 0,$$

from where follows

$$p_{11} > 0, \ p_{12} < 0, \ p_{21} > g, \ p_{22} < 0, \ p_{23} > 0. \tag{17}$$

Next, we consider the analytical and numerical transitions to the solution of the optimal parameters search problem. Let any parametric curve satisfying the statement of the problem be a solution, and a solution satisfying the criterion (15) is the optimal solution. The trajectories of the system (14), taking into account the optimal solution, are parabolas emerging from the point $(0,0)$, reaching the height h and point $(l,0)$. Next, we find the solutions for the model (14) in a general form without taking into account the optimality criterion. Let t_2 is the right end of the interval defining the motion time. Then, according to (14), we obtain

$$x_1(t_2) = \frac{t_2^2(3p_{11} + p_{12}t_2)}{6m} = l, \quad \dot{x}_1(t_2) = \frac{p_{11}t_2^2}{6m} + \frac{t_2(3p_{11} + p_{12}t_2)}{3m} = 0,$$

from where follows $p_{11} = \frac{6lm}{t_2^2}$, $p_{12} = -\frac{12lm}{t_2^3}$. Next we obtain

$$x_2(t_2) = \frac{t_2^2}{12m}(6p_{21} - 6gm + 2p_{22}t_2 + p_{23}t_2^2) = 0,$$

$$\dot{x}_2(t_2) = \frac{t_2}{6m}(6p_{21} - 6gm + 3p_{22}t_2 + 2p_{23}t_2^2) = 0,$$

$$x_2(t_1) = \frac{t_2^2}{12m\tau^4}(2p_{22}\tau + p_{23}t_2^2 + 6\tau^2(p_{21} - gm)) = h,$$

$$\dot{x}_2(t_1) = \frac{t_2}{6m\tau^3}(3p_{22}t_2\tau + 2p_{23}t_2^2 - 6\tau^2(gm - p_{21})) = 0,$$

from where follows $p_{21} = \frac{m(gt_2^2+32h)}{t_2^2}$, $p_{22} = -\frac{192hm}{t_2^3}$, $p_{23} = \frac{192hm}{t_2^4}$, $\tau = 2$. Then

$$P = \begin{pmatrix} \frac{6lm}{t_2^2} & -\frac{12lm}{t_2^3} & 0 \\ \frac{m(gt_2^2+32h)}{t_2^2} & \frac{192hm}{t_2^3} & \frac{192hm}{t_2^4} \end{pmatrix}. \tag{18}$$

The uniqueness of the trajectories that satisfy the formulation of the problem by virtue of the model (13) for each value t_2 follows from (18). In this case, the fulfillment of the criterion (15) is reduced to the choice of the optimal value t_2 depending on h, l, m. In Fig. 3 represents the trajectory of the system (14). The arrows indicate the direction of the action of vector forces PT.

The analytical form of the extrema turns out to be too cumbersome for the criterion of optimality in the form of a function. We developed program for calculating of the single minimum for the criterion (15) taking into account of the optimal value t_2.

The consideration of polynomials of degree $n > 2$ in (11) can lead to a function infinitely close to the globally optimal. In the general case, a search of the coefficients matrix reduces to the problem of finite-dimensional optimization for the systems (11). Most first-order optimization methods do not provide satisfactory solutions in this case. We note that Nelder–Meade, Powell, BFGS, Basin-hopping methods are applied to the problems of this type [19]. The method of differential evolution [20,21] in some cases is more effective than other methods. A comparative analysis of modern optimization methods is given in [22].

The differential evolution is a method of global optimization and is based on the basic ideas of genetic algorithms [23–26]. When using the differential evolution method, the data in the algorithm are not represented as binary sequences.

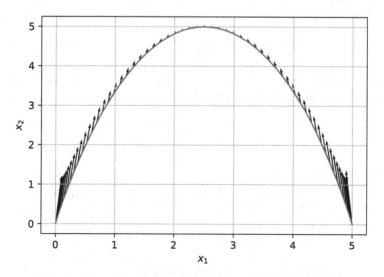

Fig. 3. The motion of the system (11) at $l = 5, h = 5, m = 1, t_2 = 3$

In this work, an algorithm is used that is a part of the library of mathematical calculations Scipy. To initialize the original population and accelerate convergence, a preliminary estimate of the region in which the desired values can lie. Taking into account (17) the preliminary estimation of the parameters for the system (14). The results are presented in Table 1.

Table 1. The values P, obtained by analytical and numerical methods

Method/values	p_{11}	p_{12}	p_{21}	p_{22}	p_{23}
Differential evolution	3.224939	−2.114714	27.008843	−33.853466	11.099497
Analytical values	3.333333	−2.222222	27.577777	−35.555555	11.851851

According to the Table 1, the differential evolution gives small absolute values for solving the optimization problem for the system (14). In this case, the objective function corresponds to a linear sum of quadratic errors in the conditions for setting the problem.

We consider the model of the form

$$m\ddot{x} = PT + mG + D, \quad P \in \Gamma, \tag{19}$$

where $x \in R^n$, Γ is a discrete map that matches the corresponding matrix of coefficients, $D = D(\dot{x}, x, t)$ is the unknown vector function that defines external perturbations.

The model (19) is the modification and generalization of the model (11). The vector of the initial conditions for the model (19) has the form $I = (a_1, b_1, a_2, b_2)$.

The coefficient matrix search, we consider the two-dimensional case. Consider the case when P has the form:

$$P = \begin{pmatrix} p_{11} & p_{12} \\ p_{21} & p_{22} \end{pmatrix}. \tag{20}$$

Let the conditions $\forall(t, x, \dot{x})$, $\|D\| = 0$ are carried out. Then the system (19) with considering (20) is written in the form

$$\begin{aligned} m\ddot{x}_1 &= p_{11} + p_{12}t, \\ m\ddot{x}_2 &= p_{21} + p_{22}t - mg. \end{aligned} \tag{21}$$

Taking into account $\dot{x}(t_2) = (0, 0)$ we obtain

$$P = \begin{pmatrix} \dfrac{2m(-2t_2 b_1 + 3l - 3a_1)}{t_2^2} & \dfrac{6m(t_2 b_1 - 2l + 2a_1)}{t_2^3} \\ mg - \dfrac{4b_2 m}{t_2} - \dfrac{6ma_2}{t_2^2} & \dfrac{6m(t_2 b_2 + 2a_2)}{t_2^3} \end{pmatrix} \tag{22}$$

We developed the following algorithm 2 which is a modification of Algorithm 1 for the case of the model (21).

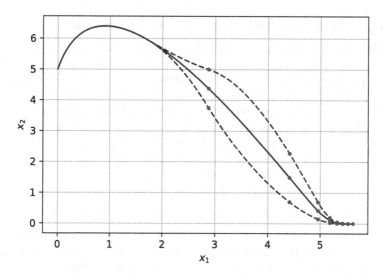

Fig. 4. Trajectories of motion with nonlinearly increasing value l

Step 1. Set the initial condition vector I, time boundary t_2, and for the specified values calculate the initial trajectory of the model (21).

Step 2. When the switching time $t_1 = \frac{t_2}{2}$ is reached, estimate the change in the position of the final point, recalculate P using (22). The current time of the system t is assumed to be zero, t_2 is equal to t_1 to simplify calculations.

Step 3. Step 2 is repeated until the final point reaches the hypersphere with a radius ρ, which characterizes the sufficient accuracy of the solution.

The difference between the algorithm 2 and algorithm 1 is as follows: (1) an additional condition is imposed on the zero finite speed; (2) more economical implementation of algorithm 2, which does not take into account the trajectory of motion, not directed to the final point at a given time; (3) the possibility of correcting the trajectory for some values $\|D\| > 0$.

This algorithm is realized in the form of a computer program. This program is written by Python. The Fig. 4 represents the trajectory of the motion with the initial conditions vector $I = (0, 1, 5, 5)$ and the initial value $l = 5$.

Under the conditions $\exists (t, x, \dot{x}) \|D\| > 0$ the efficiency of the algorithm remains high for small values $\|D\|$. As D we can consider the medium's resistance, wind effects, collisions and other unaccounted effects. Figure 4 show the trajectories of a two-dimensional system for $\|D\| = 0$ (solid line) and $\|D\| > 0$ (dotted lines) respectively, and the switching times are marked by points. Trajectories of the system motion with allowance for switching are not difficult to construct for a higher dimensionality.

2.4   Switching Model with Neural Network Controller

To modify the Algorithm 1, we introduce a regulator based on the artificial neural networks. The use of the artificial neural networks in the systems of various types is studied in [27–30] and in other works. We will assume that the matrix representation of a neural network has the form of a recursive expression. Consider a simple, fully connected network without a hidden layer. Let the inputs be denoted as $j = 1, \ldots, 4$, outputs as $k = 1, \ldots, 4$, layer number as $i = 1$. Then the weight coefficients can be represented by an adjacency matrix W_i of elements w_{kj}:

$$
W_i = \begin{pmatrix} w_{11} & \cdots & w_{14} \\ & \ddots & \\ w_{41} & \cdots & w_{44} \end{pmatrix}.
\tag{23}
$$

The values of input nodes are defined by the vector-column $Z_i = (z_1, \cdots, z_4)^T$, which specifies the scalar components of the state vector and the deviation vector from the final point $(l, 0)$ for the system (19).

For output neurons, the yield will be determined by the formula

$$
z_k = \delta \left(\sum_{l=1}^{4} w_{kj} z_j \right).
\tag{24}
$$

where δ is the activation function of single neuron. Similarly, vectorization of the layer activation function can be carried out. In this case, the derivation of the layer is reduced to the matrix expression $\bar{Z}_i = A(W_i Z_i)$, where A is the vector-function of layer activation, $\bar{Z}_i$ is output layer values. Then the general case of calculating the outputs of a neural network consisting of n layers is described by the formula

$$
Z_n = A(W_n A(W_{n-1} A(W_{n-2} \ldots A(W_1 Z_1))))
\tag{25}
$$

The advantage of a neural network representation in a vector-matrix form is high productivity at computer realization. The construction of networks of direct propagation of arbitrary topology is possible at the expense of sparse matrices of weight coefficients.

For the model (19) the following uses of the formula (25):

(1) direct control via a neural network on the right side with fixed switching times; (2) direct control on the right side with switching performed on the basis of the output of the neural network; (3) setting the coefficient matrix for (19) taking into account the output of the neural network.

In the current work, the neural network is trained using the differential evolution method for the cases of two and three layers. We implement the switching algorithm based on the neural network controller in the form of a software package.

2.5   Structure of the Software Package

Based on the obtained results and the algorithms, we offer a software package Switched System Modeling Complex (SSMC) for technical systems modeling in switching conditions of operation. The software for modeling these systems is based on the use of open platforms. The algorithms for calculating the trajectories and optimal parameters search developed for the software package are realized taking into account the modular structure of the complex. The modular structure and a combination of formal and heuristic methods make it possible to use a universal approach to the research of the studied classes of the models. One of the ultimate algorithms is based on the use of artificial neural networks. The software complex can find application in the problems of the unmanned aerial vehicles motion modeling, automated transport systems, robotic systems and in other problems.

The software complex SSMC is structurally divided into the following modules: (1) the module for constructing the state vector of the model; (2) the module for calculating the trajectory of motion; (3) the module-generator switching; (4) the module for optimizing the parameters; (5) the module for working with artificial neural networks; (6) the module of auxiliary mathematics; (7) the output module. Within the framework of the complex the following programs were developed: (1) the program for optimal model parameters search; (2) the program of the model trajectories constructing. The block-scheme and the user interface of these programs are shown in Fig. 5.

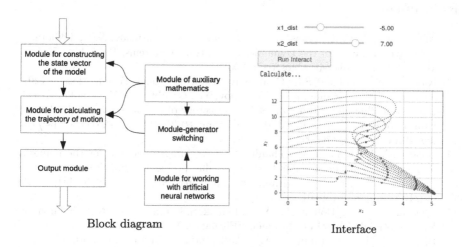

Block diagram   Interface

Fig. 5. Software package SSMC

The series of computer experiments based on the developed algorithms was carried out. The results of some computer experiments using the developed complex are given in Sect. 2.

3   Conclusions

This approach to the synthesis of the algorithms for optimal parameters search allows us to cover a fairly wide class of switchable models, in particular, the algorithms that were constructed proved to be effective for a generalized two-dimensional model with the variability of positioning, a generalized nonstationary model. The proposed algorithm for calculating the optimal trajectory taking into account the condition of minimum fuel consumption can be used in the design and improvement of technical control systems of motion. The specified algorithm is implemented as a program in the built-in language of the Octave system. Analytic and numerical methods for the parameters search are used for the case $n = 2$. The algorithms are developed for numerical optimization implemented in the form of programs in the language Python.

That is important to note that the management theory and its methods demonstrate tendencies towards convergence with cognitive-information technologies [31,32]. In this work, we use tools such as artificial neural networks and differential evolution. Prospects for the development of switched models using neural networks is to develop effective algorithms for weight coefficients matrixes search.

The results obtained in the work can be used in the problems of constructing automated transport and cargo delivery systems, in robotics, as well as in the problems of aircraft motion control.

References

1. Aleksandrov, V.V., Boltyansky, V.G., Lemak, S.S., Parusnikov, N.A., Tikhomirov, V.M.: Optimization of the Dynamics of Controlled Systems. Publishing House of Moscow State University, Moscow (2000)
2. Ivanov, A.A., Torokhov, S.L.: Control in Technical Systems. Forum, Moscow (2012)
3. Semushin, I.V., Tsiganov, A.V., Tsyganov, Yu., Golubkov, A.V., Vinokurov, D.S., et al.: Modeling and estimation of the trajectory of a moving object. Bull. South Ural State Univ. Ser. Math. Model. Program. **10**(3), 108–119 (2017). https://doi.org/10.14529/mmp170309
4. Bulatov, M.V., Lima, P.M., Thanh, D.T.: An integral method for the numerical solution of nonlinear singular boundary value problems. Bull. South Ural State Univ. Ser. Math. Model. Program. Comput. Softw. **8**(4), 5–13 (2015). https://doi.org/10.14529/mmp150401
5. Aleksandrov, V.V., Boltyanskii, V.G., Lemak, S.S., Parusnikov, N.A., Tikhomirov, V.M.: Optimal Control of Movement. Fizmatlit, Moscow (2005)
6. Blagodatskikh, V.I., Filippov, A.F.: Differential inclusions and optimal control. Proc. Steklov Inst. Math. **169**, 194–252 (1985)
7. Masina, O.N.: The problems of the motion control of transport systems. Transp. Sci. Technol. Control **12**, 10–12 (2006)
8. Cellina, A.: A view on differential inclusions. Rend. Sem. Mat. Univ. Pol. Torino **63**, 197–209 (2005)
9. Finagenko, I.A.: Differential Equations with Discontinuous Right-Hand Side. ISDCT SB RAS, Irkutsk (2013)
10. Masina, O.N., Druzhinina, O.V.: On optimal control of dynamical systems described by differential inclusions. In: Proceedings of the VII International Conference on Optimization Methods and Applications "Optimization and Applications" (OPTIMA-2016) held in Petrovac, Montenegro, 25 September–2 October, pp. 104–105. Dorodnicyn Computing Centre of FRC CSC RAS, Moscow (2016)

11. Zhang, G., Shen, Y., Wang, L.: Global anti-synchronization of a class of chaotic memristive neural networks with time-varying delays. Neural Netw. **46**(1), 195–203 (2013). https://doi.org/10.1016/j.neunet.2013.04.001

12. Wu, A., Wen, S., Zeng, Z.: Synchronization control of a class of memristor-based recurrent neural networks. Inf. Sci. **183**(1), 106–116 (2012). https://doi.org/10.1016/j.ins.2011.07.044

13. Andrzejczak, G.: Spline reproducing kernels on R and error bounds for piecewise smooth LBV problems. Appl. Math. Comput. **320**, 27–44 (2018). https://doi.org/10.1016/j.amc.2017.09.021

14. Ghasemi, M.: High order approximations using spline-based differential quadrature method: Implementation to the multi-dimensional PDEs. Appl. Math. Modell. **46**, 63–80 (2017). https://doi.org/10.1016/j.apm.2017.01.052

15. Zin, S.M.: Quartic B-spline and two-step hybrid method applied to boundary value problem. In: AIP Conference Proceedings, vol. 1522, no. 744 (2013). https://doi.org/10.1063/1.4801200

16. Druzhinina, O.V., Masina, O.N., Petrov, A.A.: Model of motion control of transport system taking into account conditions of optimality, multivaluence, and variability. Transp. Sci. Technol. Control **4**, 3–9 (2017)

17. Druzhinina, O.V., Masina, O.N., Petrov, A.A.: Approach elaboration to solution of the problems of motion control of technical systems modeled by differential inclusions. Inf. Meas. Control. Syst. **15**(4), 64–72 (2017)

18. Druzhinina, O.V., Masina, O.N., Petrov, A.A.: Models for control of technical systems motion taking into account optimality conditions. In: Proceedings of the VIII International Conference on Optimization Methods and Applications "Optimization and Application" (OPTIMA-2017) held in Petrovac, Montenegro, 2–7 October, pp. 386–391. Dorodnicyn Computing Centre of FRC CSC RAS, Moscow (2017). http://ceur-ws.org/Vol-1987/paper-56.pdf

19. Wales, D., Doye, J.: Global optimization by basin-hopping and the lowest energy structures of Lennard-Jones clusters containing up to 110 Atoms. J. Phys. Chem. A **101**, 5111–5116 (1997)

20. Das, S., Mullick, S.S., Suganthan, P.N.: Recent advances in differential evolution-an updated survey. Swarm Evol. Comput. **27**(1), 61–106 (2016). https://doi.org/10.1016/j.swevo.2016.01.004

21. Yu, W., et al.: Differential evolution with two-level parameter adaptation. IEEE Trans. Cybern. **44**(7), 1080–1099 (2014). https://doi.org/10.1109/TCYB.2013.2279211

22. Hamdy, M., Nguyen, A.-T., Hensen, J.L.M.: A performance comparison of multi-objective optimization algorithms for solving nearly-zero-energy-building design problems. Energy Build. **121**, 57–71 (2016). https://doi.org/10.1016/j.enbuild.2016.03.035

23. Arqub, O.A., Abo-Hammour, Z.: Numerical solution of systems of second-order boundary value problems using continuous genetic algorithm. Inf. Sci. **279**, 396–415 (2014). https://doi.org/10.1016/j.ins.2014.03.128

24. Elsayed, S.M., Sarker, R.A., Essam, D.L.: A new genetic algorithm for solving optimization problems. Eng. Appl. Artif. Intell. **27**, 57–69 (2014). https://doi.org/10.1016/j.engappai.2013.09.013

25. Ullah, A., Malik, S.A., Alimgeer, K.S.: Evolutionary algorithm based heuristic scheme for nonlinear heat transfer equations. PLOS ONE **13**(1) (2018). https://doi.org/10.1371/journal.pone.0191103

26. Tenenev, V.A., Rusyak, I.G., Sufiyanov, V.G., Ermolaev, M.A., Nefedov, D.G.: Construction of approximate mathematical models on results of numerical experiments. Bull. South Ural State Univ. Ser. Math. Model. Program. Comput. Softw. **8**(1), 76–87 (2015). https://doi.org/10.14529/mmp150106

27. He, W., Chen, Y., Yin, Z.: Adaptive neural network control of an uncertain robot with full-state constraints. IEEE Trans. Cybern. **46**(3), 620–629 (2016). https://doi.org/10.1109/TCYB.2015.2411285

28. He, W., Dong, Y., Sun, C.: Adaptive neural impedance control of a robotic manipulator with input saturation. IEEE Trans. Syst. Man Cybern. Syst. **46**(3), 334–344 (2016). https://doi.org/10.1109/TSMC.2015.2429555

29. Chen, C.L.P., Liu, Y.-J., Wen, G.-X.: Fuzzy neural network-based adaptive control for a class of uncertain nonlinear stochastic systems. IEEE Trans. Cybern. **44**(5), 583–593 (2014). https://doi.org/10.1109/TCYB.2013.2262935

30. Raja, M., Abbas, S., Syam, M., Wazwaz, A.: Design of neuro-evolutionary model for solving nonlinear singularly perturbed boundary value problems. Appl. Soft Comput. **62**, 373–394 (2018). https://doi.org/10.1016/j.asoc.2017.11.002

31. Vasiliev, S.N., Novikov, D.A., Bakhtadze, N.N.: Intelligent control of industrial processes. Manuf. Model. Manag. Control **1**(7), 49–57 (2013). Part 1

32. Druzhinina, O.V., Masina, O.N.: On approaches to the stability analysis of nonlinear dynamic systems with logical controllers. Modern Inf. Technol. IT-Educ. **13**(2), 40–49 (2017)

Alternative Theorem for Differential Games with Strongly Convex Admissible Control Sets

Grigorii E. Ivanov and Maxim O. Golubev$^{(\boxtimes)}$

Moscow Institute of Physics and Technology (State University), 9 Institutskiy per., Dolgoprudny, Moscow Region 141700, Russian Federation
`g.e.ivanov@mail.ru, maksimkane@mail.ru`
https://mipt.ru/education/chair/mathematics/tutors/professors/ivanovge.php,
https://mipt.ru/education/chair/mathematics/tutors/docents/golubev.php

Abstract. A linear differential game with strongly convex admissible control sets and a smooth target set is considered. For such a differential game we obtain the alternative theorem. This theorem states that for any initial position either there is a program strategy of pursuer that guarantees the capture or there is a program strategy of evader that guarantees the evasion. This result is based on the commutativity of the Minkowski sum and difference for sets with special properties of strong and weak convexity in a Banach space.

Keywords: Differential game · Strongly convex set
The Minkowski sum and difference

1 Differential Game

Foundations of the theory of zero-sum differential games were laid by Isaacs [1], Krasovskii [2], Pontryagin [3], and others. Generally, the algorithms for constructing optimal (or suboptimal) strategies in differential games have a very high computational complexity. This fact stimulates an interest in special cases of differential games when the open-loop (program) strategies are optimal. The open-loop strategies may be constructed by means of Pontryagin's maximum principle much easier than common methods of Krasovskii's stable bridge or Pontryagin's alternating integral. The question on the optimality of the open-loop strategies in linear differential games with a target set can be reduced to the question on commutativity of the Minkowski sum and difference. In general, these operations are not commutative. However, we obtain sufficient conditions of such a commutativity for weakly and strongly convex sets in a uniformly convex Banach space. In this aspect, we adapt and improve some results obtained for Hibert spaces in [4].

Supported by the Russian Foundation for Basic Research, grant 18-01-00209.

Y. Evtushenko et al. (Eds.): OPTIMA 2018, CCIS 974, pp. 321–335, 2019.
https://doi.org/10.1007/978-3-030-10934-9_23

Let X be a Banach space. Consider the differential game

$$\frac{dx}{dt} = v(t) - u(t), \qquad x(0) = x_0 \tag{1}$$

on a closed time interval $t \in [0, T]$ with the target set $M \subset X$, where $x(t)$, $u(t)$, $v(t)$ are vectors from X, $x(t)$ is the phase vector of the system, $u(t)$ is the control of the pursuer, and $v(t)$ is the control of the evader. The controls of the players obey the geometric constraints

$$u(t) \in U(t), \quad v(t) \in V(t) \quad \text{for almost all} \quad t \in [0, T], \tag{2}$$

where U, V are given multivalued mappings whose values are convex closed sets $U(t) \subset X$, $V(t) \subset X$. The goal of the pursuer is to bring the phase vector into the target set at a finite time instant: $x(T) \in M$. The goal of the evader is the opposite: $x(T) \notin M$. If the condition $x(T) \in M$ is satisfied in some realization of the game, then we say that a capture has occurred. Otherwise, we say that an evasion has occurred.

Definition 1. *By $U[0, T]$ we denote the set of admissible open-loop strategies of the pursuer consisting of all Lebesgue integrable functions $u : [0, T] \to X$ satisfying the condition $u(t) \in U(t)$ for almost all $t \in [0, T]$. By $V[0, T]$ we denote the set of admissible open-loop strategies of the evader consisting of all Lebesgue integrable functions $v : [0, T] \to X$ satisfying the condition $v(t) \in V(t)$ for almost all $t \in [0, T]$.*

Definition 2. *Given $u \in U[0, T]$, $v \in V[0, T]$ and $x_0 \in X$, we use $x_T(x_0, u, v)$ to denote the value of the phase vector at a finite time instant T for controls u, v of the players and a prescribed initial value of the phase vector x_0:*

$$x_T(x_0, u, v) = x_0 - \int_0^T u(t)\, dt + \int_0^T v(t)\, dt.$$

Definition 3. *The set of* guaranteed capture *is*

$$M_C = \{x_0 \in X \mid \exists u \in U[0, T] : \ \forall v \in V[0, T] \quad x_T(x_0, u, v) \in M\}.$$

The set of guaranteed evasion *is defined as*

$$M_E = \{x_0 \in X \mid \exists v \in V[0, T] : \ \forall u \in U[0, T] \quad x_T(x_0, u, v) \notin M\}.$$

It follows straightforward from the definition that $M_C \cap M_E = \emptyset$. Typically $M_C \cup M_E \neq X$, i.e. there exists $x_0 \in X$ which does not belong to either the set of guaranteed capture or the set of guaranteed evasion. So, for such initial position x_0 open-loop strategies are not optimal. Using properties of strong and weak convexity we obtain conditions such that $M_C \cup M_E = X$. It means that the *alternative theorem* is valid: for any initial position $x_0 \in X$ either $x_0 \in M_C$ and there is an open-loop strategy $u \in U[0, T]$ such that capture is guaranteed or $x_0 \in M_E$ and there is an open-loop strategy $v \in V[0, T]$ such that evasion is guaranteed. In this case, for any initial position $x_0 \in X$ the open-loop strategies are optimal.

2   Notation

Let X be a Banach space. Through $\langle p, x \rangle$ we denote the value of a functional $p \in X^*$ on a vector $x \in X$. For the vector $a \in X$ and the functional $p_0 \in X^*$ through $\mathfrak{B}_R(a)$ and $\mathfrak{B}_R^*(p_0)$ we denote closed balls of radius R in the spaces X and X^* respectively:

$$\mathfrak{B}_R(a) = \{x \in X : \|x - a\| \leq R\},$$

$$\mathfrak{B}_R^*(p_0) = \{p \in X^* : \|p - p_0\|_* \leq R\}.$$

Through int A, cl A, ∂A we denote the *interior*, the *closure* and the *boundary* of a set $A \subset X$ correspondingly.

The *diameter* of a set $A \subset X$ is defined as follows

$$\operatorname{diam} A = \sup_{x, y \in A} \|x - y\|.$$

Definition 4. *The* modulus of convexity *of a set $A \subset X$ is the function δ^A : $(0, \operatorname{diam} A) \to \mathbb{R}$, defined by*

$$\delta^A(\varepsilon) = \sup\left\{\delta \geq 0 \,\middle|\, \mathfrak{B}_\delta\left(\frac{x+y}{2}\right) \subset A \; \forall x, y \in A : \|x - y\| = \varepsilon\right\}$$

for all $\varepsilon \in (0, \operatorname{diam} A)$. The modulus of convexity $\delta_X(\cdot)$ of the Banach space X is the modulus of convexity of the unit ball of X: $\delta_X(\cdot) = \delta^{\mathfrak{B}_1(0)}(\cdot)$. A set $A \subset X$ is called uniformly convex if $\delta^A(\varepsilon) > 0$ for all $\varepsilon \in (0, \operatorname{diam} A)$. If the unit ball of X is uniformly convex, then the space X is called uniformly convex.

The *Hausdorff distance* between two sets $A, B \subset X$ is a function $h(A, B)$ given by the formula

$$h(A, B) = \max\left\{\sup_{a \in A} \inf_{b \in B} \|a - b\|, \quad \sup_{b \in B} \inf_{a \in A} \|a - b\|\right\}.$$

For sets $A, B \subset X$ we define

$$\varrho(A, B) = \inf_{a \in A, \, b \in B} \|a - b\|.$$

The *distance* from a point $x \in X$ to a subset $A \subset X$ is the following function

$$\varrho(x, A) = \inf_{a \in A} \|x - a\|.$$

Any element of the set

$$P_A(x) = \{a \in A : \|x - a\| = \varrho(x, A)\}$$

is called the *metric projection* of a point $x \in X$ onto a subset $A \subset X$.

The *supporting function* of a subset $A \subset X$ is defined as follows

$$s(p, A) = \sup_{x \in A} \langle p, x \rangle, \qquad \forall\, p \in X^*.$$

By the *Minkowski sum* and the *Minkowski difference* of the sets A and B in linear space X we denote the following sets

$$A + B = \{x + y \ : \ x \in A, \ y \in B\},$$

$$A \stackrel{*}{-} B = \{x \in X \ : \ x + B \subset A\}.$$

3   Prox-Regular Sets

In [5–7] the following notion of (uniformly) prox-regular sets was considered.

Definition 5. *We say that a set $A \subset X$ is R-prox-regular if for any $x \in X$ with $0 < \varrho(x, A) < R$ and $a \in P_A(x)$ one has*

$$A \bigcap \operatorname{int} \mathfrak{B}_R \left(a + \frac{R}{\|x - a\|}(x - a) \right) = \emptyset.$$

In [8] the prox-regular sets were called as sets satisfying the support condition of weak convexity.

Theorem 1. *Let $A \subset X$ be R-prox-regular and $C \subset X$ be such that for some $r \in (0, R)$ we have*

$$C \subset \mathfrak{B}_r \left(c - \frac{r}{\|x - c\|}(x - c) \right) \qquad \forall x \in X \setminus C \quad \forall c \in P_C(x). \tag{3}$$

Then $A + C$ is $(R - r)$-prox-regular.

Proof. Given $x \in X$ and $z \in P_{A+C}(x)$ with $0 < \varrho(x, A + C) < R - r$, we denote

$$\varrho = \varrho(x, A + C), \qquad q = \frac{x - z}{\varrho}, \qquad y = z + (R - r)q.$$

Note that $\varrho(A, x - C) = \inf_{a \in A, \, c \in C} \|a - x + c\| = \varrho(x, A + C) = \varrho$. Since $z \in A + C$ there exist $a \in A$, $c \in C$ such that $z = a + c$. Thus $\|a - x + c\| = \|z - x\| = \varrho = \varrho(A, x - C)$. Hence $a \in P_A(x - c)$ and $x - c \in P_{x-C}(a)$, i.e. $c \in P_C(x - a)$. Then using the equality $q = \frac{x - a - c}{\varrho}$ and assumptions for A and C we get

$$A \subset X \setminus \operatorname{int} \mathfrak{B}_R(a + Rq), \qquad C \subset \mathfrak{B}_r(c - rq).$$

So,

$$A + C \subset \left(X \setminus \operatorname{int} \mathfrak{B}_R(a + Rq) \right) + \mathfrak{B}_r(c - rq) = X \setminus \operatorname{int} \mathfrak{B}_{R-r}(a + c + (R - r)q),$$

and consequently $(A + C) \bigcap \operatorname{int} \mathfrak{B}_{R-r}(y) = \emptyset$. $\square$

4 Summand of the Ball

Definition 6. *A set $C \subset X$ is called* a summand of the ball of radius r *if there exists a convex closed subset $C_1 \subset X$ such that $C + C_1 = \mathfrak{B}_r(0)$. We use $\mathcal{SC}(r)$ to denote the class of convex closed summands of the ball of radius r.*

If $C \in \mathcal{SC}(r)$, then C is r-strongly convex set, that is C can be represented as an intersection of balls of radius r: there exists $C_1 \subset X$ such that $C = \bigcap_{c \in C_1} \mathfrak{B}_r(c)$.

In case of Hilbert space the class $\mathcal{SC}(r)$ and the class of r-strongly convex sets coincide, but in spaces L_p with $p \in (1,2) \cup (2,+\infty)$ these classes are different. Thus we consider the class $\mathcal{SC}(r)$ as a special class of strongly convex sets.

Lemma 1. *Let $C \subset X$ be a closed convex set in a reflexive Banach space X; $r > 0$. Then the condition (3) implies that $C \in \mathcal{SC}(r)$. If, in addition, X is strictly convex, then the conditions (3) and $C \in \mathcal{SC}(r)$ are equivalent.*

Proof. Let (3) holds. Let us prove

$$\forall p \in \partial \mathfrak{B}_1^*(0) \quad \forall x_p^C \in C : \quad \langle p, x_p^C \rangle = s(p, C)$$
$$\exists x_p^B \in \mathfrak{B}_1(0) : \quad \langle p, x_p^B \rangle = 1, \quad C \subset \mathfrak{B}_r(x_p^C - r \cdot x_p^B). \tag{4}$$

Let us consider a functional $p \in \partial \mathfrak{B}_1^*(0)$ and such a point $x_p^C \in C$ that $\langle p, x_p^C \rangle = s(p, C)$. Since the space X is reflexive there exists such a point $x_p^B \in \partial \mathfrak{B}_1(0)$ that $\langle p, x_p^B \rangle = 1$. We define the point $x = x_p^C + x_p^B$. Since $\langle p, x \rangle = s(p, C) + 1$, then $x \in X \setminus C$. Let us show that $x_p^C \in P_C(x)$. For any vector $c \in C$ we have

$$\|x - c\| \geq \langle p, x - c \rangle = s(p, C) + 1 - \langle p, c \rangle \geq 1 \geq \|x - x_p^C\|.$$

Therefore $x_p^C \in P_C(x)$ and from the condition (3) we have an inclusion

$$C \subset \mathfrak{B}_r \left(x_p^C - \frac{r}{\|x - x_p^C\|}(x - x_p^C) \right) = \mathfrak{B}_r(x_p^C - r \cdot x_p^B).$$

Thus we show that the condition (4) is fulfilled. Following [9, Theorem 1.1] we get $C \in \mathcal{SC}(r)$.

Now let X be strictly convex and $C \in \mathcal{SC}(r)$. As shown in [9, Theorem 1.1] the condition (4) holds. Let us prove (3). Given $x \in X \setminus C$ and $c \in P_C(x)$, our goal is to prove the following inclusion

$$C \subset \mathfrak{B}_r \left(c - \frac{r}{\|x - c\|}(x - c) \right). \tag{5}$$

Denote $\varrho = \|x - c\|$. Since $c \in C$, $x \notin C$, then $\varrho > 0$. Applying the separation theorem for the sets C and $\text{int}\,\mathfrak{B}_\varrho(x)$, we obtain such a functional $p \in \partial \mathfrak{B}_1^*(0)$ that $\langle p, c \rangle = s(p, C)$ and $\langle p, x - c \rangle = \varrho$. By virtue of condition (4) there exists a point $x_p^B \in \mathfrak{B}_1(0)$ such that $\langle p, x_p^B \rangle = 1$ and $C \subset \mathfrak{B}_r(c - r \cdot x_p^B)$. Since Banach space X is strictly convex we have $x_p^B = \frac{x-c}{\varrho} = \frac{x-c}{\|x-c\|}$. Using the inclusion $C \subset \mathfrak{B}_r(c - r \cdot x_p^B)$ we obtain the inclusion (5). $\square$

Theorem 2. *Let X be a uniformly convex Banach space. Let $A \subset X$ be closed and R-prox-regular; $C \in SC(r)$ with $r \in (0, R)$. Then $A + C$ is closed.*

Proof. Let $x \in \mathrm{cl}(A + C)$. Define the set $D = x - C$, then $\varrho(A, D) = 0$. By virtue of [10, Theorem 4.2] $\min_{a \in A,\, d \in D} \|a - d\|$ is attained. Thus there exists $a \in A \cap D$. Therefore, $a \in x - C$ and $x \in a + C \subset A + C$. $\qquad\square$

Lemma 2. *Let X be a uniformly convex Banach space and $C \in SC(r)$ with some $r > 0$. Then C is a uniformly convex set and*

$$\delta^C(\varepsilon) \geq r\delta_X\left(\frac{\varepsilon}{r}\right) \qquad \forall \varepsilon \in (0, \mathrm{diam}\, C). \tag{6}$$

Proof. Since $C \in SC(r)$ there exists $C_1 \subset X$ such that $C = \mathfrak{B}_r(0) \overset{*}{-} C_1$. Fix any $\varepsilon \in (0, \mathrm{diam}\, C)$ and denote $\delta = r\delta_X\left(\frac{\varepsilon}{r}\right)$. Let us show that $\delta^C(\varepsilon) \geq \delta$. Assume the contrary: there exists $x, y \in C$ such that $\|x - y\| = \varepsilon$ and $\mathfrak{B}_\delta\left(\frac{x+y}{2}\right) \not\subset C$. Then one can find $z \in \mathfrak{B}_\delta\left(\frac{x+y}{2}\right) \setminus C$. Since $z \notin C = \mathfrak{B}_r(0) \overset{*}{-} C_1$, we have $z + C_1 \not\subset \mathfrak{B}_r(0)$. Hence there exists $c_1 \in C_1$ such that $z + c_1 \notin \mathfrak{B}_r(0)$. In view of the inclusions $x, y \in C = \mathfrak{B}_r(0) \overset{*}{-} C_1$ we see that $x + c_1, y + c_1 \in \mathfrak{B}_r(0)$. This and Definition 4 imply that $\mathfrak{B}_{\delta_X\left(\frac{\varepsilon}{r}\right)}\left(\frac{x+y+2c_1}{2r}\right) \subset \mathfrak{B}_1(0)$ and hence $\mathfrak{B}_\delta\left(\frac{x+y}{2}\right) + c_1 \subset \mathfrak{B}_r(0)$. This contradicts $z + c_1 \notin \mathfrak{B}_r(0)$. So, the inequality $\delta^C(\varepsilon) \geq \delta = r\delta_X\left(\frac{\varepsilon}{r}\right)$ is proved. It establishes (6). This and uniform convexity of X imply uniform convexity of C. $\qquad\square$

5   Properties of Uniformly Convex Spaces

Lemma 3. *Let X be a uniformly convex Banach space with modulus of convexity δ_X. Then for any nonzero $x, y \in X$ we have*

$$\|x + y\| \leq \|x\| + \|y\| - 2\min\{\|x\|, \|y\|\} \cdot \delta_X\left(\left\|\frac{x}{\|x\|} - \frac{y}{\|y\|}\right\|\right).$$

Proof. Let us define $a = \frac{x}{\|x\|}$, $b = \frac{y}{\|y\|}$, $\lambda = \|x\|$, $\mu = \|y\|$. Without loss of generality we will suppose $\lambda \geq \mu$. Thus

$$\|x + y\| = \|\lambda a + \mu b\| = \|(\lambda - \mu)a + \mu(a + b)\| \leq \lambda - \mu + \mu\|a + b\|.$$

By the definition of modulus of convexity we have $\|a + b\| \leq 2 - 2\delta_X(\|a - b\|)$. Therefore,

$$\|x + y\| \leq \lambda + \mu - 2\mu\delta_X(\|a - b\|) = \|x\| + \|y\| - 2\min\{\|x\|, \|y\|\} \cdot \delta_X\left(\left\|\frac{x}{\|x\|} - \frac{y}{\|y\|}\right\|\right).$$

$\qquad\square$

Lemma 4. *Let X be a uniformly convex Banach space with modulus of convexity δ_X. Given positive numbers α, β, γ and vectors $x, y \in X$ such that*

$$\|x\| \leq \alpha + \gamma, \quad \|y\| \leq \beta + \gamma, \quad \|x + y\| \geq \alpha + \beta - \gamma, \quad 4\gamma \leq \min\{\alpha, \beta\},$$

one has

$$\delta_X \left(\frac{1}{2} \left\| \frac{x}{\alpha} - \frac{y}{\beta} \right\| \right) \leq \frac{3\gamma}{\min\{\alpha, \beta\}}.$$

Proof. Since $\|x\| \geq \|x + y\| - \|y\| \geq \alpha - 2\gamma$, $\|y\| \geq \|x + y\| - \|x\| \geq \beta - 2\gamma$, then $\min\{\|x\|, \|y\|\} \geq \min\{\alpha - 2\gamma, \beta - 2\gamma\}$. Hereby and from inequality $\|x\| + \|y\| - \|x + y\| \leq 3\gamma$ by Lemma 3 we obtain the inequality

$$2\delta_X \left(\left\| \frac{x}{\|x\|} - \frac{y}{\|y\|} \right\| \right) \min\{\alpha - 2\gamma, \beta - 2\gamma\} \leq 3\gamma.$$

Since $4\gamma \leq \min\{\alpha, \beta\}$, then $2\min\{\alpha - 2\gamma, \beta - 2\gamma\} \geq \min\{\alpha, \beta\}$. Therefore,

$$\delta_X \left(\left\| \frac{x}{\|x\|} - \frac{y}{\|y\|} \right\| \right) \leq \frac{3\gamma}{\min\{\alpha, \beta\}}. \tag{7}$$

From inequalities $\left\| \frac{x}{\alpha} - \frac{x}{\|x\|} \right\| = \left| 1 - \frac{\|x\|}{\alpha} \right| \leq \frac{2\gamma}{\alpha}$, $\left\| \frac{y}{\beta} - \frac{y}{\|y\|} \right\| = \left| 1 - \frac{\|y\|}{\beta} \right| \leq \frac{2\gamma}{\beta}$, we get

$$\left\| \frac{x}{\alpha} - \frac{y}{\beta} \right\| \leq \left\| \frac{x}{\|x\|} - \frac{y}{\|y\|} \right\| + \frac{4\gamma}{\min\{\alpha, \beta\}}. \tag{8}$$

If $\left\| \frac{x}{\|x\|} - \frac{y}{\|y\|} \right\| \leq \frac{4\gamma}{\min\{\alpha,\beta\}}$, then due to the inequality (8) we have $\left\| \frac{x}{\alpha} - \frac{y}{\beta} \right\| \leq \frac{8\gamma}{\min\{\alpha,\beta\}}$. Therefore, by the property of modulus of convexity $\delta_X(2t) \leq t$ for $t \in (0, 1]$ we obtain the inequality to be proved. In the opposite case $\left\| \frac{x}{\|x\|} - \frac{y}{\|y\|} \right\| > \frac{4\gamma}{\min\{\alpha,\beta\}}$ from the inequality (8) we have $\left\| \frac{x}{\alpha} - \frac{y}{\beta} \right\| \leq 2 \left\| \frac{x}{\|x\|} - \frac{y}{\|y\|} \right\|$. Hereby and from the inequality (7) by virtue of monotonicity of the modulus of convexity we obtain the inequality to be proved. $\square$

6 Efimov–Stechkin Weakly Convex Sets

Efimov and Stechkin [11] have introduced the notion of "a-convex" sets. Following the terminology of [12], we refer to these sets as to weakly convex sets in the sense of Efimov–Stechkin.

Definition 7. *$A \subset X$ is called weakly convex in the sense of Efimov–Stechkin with constant $R > 0$ (we write $A \in \mathcal{WCES}(R)$) if there exists a nonempty subset $A_1 \subset X$ such that*

$$A = \bigcap_{a \in A_1} (X \setminus \text{int } \mathfrak{B}_R(a)).$$

Also we use denotation

$$\mathcal{WCES}^-(R) = \{A \subset X : \text{cl}(X \setminus A) \in \mathcal{WCES}(R)\}.$$

Theorem 3. *Let X be a uniformly convex Banach space. Let $A \subset X$ be closed and R-prox-regular. Then $A \in \mathcal{WCES}(R)$.*

Proof. Fix a point $x \in X \setminus A$. Due to [12, Lemma 3.1] it is sufficient to find a vector $y \in X$ such that

$$x \in \operatorname{int} \mathfrak{B}_R(y) \subset X \setminus A. \tag{9}$$

If $\varrho(x, A) \geq R$, then conditions (9) are fulfilled for $y = x$. Let $\varrho(x, A) < R$. Due to [8, Lemma 4.3] there exists a point $a \in P_A(x)$. Let $y = a + \frac{R}{\|x-a\|}(x - a)$. Then $x \in \operatorname{int} \mathfrak{B}_R(y)$. Since A is R-prox-regular, we have $A \cap \operatorname{int} \mathfrak{B}_R(y) = \emptyset$. So, conditions (9) are fulfilled again. □

Lemma 5. *For a set $A \subset E$ and constant $R > 0$ the following assertions are equivalent:*

(i) $A \in \mathcal{WCES}(R)$;
(ii) $A \neq X$ and there exist $A_1 \subset X$ such that $A = A_1 \overset{*}{-} \operatorname{int} \mathfrak{B}_R(0)$;
(iii) $A \neq X$ and $A + \operatorname{int} \mathfrak{B}_R(0) \overset{*}{-} \operatorname{int} \mathfrak{B}_R(0) = A$.

Proof. The equivalence of assertions (i) and (ii) follows from the equality

$$A_1 \overset{*}{-} \operatorname{int} \mathfrak{B}_R(0) = \bigcap_{a \in X \setminus A_1} (X \setminus \operatorname{int} \mathfrak{B}_R(a)).$$

The equivalence of assertions (ii) and (iii) can be verified directly. □

Lemma 6. *Let X be a uniformly convex Banach space. Let $A \in \mathcal{WCES}(R_1) \cap \mathcal{WCES}^-(R_2)$ and $\operatorname{cl} \operatorname{int} A = A$; $a \in \partial A$. Then there exists $d \in \partial \mathfrak{B}_1(0)$ such that $\varrho(a + R_1 d, A) = R_1$ and $\varrho(a - R_2 d, X \setminus A) = R_2$.*

Proof. Since $a \in A = \operatorname{cl} \operatorname{int} A$, then there exists a sequence of points $x_k \in \operatorname{int} A$ such that $\|x_k - a\| < \frac{1}{k}$ for any $k \in \mathbb{N}$. Since $A \in \mathcal{WCES}^-(R_2)$, then for any $k \in \mathbb{N}$ there exists $y_k \in X$ such that $x_k \in \operatorname{int} \mathfrak{B}_{R_2}(y_k) \subset \operatorname{int} A$.

Since $a \in \partial A$, then there exists a sequence of points $z_n \in X \setminus A$ such that $\|z_n - a\| < \frac{1}{n}$ for any $n \in \mathbb{N}$. Since $A \in \mathcal{WCES}(R_1)$, then for any $n \in \mathbb{N}$ there exists $w_n \in X$ such that $z_n \in \operatorname{int} \mathfrak{B}_{R_1}(w_n) \subset X \setminus A$.

It follows from inclusions $\mathfrak{B}_{R_2}(y_k) \subset \operatorname{int} A$, $\mathfrak{B}_{R_1}(w_n) \subset X \setminus A$ that $\mathfrak{B}_{R_1}(w_n) \cap \mathfrak{B}_{R_2}(y_k) = \emptyset$, i.e. $\|w_n - y_k\| \geq R_1 + R_2$.

Applying Lemma 4 for $x = x_k - y_k$, $y = w_n - z_n$, $\alpha = R_2$, $\beta = R_1$, $\gamma = \frac{1}{n} + \frac{1}{k}$, for sufficiently large n, k we obtain the inequality

$$\delta_X \left(\frac{1}{2} \left\| \frac{x_k - y_k}{R_2} - \frac{w_n - z_n}{R_1} \right\| \right) \leq \frac{3}{\min\{R_1, R_2\}} \left(\frac{1}{n} + \frac{1}{k} \right).$$

Thus $\left\| \frac{x_k - y_k}{R_2} - \frac{w_n - z_n}{R_1} \right\| \to 0$ for $n, k \to \infty$. Due to completeness of the space X there exists $\lim_{k \to \infty} \frac{x_k - y_k}{R_2} = \lim_{n \to \infty} \frac{w_n - z_n}{R_1} = d \in X$. This easily implies the required statement. □

Theorem 4. *Let X be a uniformly convex Banach space. Let $A \subset X$ satisfy the equality $\operatorname{cl} \operatorname{int} A = A$. The following assertions are equivalent:*

(i) $A \in \mathcal{WCES}(R_1) \cap \mathcal{WCES}^-(R_2)$;

(ii) A and $\operatorname{cl}(X \setminus A)$ are R_1-prox-regular and R_2-prox-regular correspondingly.

Proof. Due to Theorem 3 condition (i) follows from condition (ii).

Suppose that condition (i) holds. Let us show that A is R_1-prox-regular.

Let $x \in X$, $a \in A$, $0 < \varrho(x, A) = \|x - a\| < R_1$. Since $a \in \partial A$, then due to Lemma 6 there exists a vector $d \in X$ such that $\|d\| = 1$, $\varrho(a + R_1 d, A) = R_1$, $\varrho(a - R_2 d, X \setminus A) = R_2$. Since $\operatorname{int} \mathfrak{B}_{\varrho(x,A)}(x) \cap \operatorname{int} \mathfrak{B}_{R_2}(a - R_2 d) = \emptyset$, then $\|a - R_2 d - x\| \geq R_2 + \varrho(x, A) = \|R_2 d\| + \|a - x\|$. Together with strict convexity of the space X it gives us that vectors $R_2 d$ and $x - a$ are co-directional, i.e. $d = \frac{x-a}{\|x-a\|}$. Thus $\varrho\left(a + \frac{R_1}{\|x-a\|}(x - a), A\right) = \varrho(a + R_1 d, A) = R_1$. So, A is R_1-prox-regular.

Similarly $\operatorname{cl}(X \setminus A)$ is R_2-prox-regular. $\qquad \square$

7 Calculus of Convexity Parameters with Respect to the Minkowski Operations

The following lemma follows directly from the definitions.

Lemma 7. *For any $A, B \subset X$ we have*

(i) $(\operatorname{cl} A) + B \subset \operatorname{cl}(A + B)$;

(ii) $(\operatorname{int} A) + B \subset \operatorname{int}(A + B)$;

(iii) $X \setminus (A \stackrel{*}{-} B) = (X \setminus A) + (-B)$;

(iv) $X \setminus (A + B) = (X \setminus A) \stackrel{*}{-} (-B)$.

Lemma 8. *Let X be a uniformly convex Banach space; $A \in \mathcal{WCES}(R_1) \cap \mathcal{WCES}^-(R_2)$, $\operatorname{cl} \operatorname{int} A = A$; $B \in \mathcal{SC}(r)$ with $r < R_1$. Then*

(i) $\operatorname{cl} \operatorname{int}(A + B) = A + B$;

(ii) $\operatorname{int}(A + B) = (\operatorname{int} A) + B$;

(iii) $A + B \in \mathcal{WCES}(R_1 - r)$;

(iv) $A + B \in \mathcal{WCES}^-(R_2)$.

Proof. (i). Theorem 4 implies that A and $\operatorname{cl}(X \setminus A)$ are R_1-prox-regular and R_2-prox-regular correspondingly. Due to Theorem 2 the set $A + B$ is closed. Taking into account the equality $\operatorname{cl} \operatorname{int} A = \operatorname{cl} A$ and Lemma 7(i, ii) we have

$$A + B = (\operatorname{cl} \operatorname{int} A) + B \subset \operatorname{cl}((\operatorname{int} A) + B) \subset \operatorname{cl} \operatorname{int}(A + B) \subset \operatorname{cl}(A + B) = A + B.$$

This proves the first part of the current lemma.

(ii). Suppose that the equality $\operatorname{int}(A + B) = (\operatorname{int} A) + B$ is not true. Then Lemma 7(ii) implies that there exists $x \in \operatorname{int}(A + B)$ such that $x \notin (\operatorname{int} A) + B$.

If $x \in A + \operatorname{int} B$, then there exists $b \in \operatorname{int} B$ such that $x - b \in A$. Since $b \in \operatorname{int} B$, then there exists $\varepsilon > 0$ such that $\mathfrak{B}_\varepsilon(b) \subset B$. Since $x - b \in A = \operatorname{cl} \operatorname{int} A$, then there exists $a \in \operatorname{int} A$ such that $\|x - b - a\| < \varepsilon$. Hence, $x - a \in \mathfrak{B}_\varepsilon(b) \subset B$. Therefore $x \in a + B \subset (\operatorname{int} A) + B$, which contradicts the condition $x \notin (\operatorname{int} A) + B$. We proved that $x \notin A + (\operatorname{int} B)$, i.e.

$$(x - \operatorname{int} B) \bigcap A = \emptyset. \tag{10}$$

If B is a singleton, then the desired assertion is trivially valid. Otherwise $\operatorname{int} B \neq \emptyset$ and hence $B = \operatorname{cl} \operatorname{int} B$. Since $x \in \operatorname{int}(A + B) \subset A + B$, then there exists a vector $x_0 \in (x - B) \bigcap A$. Using (10) and $B = \operatorname{cl} \operatorname{int} B$ we obtain $x_0 \in \partial A$. Due to Lemma 6 there exists a vector $d \in \partial \mathfrak{B}_1(0)$ such that $\varrho(x_0 + R_1 d, A) = R_1$, $\varrho(x_0 - R_2 d, X \setminus A) = R_2$. It follows from the equalities $\varrho(x_0 - R_2 d, X \setminus A) = R_2$ and (10) that $\varrho(x_0 - R_2 d, x - B) \geq R_2$. Taking into account inclusion $x_0 \in x - B$, we obtain that $x_0 \in P_{x-B}(x_0 - R_2 d)$. Since $B \in \mathcal{SC}(r)$, then due to Lemma 1 the inclusion $x - B \subset \mathfrak{B}_r(x_0 + rd)$ holds. Thus,

$$A \subset X \setminus \operatorname{int} \mathfrak{B}_{R_1}(x_0 + R_1 d), \qquad B \subset \mathfrak{B}_r(x - x_0 - rd).$$

Hence $A + B \subset \left(X \setminus \operatorname{int} \mathfrak{B}_{R_1}(x_0 + R_1 d) \right) + \mathfrak{B}_r(x - x_0 - rd) = X \setminus \operatorname{int} \mathfrak{B}_{R_1 - r}(x + (R_1 - r)d)$, which contradicts the inclusion $x \in \operatorname{int}(A + B)$. This proves the second part of Lemma.

(iii). Theorem 1 and Lemma 1 imply that $A + B$ is $(R_1 - r)$-prox-regular. So, the desired assertion follows from Theorem 3.

(iv). Assertion (ii) of the current lemma and Lemma 7(iv) imply that

$$\operatorname{cl}(X \backslash (A + B)) = X \setminus (\operatorname{int}(A + B)) = X \setminus ((\operatorname{int} A) + B)$$
$$= (X \setminus \operatorname{int} A) \overset{*}{-} (-B) = \operatorname{cl}(X \setminus A) \overset{*}{-} (-B).$$

So,
$$A + B \in \mathcal{WCES}^-(R_2).$$

$\square$

Lemma 9. *Let X be a uniformly convex Banach space; $A \in \mathcal{WCES}(R_1) \cap \mathcal{WCES}^-(R_2)$, $\operatorname{cl} \operatorname{int} A = A$; $C \in \mathcal{SC}(r)$ with $r < R_2$. Then*

(i) $\operatorname{int}(A \overset{*}{-} C) = (\operatorname{int} A) \overset{*}{-} C$;
(ii) $\operatorname{cl} \operatorname{int}(A \overset{*}{-} C) = A \overset{*}{-} C$;
(iii) $A \overset{*}{-} C \in \mathcal{WCES}^-(R_2 - r)$;
(iv) set $A \overset{*}{-} C \in \mathcal{WCES}(R_1)$.

Proof. It suffices to apply Lemma 8 for $A_1 = \operatorname{cl}(X \setminus A)$, $B = -C$ and Lemma 7(iii, iv).

$\square$

8   Commutativity of the Minkowski Operations

Lemma 10. *Let $A \in \mathcal{WCES}(R)$, $B \in \mathcal{SC}(r)$ with $r < R$. Then $A + B \overset{*}{-} B = A$.*

Proof. Since $B \in \mathcal{SC}(r)$, there exists $C \subset X$ such that $B + C = \mathfrak{B}_r(0)$. Denote $D = C + \operatorname{int} \mathfrak{B}_{R-r}(0)$. Then $B + D = B + C + \operatorname{int} \mathfrak{B}_{R-r}(0) = \mathfrak{B}_r(0) + \operatorname{int} \mathfrak{B}_{R-r}(0) = \operatorname{int} \mathfrak{B}_R(0)$.

Since $A \in \mathcal{WCES}(R)$, it follows that $A + \operatorname{int} \mathfrak{B}_R(0) \overset{*}{-} \operatorname{int} \mathfrak{B}_R(0) = A$. Hence $A \subset A + B \overset{*}{-} B \subset A + B + D \overset{*}{-} D \overset{*}{-} B$
$= A + \operatorname{int} \mathfrak{B}_R(0) \overset{*}{-} \operatorname{int} \mathfrak{B}_R(0) = A$. So, $A + B \overset{*}{-} B = A$.   □

Lemma 11. *Let X be a uniformly convex Banach space; $A \in \mathcal{WCES}(R) \cap \mathcal{WCES}^-(\varrho)$, $\operatorname{cl} \operatorname{int} A = A$, $B \in \mathcal{SC}(r)$ with $r < \min\{\varrho, R\}$. Then $A \overset{*}{-} B + B = A$.*

Proof. Denote $B_1 = -B$, $A_1 = \operatorname{cl}(X \setminus A)$, $A_2 = A_1 + B_1$. Then $A_1 \in \mathcal{WCES}(\varrho) \cap \mathcal{WCES}^-(R)$. Lemma 8 implies that $\operatorname{cl} \operatorname{int} A_2 = A_2$, $\operatorname{int} A_2 = (\operatorname{int} A_1) + B_1$, $A_2 \in \mathcal{WCES}(\varrho - r) \cap \mathcal{WCES}^-(R)$. Lemma 9 implies that $\operatorname{int}(A_2 \overset{*}{-} B_1) = (\operatorname{int} A_2) \overset{*}{-} B_1$. Thus $\operatorname{int}(A_1 + B_1 \overset{*}{-} B_1) = (\operatorname{int} A_2) \overset{*}{-} B_1 = (\operatorname{int} A_1) + B_1 \overset{*}{-} B_1$. Due to Lemma 10 we have $A_1 + B_1 \overset{*}{-} B_1 = A_1$. Hence $(\operatorname{int} A_1) + B_1 \overset{*}{-} B_1 = \operatorname{int} A_1$, i.e. $(X \setminus A) + B_1 \overset{*}{-} B_1 = X \setminus A$. Using Lemma 7(iii, iv) we complete the proof.   □

Theorem 5. *Let X be a uniformly convex Banach space; $A \in \mathcal{WCES}(R_1) \cap \mathcal{WCES}^-(R_2)$, $\operatorname{cl} \operatorname{int} A = A$, $B \in \mathcal{SC}(r_1)$, $C \in \mathcal{SC}(r_2)$ with $r_1 < R_1$ and $r_2 < R_2$. Then*

$$A + B \overset{*}{-} C = A \overset{*}{-} C + B \tag{11}$$

and $W = A + B \overset{*}{-} C$ possesses the following properties:

(i) $W \in \mathcal{WCES}(R_1 - r_1)$;
(ii) $\operatorname{cl}(X \setminus W) \in \mathcal{WCES}(R_2 - r_2)$;
(iii) $\operatorname{cl} \operatorname{int} W = W$.

Proof. Define a natural number n such that $\frac{r_2}{n} < \min\{R_1 - r_1, R_2 - r_2\}$. Define the subset $C_k = \frac{k}{n} C$ for any $k \in \{0, 1, \ldots, n\}$. Let us show that for any $k \in \{0, 1, \ldots, n\}$ the equality holds true

$$A \overset{*}{-} C_k + B = A + B \overset{*}{-} C_k. \tag{12}$$

If $k = 0$ then the equality (12) holds. Suppose that the equality (12) holds for $k = s \in \{0, 1, \ldots, n-1\}$. We show that it is valid for $k = s + 1$.

Lemma 9 implies that $\operatorname{cl} \operatorname{int}(A \overset{*}{-} C_s) = A \overset{*}{-} C_s$ and $A \overset{*}{-} C_s \in \mathcal{WCES}(R_1) \cap \mathcal{WCES}^-(R_2 - \frac{s}{n} r_2)$. Since the subset $C_1 = \frac{1}{n} C$ is a summand of the ball of radius $\frac{r_2}{n} < \min\{R_1, R_2 - \frac{s}{n} r_2\}$, then due to Lemma 11 we have the equality

$$A \overset{*}{-} C_s \overset{*}{-} C_1 + C_1 = A \overset{*}{-} C_s.$$

Adding to the left and right sides of this equality the subset B, we obtain the equality

$$A \overset{*}{-} C_{s+1} + B + C_1 = A \overset{*}{-} C_s + B.$$

Therefore, by the induction assumption

$$A \overset{*}{-} C_{s+1} + B + C_1 = A + B \overset{*}{-} C_s.$$

Subtracting from the left and right sides of this equality the subset C_1, we obtain the equality

$$A \overset{*}{-} C_{s+1} + B + C_1 \overset{*}{-} C_1 = A + B \overset{*}{-} C_{s+1}. \tag{13}$$

By Lemmas 8, 9 we have $\operatorname{cl int}(A \overset{*}{-} C_{s+1} + B) = A \overset{*}{-} C_{s+1} + B$ and $A \overset{*}{-} C_{s+1} + B \in \mathcal{WCES}(R_1 - r_1) \cap \mathcal{WCES}^-(R_2 - r_2)$. Since the subset C_1 is a summand of the ball of radius $\frac{r_2}{n} < \min\{R_1 - r_1, R_2 - r_2\}$, then by Lemma 10 we get

$$A \overset{*}{-} C_{s+1} + B + C_1 \overset{*}{-} C_1 = A \overset{*}{-} C_{s+1} + B.$$

This and the equality (13) imply the equality

$$A \overset{*}{-} C_{s+1} + B = A + B \overset{*}{-} C_{s+1},$$

i.e. the equality (12) holds true for $k = s + 1$. By induction, we obtain the equality (12) for any $k \in \{0, 1, \dots, n\}$. Applying the equality (12) for $k = n$, we have the equality (11).

Applying Lemma 8 to the subsets A, B, and then applying Lemma 9 to the subsets $A + B$, C, we get that W possesses the properties (i)–(iii).   □

9   Alternative Theorem for Differential Games

Definition 8. *Given a multivalued mapping* $U : [a, b] \rightrightarrows X$, *the* Aumann *integral* $\int_a^b U(t) \, dt$ *is called the set of vectors* $\int_a^b u(t) \, dt$, *where* $u : [a, b] \to X$ *are Lebesgue integrable functions satisfying the condition* $u(t) \in U(t)$ *for almost all* $t \in [a, b]$.

Theorem 6. *Assume that, in differential game* (1), (2) *the target set* $M \in \mathcal{WCES}(R_1) \cap \mathcal{WCES}^-(R_2)$ *and* $\operatorname{cl int} M = M$. *Suppose that the Aumann integrals*

$$U_0 = \int_0^T U(t) \, dt, \quad V_0 = \int_0^T V(t) \, dt \tag{14}$$

are convex closed summands of the balls of radii $r_1 \in (0, R_1)$ *and* $r_2 \in (0, R_2)$ *correspondingly. Then for any* $x_0 \in X$ *one and only one of the following conditions is satisfied:*

(i) *there exists an open-loop strategy* $u \in U[0, T]$ *of the pursuer such that, for any open-loop strategy* $v \in V[0, T]$ *of the evader, a capture occurs:* $x_T(x_0, u, v) \in M$

   or

(ii) *there exists an open-loop strategy* $v \in V[0, T]$ *of the evader such that, for any open-loop strategy* $u \in U[0, T]$ *of the pursuer, an evasion occurs:* $x_T(x_0, u, v) \notin M$.

Namely, if $x_0 \in M \overset{*}{-} V_0 + U_0$, *then the conditions (i) is satisfied; otherwise, condition (ii) is satisfied.*

Proof. Consider the case $x_0 \in M \overset{*}{-} V_0 + U_0$. In this case, there exists a vector $u_0 \in U_0$ such that $x_0 \in M \overset{*}{-} V_0 + U_0$. Hence $x_0 - u_0 + V_0 \subset M$. Therefore, for any vector $v_0 \in V_0$ we have $x_0 - u_0 + v_0 \in M$. Combining this with the definition of the Aumann integral, we see that there exists a strategy $u \in U[0, T]$ such that, for any strategy $v \in V[0, T]$, a capture occurs:

$$x_T(x_0, u, v) = x_0 - \int_0^T u(t)\, dt + \int_0^T v(t)\, dt \in M.$$

Consider the converse case: $x_0 \notin M \overset{*}{-} V_0 + U_0$. Theorem 5 implies that $M \overset{*}{-} V_0 + U_0 = M + U_0 \overset{*}{-} V_0$. Hence $x_0 \notin M + U_0 \overset{*}{-} V_0$ and therefore there exists a vector $v_0 \in V_0$ such that $x_0 + v_0 \notin M + U_0$. It means that for any $u_0 \in U_0$ one has $x_0 - u_0 + v_0 \notin M$. So, there exists a strategy $v \in V[0, T]$ such that, for any strategy $u \in U[0, T]$, an evasion occurs. □

Recall that a multivalued mapping $U : [a, b] \rightrightarrows X$ is called *Hausdorff continuous* on $[a, b]$ if

$$\forall t_0 \in [a, b] \ \forall \varepsilon > 0 \ \exists \delta > 0 : \ \forall t \in (t_0 - \delta, t_0 + \delta) \cap [a, b] \quad h(U(t), U(t_0)) < \varepsilon.$$

Lemma 12. *Let* X *be a uniformly convex Banach space. Assume that a multivalued mapping* $U : [a, b] \rightrightarrows X$ *is Hausdorff continuous on* $[a, b]$ *and* $U(t) \in \mathcal{SC}(r(t))$, *where* $r : [a, b] \to [0, +\infty)$ *is a continuous function. Then* $\int_a^b U(t)\, dt \in \mathcal{SC}(R)$ *with* $R = \int_a^b r(t)\, dt$.

Proof. By Definition 6 for any $t \in [a, b]$ there exists a closed convex subset $U_1(t) \subset X$ such that $U(t) + U_1(t) = \mathfrak{B}_{r(t)}(0)$. According to Lemma 2 for any $t \in [a, b]$ the sets $U(t)$ and $U_1(t)$ are uniformly convex with modulus of convexity $\delta^{U(t)}(\cdot)$ and $\delta^{U_1(t)}(\cdot)$ correspondingly such that

$$\min\{\delta^{U(t)}(\varepsilon), \delta^{U_1(t)}(\varepsilon)\} \geq r(t)\delta_X\left(\frac{\varepsilon}{r(t)}\right) \geq r_{\min}\delta_X\left(\frac{\varepsilon}{r_{\max}}\right) \quad \forall \varepsilon \in (0, 2r_{\max}),$$

where $r_{\min} = \min_{t\in[a,b]} r(t)$, $r_{\max} = \max_{t\in[a,b]} r(t)$. Since $U(\cdot)$ is continuous on $[a,b]$, $U_1(\cdot)$ is continuous on $[a,b]$ too. Let us show that

$$\int_a^b U(t)\,dt + \int_a^b U_1(t)\,dt = \mathfrak{B}_R(0). \tag{15}$$

Fix any $x \in \int_a^b U(t)\,dt + \int_a^b U_1(t)\,dt$. Then there exist Lebesgue integrable functions $u, u_1 : [a,b] \to X$ such that $u(t) \in U(t)$, $u_1(t) \in U_1(t)$ for almost all $t \in [a,b]$ and $x = \int_a^b u(t)\,dt + \int_a^b u_1(t)\,dt$. Hence, $\|x\| \le \int_a^b \|u(t) + u_1(t)\|\,dt \le \int_a^b r(t)\,dt = R$. It proves the inclusion

$$\int_a^b U(t)\,dt + \int_a^b U_1(t)\,dt \subset \mathfrak{B}_R(0).$$

Let us prove the reverse inclusion. Fix any $x \in \mathfrak{B}_R(0)$. Then $\|x\| \le R = \int_a^b r(t)\,dt$. Denoting $f(t) = \frac{r(t)}{R}x$, we have $x = \int_a^b f(t)\,dt$ and $f(t) \in \mathfrak{B}_{r(t)}(0) = U(t) + U_1(t)$ for all $t \in [a,b]$. Using uniform convexity of $U(t)$ and $U_1(t)$ according to [13, Theorem 3.1] we obtain that the multivalued mapping $\hat{U} : [a,b] \rightrightarrows X$ defined as $\hat{U}(t) = U(t) \cap (f(t) - U_1(t))$ for all $t[a,b]$ is Hausdorff continuous on $[a,b]$. By the Michael selection theorem there exists a continuous selection $u : [a,b] \to X$ of the multivalued mapping $\hat{U}(\cdot)$. Hence, $u(t) \in U(t)$ and $f(t) - u(t) \in U_1(t)$ for all $t \in [a,b]$. So,

$$x = \int_a^b f(t)\,dt = \int_a^b u(t)\,dt + \int_a^b (f(t) - u(t))\,dt \in \int_a^b U(t)\,dt + \int_a^b U_1(t)\,dt.$$

This concludes the proof of (15). From (15), convexity and closedness of $\int_a^b U(t)\,dt$ we see that $\int_a^b U(t)\,dt \in SC(R)$. $\qquad\square$

Lemma 12 implies that if the multivalued mappings $U, V : [0,T] \rightrightarrows X$ are Hausdorff continuous on $[0,T]$ and the admissible control sets $U(t)$ and $V(t)$ satisfy the conditions $U(t) \in SC(r_1(t))$ and $V(t) \in SC(r_2(t))$ for all $t \in [0,T]$ with continuous functions $r_1, r_2 : [0,T] \to [0,+\infty)$ such that $\int_0^T r_i(t)\,dt < R_i$, $i = 1,2$, then the Aumann integrals (14) are convex closed summands of the balls of radii $r_1 \in (0, R_1)$ and $r_2 \in (0, R_2)$ correspondingly as needed in Theorem 6.

Example 1. Assume that, in differential game (1), (2) the target set $M = \mathfrak{B}_R(c_0)$, the admissible control sets are $U(t) = \mathfrak{B}_{r_1(t)}(c_1(t))$ and $V(t) = \mathfrak{B}_{r_2(t)}(c_2(t))$ for all $t \in [0,T]$ with some continuous functions $r_1, r_2 : [0,T] \to [0,+\infty)$, $c_1, c_2 : [0,T] \to X$ and a vector $c_0 \in X$. Suppose that $\int_0^T r_2(t)\,dt < R$. Then the assumptions of Theorem 6 are satisfied.

References

1. Isaacs, R.: Differential Games. Wiley, New York (1965)
2. Krasovskii, N.N.: Control of a Dynamic System. Nauka, Moscow (1985)
3. Pontryagin, L.S.: Linear differential pursuit games. Mat. Sb. **112**(3), 307–330 (1980)
4. Ivanov, G.E.: Weakly convex sets and their properties. Math. Notes **79**(1), 55–78 (2006)
5. Bernard, F., Thibault, L., Zlateva, N.: Characterization of proximal regular sets in super reflexive Banach spaces. J. Convex Anal. **13**(3–4), 525–559 (2006)
6. Bernard, F., Thibault, L., Zlateva, N.: Prox-regular sets and epigraphs in uniformly convex Banach spaces: various regularities and other properties. Trans. Amer. Math. Soc. **363**, 2211–2247 (2011)
7. Ivanov, G.E.: Weak convexity of sets and functions in a banach space. J. Convex Anal. **22**(2), 365–398 (2015)
8. Balashov, M.V., Ivanov, G.E.: Weakly convex and proximally smooth sets in Banach spaces. Izv. Math. **73**(3), 455–499 (2009)
9. Balashov, M.V., Polovinkin, E.S.: M-strongly convex subsets and their generating sets. Sbornik: Math. **191**(1), 25–60 (2000)
10. Ivanov, G.E., Lopushanski, M.S.: Well-posedness of approximation and optimization problems for weakly convex sets and functions. J. Math. Sci. **209**(61), 66–87 (2015)
11. Efimov, N.V., Stechkin, S.B.: Support properties of sets in Banach spaces. Dokl. Akad. Nauk SSSR **127**(2), 254–257 (1959). (in Russian)
12. Ivanov, G.E.: Weak convexity in the senses of Vial and Efimov-Stechkin. Izv. Math. **69**(6), 1113–1135 (2005)
13. Balashov, M.V., Repovš, D.: Uniform convexity and the splitting problem for selections. J. Math. Anal. Appl. **360**(1), 307–316 (2009)

On Optimal Selection of Coefficients of Path Following Controller for a Wheeled Robot with Constrained Control

Alexander Pesterev[(⊠)] [iD]

Institute of Control Sciences, Moscow 117997, Russia
alexanderpesterev.ap@gmail.com

Abstract. Stabilization of motion of a wheeled robot with constrained control resource by means of a continuous feedback linearizing the closed-loop system in a neighborhood of the target path is considered. The problem of selection of the feedback coefficients is set and discussed. In the case of a straight target path, the desired feedback coefficients are defined to be those that result in the partition of the phase plane into two invariant sets of the nonlinear closed-loop system while ensuring the greatest asymptotic rate of the deviation decrease. A hybrid control law is proposed that ensures the desired properties of the phase portrait and minimal overshooting and is stable to noise. The proposed techniques are extended to the case of circular target paths.

Keywords: Wheeled robot · Path following problem
Saturated control · Optimal feedback coefficients

1 Introduction

There exist many applications (for example, in agriculture [1,2]) where a vehicle is to be driven along some target path with a high level of accuracy. Such tasks are performed by automatic vehicles (further referred to as *wheeled robots* (WRs), or simply *robots*) equipped with navigational and inertial tools and satellite antennas. The problem of bringing the robot from an initial state to a preassigned target path and stabilizing its motion along the path is called *path stabilization problem* (or *path following problem*); it was discussed in a great number of publications. Various models (e.g., monocycle, simple car, car-like model with and without drive actuator, tractor with trailers, etc.) and target curves (e.g., straight lines, circles, general-form curvilinear paths) were considered (see, e.g., [1–7] and references therein).

This work was supported by the Russian Foundation for Basic Research, project no. 18-08-00531, and by the Presidium of Russian Academy of Sciences, Program no. 29 "Advanced Topics of Robotic Systems."

© Springer Nature Switzerland AG 2019
Y. Evtushenko et al. (Eds.): OPTIMA 2018, CCIS 974, pp. 336–350, 2019.
https://doi.org/10.1007/978-3-030-10934-9_24

One of the commonly accepted approaches frequently and successfully used for solving the path stabilization problem is based on the exact linearization of the equations of motion in a part of variables by applying an appropriate nonlinear feedback (partial feedback linearization) [1–5]. In the framework of this approach, the original affine system is transformed into a normal form, from which the desired feedback linearizing the system in stabilizable variables is easily found. If the vehicle speed varies in time, in order to transform to the normal form, it is required first to change the independent variable. By applying different time scales, one can obtain different normal forms and, accordingly, different feedbacks. A detailed discussion of this issue, as well as comparison of different linearizing feedbacks can be found in [3].

The presence of control constraints makes the path stabilization problem much more complicated since it cannot be linearized in the entire state space. One of the commonly accepted techniques, in this case, is to use the saturated linearizing feedback, which yields linearity only in the vicinity of the target path. The goal of this study is to find out how to select coefficients of the saturated linearizing feedback to improve the efficiency of stabilization when the control resource is bounded.

2   Problem Statement

We consider the *kinematic model* of a wheeled robot, which describes the simplest vehicle moving without lateral slippage with two rear driving wheels and front wheels responsible for steering the platform. In the planar case, the robot position is described by two coordinates (x_c, y_c) of some point of the platform, the so-called *target point,* and one angle describing the orientation of the platform with respect to a fixed reference system Oxy. For the target point, the point located in the middle of the rear axle is taken, and for the angle, the angle θ between the central line of the platform (which coincides with the direction of the velocity vector) and the x-axis. The kinematic equations of such a robot are well known to be (see, for example, [1–7])

$$
\begin{aligned}
\dot{x}_c &= v \cos \theta, \\
\dot{y}_c &= v \sin \theta, \\
\dot{\theta} &= v \tan \phi / L.
\end{aligned}
\tag{1}
$$

Here, the dot over a symbol denotes differentiation with respect to time, $v \equiv v(t)$ is a scalar linear velocity of the target point, and L is the distance between the front and rear axles. The vehicle is controlled by turning the front wheels through an angle ϕ, $|\phi| \leq \phi_{\max} < \pi/2$. Since the angle ϕ in the above range and the instant curvature u of the curve described by the target point are related by the one-to-one relationship $u = \tan \phi / L$, it is convenient to take u to be the control variable, which satisfies the two-sided constraints

$$
-\bar{u} \leq u \leq \bar{u},
\tag{2}
$$

where $\bar{u} = \tan \phi_{\max} / L$ is the maximal possible curvature of the actual trajectory described by the target point.

In the path stabilization problem, it is required to synthesize a control law u that brings the robot to a given target path and stabilizes its motion along the curve. The target path is given in a parametric form by a pair of functions $(X(s), Y(s))$, where s is a natural parameter (arc length), and is assumed to be feasible. The latter means that the functions $X(s)$ and $Y(s)$ are twice differentiable [4] everywhere except for a finite number of points (and, hence, curvature $k(s)$ of the target curve is a piecewise continuous function), and the maximum curvature $\bar{k} = \max_s k(s)$ of the target path satisfies the constraint $\bar{k} < \bar{u}$.

It has been shown in [5] that, by changing state variables and applying time scaling, the path following problem can be written in the canonical form [3], as the problem of finding a feedback that stabilizes the zero solution of the system

$$z_1' = z_2, \quad z_2' = (1 + z_2^2)^{3/2} u - \frac{k(1 + z_2^2)}{1 - kz_1}. \tag{3}$$

In (3), z_1 is the deviation of the robot from the target path, $z_2 = \tan\psi$, where ψ is the angle between the direction of the velocity vector and the tangent line to the target curve at the point closest to the robot, and the prime denotes differentiation with respect to the new independent variable ξ, which satisfies the equation $\dot\xi = v \cos\psi$.

If the control resource is not bounded ($\bar{u} = \infty$), then closing system (3) by the feedback

$$u = -\frac{\sigma(z)}{(1 + z_2^2)^{3/2}} + \frac{k}{\sqrt{1 + z_2^2}(1 - kz_1)}, \tag{4}$$

where $z = [z_1, z_2]^T$ and $\sigma(z)$ is a linear function with positive coefficients, we obtain the linear system

$$z_1' = z_2, \quad z_2' = -\sigma(z), \tag{5}$$

the zero solution of which is globally asymptotically stable. For convenience of calculations, without loss of generality, we will represent function $\sigma(z)$ in the form

$$\sigma(z) = \lambda^2 z_1 + 2\lambda\gamma z_2, \quad \lambda > 0, \gamma > 0. \tag{6}$$

In the case of the constrained control resource, applying the saturation function to the linearizing control law, i.e., selecting the feedback in the form

$$u = -\mathrm{sat}_{\bar{u}}\left(\frac{\sigma(z)}{(1 + z_2^2)^{3/2}} - \frac{k}{\sqrt{1 + z_2^2}(1 - kz_1)}\right), \tag{7}$$

we get a hybrid system given by the linear Eq. (5) in the set where $|u| < \bar{u}$ and by the nonlinear equations

$$z_1' = z_2, \quad z_2' = -\frac{k(1 + z_2^2)}{1 - kz_1} - \mathrm{sign}(\sigma(z))(1 + z_2^2)^{3/2}\bar{u} \tag{8}$$

in the set where the control reaches saturation.

As can be seen, the properties of the system under study are determined by the four parameters: the feedback coefficients λ and γ, path curvature k, and the control resource $\bar{u}$. Let us turn to an equivalent system of equations in dimensionless variables, which will allow us to get rid of one parameter. It is easy to verify that all above equations are invariant with respect to the transformation

$$\tilde{u} = u/\bar{u}, \ \tilde{\xi} = \xi\bar{u}, \ \tilde{z}_1 = z_1\bar{u}, \ \tilde{z}_2 = z_2, \ \tilde{\lambda} = \lambda/\bar{u}, \ \tilde{\gamma} = \gamma, \ \tilde{k} = k/\bar{u}. \tag{9}$$

Thus, study of the behavior of an arbitrary WR reduces to studying a dimensionless WR with the control resource equal to one. To simplify subsequent calculations and formulas, we will use the same notation (without tilde) as in the dimensional case to denote dimensionless quantities and parameters. Thus, all above equations remain valid for the dimensionless variables by setting $\bar{u} = 1$ in Eqs. (7) and (8).

Clearly, the efficiency of the stabilization directly depends on the coefficients of the linear function $\sigma(z)$. However, the author failed to find any publications on wheeled robots where the problem of finding optimal (in one or another sense) feedback coefficients is solved or even posed. In works [1,2,5], when discussing the practical implementation of the proposed linearizing feedbacks, the values of the coefficients are either set arbitrary (in numerical experiments with WR models) or selected experimentally (when the case in point are real automated vehicles). For example, in [1,2], to control a farm tractor, the authors used a linearizing feedback depending on one parameter, the value of which was selected experimentally from the condition that stabilization in a strip of width 15 m is achieved without reaching the control constraint (without getting into the saturation mode). Such an approach is not only badly justified but also is too cumbersome, since the results obtained for one WR cannot be used for another WR with different geometric characteristics and/or different control resource, which brings us to the necessity of development of a theoretically justified approach to solving the problem of selection of the feedback coefficients.

In the particular case of a straight target path and one-parameter family of the coefficients with the fixed $\gamma = 1$ (in this case, system (5) has one repeated pole $-\lambda$), this problem was studied in [8]. The desired value of the parameter was defined in [8] as the greatest λ for which there exists a partition of the phase plane into two invariant half-planes (i.e., any trajectory of the closed-loop system completely lies in one of the half-planes). The existence of two invariant half-spaces implies that the phase portrait of the nonlinear system (3), (7) is topologically equivalent to the phase portrait of the linear system (5) in the entire plane R^2 (rather than in a neighborhood of the origin). It was proved that the optimal in this sense value is $\lambda_{\text{opt}} = 3\sqrt{3}\bar{u}/2$ and that the phase plane is partitioned into the desired half-planes by the asymptote $z_2 = -\lambda_{\text{opt}}z_1$.

In this paper, we study the more general case where the roots of the characteristic polynomial of the linear system (5) are different. First, we pose the same problem as in [8]:

Problem 1. Determine feedback coefficients for which (i) there exists a partition of the phase plane into two invariant sets and (ii) the asymptotic rate of approaching the target path is as high as possible.

We will also study how the results obtained for the straight paths can be extended to the case of circular paths.

It should be emphasized that the above criterion of the selection of the feedback coefficients is quite natural. The fulfillment of this criterion means that the WR approaches the target path in a non-oscillatory way: the trajectory of the WR intersects the target curve at most once.

To get an idea of how the feedback coefficients affect system behavior, we first consider the phase portrait of the linear system (5), which governs the closed-loop system behavior when the control resource is unbounded.

3   Phase Portrait of the Linear System

The roots of the characteristic equation $\mu^2 + 2\gamma\lambda\mu + \lambda^2 = 0$ of the linear system (5) are easily found to be

$$\mu_{1,2} = -\lambda_{1,2}, \; \lambda_1 = \gamma\lambda(1 - \sqrt{1 - 1/\gamma^2}), \; \lambda_2 = \gamma\lambda(1 + \sqrt{1 - 1/\gamma^2}). \qquad (10)$$

For $\gamma \geq 1$, the roots are real negative numbers and the linear system has a stable node at the origin. If $\gamma < 1$, μ_1 and μ_2 are complex conjugate numbers, and $z = 0$ is a focus of the linear system. Clearly, in the latter case, no entire trajectory of the system can lie from the one side of a straight line passing through the origin; i.e., the above-formulated criterion certainly cannot be satisfied. Therefore, in what follows, we assume that $\gamma \geq 1$. Note that, for $\gamma = 1$, we arrive at the case of a degenerate node (repeated root $\mu_{1,2} = -\lambda$) considered in [8].

A typical phase portrait of a system with a stable node is shown in Fig. 1. Here, $\gamma = 1.1$, $\lambda \approx 4.0$, $\lambda_1 \approx 2.6$, and $\lambda_2 \approx 6.3$. The system has two eigenvectors collinear to the straight lines $z_2 = -\lambda_1 z_1$ and $z_2 = -\lambda_2 z_1$. Any system trajectory, except those beginning at points on the straight line corresponding to the larger eigenvalue (λ_2), touches the asymptote $z_2 = -\lambda_1 z_1$ at the origin. With regard to (10), the equation of the asymptote can be written as $z_2 + \gamma\lambda(1 - \sqrt{1 - 1/\gamma^2})z_1 = 0$. Multiplying this equation by $\gamma\lambda(1 + \sqrt{1 - 1/\gamma^2})$, we obtain the asymptote equation in the form

$$\gamma\lambda(1 + \sqrt{1 - 1/\gamma^2})z_2 + \lambda^2 z_1 = 0. \qquad (11)$$

The asymptote divides the phase plane into two half-planes A_- (below the asymptote) and A_+ (above the asymptote), where the left-hand side of (11) is less or greater than zero, respectively. Clearly, A_- and A_+ are invariant sets of system (5), i.e., any trajectory completely lies in one of these half-planes and may intersect the target path not more than once. The deviation decreases exponentially with the exponent equal to the lesser eigenvalue $\mu_1 = -\lambda_1$.

4   Stabilization of a Robot with Constrained Control Along a Straight Path

The system with a constrained control ceases being linear when it comes to the "saturation" region, the set where the inequality

$$|\sigma(z)| \geq (1 + z_2^2)^{3/2} \tag{12}$$

holds. Clearly, the saturation region is a disconnected set consisting of two non-intersecting sets lying from both sides of the straight line $\sigma(z) = 0$. It is easy to see that the system moves along an integral curve in the direction of increasing (decreasing) variable z_2 in the left (right) saturation region.

Any trajectory of the nonlinear system (3), (7) completely lies in the domain A_- or A_+ if and only if the asymptote $z_2 = -\lambda z_1$ does not intersect the saturation regions, since a system trajectory can intersect the asymptote only in the saturation region (where the system is nonlinear). Let us find conditions the fulfillment of which guarantees that the asymptote does not intersect the saturation region. In view of symmetry, it will suffice to consider one (say, left) component of the saturation region. Let us rewrite inequality (12) holding in the saturation region as

$$\lambda^2 z_1 + \lambda \gamma z_2 \leq -(1 + z_2^2)^{3/2} - \lambda \gamma z_2.$$

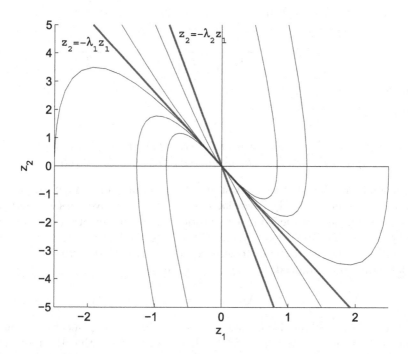

Fig. 1. Phase portrait of a linear system with stable node at the origin.

Adding $\gamma\lambda\sqrt{1 - 1/\gamma^2}z_2$ to the both sides of this inequality, we obtain

$$\lambda^2 z_1 + \lambda\gamma(1 + \sqrt{1 - 1/\gamma^2})z_2 \leq -(1 + z_2^2)^{3/2} - \lambda\gamma(1 - \sqrt{1 - 1/\gamma^2})z_2. \quad (13)$$

In order that the entire component of the saturation region lies under the asymptote, the right-hand side of the inequality must be negative in it. Indeed, in this case, the left-hand side of the inequality is also negative in the saturation region, i.e., in view of (11), belongs to the set A_-. The condition of negativeness of the right-hand side of the inequality can be written with regard to (10) as $(1 + z_2^2)^{3/2} \geq -\lambda_1 z_2$. Replacing the inequality sign by the equality sign, squaring both sides of the equality obtained and introducing the notation $x = z_2^2$, we arrive at the equation $(1 + x)^3 = \lambda_1^2 x$. This equation can be viewed as the equation in the unknown point of tangency of the cubic and linear (with an unknown coefficient) functions of x. Equating derivatives of both functions at the point of tangency, $\lambda_1^2 = 3(1 + x)^2$, and substituting the right-hand side of the last equation for λ_1^2 into the previous equation, we obtain $x = 1/2$. Substituting this into the last equation, we find that the asymptote touches the saturation region when $\lambda_1 = 3\sqrt{3}/2$ and that the ordinate of the point of tangency is $z_2 = -\sqrt{1/2}$. Thus, for the smaller eigenvalue λ_1, we obtained the same value λ_{opt} that was obtained in [8] for λ in the case of the multiple roots. The corresponding λ is easily found from the relations (10): $\lambda = \lambda_{opt}/(\gamma(1 - \sqrt{1 - 1/\gamma^2}))$. Thus, we have proved the following assertion valid for a straight target path.

Theorem 1. *Let the coefficients of feedback (7), (6) satisfy the condition*

$$\gamma \geq 1, \quad \lambda(\gamma) = \frac{\lambda_{opt}}{\gamma(1 - \sqrt{1 - 1/\gamma^2})}, \quad \lambda_{opt} = \frac{3\sqrt{3}}{2}. \quad (14)$$

Then, the half-planes A_- or A_+ lying from the two sides of the straight line $a(z) = 0$, where

$$a(z) = \lambda_{opt} z_1 + z_2, \quad (15)$$

are invariant sets of the closed-loop system (3), (7), (6), and any solution of the system asymptotically tends to the origin with the exponential rate $(-\lambda_{opt})$.

Theorem 1 implies that there exist an infinite number of pairs of the parameters γ and $\lambda(\gamma)$ related by the condition (14) for which we have the same partition of the phase plane and the same asymptotic rate of the deviation decrease. On the asymptote, the closed-loop system (3), (7), (6) is linear for any $\gamma \geq 1$, and any solution of the system tends to zero exponentially with the exponent $(-\lambda_{opt})$ not depending on γ. Indeed, for any γ, we have

$$\sigma(z) = \lambda^2 z_1 + 2\gamma\lambda z_2 = \lambda_1\lambda_2 z_1 + (\lambda_1 + \lambda_2)z_2 = \lambda_2(\lambda_1 z_1 + z_2) + \lambda_1 z_2,$$

where $\lambda_1 = \lambda_{opt}$. On the asymptote, $\lambda_1 z_1 + z_2 = 0$, and the linearizing control (4) takes the form $u = \lambda_1 z_2/(1 + z_2^2)^{3/2}$. It is easy to check that the right-hand side of the last expression does not exceed one, with the extreme values $u = \pm 1$

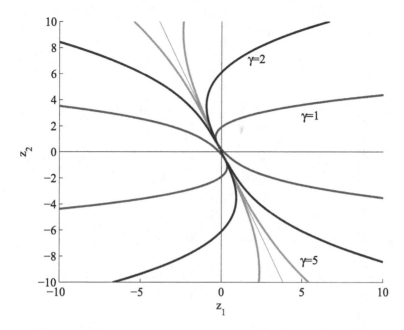

Fig. 2. Saturation regions of system (3) with $k = 0$ closed by feedback (7), (6) for $\gamma = 1$, $\gamma = 2$, and $\gamma = 5$.

being achieved only at $z_2 = \pm 1/\sqrt{2}$. Hence, on the asymptote, the closed-loop system (3), (7) takes the form $z_1' = z_2$, $z_2' = -\lambda_{\text{opt}} z_2$.

Let us find out in what way the selection of the value of γ affects the behavior of the closed-loop system. First, the greater the value of γ, the larger the saturation region. Figure 2 shows the saturation sets for three values of γ: $\gamma = 1$, $\gamma = 2$, and $\gamma = 5$. The thin line depicts the asymptote, which separates the two components of the saturation region for any $\gamma \geq 1$ and touches them at the points with the coordinates $z_2 = \pm 1/\sqrt{2}$.

Let $z(\xi, z^0, \gamma)$ denote the trajectory of the system (3) closed by the feedback (7), (6) with the initial condition $z(0) = z^0$. From Theorem 1, it follows that deviation $z_1(\xi, z^0, \gamma)$ either monotonically tends to zero or has one local extremum at the point of intersection with the axis z_1. In the latter case, the quality of the control can additionally be characterized by the magnitude of this extremum, which will we referred to as "overshooting" and denoted as $M(z^0, \gamma)$,

$$M(z^0, \gamma) = \max_{\xi}[z_1(\xi)\text{sign}(z_2(0))]. \tag{16}$$

If the deviation tends to zero monotonically, the overshooting is zero.

It can be shown that the minimum of the overshooting is achieved on the limit trajectories, the trajectories to which trajectories $z(\xi, z^0, \gamma)$ tend as $\gamma \to \infty$. The corresponding limit feedback (7), (6) is given by the discontinuous function

$$u(z) = \begin{cases} -1, z \in A_+, \\ 1, \ z \in A_-. \end{cases} \tag{17}$$

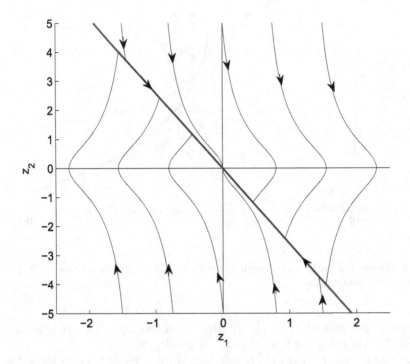

Fig. 3. Phase portrait of the closed-loop system (3), (17).

Figure 3 shows the phase portrait of system (3) closed by the limit control law (17). In the sets A_+ and A_-, the system moves along integral curves of equation (8) given by

$$z_1 = z_1(0) \mp \left(\frac{1}{\sqrt{1 + z_2^2(0)}} - \frac{1}{\sqrt{1 + z_2^2}} \right). \tag{18}$$

Having reached the asymptote, the system moves ("slides") along it to the origin.

However, in practice, the discontinuous control law (17) is not applicable because of the chattering arising when the system moves along the asymptote. Indeed, since the set of points belonging to the asymptote has zero measure, the control will alternately take limit values ± 1 when moving along it.

Is it possible to get rid of the above drawback and still preserve minimality of the overshooting? To answer this question, let us analyze the character of the trajectories of the closed-loop system. It can be seen from Fig. 3 that motion of the system closed by the limit control law (17) consists of the following two stages: motion along an integral curve (18) with the limit control $+1$ or (-1)

and motion along the asymptote. In terms of the original system, on the first stage, the robot makes a turn moving with the front wheels turned through the maximal angle until the angle ψ takes the value $\psi = -\arctan(\lambda_{\mathrm{opt}} z_1)$, after which the second stage starts when the system "slides" along the asymptote. It is on the second stage where chattering arises. Note also that, if the overshooting $M(z^0)$ is positive, then it is achieved on the first stage, when the system moves with the limit control.

The above observation makes us think of a combined (hybrid) two-stage control law: to apply the limit (saturated) control on the first stage, like in the discontinuous control law (17), and, after hitting the asymptote, to use the continuous feedback (7), (6) with the minimal $\gamma = 1$ on the second stage. Such a strategy makes it possible to combine advantages of the limit discontinuous control law and the continuous feedback with small γ and to get rid of disadvantages of both. The discontinuous law (17) brings the system from an initial state z^0 to the asymptote in a minimal time and with the minimal overshooting $M(z^0)$. The control on this stage takes only one value $u = 1$ or $u = -1$ and is insensitive to noise. Switching to the continuous feedback on the second stage allows the system to avoid chattering. When moving along the asymptote, the system is linear, and the deviation decreases exponentially. Since the rate of convergence does not depend on γ, the use of the minimal $\gamma = 1$ ensures the least sensitivity to measurement noise without sacrificing the convergence rate.

In practice, the exact hit of the asymptote is impossible because of measurement noise and approximation errors. Therefore, switching from the discontinuous to continuous feedback should occur upon entering some neighborhood of the asymptote, the so-called *control switching set* $\varPi$, so that the control law takes the form

$$
u(z) = \begin{cases} -1, & z \notin \varPi,\, z \in A_+, \\ 1, & z \notin \varPi,\, z \in A_-, \\ -\mathrm{sat}_1[(\lambda_{\mathrm{opt}}^2 z_1 + 2\lambda_{\mathrm{opt}} z_2)/(1 + z_2^2)^{3/2}], & z \in \varPi. \end{cases} \tag{19}
$$

How to select the asymptote neighborhood depends on a particular implementation of the control law (19). The answer to this question may depend on many factors, such as the accuracy of measurements of the state variables, digitization frequency, robot's velocity, and so on. Here, we would only like to emphasize that it is important that the switching set be invariant. This property guarantees that have occurred in the set, the system will never leave it.

5 Stabilization of a Robot with Constrained Control Along a Circular Path

The case of curvilinear target paths is much more complicated, and the optimality criterion adopted in the case of a straight path cannot be satisfied. In this section, we confine our consideration to only paths with constant curvature, i.e., circles, and will show, first, why this criterion is not applicable and, second, how to modify it.

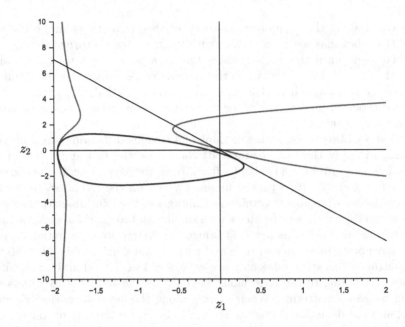

Fig. 4. Saturation regions U_+ (blue boundary) and U_- (green boundary) of system (3) with $k = -0.5$ closed by feedback (7), (6) with $\gamma = 1$ and $\lambda = 3.54$. (Color figure online)

For definiteness, we assume that the robot moves in the clockwise direction, which implies that the path curvature is negative, $k = \text{const} < 0$. Unlike in the previous case, the domain of the system is a half-plane $z_1 > 1/k$ rather than R^2, and the saturation regions depend not only on the feedback coefficients but also on the path curvature k. Let us denote the sets where the control takes values $+1$ and -1 as $U_+(k)$ and $U_-(k)$, respectively,

$$U_+(k) = \left\{ z : -\frac{\sigma(z)}{(1 + z_2^2)^{3/2}} + \frac{k}{\sqrt{1 + z_2^2}(1 - kz_1)} \geq 1 \right\}, \qquad (20)$$

$$U_-(k) = \left\{ z : -\frac{\sigma(z)}{(1 + z_2^2)^{3/2}} + \frac{k}{\sqrt{1 + z_2^2}(1 - kz_1)} \leq -1 \right\}. \qquad (21)$$

Taking into account the negativeness of k, it is not difficult to see that $U_+(k)$ is a bounded set and $U_+(k) \subset U_+(0)$, whereas $U_-(k) \supset U_-(0)$ is unbounded and may consist of two disconnected subsets (see, e.g. Fig. 4), where $U_+(0)$ and $U_-(0)$ are the saturation regions for a straight target path. A typical picture of the saturation regions for a circular path is shown in Fig. 4. Here, $k = -0.5$, $\lambda = 3.54$, and $\gamma = 1$. The boundary of the bounded set U_+ is shown by the thick blue line, while the boundaries of the two unbounded components of U_- are depicted by the thick green lines.

Since the term $k/\sqrt{1+z_2^2}(1-kz_1)$ tends to $-\infty$ when $z_1 \to 1/k$ for any z_2, any straight line passing through the origin (including the asymptote for any feedback coefficients), necessarily intersects the set $U_-(k)$. Let, similar to the straight-path case, A_+ and A_- denote the intersection of the system domain $z_1 > 1/k$ with the half-planes $z_2 + \lambda z_1 > 0$ and $z_2 + \lambda z_1 < 0$, respectively. Taking into account that trajectories can intersect the asymptote from the lower set A_- to the upper one A_+ in the region $U_-(k)$, the both sets cannot be invariant sets of the system whatever feedback coefficients are; i.e., the criterion formulated in Sect. 2 cannot be satisfied.

On the other hand, a trajectory beginning in the upper half-plane A_+ can intersect the asymptote and enter the lower half-plane A_- only in the set $U_+(k)$. Hence, if the asymptote does not intersect the set $U_+(k)$, no trajectories can come from A_+ to A_-; i.e., A_+ is a *positive invariant set* of the system in this case. Taking into account that our goal is to obtain the closed-loop system with the phase portrait similar to that of the corresponding linear system and the greatest convergence rate, this brings us at the following optimization problem statement.

Problem 2. Find feedback coefficients that guarantee the existence of a positive invariant half-plane while ensuring the greatest rate of deviation decrease.

Similar to the case of a straight path, given the curvature k, we seek for λ_1 such that the line $z_2 = -\lambda_1 z_1$ touches the boundary of the set $U_+(k)$, which is found similar to that in the case of a straight target path. To this end, like in Sect. 4, we rewrite the inequality defining the set $U_+(k)$ such that its left-hand side coincide with the left-hand side of the asymptote Eq. (11) (cf. (13))

$$\lambda^2 z_1 + \lambda\gamma(1 + \sqrt{1 - 1/\gamma^2})z_2 \leq -f(z, k, \lambda_1)z_2, \tag{22}$$

where

$$f(z, k, \lambda_1) = (1 + z_2^2)^{3/2} - \frac{k(1 + z_2^2)}{1 - kz_1} + \lambda_1 z_2. \tag{23}$$

From (22), it follows that the set $U_+(k)$ completely lies under the asymptote when $f(z, k, \lambda_1) \geq 0 \; \forall z \in U_+$.

To find the value of λ_1 for which the asymptote touches the boundary of the set $U_+(k)$, we consider the restriction

$$\bar{f}(z_2, k, \lambda_1) = (1 + z_2^2)^{3/2} + \frac{\lambda_1(\lambda_1 z_2 - k)}{\lambda_1 + kz_2} \tag{24}$$

of the function $f(z, k, \lambda_1)$ to the asymptote, which is obtained by substituting $z_1 = -z_2/\lambda_1$ into (23). The desired $\lambda_1(k)$ is a solution to the nonlinear equation

$$F(k, \lambda_1) = 0, \tag{25}$$

where

$$F(k, \lambda_1) = \min_{z_2 < 0} \bar{f}(z_2, k, \lambda_1). \tag{26}$$

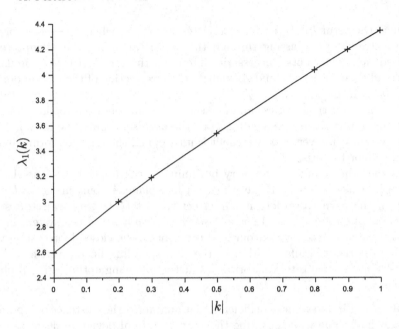

Fig. 5. Dependence of the asymptotic convergence rate λ_1 on the circular path curvature.

For a given k, the solution to problem (24)–(26) is easily found numerically. Simple analysis shows that $\lambda_1(k)$ is a monotonically increasing function of $|k|$. Figure 5 shows dependence of the convergence rate λ_1 on the path curvature obtained by numerical solving problem (24)–(26). For example, for the case of the circular path with $k = -0.5$, the saturation regions for which are depicted in Fig. 4, $\lambda_1 \approx 3.54$. As can be seen, the corresponding asymptote $z_2 = -3.54z_1$ shown in Fig. 4 by the inclined thin line actually touches the set U_+.

The results of this section are summarized in the following theorem.

Theorem 2. *Let $k = const$ and the coefficients of feedback (7), (6) satisfy the condition*

$$\gamma \geq 1, \ \ \lambda(\gamma) = \frac{\lambda_1(k)}{\gamma(1 - \sqrt{1 - 1/\gamma^2})}, \tag{27}$$

where $\lambda_1(k)$ is the solution of problem (24)–(26). Then, the set $A_+ = \{(z_1, z_2) : z_1 > 1/k, z_2 > -\lambda_1(k)z_1\}$ is a positive invariant set of the closed-loop system (3), (7), (6), and any solution of the system asymptotically tends to the origin with the exponential rate $-\lambda_1(k)$.

Thus, like in the case of a straight target path, there exist an infinite number of pairs $(\gamma, \lambda(\gamma))$ resulting in the same partition of the phase plane and the same convergence rate. All reasonings regarding the selection of a particular pair presented in Sect. 4 remain valid for the circular paths. The best option is the combined (hybrid) strategy: to apply the limit (saturated) control to bring the system from an initial state $z(0)$ to the asymptote (which is equivalent to using infinitely large γ and λ) and, then, to switch to the saturated linearizing feedback with the minimal $\gamma = 1$ to slide to the origin along the asymptote. The only difference compared to the straight-path case is that in the region of negative z_1 (inside the circle), where the asymptote intersects the set U_- (see Fig. 4), the system can leave the asymptote staying in the positive invariant set A_+. For any initial condition, the WR can intersect the target path at most twice.

6 Conclusions

In the paper, stabilization of a wheeled robot along a target path has been discussed. In the case of an unlimited control resource, the problem is easily solved by applying the feedback linearization technique. If the control is bounded, the application of the saturated linearizing feedback results in a nonlinear closed-loop system and brings one to the problem of selecting the feedback coefficients to optimize the performance of the stabilization. For a straight target path, the desired feedback coefficients are defined to be those that result in the partition of the phase plane into two invariant sets of the nonlinear closed-loop system while ensuring the greatest asymptotic rate of the deviation decrease. The use of the feedback law with such coefficients guarantees that the robot intersects the target path not more than once. It has been proved that there exists a family of the optimal coefficients. A hybrid control law has been proposed that ensures the desired properties of the phase portrait and minimal overshooting and is stable to noise. Such a partition has been shown to be impossible for circular target paths. In this case, optimal feedback coefficients are defined to be those that guarantee the existence of a positive invariant half-plane while ensuring the greatest rate of deviation decrease. The problem of numerical finding the optimal coefficients has been solved.

In the future, we plan to study the problem of finding optimal feedback coefficients for the problem of stabilizing robot's motion along general-form target paths.

References

1. Thuilot, B., Cariou, C., Martinet, P., Berducat, M.: Automatic guidance of a farm tractor relying on a single CP-DGPS. Auton. Robots **13**, 53–61 (2002)
2. Thuilot, B., Lenain, R., Martinet, P., Cariou, C.: Accurate GPS-based guidance of agricultural vehicles operating on slippery grounds. In: Liu, J.X. (ed.) Focus on Robotics Research. Nova Science, New York (2005)
3. Pesterev, A.V., Rapoport, L.B.: Canonical representation of the path following problem for wheeled robots. Autom. Remote Control **74**(5), 785–801 (2013)
4. De Luca, A., Oriolo, G., Samson, C.: Feedback control of a nonholonomic car-like robot. In: Laumond, J.-P. (ed.) Robot Motion Planning and Control, pp. 170–253. Springer, New York (1998)

5. Pesterev, A.V.: Synthesis of a stabilizing control for a wheeled robot following a curvilinear path. Autom. Remote Control **73**(7), 1134–1144 (2012)
6. Morin, P., Samson, C.: Motion control of wheeled mobile robots. In: Siciliano, B., Khatib, O. (eds.) Springer Handbook of Robotics. Springer, Heidelberg (2008). https://doi.org/10.1007/978-3-540-30301-5_35
7. Rapoport, L.B.: Estimation of attraction domain in a wheeled robot control problem. Autom. Remote Control **67**(9), 1416–1435 (2006)
8. Pesterev, A.: Stabilizing control for a wheeled robot following a curvilinear path. In: Proceedings of the 10th International IFAC Symposium on Robot Control, Dubrovnik, pp. 643–648 (2012)

The Space-Time Representation for Optimal Impulsive Control Problems with Hysteresis

Olga N. Samsonyuk$^{(\boxtimes)}$ [ID]

Matrosov Institute for System Dynamics and Control Theory SB RAS,
Irkutsk, Russia
samsonyuk.olga@gmail.com

Abstract. An optimal control problem for a sweeping process driven by impulsive controls is considered. The control system we study is described by both a measure-driven differential equation and a differential inclusion. This system is the impulsive-trajectory relaxation of an ordinary control system with nonlinearity of hysteresis type, in which the hysteresis is modeled by the play operator and considered as a particular case of a nonconvex sweeping process. The concept of a sweeping process for the so-called graph completions of functions of bounded variation, defining the corresponding moving set, is developed. The space-time representation based on the singular space-time transformation and a method to obtain optimality conditions for impulsive processes are proposed. By way of motivation, an example from mathematical economics is considered.

Keywords: Measure-driven differential equations · Sweeping process
Rate independent hysteresis · Impulsive control
Space-time representation · Optimal control

1 Introduction

In this paper, we address an optimal impulsive control problem with hysteresis. This problem is considered within the framework of the sweeping process introduced and deeply studied by Moreau [31].

Let $T = [a, b]$ be a given time interval, $\text{comp}(\mathbb{R}^r)$ be the collection of all nonempty compact subsets from $\mathbb{R}^r$, and $AC(T, \mathbb{R}^n)$ be the set of absolutely continuous functions from T to $\mathbb{R}^n$.

Consider the problem: given $t \to C(t)$, a Lipschitz continuous set-valued map from T to $\text{comp}(\mathbb{R}^r)$, and $y_0 \in \mathbb{R}^r$, find $y \in AC(T, \mathbb{R}^r)$ such that

$$- \dot{y}(t) \in N_{C(t)}\big(y(t)\big) \qquad \text{for a.e. } t \in T, \tag{1}$$

$$y(t) \in C(t) \qquad \text{for all } t \in T, \tag{2}$$

$$y(0) = y_0 \in C(a), \tag{3}$$

Supported by the Russian Foundation for Basic Research, Project no. 18-01-00026.

Y. Evtushenko et al. (Eds.): OPTIMA 2018, CCIS 974, pp. 351–366, 2019.
https://doi.org/10.1007/978-3-030-10934-9_25

where $N_C(y)$ is some appropriate normal cone of the set C at the point y.

The differential inclusion (1)–(3) describes a sweeping process. The point $y(t)$ remains at rest when it does not touch the boundary of the moving set $C(t)$, or otherwise, it is swept by the boundary of $C(t)$ towards its interior. Sweeping process appears in many applications, for example, elastoplasticity, crowd motion, mathematical economy, hysteresis, and others (see, e.g., [4,17,23, 27,31,32,40,42]).

The existence and uniqueness of a solution to Problem (1)–(3) are studied in [31] for the case when the set-valued map $t \to C(t)$ is absolutely continuous (with respect to the Hausdorff metric) and has nonempty closed convex values for every $t \in T$. For the convex sweeping process the normal cone $N_C(y)$ is understood in the sense of the convex analysis.

In [1,7,8,42,44] Problem (1)–(3) is investigated for the nonconvex moving set $C(t)$ under the assumption that $C(t)$ is a uniformly proximal regular set. The latter means that for every point from some neighborhood of the set $C(t)$ there exists a unique projection of this point. For this case, $N_C(y)$ is the Clarke normal cone to the set C at the point y. The existence and uniqueness of a solution to the sweeping process (1)–(3), where $C(t)$ is a uniformly proximal regular set and moves in an absolutely continuous way, are established in [42].

We note that for a uniformly proximal regular set C the Clarke normal cone $N_C(y)$ coincides with the proximal normal cone. We recall that a vector $p \in \mathbb{R}^r$ is a proximal normal vector to a closed set $C \subset \mathbb{R}^r$ at a point $y \in C$ if and only if there exists $\alpha > 0$ such that $d_C(y + \alpha p) = \alpha ||p||_2$, where $|| \cdot ||_2$ is the Euclidean norm and $d_C(\cdot)$ is the distance function defined by the rule $d_C(z) = \inf_{y \in C} ||z - y||_2$.

The proximal normal cone to C at y, written $N_C^P(y)$, is the set of all proximal normal vectors to C at y.

In this paper, we concern with the moving set $C(t) \doteq q(t) - Z(t)$, $t \in T$, where $q : T \to \mathbb{R}^r$ is a Lipschitz continuous function and $Z : T \to \mathrm{comp}(\mathbb{R}^r)$ is a given set-valued function. We assume that $t \to Z(t)$ is Lipschitz continuous with respect to the Hausdorff metric and for every $t \in T$ the set $Z(t)$ is a δ-proximal regular set (see Definition 1 below). We denote by $\mathcal{F} : (q(\cdot), y_0) \mapsto y(\cdot)$ the solution operator for our sweeping process. This operator is characterized by the rate independence property:

$$\mathcal{F}(q \circ \varphi, y_0) = \mathcal{F}(q, y_0) \circ \varphi$$

whenever $\varphi : T \to T$ is an increasing surjective Lipschitz reparametrization of time. We note that for the case when the moving set $C(t)$ has the constant shape, namely $Z(t) = Z$ for all $t \in T$, where Z is a closed convex set from $\mathbb{R}^r$, the operator $\mathcal{F}$ is also called the play operator. The terms input and output are used for q and y, respectively. The play operator has been investigated from various points of view and its properties have been thoroughly studied. We refer the reader to [4,22,23,40,45] for the properties and applications of the play operator.

We consider a sweeping process driven by impulsive controls, for which the dynamics is described by both the differential inclusion (1)–(3) and some measure-driven differential equation. This sweeping process corresponds to a relaxation of (1)–(3) for discontinuous inputs $q(\cdot)$ with bounded total variation on T. This paper continues the research initiated in [37,38].

Consider the following optimal control problem (P_0):

$$J_0 = l\big(x(b)\big) \to \inf \tag{4}$$

subject to the dynamics

$$\dot{x}(t) = f\big(t, x(t), y(t)\big) + G\big(t, x(t)\big)v(t), \qquad x(a) = x_0, \tag{5}$$

$$v(t) \in K \qquad \text{for a.e. } t \in T, \tag{6}$$

$$q(t) = Ax(t) \qquad \text{for all } t \in T. \tag{7}$$

Here, the control constraints set K is a closed convex cone in $\mathbb{R}^m$, $x \in AC(T, \mathbb{R}^n)$, $y \in AC(T, \mathbb{R}^r)$, $v \in L^\infty(T, \mathbb{R}^m)$, A is a constant matrix of dimension $(r \times n)$. Furthermore, l, f, G are given functions satisfying the assumptions (H_1), (H_2) below. The function $y(\cdot)$ on the right-hand side of the Eq. (5) is a solution to the differential inclusion:

$$- \dot{y}(t) \in N^P_{C(t)}\big(y(t)\big) \qquad \text{for a.e. } t \in T, \tag{8}$$

$$y(a) = y_0 \in C(a), \tag{9}$$

where $C(t) = q(t) - Z(t)$ for all $t \in T$, $t \to Z(t)$ is a given set-valued function, and $N^P_C(y)$ is the proximal normal cone to the set C at the point y. We denote by σ_0 a feasible process to (P_0), i.e. $\sigma_0 = (x(\cdot), q(\cdot), y(\cdot); v(\cdot))$, where the components satisfy the relations (5)–(9). Let Σ_0 be the collection of all feasible processes.

We posit the following assumptions on the functions $l(x)$, $f(t, x, y)$, $G(t, x)$, and $Z(t)$:

(H_1) l, f, G are continuous.

(H_2) f, G are locally Lipschitz continuous in x, y, i.e. for any compact set $Q \subset \mathbb{R}^n \times \mathbb{R}^r$ there exist constants $L_1(Q)$, $L_2(Q) > 0$ such that

$$\|f(t, x_1, y_1) - f(t, x_2, y_2)\| \le L_1(Q) \left(\|x_1 - x_2\| + \|y_1 - y_2\|\right),$$
$$\|G(t, x_1) - G(t, x_2)\| \le L_2(Q) \left(\|x_1 - x_2\|\right)$$
$$\text{for all } (x_1, y_1), (x_2, y_2) \in Q \quad \text{and all } t \in T.$$

Moreover, there exist constants c_1, $c_2 > 0$ such that

$$\|f(t, x, y)\| \le c_1 \left(1 + \|x\| + \|y\|\right), \quad \|G(t, x)\| \le c_2 \left(1 + \|x\|\right)$$
$$\text{for all } (t, x, y) \in T \times \mathbb{R}^n \times \mathbb{R}^r.$$

Here $\| \cdot \|$ denotes the vector norm defined by $\|x\| = \sum\limits_{j=1}^{n} |x_j|$ or a consistent matrix norm of the proper dimension.

(H_3) The set valued function $Z : T \to \mathrm{comp}(\mathbb{R}^r)$ is L-Lipschitz continuous with respect to the Hausdorff metric, i.e.,

$$d_H\big(C(t), C(s)\big) \le L|t - s| \quad \text{for all } t, s \in T,$$

where $d_H(\cdot,\cdot)$ is the Hausdorff distance. Furthermore, there exists $\delta > 0$ such that for every $t \in T$ the set $Z(t)$ is a δ-prox-regular set. The latter is understood in the sense of the following definition:

Definition 1 ([8]). *The closed set $Z \subset \mathbb{R}^r$ is called δ-prox-regular if and only if each point y in the δ-enlargement of Z, i.e. $U_\delta(Z) = \{w \in \mathbb{R}^r \mid d_Z(w) < \delta\}$, has a unique nearest point $proj_Z(y)$ and the mapping $proj_Z(\cdot)$ is continuous in $U_\delta(Z)$.*

We note that the assumption (H_3) for $Z(t)$ implies the same properties for $C(t) = q(t) - Z(t)$ whenever $q(\cdot)$ is a Lipschitz continuous function.

In general, Problem (P_0) does not have an optimal solution within absolutely continuous trajectories and Lebesgue measurable controls. Due to the fact that the right-hand side of (5) is not bounded, there may exist a sequence of controls $\{v_k\}$ converging in the sense of distributions to the Dirac delta function such that the corresponding trajectories $\{x_k\}$ and the outputs $\{y_k\}$ tend pointwise to discontinuous functions. Optimization problems of this kind are singular in the sense of [11,16,18,29,33,39] and their relaxations lead to control vector measures (impulsive controls) and trajectories of bounded variation. In this paper, we study an optimal impulsive control problem relaxing the singular problem (P_0).

Extensions of the play operator and the corresponded sweeping process to include the input of bounded variation are considered in [21,24–26,34,35]. We note that the extensions mentioned above are applicable for the trajectory relaxation of Problem (P_0) only in the case of scalar nonnegative (or nonpositive) controls v. In this paper, we propose an alternative extension suitable for K-valued controls.

Extensions of the sweeping processes to the inputs of bounded variation arise, for example, in models with material softening due to fatigue, also in modeling thermoplasticity with temperature dependent yield surface [20,21]. Such problems may also appear in many other applications. By way of motivation, consider an example from mathematical economics.

Example 1. Consider a model of advertising expenses optimization for two mutually complementary products. This model is described as follows:

$$\int_0^{t_1} \big(p_1 x_1(t) + p_2 x_2(t) - v_1(t) - v_2(t)\big)dt \to \sup, \tag{10}$$

$$\begin{aligned}
\dot{x}_1 &= (a v_1 + k_1 y_2)(1 - x_1) - b x_1, & x_1(0) &= x_{10}, \\
\dot{x}_2 &= (c v_2 + k_2 y_1)\left(1 - \tfrac{x_2}{x_1}\right) - d x_2, & x_2(0) &= x_{20},
\end{aligned} \tag{11}$$

$$\dot{y}(t) \in -N_{C(t)}\big(y(t)\big) \qquad \text{for a.e. } t \in [0, t_1], \quad y(0) = x_0, \tag{12}$$

$$v_1(t) \ge 0, \quad v_2(t) \ge 0 \quad \text{a.e. } t \in [0, t_1]. \tag{13}$$

Here, $x_1(t)$ and $x_2(t)$ are the market shares controlled by a company at the instant of time t by way of selling the first and the second products, respectively; $x_0 = (x_{10}, x_{20})$, $x_{10} \in (0,1)$ and $x_{20} \in (0, x_{10})$ are given initial market shares; $v_1(t)$ and $v_2(t)$ are the current advertising costs. The sweeping process is defined by the moving set $C(t) = x(t) - Z(t)$, where $Z(t) = \alpha(t)B$, $B \doteq \{z \in \mathbb{R}^2 \mid z_1^2 + z_2^2 \leq 1\}$, $\alpha : [0, t_1] \to (0, 1)$ is a given Lipschitz continuous function. The sweeping process here means that there exists some delay in the reaction of the customers on changing in the volume of sales of the products. The parameters a, b, c, d, p_1, p_2, k_1, and k_2 are nonnegative, with $b < d$. The first product being considered as the primary or leading one, the other is secondary in the sense that its consumption depends on the consumption of the first product. In the market model, the market share of the second product is bounded by the controlled market share of the first product: if $x_{10} \in (0, 1)$ and $x_{20} \in (0, x_{10})$, then, in view of (11), $0 < x_1(t) < 1$, $0 < x_2(t) < x_1(t)$ for the trajectories of the model. The parameters k_1, k_2 characterize the customer's preferences towards the complementary product. Therefore, the sales amounts of the products may increase even without advertisement. The objective functional is interpreted as the total profit over the time interval $[0, t_1]$. Moreover, the value $\int_0^t (v_1(s) + v_2(s)) ds$ describes the total advertising expenses over the period $[0, t]$. The current advertising expenses are unbounded, which means that an aggressive advertising campaign is possible along with a sharp increase in the volume of sales for both types of products. When $k_1 = k_2 = 0$ and $b = d = 0$, the corresponding model was studied in [10] by using the dynamic programming method, and in [12], where the maximum principle for impulsive processes was used. Problem (10)–(13) does not have an optimal measurable control. In order to cope with Problem (10)–(13) we need to relax it in a suitable sense.

The paper is organized as follows. In Sect. 2, an extension of the operator $\mathcal{F}$ for the so-called graph completions of BV–functions is proposed. In Sect. 3, the optimal impulsive control problem (P) is stated. In Sect. 4, we introduce the so-called space-time representation for the sweeping process driven by impulsive control and discuss the relationship between Problems (P) and (P_0).

Throughout this paper we use the following notation:

- $BV(T, \mathbb{R}^n)$ is the space of functions of bounded variation (BV–functions). We recall that the total variation of $w \in BV(T, \mathbb{R})$ is defined by the rule

$$\operatorname*{var}_{[a,b]} w(\cdot) = \sup_\rho \sum_{i=1}^l |w(t_i) - w(t_{i-1})|,$$

where the supremum is taken over all $\rho = \{t_0, t_1, \ldots, t_l\}$ such that $a = t_0 \leq t_1 \leq \ldots \leq t_l = b$ is a partition of the interval T. Moreover, for $w \in BV(T, \mathbb{R}^n)$

$$\operatorname*{var}_{[a,b]} w(\cdot) = \sum_{j=1}^n \operatorname*{var}_{[a,b]} w_j(\cdot).$$

- $BV_r(T, \mathbb{R}^n)$ is the set of BV-functions right continuous on $(a, b]$.
- Given $w \in BV_r(T, \mathbb{R}^n)$, $S_d(w)$ denotes the set of jumps of w, i.e., $S_d(w) = \{s \in [a, b] \mid w(s) - w(s-) \neq 0\}$.

2 Sweeping Process for Graph Completions of *BV* Functions

2.1 Graph Completions for *BV* Functions

Let $\tau_1 > 0$ be given. A nondecreasing Lipschitz continuous function $\eta(\cdot)$ from $[0, \tau_1]$ to $[a, b]$ such that $\eta(0) = a$, $\eta(\tau_1) = b$ is called a time reparametrization.

Given a time reparametrization $\eta(\cdot)$, define the pseudoinverse $\theta : [a, b] \to [0, \tau_1]$ by the rule:

$$\theta(t) = \inf\{\tau \in [0, \tau_1] \mid \eta(\tau) > t\}, \quad t \in (a, b], \qquad \theta(a) = 0. \tag{14}$$

Then, $\theta(\cdot)$ is increasing and $\theta \in BV_r([a, b], [0, \tau_1])$. Moreover,

$$\eta(\theta(t)) = t \quad \text{for all } t \in [a, b].$$

Given $\eta(\cdot)$ and its pseudoinverse $\theta(\cdot)$, let $S^\eta \doteq S_d(\theta)$, $d_s^\eta \doteq \theta(s) - \theta(s-)$.

Definition 2. *Let* $x \in BV_r([a, b], \mathbb{R}^n)$, $\eta(\cdot)$ *be a time reparametrization, and* $\theta(\cdot)$ *be the pseudoinverse for* $\eta(\cdot)$. *The time reparametrization* $\eta(\cdot)$ *is said to be consistent with* $x(\cdot)$ *if* $S_d(x) \subseteq S_d(\theta)$ *and the function* $\tau \to x(\eta(\tau))$ *is a Lipschitz continuous function.*

Let $\mathcal{T}_x([0, \tau_1], [a, b])$ denote the set of all time reparametrizations consistent with $x(\cdot)$.

Definition 3. *Let* $x \in BV_r([a, b], \mathbb{R}^n)$, $\eta \in \mathcal{T}_x([0, \tau_1], [a, b])$, *and* $\theta(\cdot)$ *be the pseudoinverse for* $\eta(\cdot)$. *In addition, let* $z^s : [0, d_s^\eta] \to \mathbb{R}^n$, $s \in S^\eta$ *be the family of Lipschitz continuous functions with uniformly bounded Lipschitz constants. Suppose that* $z^s(0) = x(s-)$, $z^s(d_s^\eta) = x(s)$, $s \in S^\eta$. *Then, the tuple* $(x(\cdot), \{z^s(\cdot)\}_{s \in S^\eta})$, *denoted by* x_η, *is said to be a graph completion corresponding to* $\eta(\cdot)$ *for* $x(\cdot)$.

With $x_\eta = (x(\cdot), \{z^s(\cdot)\}_{s \in S^\eta})$ we associate a Lipschitz continuous function $\xi : [0, \tau_1] \to \mathbb{R}^n$ such that

$$\xi(\tau) = x(\eta(\tau)) \quad \text{for all } \tau \in [0, \tau_1] \setminus \bigcup_{s \in S^\eta} \Delta_s, \tag{15}$$

$$\xi(\tau) = z^s(\tau - \theta(s-)) \quad \text{for all } \tau \in \Delta_s \text{ and all } s \in S^\eta, \tag{16}$$

where $\Delta_s \doteq [\theta(s-), \theta(s)]$, $s \in S^\eta$. It is easy to see that

$$x(t) = \xi(\theta(t)) \quad \text{for all } t \in [a, b],$$
$$z^s(\tau) = \xi(\tau + \theta(s-)) \quad \text{for all } \tau \in [0, d_s^\eta] \text{ and all } s \in S^\eta.$$

Then, $(\xi(\cdot), \eta(\cdot))$ is a parametric representation of x_η. We denote this parametric representation by $x_{\eta, \tau}$.

We denote by $BV^{gr}([a, b], \mathbb{R}^n)$ the set of graph completions corresponding to $BV_r([a, b], \mathbb{R}^n)$, namely

$$BV^{gr}([a, b], \mathbb{R}^n) \doteq \{x_\eta \mid x \in BV_r([a, b], \mathbb{R}^n), \ \eta \in \mathcal{T}_x([0, \tau_1], [a, b]), \ \tau_1 \geq b - a\}.$$

2.2 Sweeping Process for Graph Completions of BV Functions

Let $q \in BV_r([a,b], \mathbb{R}^r)$, $\eta \in \mathcal{T}_q([0, \tau_1], [a, b])$, $\theta(\cdot)$ be the pseudoinverse for $\eta(\cdot)$ defined by (14), and $q_\eta = (q(\cdot), \{z_q^s(\cdot)\}_{s \in S^\eta})$ be a graph completion for $q(\cdot)$.

As above, with q_η we associate the function $\zeta : [0, \tau_1] \to \mathbb{R}^r$ defined by (15), (16) (with ξ, x, and z^s replaced by ζ, q, and z_q^s, respectively). We note that the parametric representation $q_{\eta,\tau} = (\zeta(\cdot), \eta(\cdot))$ is a Lipschitz continuous function acting from $[0, \tau_1]$ to $\mathbb{R}^r \times [a, b]$.

Define the moving set $\mathcal{C}(\tau)$, $\tau \in [0, \tau_1]$, by the rule:

$$\mathcal{C}(\tau) = \zeta(\tau) - Z(\eta(\tau)), \qquad \tau \in [0, \tau_1].$$

Then, the set-valued map $\tau \to \mathcal{C}(\tau)$ is Lipschitz continuous. Moreover, the assumption (H_3) implies that for every $\tau \in [0, \tau_1]$ $\mathcal{C}(\tau)$ is a δ-prox-regular set. We consider the sweeping process defined by the differential inclusion:

$$-\dot{\nu}(\tau) \in N_{\mathcal{C}(\tau)}^P(\nu(\tau)) \qquad \text{for a.e. } \tau \in [0, \tau_1], \tag{17}$$

$$\nu(0) = y_0 \in \mathcal{C}(0), \tag{18}$$

where $N_{\mathcal{C}(\tau)}^P(\nu(\tau))$ is the proximal normal cone to the set $\mathcal{C}(\tau)$ at the point $\nu(\tau)$. Following to [1,7,8,42], we get that for given $q_{\eta,\tau}$ there exists a unique solution to (17), (18).

We denote by $\mathcal{F}_\eta : (\zeta, y_0) \mapsto \nu$ the solution operator to (17), (18). Moreover, define the operator $\mathcal{P} : (q_\eta, y_0) \mapsto y_\eta$ acting on $BV^{gr}([a, b], \mathbb{R}^r)$ such that

$$\mathcal{P}(q_\eta, y_0) = y_\eta,$$

where $y_\eta = (y(\cdot), \{z_y^s(\cdot)\}_{s \in S^\eta})$ satisfies the relation:

$$y(t) = \nu(\theta(t)) \qquad \text{for all } t \in [a, b], \tag{19}$$

$$z_y^s(\tau) = \nu(\tau + \theta(s-)) \qquad \text{for all } \tau \in [0, d_s^\eta], \ s \in S^\eta, \tag{20}$$

$$\nu(\cdot) \text{ is a solution corresponding to } q_{\eta,\tau} \text{ to } (17), (18). \tag{21}$$

It is easy to see that $y(\cdot) \in BV_r([a, b], \mathbb{R}^r)$, y_η is a graph completion (corresponding to $\eta(\cdot)$) for $y(\cdot)$, and $y_{\eta,\tau} = (\nu(\cdot), \eta(\cdot))$, $\tau \in [0, \tau_1]$, is the parametric representation defined similarly to (15), (16).

Note that using time reparametrization allows us to consider the sweeping process in an auxiliary time, in which the moving set becomes Lipschitz continuous and the instants of jumps of $q(\cdot)$ are transformed to some nonempty intervals.

The idea of consistent time reparametrization is closely related to the so-called singular space-time transformation of impulsive processes, which is the most popular tool in the impulsive control theory. We mention the papers [28, 29,33,39], where the singular time reparametrization for impulsive processes was introduced and thoroughly studied. There are many papers, where different modifications for this singular time reparametrization were considered (see, e.g., [13,14,18,19,30,36,41]).

3 Statement of the Problem

Let $K_1 \doteq \{v \in K \mid ||v|| = 1\}$, where $||v|| \doteq \sum_{j=1}^m |v_j|$, and $co\,A$ be the convex hull of a set A. Given μ, a bounded Borel measure on T, we denote by μ_{c}, $|\mu_{\mathrm{c}}|$, and $S_{\mathrm{d}}(\mu)$ the continuous component in the Lebesgue decomposition of the measure μ, the total variation of μ_{c}, and the set on which the discrete component of μ is concentrated, i.e., $S_{\mathrm{d}}(\mu) \doteq \{s \in T \mid \mu(\{s\}) \neq 0\}$, respectively.

By an impulsive control $\pi(\mu)$ we mean a tuple

$$\pi(\mu) = \big(\mu, S, \{d_s, \omega_s(\cdot)\}_{s \in S}\big)$$

satisfying the following conditions:

(i) μ is a K-valued bounded Borel measure on T,
(ii) the set $S \subset T$ is at most countable subset of T, and $S_{\mathrm{d}}(\mu) \subseteq S$,
(iii) for every $s \in S$, $d_s \in \mathbb{R}$, ω_s is a measurable function from $[0, d_s]$ to $co\,K_1$
   such that

$$d_s \geq ||\mu(\{s\})||, \quad \int_0^{d_s} \omega_s(\tau)d\tau = \mu(\{s\}),$$

(iv) $\displaystyle\sum_{s \in S} d_s < \infty$.

We denote by $\mathcal{W}(T, K)$ the set of all impulsive controls $\pi(\mu)$. With every $\pi(\mu) \in \mathcal{W}(T, K)$ we associate the following functions: $V : T \to \mathbb{R}$, $\theta_\mu : [a, b] \to [0, \tau_{1,\mu}]$, where $\tau_{1,\mu} \doteq b - a + V(b)$, and $\eta_\mu : [0, \tau_{1,\mu}] \to [a, b]$ such that

$$V(t) = |\mu_{\mathrm{c}}|([a, t]) + \sum_{s \in S,\, s \leq t} d_s, \quad t \in (a, b], \quad V(a) = 0,$$

$$\theta_\mu(t) = t - a + V(t) \quad \text{for all } t \in T,$$

$$\theta_\mu(\cdot) \text{ is the pseudoinverse function for } \eta_\mu(\cdot).$$

Consider the optimal impulsive control problem (P):

$$\text{Minimize } J = l\big(x(b)\big)$$

subject to the dynamics, denoted $(\mathcal{D})$,

$$x(t) = x_0 + \int_a^t f\big(\tau, x(\tau), y(\tau)\big)d\tau + \int_a^t G\big(\tau, x(\tau)\big)\mu_{\mathrm{c}}(d\tau)$$

$$+ \sum_{s \in S,\, s \leq t} \big(z^s(d_s) - x(s-)\big), \quad t \in (a, b], \quad x(a) = x_0, \tag{22}$$

$$q(t) = Ax(t), \quad t \in [a, b], \tag{23}$$

$$\frac{dz^s(\tau)}{d\tau} = G\big(s, z^s(\tau)\big)\omega_s(\tau), \quad z^s(0) = x(s-), \tag{24}$$

$$z_q^s(\tau) = Az^s(\tau), \quad \tau \in [0, d_s], \quad s \in S, \tag{25}$$

$$\big(y(\cdot), \{z_y^s(\cdot)\}_{s \in S}\big) = \mathcal{P}\big(q(\cdot), \{z_q^s(\cdot)\}_{s \in S}, y_0\big), \tag{26}$$

$$\pi(\mu) = \big(\mu, S, \{d_s, \omega_s(\cdot)\}_{s \in S}\big) \in \mathcal{W}(T, K). \tag{27}$$

Here, $x \in BV_r(T, \mathbb{R}^n)$, $q, y \in BV_r(T, \mathbb{R}^r)$. We note that $x_{\eta_\mu} = (x(\cdot), \{z^s(\cdot)\}_{s \in S})$, $q_{\eta_\mu} = (q(\cdot), \{z_q^s(\cdot)\}_{s \in S})$, and $y_{\eta_\mu} = (y(\cdot), \{z_y^s(\cdot)\}_{s \in S})$ are graph completions for $x(\cdot)$, $q(\cdot)$, and $y(\cdot)$, respectively.

Let $\sigma = \big(x_{\eta_\mu}, q_{\eta_\mu}, y_{\eta_\mu}; \pi(\mu)\big)$ be a feasible process to (P) and Σ be the collection of all processes σ.

4   The Space-Time Representation for Problem (P)

In this section, we describe the so-called space-time representation for Problem (P). This representation is based on the singular time reparametrization and leads to the transformation of Problem (P) to an auxiliary optimal control problem whose controls are Lebesgue measurable functions.

Let $\{\sigma_{0k}\} = \big\{\big(x_k(\cdot), q_k(\cdot), y_k(\cdot); v_k(\cdot)\big)\big\}$ be a sequence of feasible processes of the system (5)–(9) such that $\sup\limits_{k} \|v_k\|_{L^1} < \infty$. Define the function $V(\cdot)$ by the rule $V_k(t) = \int_a^t \|v_k(s)\| ds$. Then, there exists $M > 0$ such that

$$V_k(b) \le M \quad \text{for all } k = 1, 2, \ldots. \tag{28}$$

Proposition 1. *Let the assumptions (H_1)–(H_3) hold. Then, the estimate (28) implies that the functions $\{x_k(\cdot), q_k(\cdot), y_k(\cdot)\}$ and their total variations on $[a, b]$ are uniformly bounded.*

Proof. First, we derive some inequalities for solutions of the differential inclusion (8), (9). Let $q : [a, b] \to \mathbb{R}^r$ be a given Lipschitz continuous function. Let $t_1, t_2 \in [a, b]$ such that $t_2 > t_1$. From the assumption (H_3) and the estimate for the norms $\|\cdot\|$ and $\|\cdot\|_2$:

$$\|p\|_2 \le \|p\| \le \sqrt{r}\|p\|_2 \quad \text{for all } p \in \mathbb{R}^r,$$

we have

$$
\begin{aligned}
d_H\big(C(t_2), C(t_1)\big) &\le \|q(t_2) - q(t_1)\|_2 + d_H\big(Z(t_2), Z(t_1)\big) \\
&\le \|q(t_2) - q(t_1)\| + d_H\big(Z(t_2), Z(t_1)\big) \\
&\le \operatorname*{var}_{[a, t_2]} q(\cdot) - \operatorname*{var}_{[a, t_1]} q(\cdot) + L(t_2 - t_1),
\end{aligned}
$$

where $\operatorname*{var}_{[a, t]} q(\cdot) \doteq \int_a^t \|\dot{q}(s)\| ds$. Define the function $u : [a, b] \to \mathbb{R}$ by the rule

$$u(t) = \operatorname*{var}_{[a, t]} q(\cdot) + L(t - a).$$

Then,

$$d_H\big(C(t_2), C(t_1)\big) \le |u(t_2) - u(t_1)| \quad \text{for all } t_1, t_2 \in [a, b].$$

Hence, in view of ([43], p. 36, Corollary 4.2), we obtain

$$\|y(t_2) - y(t_1)\|_2 \le |u(t_2) - u(t_1)|.$$

Therefore, for $t_2 > t_1$ we have

$$||y(t_2) - y(t_1)|| \leq \sqrt{r}(u(t_2) - u(t_1)) \leq \sqrt{r}(\operatorname*{var}_{[a,t_2]} q(\cdot) + L(t_2 - t_1)).$$

Hence, we obtain the following estimates:

$$||y(t)|| \leq ||y_0|| + \sqrt{r}\left(L(t - a) + \int_a^t ||\dot{q}(s)||ds\right) \quad \text{for all } t \in [a, b], \tag{29}$$

$$\operatorname*{var}_{[a,t]} y(\cdot) \leq \sqrt{r}\left(L(t - a) + \operatorname*{var}_{[a,t]} q(\cdot)\right) \quad \text{for all } t \in [a, b]. \tag{30}$$

By using the estimate (29) and (H_2), we have

$$||x_k(t)|| \leq ||x_0|| + \int_a^t ||f(s, x_k(s), y_k(s))||ds$$

$$+ \int_a^t ||G(s, x_k(s))v_k(s)||ds$$

$$\leq ||x_0|| + c_1(t - a) + c_2 M + \int_a^t c_1||y_k(s)||ds$$

$$+ \int_a^t ||x_k(s)|| (c_1 + c_2||v_k(s)||)ds. \tag{31}$$

$$||y_k(t)|| \leq ||y_0|| + \sqrt{r}L(t - a) + c_3 \int_a^t ||x_k(s)||ds$$

where $c_3 > 0$ is defined by r and a consistent norm of the matrix A (c_3 does not depend on k). Let $\alpha_k(t) = ||x_k(t)|| + ||y_k(t)||$, $\beta(t) = ||x_0|| + ||y_0|| + (c_1 + \sqrt{r}L)(t - a) + c_2 M$, $t \in [a, b]$. Then, we have

$$\alpha_k(t) \leq \beta(t) + c_1 \int_a^t \alpha_k(s)ds + c \int_a^t \alpha_k(s)dV_k(s)$$

$$\leq \beta(t) + c \int_a^t \alpha_k(s)d(s + V_k(s)),$$

where $c = \max\{c_1, c_2, c_3\}$. Invoking Gronwall's lemma the last inequality yields

$$\alpha_k(t) \leq (||x_0|| + ||y_0|| + (c_1 + \sqrt{r}L)(b - a) + c_2 M)e^{c(b-a+M)} \doteq K.$$

Therefore, we have

$$||x_k(t)|| \leq K, \quad ||q_k(t)|| \leq c_3 K, \quad ||y_k(t)|| \leq K \quad \text{for all } t \in [a, b].$$

The uniform boundedness of $\{x_k(\cdot), q_k(\cdot), y_k(\cdot)\}$ is thus proven.

Next, we have the inequality

$$\operatorname*{var}_{[a,t]} x_k(\cdot) = \int_a^t ||\dot{x}_k(s)|| ds$$
$$\leq \int_a^t ||f(s, x_k(s), y_k(s))|| ds + \int_a^t ||G(s, x_k(s))|| dV_k(s).$$

Hence, in view of the uniform boundedness of $\{x_k(\cdot), y_k(\cdot)\}$ and the assumptions (H_1)–(H_3) we see that $\operatorname*{var}_{[a,t]} x_k(\cdot)$, $\operatorname*{var}_{[a,t]} q_k(\cdot)$ are uniformly bounded. Therefore, from (30) it follows that the total variations $\operatorname*{var}_{[a,t]} y_k(\cdot)$ are also uniformly bounded and the proposition thus follows.

For every k consider the function

$$\theta_k(t) = t - a + V_k(t),$$

which is Lipschitz continuous and strictly increasing. Let $\eta_k(\cdot)$ be the inverse of $\theta_k(\cdot)$, i.e. $\eta_k(\tau) = \theta_k^{-1}$. We set $\tau_{1k} = b - a + V_k(b)$.

Our aim now is to change the time $t = \eta_k(\tau)$ in $\{\sigma_{0k}\}$ and obtain differential relations characterizing the components of $\{\sigma_{0k}\}$ in the auxiliary time $\tau \in [0, \tau_{1k}]$.

First, we introduce the functions $\xi_k(\cdot)$, $\zeta_k(\cdot)$, $\nu_k(\cdot)$, $\omega_{0k}(\cdot)$, $\omega_k(\cdot)$ defined on $[0, \tau_{1k}]$ by the rule:

$$\xi_k(\tau) = x_k(\eta_k(\tau)), \quad \zeta_k(\tau) = q_k(\eta_k(\tau)), \quad \nu_k(\tau) = y_k(\eta_k(\tau)),$$

$$\omega_{0k}(\tau) = \frac{1}{1 + ||v_k(\eta_k(\tau))||}, \quad \omega_k(\tau) = \frac{v_k(\eta_k(\tau))}{1 + ||v_k(\eta_k(\tau))||}.$$

We note that the rate independence property of the sweeping process implies that

$$\mathcal{F}(q_k, y_0) \circ \eta_k = \mathcal{F}_{\eta_k}(q_k \circ \eta_k, y_0),$$

where $\mathcal{F}$ is defined by (8), (9) and $\mathcal{F}_{\eta_k}$, by (17), (18). Consequently, $\nu_k = \mathcal{F}_{\eta_k}(\zeta_k, y_0)$.

Then, it is easy to see that the functions $\xi_k(\cdot)$, $\zeta_k(\cdot)$, $\nu_k(\cdot)$, $\omega_{0k}(\cdot)$, $\omega_k(\cdot)$ satisfy the following relations:

$$\eta_k'(\tau) = \omega_{0k}(\tau), \tag{32}$$
$$\xi_k'(\tau) = f(\eta_k(\tau), \xi_k(\tau), \nu_k(\tau))\omega_{0k}(\tau) + G(\eta_k(\tau), \xi_k(\tau))\omega_k(\tau), \tag{33}$$
$$\zeta_k(\tau) = A\xi_k(\tau) \quad \text{for all } \tau \in [0, \tau_{1k}], \tag{34}$$
$$\nu_k = \mathcal{F}_{\eta_k}(\zeta_k, y_0), \tag{35}$$

$$\omega_{0k}(\tau) > 0, \quad \omega_k(\tau) \in K, \tag{36}$$

$$\omega_{0k}(\tau) + ||\omega_k(\tau)|| = 1 \quad \text{for a.e. } \tau \in [0, \tau_{1k}] \tag{37}$$

with the initial and terminal conditions: $\eta_k(0) = a$, $\eta_k(\tau_1) = b$, $\xi_k(0) = x_0$, $\zeta_k(0) = q_0$, $\nu_k(0) = y_0$. Here, the prime denotes differentiation with respect to τ.

From the assumption (H_1)–(H_3) and Proposition 1 it follows that $\xi_k(\cdot)$, $\zeta_k(\cdot)$, and $\nu_k(\cdot)$ are uniformly bounded and equicontinuous. The Arzelà-Ascoli theorem implies then that there exists a subsequence $\{\xi_{kj}(\cdot), \zeta_{kj}(\cdot), \nu_{kj}(\cdot)\}$ converging uniformly to some functions $(\bar{\xi}(\cdot), \bar{\zeta}(\cdot), \bar{\nu}(\cdot))$ on the interval $[0, \bar{\tau}_1]$, where $\bar{\tau}_1 = \lim_{j\to\infty} \tau_{kj}$.

Consider the dynamics:

$$\eta_k'(\tau) = \omega_{0k}(\tau), \qquad \eta_k(0) = a, \quad \eta_k(\tau_{1k}) = b, \qquad (38)$$
$$\gamma_k'(\tau) = \omega_k(\tau), \qquad \gamma_k(0) = 0 \qquad (39)$$

coupled with the control constraints (36), (37). Passing to a subsequence if necessary, let $\{\eta_k(\cdot), \gamma_k(\cdot)\}$ converges uniformly to $(\bar{\eta}(\cdot), \bar{\gamma}(\cdot))$. Then, from Filippov's lemma it follows that there exists a control $(\bar{\omega}_0(\cdot), \bar{\omega}(\cdot))$ such that:

$$\bar{\eta}'(\tau) = \bar{\omega}_0(\tau), \qquad \bar{\eta}(0) = a, \quad \bar{\eta}(\bar{\tau}_1) = b,$$
$$\bar{\gamma}'(\tau) = \bar{\omega}(\tau), \qquad \bar{\gamma}(0) = 0,$$
$$(\bar{\omega}_0(\tau), \bar{\omega}(\tau)) \in co\,\tilde{K}_1 \quad \text{for a.e. } \tau \in [0, \bar{\tau}_1],$$

where $\tilde{K}_1 \doteq \{(\omega_0, \omega) \in [0, 1] \times K \mid \omega_0 + \|\omega\| = 1\}$.

Using Proposition 1, the local Lipschitz continuity in x, y of f and G, and Gronwall's lemma, one can show that $(\bar{\xi}(\cdot), \bar{\zeta}(\cdot), \bar{\nu}(\cdot))$ is a solution corresponding to $\bar{\omega}_0(\cdot)$, $\bar{\omega}(\cdot)$ of the following system:

$$\eta'(\tau) = \omega_0(\tau), \qquad (40)$$
$$\xi'(\tau) = f(\eta(\tau), \xi(\tau), \nu(\tau))\omega_0(\tau) + G(\eta(\tau), \xi(\tau))\omega(\tau), \qquad (41)$$
$$\zeta(\tau) = A\xi(\tau) \quad \text{for all } \tau \in [0, \tau_1], \qquad (42)$$
$$\nu = \mathcal{F}_\eta(\zeta, y_0), \qquad (43)$$
$$(\omega_0(\tau), \omega(\tau)) \in co\,\tilde{K}_1 \quad \text{for a.e. } \tau \in [0, \tau_1] \qquad (44)$$

with the initial and terminal conditions

$$\eta(0) = a, \ \eta(\tau_1) = b, \ \xi(0) = x_0, \ \zeta(0) = q_0, \ \nu(0) = y_0. \qquad (45)$$

Here, $\tau_1 \in [b - a, b - a + M]$. We denote by $\mathcal{G}$ the collection of all processes $g = (\eta(\cdot), \xi(\cdot), \zeta(\cdot), \nu(\cdot); \omega_0(\cdot), \omega(\cdot))$ for (40)–(45).

Now, we establish a relationship between the systems $(\mathcal{D})$ and (40)–(45).

First, let $\sigma = (x(\cdot), q(\cdot), y(\cdot); \pi(\mu))$ coupled with $\{z^s(\cdot), z_q^s(\cdot), z_y^s(\cdot)\}_{s\in S}$ be a process of $(\mathcal{D})$. Consider the graph completions x_{η_μ}, q_{η_μ}, and y_{η_μ} and define their parametric representations as in Sect. 2. Namely, let $\eta_\mu(\cdot)$, $\xi(\cdot)$, $\zeta(\cdot)$, $\nu(\cdot)$ be such that

$$x_{\eta_\mu, \tau} = (\xi(\cdot), \eta_\mu(\cdot)), \quad q_{\eta_\mu, \tau} = (\zeta(\cdot), \eta_\mu(\cdot)), \quad y_{\eta_\mu, \tau} = (\nu(\cdot), \eta_\mu(\cdot)).$$

Then, $\big(\eta_\mu(\cdot), \xi(\cdot), \zeta(\cdot), \nu(\cdot)\big)$ is a solution to (40)–(45) corresponding to some control $\big(\omega_0(\cdot), \omega(\cdot)\big)$. Given $\pi(\mu) = \big(\mu, S, \{d_s, \omega_s(\cdot)\}_{s \in S}\big)$, the control $\big(\omega_0(\cdot), \omega(\cdot)\big)$ can be defined by the following rule:

(a) Define auxiliary functions $w(\cdot)$, $\{z_w^s(\cdot)\}_{s \in S}$ such that

$$w(t) = \int_a^t \mu_c(dt) + \sum_{s \leq t,\, s \in S} \big(z_w^s(d_s) - w(s-)\big), \quad t \in (a, b], \quad w(a) = 0,$$

$$\frac{dz_w^s(\tau)}{d\tau} = \omega_s(\tau), \quad z_w^s(0) = w(s-) \text{ for a.e. } \tau \in [0, d_s], \text{ and all } s \in S.$$

Note that $w_{\eta_\mu} = \big(w(\cdot), \{z_w^s(\cdot)\}_{s \in S}\big)$ is a graph completion for $w(\cdot)$.
(b) Find the parametric representation $w_{\eta_\mu, \tau} = \big(\gamma(\cdot), \eta_\mu(\cdot)\big)$.
(c) Then,
$$\omega_0(\tau) = \eta_\mu'(\tau), \quad \omega(\tau) = \gamma'(\tau) \text{ for a.e. } \tau \in [0, \tau_1].$$

Second, for any $g = \big(\eta(\cdot), \xi(\cdot), \zeta(\cdot), \nu(\cdot); \omega_0(\cdot), \omega(\cdot)\big) \in \mathcal{G}$ there exists an impulsive process $\sigma \in \Sigma$ such that

$$x_{\eta_\mu, \tau} = \big(\xi(\cdot), \eta(\cdot)\big), \quad q_{\eta_\mu, \tau} = \big(\zeta(\cdot), \eta(\cdot)\big), \quad y_{\eta_\mu, \tau} = \big(\nu(\cdot), \eta(\cdot)\big),$$

where $\big(x_{\eta_\mu, \tau}, q_{\eta_\mu, \tau}, y_{\eta_\mu, \tau}\big)$ is the trajectory parametric representation corresponding to σ. The proof of this fact follows closely the lines of the proof of a similar statement from [29]. Now, given $\big(\omega_0(\cdot), \omega(\cdot)\big)$, the impulsive control $\pi(\mu) = \big(\mu, S, \{d_s, \omega_s(\cdot)\}_{s \in S}\big)$ is defined by the following rule:

(a') Let $\theta(\cdot)$ be the pseudoinverse for $\eta(\cdot)$ defined by (14) and

$$w(t) = \int_0^{\theta(t)} \omega(\tau)d\tau, \quad t \in (a, b], \quad w(a) = 0.$$

(b') Then, the components of $\pi(\mu)$ are such that:
    (i) μ is the Stieltjes-Lebesgue measure generated by $w(\cdot)$,
    (ii) $S = S_d(\theta)$,
    (iii) for every $s \in S$, $d_s = \theta(s) - \theta(s-)$, $\omega_s(\cdot)$ is defined by the rule

$$\omega_s(\tau) = \omega(\tau + \theta(s-)) \quad \text{a.e. } \tau \in [0, d_s].$$

In the common terminology of impulsive control theory (see [30,33] and the references therein) the system (40)–(45) is called a space-time representation for the impulsive control system $(\mathcal{D})$.

Consider the auxiliary optimal control problem (P_a):

$$\text{Minimize} \quad J_a(g) = l\big(\xi(\tau_1)\big)$$

over processes $g = \big(\eta(\cdot), \xi(\cdot), \zeta(\cdot), \nu(\cdot); \omega_0(\cdot), \omega(\cdot)\big) \in \mathcal{G}$.

Problem (P_a) is said to be the space-time representation for Problem (P). It is easy to show that if processes $\sigma \in \Sigma$ and $g \in \mathcal{G}$ correspond to each other, then the corresponding trajectories satisfy the equality $x(b) = \xi(\tau_1)$.

We thus have the following result.

Theorem 1. $\bar{\sigma} \in \Sigma$ *yields the optimal value to Problem* (P) *if and only if* $\bar{g} \in \mathcal{G}$ *(corresponding to $\bar{\sigma}$) yields the optimal value to Problem* (P_a). *Moreover,*

$$\min_{\sigma \in \Sigma} J(\sigma) = \min_{g \in \mathcal{G}} J_a(g), \quad i.e. \ J(\bar{\sigma}) = J_a(\bar{g}).$$

We note that the space-time representation for impulsive processes allows us to obtain optimality conditions for Problem (P) by resorting to Problem (P_a). We mention [2,3,5,6,9,15], where optimal control problems close to Problem (P_a) were studied.

Acknowledgements. The work is partially supported by the Russian Foundation for Basic Research, Project no. 18-01-00026.

References

1. Benabdellah, H.: Existence of solutions to the nonconvex sweeping process. J. Differential Equ. **164**, 286–295 (2000). https://doi.org/10.1006/jdeq.1999.3756
2. Bensoussan, A., Turi, J.: Optimal control of variational inequalities. Commun. Inf. Syst. **10**(4), 203–220 (2010). https://doi.org/10.4310/CIS.2010.v10.n4.a3
3. Brokate, M.: Optimal Streuerungen von gewöhnlichen Differentialgleichungen mit Nichtlinearitäten vom Hysteresis-Typ.Peter D. Lang Verlag, Frankfurt am Main (1987)
4. Brokate, M., Sprekels, J.: Hysteresis and Phase Transitions. Series of Applied Mathematical Sciences, vol. 121. Springer, New York (1996). https://doi.org/10.1007/978-1-4612-4048-8
5. Brokate, M., Krejčí, P.: Optimal control of ODE systems involving a rate independent variational inequality. Disc. Cont. Dyn. Syst. Ser. B. **18**(2), 331–348 (2013). https://doi.org/10.3934/dcdsb.2013.18.331
6. Cao, T.H., Mordukhovich, B.S.: Optimality conditions for a controlled sweeping processswith applications to the crowd motion model. Discrete Contin. Dyn. Syst. Ser. B **21**, 267–306 (2017). https://doi.org/10.3934/dcdsb.2017014
7. Castaing, C., Monteiro Marques, M.D.P.: Evolution problems associated with nonconvex closed moving sets with bounded variation. Port. Math. **53**, 73–87 (1996). ftp://ftp4.de.freesbie.org/pub/EMIS/journals/PM/53f1/pm53f106.pdf
8. Colombo, G., Goncharov, V.V.: The sweeping processes without convexity. Set-Valued Anal. **7**, 357–374 (1999). https://doi.org/10.1023/A:100877452
9. Colombo, G., Henrion, R., Hoang, N.D., Mordukhovich, B.S.: Optimal control of the sweeping process over polyhedral controlled sets. J. Differential Equ. **260**, 3397–3447 (2016). https://doi.org/10.1016/j.jde.2015.10.039
10. Dorroh, J.R., Ferreyra, G.: A multistate, multicontrol problem with unbounded controls. SIAM J. Control Optim. **32**, 1322–1331 (1994). https://doi.org/10.1137/S0363012992229823
11. Dykhta, V.A., Samsonyuk, O.N.: Optimal Impulsive Control with Applications, 2nd edn. Fizmatlit, Moscow (2003)
12. Dykhta, V.A., Samsonyuk, O.N.: A maximum principle for smooth optimal impulsive control problems with multipoint state constraints. Comput. Math. Math. Phys. **49**, 942–957 (2009). https://doi.org/10.1134/S0965542509060050
13. Dykhta, V.A., Samsonyuk, O.N.: Hamilton-Jacobi Inequalities and Variational Optimality Conditions. ISU, Irkutsk (2015)

14. Goncharova, E., Staritsyn, M.: On BV-extension of asymptotically constrained control-affine systems and complementarity problem for measure differential equations. Disc. Cont. Dyn. Syst. Ser. **11**(6), 1061–1070 (2018). https://doi.org/10.3934/dcdss.2018061

15. Gudovich, A., Quincampoix, M.: Optimal control with hysteresis nonlinearity and multidimensional play operator. SIAM J. Control. Optim. **49**(2), 788–807 (2011). https://doi.org/10.1137/090770011

16. Gurman, V.I.: The Extension Principle in Optimal Control Problems, 2nd edn. Fizmatlit, Moscow (1997)

17. Henry, C.: An existence theorem for a class of differential equations with multivalued right-hand side. J. Math. Anal. Appl. **41**, 179–186 (1973)

18. Karamzin, D.Yu., Oliveira, V.A., Pereira, F.L., Silva, G.N.: On some extension of optimal control theory. Eur. J. Control **20**(6), 284–291 (2014). https://doi.org/10.1016/j.ejcon.2014.09.003

19. Karamzin, D.Yu., Oliveira, V.A., Pereira, F.L., Silva, G.N.: On the properness of the extension of dynamic optimization problems to allow impulsive controls. ESAIM Control Optim. Calculus Var. **21**(3), 857–875 (2015). https://doi.org/10.1051/cocv/2014053

20. Kopfová, J.: BV-norm continuity of the play operator. In: 8th Workshop on Multi-Rate Processes and Hysteresis and the HSFS Workshop (Hysteresis and Slow-Fast Systems) (2016)

21. Kopfova, J., Recupero, V.: BV-norm continuity of sweeping processes driven by a set with constant shape. J. Differential Equ. **261**(10), 5875–5899 (2016). https://doi.org/10.1016/j.jde.2016.08.025

22. Krasnoselskii, M.A., Pokrovskii, A.V.: Systems with Hysteresis. Springer, Heidelberg (1989). https://doi.org/10.1007/978-3-642-61302-9

23. Krejčí, P.: Vector hysteresis models. Eur. J. Appl. Math. **2**, 281–292 (1991). https://doi.org/10.1017/S0956792500000541

24. Krejčí, P., Liero, M.: Rate independent Kurzweil process. Appl. Math. **54**, 117–145 (2009). https://doi.org/10.1007/s10492-009-0009-5

25. Krejčí, P., Recupero, V.: Comparing BV solutions of rate independent processes. J. Convex. Anal. **21**, 121–146 (2014)

26. Krejčí, P., Roche, T.: Lipschitz continuous data dependence of sweeping processes in BV spaces. Disc. Cont. Dyn. Syst. Ser. B. **15**, 637–650 (2011). http://ncmm.karlin.mff.cuni.cz/preprints/1042115110pr12.pdf

27. Kunze, M., Marques, M.M.: An introduction to Moreau's sweeping process. In: Brogliato, B. (ed.) Impacts in Mechanical Systems. Lecture Notes in Physics, vol. 551, pp. 1–60. Springer, Berlin (2000). https://doi.org/10.1007/3-540-45501-9_1

28. Miller, B.M.: The generalized solutions of nonlinear optimization problems with impulse control. SIAM J. Control Optim. **34**, 1420–1440 (1996). https://doi.org/10.1137/S0363012994263214

29. Miller, B.M., Rubinovich, E.Ya.: Impulsive Controls in Continuous and Discrete-Continuous Systems. Kluwer Academic Publishers, New York (2003). https://doi.org/10.1007/978-1-4615-0095-7

30. Miller, B.M., Rubinovich, E.Ya.: Discontinuous solutions in the optimal control problems and their representation by singular space-time transformations. Autom. Remote Control **74**, 1969–2006 (2013). https://doi.org/10.1134/S0005117913120047

31. Moreau, J.-J.: Evolution problem associated with a moving convex set in a Hilbert space. J. Differential Equ. **26**, 347–374 (1977). https://doi.org/10.1016/0022-0396(77)90085-7

32. Moreau, J.-J.: Numerical aspects of the sweeping process. Comput. Methods Appl. Mech. Eng. **177**(3–4), 329–349 (1999). https://doi.org/10.1016/S0045-7825(98)00387-9

33. Motta, M., Rampazzo, F.: Space-time trajectories of nonlinear systems driven by ordinary and impulsive controls. Differential Integral Equ. **8**, 269–288 (1995)

34. Recupero, V., Santambrogio, F.: Sweeping processes with prescribed behavior on jumps. ArXiv:1707.09765 (2017)

35. Recupero, V.: BV continuous sweeping processes. J. Differential Equ. **259**, 4253–4272 (2015). https://doi.org/10.1016/j.jde.2015.05.019

36. Samsonyuk, O.N.: Invariant sets for nonlinear impulsive control systems. Autom. Remote Control **76**(3), 405–418 (2015). https://doi.org/10.1134/S0005117915030054

37. Samsonyuk, O.N., Timoshin, S.A.: Optimal impulsive control problems with hysteresis. In: Constructive Nonsmooth Analysis and Related Topics (dedicated to the Memory of V.F. Demyanov), CNSA–2017, pp. 276–280 (2017). https://doi.org/10.1109/CNSA.2017.7974010

38. Samsonyuk, O.N., Tolkachev, D.E.: Approximation results for impulsive control systems with hysteresis. In: Tkhai, V.N. (ed.) 14th International Conference "Stability and Oscillations of Nonlinear Control Systems" (Pyatnitskiy's Conference) (STAB) (2018). https://doi.org/10.1109/STAB.2018.8408396

39. Sesekin, A.N., Zavalishchin, S.T.: Dynamic Impulse Systems: Theory and Applications. Kluwer Academic Publishers, Dordrecht (1997). https://doi.org/10.1007/978-94-015-8893-5

40. Siddiqi, A.H., Manchanda, P., Brokate, M.: On some recent developments concerning Moreau's sweeping process. In: Siddiqi, A.H., Kocvara, M. (eds.) Trends in Industrial and Applied Mathematics. Applied Optimization, vol. 72, pp. 339–354. Springer, Boston (2002). https://doi.org/10.1007/978-1-4613-0263-6_15

41. Staritsyn, M.: On "discontinuous" continuity equation and impulsive ensemble control. Syst. Control Lett. **118**, 77–83 (2018). https://doi.org/10.1016/j.sysconle.2018.06.001

42. Thibault, L.: Regularization of nonconvex sweeping process in Hilbert space. Set-Valued Anal. **16**, 319–333 (2008). https://doi.org/10.1007/s11228-008-0083-y

43. Thibault, L.: Moreau sweeping process with bounded truncated retraction. J. Convex Anal. **23**(4), 1051–1098 (2016). http://www.heldermann.de/JCA/JCA23/JCA234/jca23039.htm

44. Valadier, M.: Lipschitz approximation of the sweeping (or Moreau) process. J. Differential Equ. **88**, 248–264 (1990). https://doi.org/10.1016/0022-0396(90)90098-A

45. Visintin, A.: Differential Models of Hysteresis. Series in Applied Mathematical Sciences, vol. 111. Springer, Berlin (1994). https://doi.org/10.1007/978-3-662-11557-2

Impulsive Relaxation of Continuity Equations and Modeling of Colliding Ensembles

Maxim Staritsyn$^{(\boxtimes)}$ and Nikolay Pogodaev

Matrosov Institute for System Dynamics and Control Theory of Siberian Branch
of the Russian Academy of Sciences, Irkutsk, Russia
starmaxmath@gmail.com, nickpogo@gmail.com

Abstract. The paper promotes a relatively novel class of multi-agent
control systems named "impulsive" continuity equations. Systems of this
sort, describing the dynamics of probabilistically distributed "crowd" of
homotypic individuals, are intensively studied in the case when the driving vector field is bounded and sufficiently regular. We, instead, consider
the case when the vector field is unbounded, namely, affine in a control parameter, which is only integrally constrained. This means that
the "crowd" can be influenced by "shock" impacts, i.e., actions of small
duration but very high intensity. For such control continuity equations,
we design an impulsive relaxation by closing the set of solutions in a
suitable coarse topology. The main result presents a constructive form
of the relaxed system. A connection of the obtained results to problems
of contact dynamics is also discussed along with applications to optimal
ensemble control and other promising issues.

Keywords: Multi-agent systems · Ensemble control
Mean-field type control · Continuity equation · Impulsive control

1 Introduction

This study is primarily inspired by the following two connected issues:

(i) impulsive control under uncertainty, and
(ii) impulsive control of dynamic ensembles (multi-agent systems)/mean-field
impulsive control.

Similar subjects, which draw their practical motivation in the variety of real-
life cases [14,18,19,26], are well-recognized within the ordinary (non-impulsive)
control setup and became a trendy and rapidly developing field of research in the
recent years [11,12,21,27,28]. Meanwhile, though such questions are naturally
posed in the impulsive control framework, we lose to find a noticeable record

The work is supported by the Russian Foundation for Basic Research, grants 18-31-
20030, 18-31-00425, 18-01-00026.

related to the actual impulsive case, excepting some fragmentarily connected works. Regarding continuity equations with velocity fields of a low regularity, we mention papers [3,4]; one can also revise [1,2] and citations therein, where a basic framework for measure-driven evolution equations is developed.

Control problems for multi-agent systems typically look like "how to steer a crowd (collection, ensemble) of individuals (agents, objects) by a single, common for all the individuals, control influence?". If the cardinality of the crowd is relatively large (up to continuum), it is not more reasonable (or, even, possible) to trace – not to say "to control" – each an agent personally. In such a case, it is natural to regard the state of the crowd as a probability measure, saying us, how many objects currently occupy a subset of the state space. The time evolution of this measure in the simplest case (under no diffusion etc.) can be described by the continuity equation. A similar representation also comes in mind, if we deal with a single object, whose parameters or initial position are uncertain [11,12,21,27]. As an illustration, we propose the following toy example from the contact dynamics (some practically relevant models of this sort can be found in control of quantum systems, see, e.g., [27] and the references therein).

Example 1. Consider an inverted pendulum of the unit mass and length, posed on a cart of the unit mass, which moves linearly without acceleration. Assume that the cart hits on an elastic curb, placed at the origin, that does not constraint the motion of the pendulum. Following the approach [8,23], we approximate the purely elastic contact by a viscoelastic one with elasticity $u \to +\infty$ (a "penalty for entering the obstacle") and accept the restoration law [15] modeled by a linear spring with damping (for simplicity, with the unit natural frequency and damping coefficient)[1]:

$$2\ddot{x} - \ddot{\theta}\cos\theta + \dot{\theta}^2\sin\theta = (-x - 2\dot{x})\,u\chi_{\{x \le 0\}},$$

$$\ddot{\theta} - g\sin\theta = \ddot{x}\cos\theta.$$

Here, x is the linear coordinate of the cart, θ is the angular position of the pendulum, and g is the acceleration due to gravity; $\dot{x} \doteq \frac{dx}{dt}$, and χ_A denotes the characteristic function of a set A. We consider the system evolution on a small interval $[0, T]$ containing the moment $0 < \tau < T$ of collision. Suppose, we know the values $x(0) > 0$ and $\dot{x}(0) < 0$, while $\theta(0)$ can not be accurately measured by the available tools; instead, we are given the probability ϑ of finding it in a subset of $\mathbb{R}$. Passing to a first-order ODE with an *u-affine* velocity field $v(x, u) \in \mathbb{R}^4$, $x = (x_1, x_2, x_3, x_4) \doteq (x, \dot{x}, \theta, \dot{\theta})$, we come to a control system with a fuzzy initial state (the exact representation of this system can be easily deduced by

---

[1] The restoration law represents a mechanical specification of the obstacle (in our interpretation, of the "curb") due to concrete properties of its material; we deal with an "idealized", academic model. Recall that, in general, the intensity of impact depends on mechanical properties of colliding objects, and restoration laws can be different, depending on the law of interaction [24]. Practically, computation of the actual restoration law is a complicated problem with a number of pitfalls such as the famous Painlevé paradox [22].

the reader, and we skip it here for brevity). As it was discussed above, the time evolution of measure ϑ on $[0, T]$ can be described by the continuity equation

$$\mu_0 = \vartheta; \quad \partial_t \mu_t + \nabla \cdot (\mu_t\, v(x, u)) = 0. \tag{1}$$

Now, attempted to extend the picture towards the case of elastic collision, we are to answer, what happens with the solution $t \mapsto \mu_t[u]$ of (1) as $u \to \infty$. Note that a similar issue arises in the control-theoretical context, i.e. when considering systems of the form (1) driven by a control-affine vector field $v(x, u)$, as soon as control inputs are not a priori constrained in the pointwise sense. To answer such questions, we propose to recruit the theory of finite-dimensional impulsive control and measure differential equations [7,9,13,16,17,23,25,30,33,34] combined with the apparatus of continuity equations [4,5].

This paper follows our recent work [31] and essentially relies on [27]. In [31], we consider Eq. (1) driven by a control-affine function

$$v(x, u) \doteq f(x) + g(x)\, u$$

with a drift $f : \mathbb{R}^n \mapsto \mathbb{R}^n$ and a control vector field $g : \mathbb{R}^n \mapsto \mathbb{R}^n$ weighted by a measurable scalar-valued signal u. Now we extend some results of [31] to systems driven by vector fields of the form

$$v(x, w, u) \doteq f(x, w) + G(x, w)\, u, \tag{2}$$

with "regular" signals $w = (w_j)_{j=\overline{1,k}}$ ranged in a given compact subset of $\mathbb{R}^k$ and "shock", "unbounded" *vector-valued* inputs $u = (u_j)_{j=\overline{1,m}}$, under *no commutativity/involutivity assumptions* related to the associated control vector fields G_j, $j = \overline{1, m}$. Formally, the addressed case is substantially different – compared to the case of scalar-valued controls – due to a complex (less intuitive) notion of impulsive control [7] and the associated representation of characteristics. In view of this (though, as one could conclude at the end of this manuscript, the undertaken generalization is not very complicated in the technical sense), the present work essentially broadens the sphere of application of the previously derived results. Finally, we believe that this paper could contribute towards dissemination of the problematics of multi-agent impulsive control systems, and attraction of one's interest to this very promising area of the modern control theory.

2   Notations and Preliminaries

In this paragraph, we collect some necessary notations, definitions, and useful facts from measure theory and functions of bounded variation.

Probability Measures on $\mathbb{R}^n$. The states of our distributed dynamical system are measures on the vector space $\mathbb{R}^n$. Among them, we distinguish the n-dimensional Lebesgue measure denoted by λ^n.

By $\mathcal{P} = \mathcal{P}(\mathbb{R}^n)$ we denote the set of *probability* measures defined on Borel subsets of $\mathbb{R}^n$, and let $\mathcal{P}_1 \subset \mathcal{P}$ be composed of all such measures with a bounded first moment, i.e., such that

$$\int_{\mathbb{R}^n} |\eta|\, d\mu(\eta) < \infty.$$

As is known, $\mathcal{P}_1$ admits a natural structure of a (complete separable [5]) metric space $(\mathcal{P}_1, \varrho)$, while it is endowed with the Kantorovich metric (which is frequently unfairly called "the Wasserstein distance"):

$$\varrho(\mu, \nu) \doteq \sup \left\{ \int_{\mathbb{R}^n} \varphi\, d(\nu - \mu) \,\middle|\, \begin{array}{c} \varphi \in C(\mathbb{R}^n, \mathbb{R}), \\ \mathrm{Lip}(\varphi) \le 1 \end{array} \right\}.$$

Here, $\mathrm{Lip}(f)$ is the minimal Lipschitz constant of f. Recall that the metric ϱ corresponds to the topology of narrow convergence in $\mathcal{P}_1$. In what follows, the convergence of measures in $\mathcal{P}$ is always understood in the narrow sense.

Given a measure $\mu \in \mathcal{P}$ and a Borel measurable map $F : \mathbb{R}^n \mapsto \mathbb{R}^n$, one defines the operator $\mathcal{P} \mapsto \mathcal{P}$, called the *push-forward of μ through F*, by the relation:

$$F_\sharp \mu(E) \doteq \mu\big(F^{-1}(E)\big) \quad \text{for any Borel } E \subseteq \mathbb{R}^n.$$

Recall the following useful properties of the operator $\sharp$ [27]:

(a) for any measurable function $\varphi : \mathbb{R}^n \mapsto \mathbb{R}$, Borel measurable F, and $\mu \in \mathcal{P}$, one has:

$$\int_{\mathbb{R}^n} \varphi(x)\, dF_\sharp \mu(x) = \int_{\mathbb{R}^n} (\varphi \circ F)(x)\, d\mu(x);$$

(b) the continuity of F implies the continuity of $F_\sharp$, and

(c) given $\mu \in \mathcal{P}$, a map $F : \mathbb{R}^n \mapsto \mathbb{R}^n$ and a family of *continuous* maps $F^k : \mathbb{R}^n \mapsto \mathbb{R}^n$, $k \ge 0$, such that $F^k(x) \to_{k \to \infty} F(x)$ for μ-almost all $x \in \mathbb{R}^n$, it holds that $F^k_\sharp \mu \to_{k \to \infty} F_\sharp \mu$; furthermore, the pointwise convergence $F^k \to_{k \to \infty} F$ implies that $F^k_\sharp \mu \to_{k \to \infty} F_\sharp \mu$, for any $\mu \in \mathcal{P}$.

Functions of Bounded Variation. Let $\mathcal{X} \doteq (\mathcal{X}, d)$ be a metric space. Given a closed interval $\mathcal{I} \doteq [\underline{t}, \overline{t}] \subseteq \mathbb{R}$, a function $\mathcal{F}_{(\cdot)} : \mathbb{R} \ni t \mapsto \mathcal{F}_t \in X$ is said to have bounded variation on $\mathcal{I}$, if

$$\mathrm{Var}_{\mathcal{I}}\, \mathcal{F}_{(\cdot)} \doteq \sup \sum_{i=1}^{\mathrm{card}(\pi)-1} d\big(\mathcal{F}_{t_i}, \mathcal{F}_{t_{i+1}}\big) < \infty,$$

where sup is taken over all finite partitions $\pi = \{t_i\} \subset \mathcal{I}$, $t_i < t_{i+1}$, of the interval $\mathcal{I}$. The set of functions $\mathcal{I} \mapsto \mathcal{X}$ of bounded variation ("*BV*-functions") is denoted by $BV(\mathcal{I}, \mathcal{X})$. Note that, for any *BV*-function $\mathcal{F}_{(\cdot)} : \mathcal{I} \doteq [\underline{t}, \overline{t}] \ni t \mapsto \mathcal{F}_t \in \mathcal{X}$, the set $\Delta_{\mathcal{F}} \subset \mathcal{I}$ of its discontinuity points ("jump points") is finite or countable.

The set of all $BV(\mathcal{I}, \mathcal{X})$-functions, which are *right continuous* on $[\underline{t}, \overline{t})$, is denoted by $BV^+(\mathcal{I}, \mathcal{X})$.

**Vector Measures on the Real Line and Functions of Bounded Varia-
tion.** In the finite-dimensional case, namely, if $\mathcal{X} = \mathbb{R}^n$, the set of BV^+-functions
is notorious to be isomorphic to the space of so-called Lebesgue-Stieltjes mea-
sures, which are defined as (the unique) extensions of vector-valued Borel measures on $\mathcal{I}$. In other words, any such a measure U can be viewed as a generalized derivative of a function F_U of bounded variation, and the correspondence
$U \leftrightarrow F_U$ is one-to-one. Therefore, in this case, one can look at $BV^+(\mathcal{I}, \mathbb{R}^n)$ as
at the dual $C^*(\mathcal{I}, \mathbb{R}^n)$ of the space of continuous functions $\mathcal{I} \mapsto \mathbb{R}^n$ (with the
topology of uniform convergence).

The space $BV^+ \doteq BV^+(\mathcal{I}, \mathbb{R}^n)$ has a natural structure of the Banach space
with a norm, defined by the total variation; this norm corresponds to the standard, strong topology of BV^+. Another useful topology (the one, we shall employ
throughout the remaining part of our paper) can be defined through the convergence $\rightharpoonup$ at all continuity points of the limit function and at the boundary
points of the interval $\mathcal{I}$. This topology is weaker than the natural topology of
the norm, while (for the case of signed measures) it is stronger than the weak*
topology of C^*—the topology of narrow convergence of the respective measures.

Vector measures on $\mathbb{R}$ play an important role in our further investigation. Any
$\mathbb{R}^n$-valued measure U admits a unique Lebesgue decomposition $U = U_c + U_d \doteq
U_{ac} + U_{sc} + U_d$, where the components U_{ac} and U_{sc} are the absolutely continuous
and singular continuous (with respect to $\lambda = \lambda^1$) parts, respectively, while U_d
is a discrete (purely atomic) measure being a series of Dirac type point-mass
distributions concentrated on an at most countable set; Δ_U stands for the set of
atoms of U, and the total variation of U is denoted by $|U|$.

3   An Insight to Finite-Dimensional Impulsive Control

In this section, we recall the needed mathematical framework related to the
theory of finite-dimensional impulsive systems.

Given a time interval $\mathcal{T} \doteq [0, T]$, a real $M > 0$, a compact set $W \subset \mathbb{R}^k$ and
functions $f : \mathbb{R}^n \mapsto \mathbb{R}^n$ and $G : \mathbb{R}^n \mapsto \mathbb{R}^{n \times m}$, consider the following control
system on $\mathbb{R}^n$:

$$\dot{x} = v(x, w, u) \doteq f(x, w) + G(x, w)\, u, \qquad t \in \mathcal{T}, \tag{3}$$

whose inputs $(w, u) : \mathcal{T} \mapsto \mathbb{R}^k \times \mathbb{R}^m$ subject to the constraints:

$$w \in \mathcal{W}(\mathcal{T}) \doteq \left\{ \omega \in L_\infty : \ \omega(t) \in W \ \lambda\text{-a.e. on } \mathcal{T} \right\}, \tag{4}$$

$$u \in \mathcal{U} = \mathcal{U}(M) \doteq \left\{ v \in L_\infty : \ \|v\|_{L_1} \leq M \right\}. \tag{5}$$

From now on, "U-a.e." abbreviates "almost everywhere with respect to a given
measure U".

As is known (and simply checked), the tube of admissible arcs of (3), (4)
and (5) fails to be closed in C due to the affine dependence on u, which is

"unbounded". This, in particular, implies the generic ill-posedness of associated optimization problems. The latter fact brings us to the problem of trajectory relaxation (compactification of the tube of admissible arcs in an appropriate weak topology), which is one of the actual theoretical roots of the impulsive control theory.

Under some typical convexity assumptions to be introduced below, the desired relaxation can be designed in the coarse topology of the space BV^+, defined by the convergence at continuity points and at the boundary points of the interval $\mathcal{I}$,—through extending the notions of control input and the associated state. An obvious (naive but rather profitable) idea is to extend the original set $\mathcal{U}(M)$ up to the set of $\mathbb{R}^m$-valued measures U, whose total variations $|U|$ are uniformly bounded by M, through the formal embedding $u \to u\,\lambda$. Following this idea, Eq. (3) can be mechanically rewritten as a measure differential equation (MDE):

$$dx = f(x, w)\, dt + G(x, w)\, U(dt), \tag{6}$$

and the constraint $\|u\|_{L_1} \le M$ turns into

$$|U|(\mathcal{T}) \le M. \tag{7}$$

One immediately observes that Eq. (6) does not make a literal sense, since it contains a product of a generalized function and a discontinuous (just measurable) function, and its actual meaning requires specification. In fact, (6) performs a correct mathematical object only if U is continuous, and – with a slight abuse of notation – still remains somehow sensible in the case, when the matrix function G is independent of w, while it enjoys the well-known Frobenius commutativity property [29]. In general, unlike the case of scalar-valued inputs, the presently addressed model requires a more accurate understanding of impulsive control and a particular concept of the solution to MDE (6).

Impulsive Controls. Following [6,7], by impulsive control of system (6), (7) we agree to mean a collection

$$\mathbf{u} \doteq (U, V, \{w_\tau, u_\tau\}_{\tau \in \Delta_V}),$$

where U and V are Borel measures of dimensions m and 1, respectively, such that

$$|U|(B) \le V(B) \ \ \forall \text{ Borel } B \subseteq \mathcal{T}, \quad |U|_c = V_c, \quad V(\mathcal{T}) \le M; \tag{8}$$

$w_\tau : \mathcal{T}_\tau \mapsto \mathbb{R}^k$ and $u_\tau : \mathcal{T}_\tau \mapsto \mathbb{R}^m$, $\mathcal{T}_\tau \doteq [0, T_\tau]$, $T_\tau \doteq V(\{\tau\})$, are Borel functions, parameterized by atoms of V, such that, for all $\tau \in \Delta_V$, it holds

$$\int_{\mathcal{T}_\tau} u_\tau(\varsigma)\, d\varsigma = U(\{\tau\}), \tag{9}$$

$$w_\tau(\varsigma) \in W \text{ and } |u_\tau(\varsigma)| = 1 \quad \lambda\text{-a.e. on } \mathcal{T}_\tau. \tag{10}$$

The total variation $|\mathbf{u}|$ of an impulsive control $\mathbf{u}$ is defined formally as $|\mathbf{u}| \doteq V$.

Let $\mathbf{U} \doteq \mathbf{U}(M)$ be the set of all $\mathbf{u}$ enjoying (8)–(10). The set $\mathbf{U}$ can be turned into a metric space while endowed with a specific distance function d like in [6]. Then, the input-output map $\mathbf{u} \mapsto x[\mathbf{u}]$ is proved to be continuous as a function $(\mathbf{U}, d) \mapsto BV^+(\mathcal{T}, \mathbb{R}^n)$.

Measure Differential Equations. We make the following (restrictive but rather standard) regularity assumptions (**H**):

(1) There is a constant $L > 0$ such that

$$\big|f(x,w) - f(y,w)\big| + \|G(x,w) - G(y,w)\| \leq L\,|x - y|,$$

for all $x, y \in \mathbb{R}^n$ and $w \in W$; here, $|\cdot|$ and $\|\cdot\|$ are fixed (agreed) vector and matrix norms.

(2) The set $\{\alpha f(x,w) + G(x,w)\beta : (\alpha, \beta) \in K,\ w \in W\}$ is convex in $\mathbb{R}^n$ for all $x \in \mathbb{R}^n$. Here,

$$K \doteq \big\{(\alpha, \beta) \in \mathbb{R}^{1+m}\,|\, \alpha \geq 0,\ \alpha + |\beta| \leq 1\big\}. \tag{11}$$

By a solution to the MDE (6) with the initial condition $x(0^-) = \eta \in \mathbb{R}^n$ ($x(t^-)$ always denotes the left one-sided limit of a function x at a point t) under an impulsive control $\mathbf{u} \doteq (U, V, \{w_\tau, u_\tau\}_{\tau \in \Delta_V}) \in \mathbf{U}$, we mean a function $t \mapsto X^t[\mathbf{u}](\eta)$ of the class $BV^+(\mathcal{T}, \mathbb{R}^n)$, turning into identity the following equation:

$$x(t) = \eta + \int_0^t f\big(x(\varsigma), w(\varsigma)\big)\, d\varsigma + \int_0^t G\big(x(\varsigma), w(\varsigma)\big)\, dU_c(\varsigma)$$

$$+ \sum_{\tau \in [0,t],\ \tau \in \Delta_V} \Big[\varkappa_\tau(T_\tau) - x(\tau^-)\Big]. \tag{12}$$

Here, for each $\tau \in \Delta_V$, the function $\varsigma \mapsto \varkappa_\tau(\varsigma)$ is the solution to the following Cauchy problem for the auxiliary ODE, called the "limit system":

$$\frac{d}{d\varsigma}\varkappa(\varsigma) = G\big(\varkappa(\varsigma), w_\tau(\varsigma)\big)\, u_\tau(\varsigma), \quad \varkappa(0) = x(\tau^-). \tag{13}$$

Note that the adapted concept of solution to MDE is correct, i.e., under assumptions (**H**), a solution defined this way does exist and is unique [23], while if the measure U is absolutely continuous with respect to λ, then (12) reduces to (3).

Discontinuous Time Change. The main arguments behind the accepted optimal control framework are due to the well-known technical trick called the discontinuous time reparameterization approach, see, e.g. [10, 23, 30, 33]. This technique suggests an equivalent (in the sense to be specified just below) transformation of our MDE to an ordinary control system of the form:

$$\xi' \doteq \frac{d}{ds}\xi(s) = \alpha(s), \tag{14}$$

$$y' = \hat{v}_s(y) \doteq \hat{v}\big(y, \omega(s), \alpha(s), \beta(s)\big) \doteq \alpha(s) f\big(y, \omega(s)\big) + G\big(y, \omega(s)\big)\beta(s), \quad (15)$$

subject to

$$\omega \in \mathcal{W}(\mathcal{S}); \quad (\alpha, \beta) \in \mathcal{A}, \quad (16)$$

where $\mathcal{W}(\mathcal{S})$ is defined as above due to replacing $\mathcal{T}$ with $\mathcal{S} \doteq [0, S]$, $T \leq S \leq T + M$,

$$\mathcal{A} = \mathcal{A}(S, T) \doteq \left\{ (\alpha, \beta) \in L_\infty(\mathcal{S}, \mathbb{R}^{1+m}) \; \middle| \; \begin{array}{c} (\alpha, \beta)(s) \in K \;\; \lambda\text{-a.e. on } \mathcal{S} \\ \|\alpha\|_{L_1} = T \end{array} \right\},$$

and K is given by (11).

Given $(w, u) \in \mathcal{W}(\mathcal{T}) \times \mathcal{U}$ and an initial position $x(0) = \eta$, one embeds (3)–(5) into the ODE (14)–(16) with $(\xi, y)(0) = (0, \eta)$ by defining controls $\omega \in \mathcal{W}(\mathcal{S})$ and $(\alpha, \beta) \in \mathcal{A}$ through the formulas:

$$\omega = w \circ \xi; \quad (\alpha(s), \beta(s)) \doteq \left(\frac{1}{1 + |u(t)|}, \frac{u(t)}{1 + |u(t)|} \right) \Bigg|_{t = \xi(s)} \quad \lambda\text{-a.e. on } \mathcal{S}, \quad (17)$$

where $\xi = \xi[\alpha] : \mathcal{S} \mapsto \mathcal{T}$ is a solution to the Cauchy problem $\xi' = \alpha(s)$, $\xi(0) = 0$, and $\circ$ stands for the composition of functions. One observes that, as soon as (α, β) is of the form (17), the respective $\xi = \xi[\alpha]$ is absolutely continuous and strictly increasing (so, its inverse $t \mapsto \xi^{-1}(t)$ is correctly defined). Meanwhile, taken an arbitrary control $(\alpha, \beta) \in \mathcal{A}$, the associated trajectory component $\xi = \xi[\alpha]$ is not necessarily strictly monotone and its inverse ξ^{-1} does not, generally, exist. We define the *pseudo-inverse* of ξ as

$$\xi^{\leftarrow}(t) = \inf\big\{ s \in \mathcal{S} : \xi(s) > t \big\}, \; t \in [0, T); \quad \xi^{\leftarrow}(T) = S.$$

The function $t \mapsto \xi^{\leftarrow}(t)$ can be discontinuous, while it is still strictly increasing, right continuous on $[0, T)$ and has bounded variation on $\mathcal{T}$.

Note that, under assumption (**H**), the Eq. (15) has a solution $s \mapsto Y_\theta^s[\omega; \alpha, \beta](\eta)$ globally defined on $\mathcal{S}$, for any initial condition $\eta \in \mathbb{R}^n$ and any input $(\omega; \alpha, \beta) \in \mathcal{W}(\mathcal{S}) \times \mathcal{A}$.

Here are some facts clarifying the connection between systems (12) and (14), (15) under the assumption (**H**):

(i) Given $\eta \in \mathbb{R}^n$, let $t \mapsto X^t(\eta)$ be a "$\rightharpoonup$"-limit in $BV^+(\mathcal{T}, \mathbb{R}^n)$ of a sequence $\{t \mapsto X^t[w^k, u^k](\eta)\}_{k \geq 0}$ of solutions to the ODE (15), produced by a control sequence $\{(w^k, u^k)\}_{k \geq 0} \subset \mathcal{W}(\mathcal{T}) \times \mathcal{U}$. Then, there exists (see [23, Theorem 2.13]) a control $(\omega; \alpha, \beta) \in \mathcal{W}(\mathcal{S}) \times \mathcal{A}$ such that the following relation holds:

$$X^t(\eta) = Y^{\xi^{\leftarrow}[\alpha](t)}[\omega; \alpha, \beta](\eta) \text{ for all } t \in \mathcal{T}. \quad (18)$$

(ii) By the same arguments as in [23, Lemma 5.2], for any $(\omega; \alpha, \beta) \in \mathcal{W}(\mathcal{S}) \times \mathcal{A}$, there exists a sequence of ordinary controls $\{w^k, u^k\}_{k \geq 0} \subset \mathcal{W}(\mathcal{T}) \times \mathcal{U}$ such that, for any $\eta \in \mathbb{R}^n$, the function $t \mapsto X^t(\eta)$, defined by (18), is approximated in the topology of "$\rightarrow$"-convergence by the sequence $\{X^{(\cdot)}[w^k, u^k](\eta)\}_{k \geq 0}$ of the associated solutions to the ODE (3), (4) and (5) with the initial condition $x(0) = \eta$, i.e. $X^t[w^k, u^k](\eta) \rightarrow_{k \to \infty} X^t(\eta)$; the choice of the approximating sequence is independent of η.

(iii) Similarly to [23, Theorem 4.7], one can prove that, for any $(\omega; \alpha, \beta) \in \mathcal{W}(\mathcal{S}) \times \mathcal{A}$, there exists an impulsive control $\mathbf{u} \doteq \mathbf{u}[\omega; \alpha, \beta] \in \mathbf{U}$ such that, for all $\eta \in \mathbb{R}^n$, the map $t \mapsto X^t(\eta)$, defined by (18), coincides with the solutions $t \mapsto X^t[\mathbf{u}](\eta)$ to the MDE (6) in the sense (12).

(iv) For any $\mathbf{u} \in \mathbf{U}$, one can define (see, e.g., [6, Proof of Theorem 3.1]) a control $(\omega; \alpha, \beta) \doteq (\omega; \alpha, \beta)[\mathbf{u}] \in \mathcal{W}(\mathcal{S}) \times \mathcal{A}$ such that, for all $\eta \in \mathbb{R}^n$, the solution $t \mapsto X^t(\eta) \doteq X^t[\mathbf{u}](\eta)$ to (12) meets relation (18).

4   Impulsive Continuity Equation

Now we return to the framework of multi-agent systems.

4.1   Prototypic Continuity Equation

We study the time evolution of a given measure $\vartheta \in \mathcal{P}_1(\mathbb{R}^n)$ under the action of the vector field

$$v_t(x) \doteq v\big(x, w(t), u(t)\big)$$

with v having the form (2), and inputs (w, u) satisfying (4), (5). This evolution can be described by a solution $t \mapsto \mu_t$ of the continuity equation

$$\mu_0 = \vartheta; \quad \partial_t \mu_t + \nabla \cdot (\mu_t v_t) = 0, \quad t \in \mathcal{T}, \tag{19}$$

which, as usual, should be understood in the weak sense, i.e., as the equality

$$\int_{\mathcal{T}} \int_{\mathbb{R}^n} \big(\partial_t \varphi(\eta) + \langle f_t(\eta), \nabla_\eta \varphi(t, \eta) \rangle\big) \, d\mu_t(\eta) \, dt = 0$$

to be hold for all smooth compactly supported test functions $\varphi : (0, T) \times \mathbb{R}^n \mapsto \mathbb{R}$. The set of weak solutions to (4), (5) and (19) is denoted by $\mathcal{M} \doteq \mathcal{M}(\mathcal{U})$.

As is known, the (weak) solution of the continuity equation can be constructed using characteristics being solutions to the ODE (3): given $(w, u) \in \mathcal{W}(\mathcal{T}) \times \mathcal{U}$, $\eta \in \mathbb{R}^n$ and $\theta \in [0, T)$, let $t \mapsto X_\theta^t[w, u](\eta)$, $t \in [\theta, T]$, denote the solution to (3) with the initial condition $x(\theta) = \eta$. Then, for any narrowly continuous family $\{\mu_t\}_{t \in \mathcal{T}} \subset \mathcal{P}_1$ such that $t \mapsto \mu_t$ solves (19) under the input (w, u), one has:

$$\mu_t = \big(X^t[w, u]\big)_\sharp \vartheta \quad \forall t \in \mathcal{T}, \tag{20}$$

where $X^t[w, u](\eta) \doteq X_0^t[w, u](\eta)$.

4.2   Relaxation of the Continuity Equation

As in the finite-dimensional case, the relaxation of continuity equation (19) relies on the notion [31] of *generalized solution*, which is somehow analogous to [23]:

Definition 1. *A map $\mu_{(\cdot)} \in BV(T, \mathcal{P}_1)$ is called a generalized solution of (4), (5) and (19) iff it is right continuous on $[0, T)$, and there exists a sequence $\{\mu_{(\cdot)}^k\}_{k \geq 0} \subset \mathcal{M}$ of solutions to the continuity equation (4), (5) and (19), such that $\mu_t^k \to_{k \to \infty} \mu_t$ at all continuity points $t \in T$ and at $t = T$ (similar to the finite dimensional case, we write $\mu^k \to_{k \to \infty} \mu$ to identify this type of convergence).*

The reasoning for this definition is clarified by the following

Proposition 1. *Given a sequence $(w^k, u^k)_{k \geq 0} \in \mathcal{W}(T) \times \mathcal{U}$, let $\{\mu_{(\cdot)}^k\}_{k \geq 0} \subset C(T, \mathcal{P})$ be the family of the associated solutions to the continuity equation (19). Then, the convergence $u^k \lambda \to_{k \to \infty} \mathbf{u}$ in $(\mathbf{U}, d)$ to some impulsive control $\mathbf{u} \doteq (U, V, \{w_\tau, u_\tau\}) \in \mathbf{U}$ implies the existence of a map $\mu \in BV^+(T, \mathcal{P})$ such that $\mu_t^k \to \mu_t$ at all continuity points $t \in T$ and at $t = T$.*

The proof of this theorem, being a simple combination of arguments from [6] and [31], is skipped in order to avoid repetitions.

4.3   Discontinuous Time Reparameterization of the Continuity Equation

The main idea behind our approach is to adapt for the continuity equation (19) the same trick, that we used to recruit in the finite dimensional case, namely, the discontinuous time reparameterization. In fact, one can reduce (19) to the following continuity equation with a *bounded* velocity field $s \mapsto \hat{v}_s(y)$ defined as in (15) for inputs $(\omega; \alpha, \beta) \in \mathcal{W}(\mathcal{S}) \times \mathcal{A}$:

$$\nu_0 = \vartheta; \quad \partial_s \nu_s + \nabla \cdot (\nu_s \, \hat{v}_s) = 0, \quad s \in \mathcal{S}. \tag{21}$$

Again, given a narrowly continuous family $\{\nu_s\}_{s \in \mathcal{S}} \subset \mathcal{P}_1$ such that $s \mapsto \nu_s$ is the weak solution of (21), one has the representation:

$$\nu_s = (Y^s)_\sharp \, \vartheta \quad \text{for all } s \in \mathcal{S},$$

where $Y^s(\eta) \doteq Y_0^s[\omega; \alpha, \beta](\eta)$, and $Y_0^s[\omega; \alpha, \beta](\eta)$ satisfies the characteristic system (15) with the initial condition $y(\theta) = \eta$. Given $(\omega; \alpha, \beta)$ defined by (17) and

$$\hat{v} \doteq \xi' \, (v \circ \xi),$$

where $\xi = \xi[\alpha]$ is a solution to (14) with $\xi(0) = 0$, one concludes, by [5, Lemma 2.7], that the pair (μ_t, v_t) satisfies (19) in the distributional sense if $(\nu_s, \hat{v}_s)$ satisfies (21), and

$$\mu_t = \nu_{\xi^{-1}(t)}. \tag{22}$$

Thus, the transformed control continuity equation (16), (21) describes solutions to (4), (5) and (19), up to a suitable change of variable. However, the transformed system is richer than the original one: in fact, it represents all generalized solutions of (4), (5) and (19).

4.4   Relaxed Model and Its Representation

We proceed with an auxiliary result, which generalizes [31, Lemmas 2, 3], while is proved by the very same arguments.

Lemma 1. *Assume* (**H**). *Then, for any* $\mathbf{u} \in \mathbf{U}$,

- *for all* $t \in \mathcal{T}$, *the function* $\mathbb{R}^n \ni \eta \mapsto X^t[\mathbf{u}](\eta) \in \mathbb{R}^n$ *is Lipschitz continuous, and* $\mathrm{Lip}\big(X^t[\mathbf{u}](\cdot)\big) \le e^{L(T+M)}$;
- *the function* $\mathbb{R}^n \ni \eta \mapsto \mathrm{Var}_{\mathcal{T}} X^{(\cdot)}[\mathbf{u}](\eta) \in \mathbb{R}_+$ *is Lipschitz continuous with a Lipschitz constant depending only on the data* L, T *and* M.

Corollary 1. *Assume that* (**H**) *hold and let* $\vartheta \in \mathcal{P}_1(\mathbb{R}^n)$. *Then, for any* $\mathbf{u} \in \mathbf{U}$, *the function* $\eta \mapsto \mathrm{Var}_{\mathcal{T}} X^{(\cdot)}[\mathbf{u}](\eta)$ *belongs to* $L_1(\mathbb{R}^n, \mathbb{R}; \vartheta)$.

Given $(\omega; \alpha, \beta) \in \mathcal{W}(\mathcal{S}) \times \mathcal{A}$, we introduce the map $\mathcal{T} \ni t \mapsto \mu_t \doteq \mu_t[\omega; \alpha, \beta] \in \mathcal{P}_1$ as

$$\mu_t[\omega; \alpha, \beta] \doteq \nu_{\xi^{\leftarrow}(t)} \doteq \left(Y^{\xi^{\leftarrow}[\alpha](t)}[\omega; \alpha, \beta]\right)_\sharp \vartheta, \quad t \in \mathcal{T}, \tag{23}$$

and denote

$$\overline{\mathcal{M}} \doteq \big\{\mu_{(\cdot)}[\omega; \alpha, \beta] : (\omega; \alpha, \beta) \in \mathcal{W}(\mathcal{S}) \times \mathcal{A}\big\}.$$

In view of representation (12), one deduces the following

Proposition 2. $\overline{\mathcal{M}} = \{\mu_{(\cdot)}[\mathbf{u}] | \mathbf{u} \in \mathbf{U}\}$, *where measure-valued functions* $\mu_{(\cdot)} \doteq \mu_{(\cdot)}[\mathbf{u}]$ *are defined through the relation:*

$$\mu_t = \big(X^t[\mathbf{u}]\big)_\sharp \vartheta, \quad t \in \mathcal{T}, \tag{24}$$

with $t \mapsto X^t[\mathbf{u}](\eta)$ *being solutions of* (12) *with the input* $\mathbf{u}$ *and the initial state* $x(0^-) = \eta \in \mathbb{R}^n$.

Proposition 3. *Let* $\mu_{(\cdot)} = \mu_{(\cdot)}[\mathbf{u}] \in \overline{\mathcal{M}}$, $\mathbf{u} \in \mathbf{U}$. *Then,*
 (1) $\mu_{(\cdot)}$ *belongs to the class* $BV^+(\mathcal{T}, \mathcal{P})$, *i.e. it is right continuous on* $[0, T)$ *and has bounded variation; furthermore, the total variation of* $\mu_{(\cdot)}$ *on* $\mathcal{T}$ *is estimated as*

$$\mathrm{Var}_{\mathcal{T}} \mu_{(\cdot)}[\mathbf{u}] \le \int_{\mathbb{R}^n} \mathrm{Var}_{\mathcal{T}} X^{(\cdot)}[\mathbf{u}](\eta) \, d\vartheta(\eta);$$

(2) Discontinuity points of the measure-valued map $\mu_{(\cdot)}[\mathbf{u}]$ *on* $(0, T)$ *are concentrated within the set* $\Delta_{|\mathbf{u}|} \doteq \Delta_V$.

Corollary 2. *One has the uniform estimate:*

$$\mathrm{Var}_{\mathcal{T}} \mu_{(\cdot)}[\mathbf{u}] \le CS\big[1 + e^{CS}\big(1 + \mathrm{m}_1(\vartheta)\big)\big] \quad \forall \, \mathbf{u} \in \mathbf{U}.$$

The following assertion, which is a direct extension of the main result of [31], proposes a constructive representation of generalized solutions through a finite-dimensional impulsive system—the measure differential equation (6).

Theorem 1. *Assume that* (**H**) *holds. The set of all generalized solutions of* (4), (5) *and* (19), *coincides with* $\overline{\mathcal{M}}$.

The proof of this result is, again, a technical modification of the proof of [31, Theorem 1]. To avoid repetitions, we drop the technicalities and present only the main arguments.

A simple part is the implication "$\mu_{(\cdot)} \in \overline{\mathcal{M}} \Rightarrow \mu_{(\cdot)}$ is a generalized solution". The existence of a desired sequence of processes $(\mu^k_{(\cdot)}; u^k, w^k)$ of the prototypic distributed control system (4), (5) and (19), immediately follows from the properties (iv) and (ii) of the characteristic systems, presented in the concluding part of Sect. 3, representations (20) and (24), and the property (c) of the operator $\sharp$; the right continuity of $t \mapsto \mu_t$ is implied by Proposition 3.

To prove the assertion "$\mu_{(\cdot)}$ is a generalized solution $\Rightarrow \mu_{(\cdot)} \in \overline{\mathcal{M}}$", one considers an approximating sequence $(\mu^k_{(\cdot)}; u^k, w^k)$ from the definition along with its time-reparameterized counterpart $(\nu^k_{(\cdot)}; \omega^k, \alpha^k, \beta^k)$. By the Arzela-Ascoli selection principle, one extracts from $(\omega^k, \alpha^k, \beta^k)$ a subsequence weakly converging in $L_1(\mathcal{S}, \mathbb{R}^{1+m+k})$ to some $(\omega, \alpha, \beta) \in \mathcal{W}(\mathcal{S}) \times \mathcal{A}$. The respective solutions to the reduced characteristic system (15) then converge pointwise to $Y^{(\cdot)}[\eta] = Y^{(\cdot)}[\omega, \alpha, \beta; \eta]$, the solution of (15) under the input (α, β, ω) with the initial condition $y(0) = \eta$, for any $\eta \in \mathbb{R}^n$. By property (iii) of the characteristic systems, the discontinuous time reparameterization of $Y^{(\cdot)}[\eta]$ gives a solution to the measure-driven system (6) under some admissible impulsive control **u**, common for all η. Thanks to representation (20) and the property (c) of $\sharp$, one asserts that the pushforward of the initial measure ϑ along the flow, generated by the time-reparameterized characteristics $Y^{(\cdot)}[\eta]$, is also a generalized solution. Finally, due to the density of the continuity points of $t \mapsto \mu_t$ in $\mathcal{T}$, one proves that the above defined pushforward measure coincides with $\mu_{(\cdot)}$, i.e., $\mu_{(\cdot)}$ does admit representation (24). By Proposition 2, $\mu_{(\cdot)} \in \overline{\mathcal{M}}$, as desired.

4.5   Extremal Problem

As an application of the exhibited results, we present an impulsive extension of an optimal ensemble control problem [21,27]. Given a cost function $\ell : \mathbb{R}^n \mapsto \mathbb{R}$, consider the following optimization problems:

$$(P) \quad \int_{\mathbb{R}^d} \ell(x) \, d\mu_T(x) \to \inf \ \text{over solutions } t \mapsto \mu_t \text{ of system } (4), (5), (19)$$

and

$$(\overline{P}) \quad \int_{\mathbb{R}^d} \ell(x) \, d\mu_T(x) \to \min, \quad \mu_{(\cdot)} \in \overline{\mathcal{M}}.$$

One of prominent practical interpretations here is the problem of focusing the crowd in a given target area (when ℓ is a characteristic function of this area, or its approximation).

Theorem 2. *Assume that* (**H**) *hold, and the function ℓ is lower semicontinuous. Then, the minimum in* $(\overline{P})$ *is achieved, and* $\inf(P) = \min(\overline{P})$.

In other words, $(\overline{P})$ is a relaxation of (P) in the sense that the problems coincide, as soon as they are stated "on minimizing sequences", while the solution of $(\overline{P})$ does exist in the respective class of admissible inputs. The assertion is, in fact, an immediate corollary of Theorem 1.

5   Conclusions and Open Problems

The main result of this paper can be summarized as follows: the proposed trajectory relaxation and the superposition operator, actually, "commute" in the sense that, on the one hand, we can first relax the characteristic system and then make the superposition of the relaxed characteristics resulting in a relaxed solution of the continuity equation or, alternatively, one can relax directly the continuity equation, and then represent the solution though the pushforward along the flow of the relaxed ODE.

One observes that the proposed generalization of the results [31] (i.e. their extension to the case of vector-valued controls) is rather straightforward. We shall mention two natural directions for future research, which are not so simple. The first issue concerns the optimal control problems similar to $(\overline{P})$. Here, the primal question (whose elaboration is, by now, rather plain for us) is the necessary optimality condition of the type of Pontryagin Maximum Principle. Another very natural and challenging problem is due to modeling rather than optimization, and this is a point of the actual connection with the area of mean-field control. Recall that our present model (based on the continuity equation) assumes that the individuals do not interact. Now, we wonder, how to describe the behavior of the ensemble in the case, when the agents do influence one another. This kind of models was recently raised by [20, 32].

Acknowledgements. The authors are grateful to Krzysztof Konewski for inspiration and careful attention to this work.

References

1. Ahmed, N.U.: Vector and operator valued measures as controls for infinite dimensional systems: optimal control. Discuss. Math. Differ. Incl. Control Optim. **28**, 95–131 (2008)
2. Ahmed, N.U., Teo, K.L., Hou, S.H.: Nonlinear impulsive systems on infinite dimensional spaces. Nonlinear Anal. **54**(5), 907–925 (2003)
3. Ambrosio, L.: Metric space valued functions of bounded variation. Ann. Scuola Norm. Sup. Pisa Cl. Sci. **17**(3), 439–478 (1990)
4. Ambrosio, L., Fusco, N., Pallara, D.: Functions of Bounded Variation and Free Discontinuity Problems. Oxford Mathematical Monographs. The Clarendon Press, Oxford University Press, New York (2000)
5. Ambrosio, L., Savaré, G.: Gradient flows of probability measures. In: Handbook of Differential Equations: Evolutionary Equations, vol. III, pp. 1–136. Elsevier/North-Holland, Amsterdam (2007)

6. Arutyunov, A.V., Karamzin, D.Y., Pereira, F.L.: On constrained impulsive control problems. J. Math. Sci. **165**(6), 654–688 (2010). https://doi.org/10.1007/s10958-010-9834-z
7. Arutyunov, A., Karamzin, D., Pereira, F.L.: On a generalization of the impulsive control concept: controlling system jumps. Discrete Contin. Dyn. Syst. **29**(2), 403–415 (2011). https://doi.org/10.3934/dcds.2011.29.403
8. Bentsman, J., Miller, B.M.: Dynamical systems with active singularities of elastic type: a modeling and controller synthesis framework. IEEE Trans. Automat. Control **52**(1), 39–55 (2007). https://doi.org/10.1109/TAC.2006.887899
9. Bressan Jr., A., Rampazzo, F.: Impulsive control systems without commutativity assumptions. J. Optim. Theory Appl. **81**(3), 435–457 (1994). https://doi.org/10.1007/BF02193094
10. Bressan, A., Rampazzo, F.: On differential systems with quadratic impulses and their applications to Lagrangian mechanics. SIAM J. Control Optim. **31**(5), 1205–1220 (1993). https://doi.org/10.1137/0331057
11. Brockett, R.: Notes on the control of the Liouville equation. In: Alabau-Boussouira, F., Brockett, R., Glass, O., Le Rousseau, J., Zuazua, E. (eds.) Control of Partial Differential Equations. LNM, vol. 2048, pp. 101–120. Springer, Heidelberg (2012). https://doi.org/10.1007/978-3-642-27893-8_2
12. Colombo, R.M., Garavello, M., Lécureux-Mercier, M., Pogodaev, N.: Conservation laws in the modeling of moving crowds. In: Hyperbolic Problems: Theory, Numerics, Applications, AIMS Series in Applied Mathematics, vol. 8, pp. 467–474. American Institute of Mathematical Sciences (AIMS), Springfield (2014)
13. Dykhta, V.A., Samsonyuk, O.N.: Optimalnoe impulsnoe upravlenie s prilozheniyami. Fizmatlit "Nauka", Moscow (2000)
14. Fornasier, M., Solombrino, F.: Mean field optimal control. ESAIM Control Optim. Calc. Var. (2014). https://doi.org/10.1051/cocv/2014009
15. Fraga, S.L., Gomes, R., Pereira, F.L.: An impulsive framework for the control of hybrid systems. In: 2007 46th IEEE Conference on Decision and Control, pp. 3202–3207 (2007). https://doi.org/10.1109/CDC.2007.4434895
16. Gurman, V.: Extensions and global estimates for evolutionary discrete control systems. In: Kurzhanski, A., Lasiecka, I. (eds.) Modelling and Inverse Problems of Control for Distributed Parameter Systems. LNCIS, vol. 154, pp. 16–21. Springer, Berlin, Heidelberg (1991). https://doi.org/10.1007/BFb0044479
17. Karamzin, D.Y., de Oliveira, V.A., Pereira, F.L., Silva, G.N.: On the properness of an impulsive control extension of dynamic optimization problems. ESAIM Control Optim. Calc. Var. **21**(3), 857–875 (2015). https://doi.org/10.1051/cocv/2014053
18. Li, J.S.: Ensemble control of finite-dimensional time-varying linear systems. IEEE Trans. Autom. Control **56**(2), 345–357 (2011)
19. Lions, J.L.: Exact controllability, stabilization and perturbations for distributed systems. SIAM Rev. **30**(1), 1–68 (1988)
20. Maltugueva, N., Pogodaev, N., Sorokin, S., Staritsyn, M.: On impulsive control of dynamical systems with network structure modeling the spread of political influence. In: 2018 6th International Conference "Nonlinear Analysis and Extremal Problems", pp. 90–91 (2018)
21. Marigonda, A., Quincampoix, M.: Mayer control problem with probabilistic uncertainty on initial positions. J. Differ. Equ. **264**(5), 3212–3252 (2018). https://doi.org/10.1016/j.jde.2017.11.014
22. Miller, B.M., Rubinovich, E.Y., Bentsman, J.: Singular space-time transformations: towards one method for solving the Painleve problem. J. Math. Sci. **219**(2), 208–219 (2016). https://doi.org/10.1007/s10958-016-3098-1

23. Miller, B.M., Rubinovich, E.Y.: Impulsive Control in Continuous and Discrete-Continuous Systems. Kluwer Academic/Plenum Publishers, New York (2003). https://doi.org/10.1007/978-1-4615-0095-7

24. Monteiro-Marques, M.D.P.: Differential Inclusions in Nonsmooth Mechanical Problems: Shocks and Dry Friction. Birkhauser, Boston (1993)

25. Motta, M., Rampazzo, F.: Space-time trajectories of nonlinear systems driven by ordinary and impulsive controls. Differ. Integr. Equ. **8**(2), 269–288 (1995)

26. Ovseevich, A.I., Fedorov, A.K.: Asymptotically optimal control of a simplest distributed system. Dokl. Akad. Nauk **473**(5), 525–528 (2017)

27. Pogodaev, N.: Optimal control of continuity equations. NoDEA Nonlinear Differ. Eq. Appl. **23**(2), 21 (2016)

28. Propoĭ, A.I.: Problems of the optimal control of mixed states. Avtomat. i Telemekh. **3**, 87–98 (1994)

29. Rampazzo, F.: Lie brackets and impulsive controls: an unavoidable connection. In: Differential Geometry and Control (Boulder, CO, 1997), Proceedings of Symposium on Pure Mathematics, vol. 64, pp. 279–296. American Mathematical Society, Providence (1999). https://doi.org/10.1090/pspum/064/1654552

30. Rishel, R.W.: An extended Pontryagin principle for control systems whose control laws contain measures. J. Soc. Indust. Appl. Math. Ser. A. Control **3**, 191–205 (1965)

31. Staritsyn, M.: On "discontinuous" continuity equation and impulsive ensemble control. Syst. Control Lett. **118**, 77–83 (2018)

32. Staritsyn, M., Maltugueva, N., Pogodaev, N., Sorokin, S.: Impulsive control of systems with network structure describing spread of political influence. Izvestiya Irkutskogo Gosudarstvennogo Universiteta. Seriya "Matematika" (Bulletin of Irkutsk State Univ. Ser. "Math.") **25**, 126–143 (2018). (in Russian)

33. Warga, J.: Variational problems with unbounded controls. J. Soc. Indust. Appl. Math. Ser. A Control **3**, 424–438 (1965)

34. Zavalishchin, S.T., Sesekin, A.N.: Dynamic Impulse Systems: Theory and applications. Mathematics and Its Applications, p. 394. Kluwer Academic Publishers Group, Dordrecht (1997). https://doi.org/10.1007/978-94-015-8893-5

Optimization in Economy, Finance and Social Sciences

Analysis of Indicators of High-Technology Production Using Optimization Models, Taking into Account the Shortage of Working Capital

Damir Alimov[1,2]([✉]) [iD], Nataliia Obrosova[2,3] [iD], and Alexander Shananin[2,3] [iD]

[1] Lomonosov Moscow State University,
GSP-1, Leninskie Gory, Moscow 119991, Russian Federation
`alimov2007d@gmail.com`
[2] Moscow Institute of Physics and Technology (State University),
9 Institutskiy per., Dolgoprudny, Moscow Region 141701, Russian Federation
`nobrosova@yandex.ru, alexshan@yandex.ru`
[3] Dorodnicyn Computing Centre of FRC CSC RAS,
40 Vavilov st., Moscow 119333, Russian Federation

Abstract. We present a new approach to an estimate of company's value in the high-tech sector, based on the results of research of the mathematical model of production, taking into account the crediting of working capital in an unstable demand. The model is formalized in the form of the Bellman equation, which solution gives an estimate of the enterprise value depending on its performance indicators and external economic conditions. The method develops the income approach to an estimate of company's value and takes into account the specifics of the economic regulation of production in the industry and provides an opportunity for an operative analysis of production indicators when external conditions change. The paper presents the results of analysis of the capitalization dynamics of the large Russian company Kamaz using the developed model. The results of the study of the modified model that takes into account the impact of the accumulated debt burden on the enterprise's indicators are proposed. The solution of the corresponding Bellman equation makes it possible to estimate the magnitude of the critical debt of the company corresponding to the bankruptcy boundary and the debt accounting parameter for the bankruptcy of the company. In the paper, the properties of the company's capitalization in terms of a model that takes into account the debt burden and infrastructure constraint are also given.

Keywords: Bellman equation · Production model
Unstable demand · Current assets deficit · Debt burden
Identification · Capitalization

The research was supported by RSF, project 16-11-10246.

Y. Evtushenko et al. (Eds.): OPTIMA 2018, CCIS 974, pp. 385–398, 2019.
https://doi.org/10.1007/978-3-030-10934-9_27

1   Introduction

Prospects for the development of countries with a catching-up economy largely depend on the success of the modernization processes of enterprises in the high-tech sector. Under the current conditions, the manufacturing sector enterprises in the catching-up economies are at a lower technological level than the industrial development leaders (for ex. Germany, USA, Great Britain, Japan). Products of the sector lose in the domestic market to better import analogs and the enterprises have problems with sales. As a result, manufacturers face a deficit of working capital and the need of advancing them by various forms of credit. Thus, the functioning of production depends on the availability of credit resources. In an unstable financial and economic situation, fluctuations in interest rates may negatively affect the performance of the company and affect its credit rating. The decrease in the credit rating, in turn, causes an increase in interest rates when lending to the company's current assets. There is a trap of high-interest rates [1]. This situation is typical for the industry of a number of Eurozone countries (Italy, Spain, Portugal, Greece), as well as for Russian manufacturers of high-tech products.

One of the most important indicators affecting the credit rating is a company's capitalization. The value of capitalization plays the role of a collateral when a company applies for a loan. However, the value of market capitalization (valuation based on the results of trading in company shares on the exchange) takes into account speculative factors and does not always objectively reflect the real value of the enterprise. Therefore a valuation of the company based on real indicators of its activity is in demand. At present three main approaches to business valuation are recognized: income (methods for estimating the company's expected revenues), costly (assessing the market value of the company's assets or the costs of building a new comparable facility) and comparative (comparative analysis of the valuation object with similar objects). All these approaches have certain drawbacks. The values obtained by different methods diverge with each other. Therefore a combination of approaches is often used, and the choice of the final variant is determined subjectively by experts, taking into account the type of activity of the company and its position on the market. A common drawback of the classical approaches is that they initially do not take into account the listed features of the activities of enterprises in the high-tech sector, so they require individual adjustments by experts for each particular enterprise.

Classical methods do not allow to take into account promptly the changing of market conditions in which the company operates (for example, the structure of prices for raw materials and products, interest rates on borrowed funds of the company), and also do not allow to reflect promptly the influence of large management decisions on the company's performance indicators. In conditions of financial instability, which are typical, for example, for the Russian economy, carrying out such research is important for assessing the real value of a business. Therefore, the actual task is the development of modified methods that allow one to quickly assess the real value of an enterprise depending on the changing parameters of the economic situation, taking into account the specificity of high-tech industries in countries with catching-up economies. An adequate tool for developing modified methods for estimating the value of a company in the high-tech sector are the mathematical models of the enterprise, taking into account the industry specifics of the financial and economic mechanisms for regulating production.

Our group since the end of the 1990s has successfully developed mathematical methods for describing the production process in the manufacturing sector of the Russian economy, taking into account the shortage of working capital and the instability of product sales. Models are based on the Houthakker-Johansen approach [2]. Unlike the classical approach, we take into account financial and economic mechanisms of production regulation. We built a system of models of production in the manufacturing sector, which describes the mechanisms for regulating production at different stages of the evolution of the Russian economy [3].

The current version of the model takes into account the limitations of the trading infrastructure. The model is formalized in the form of the Bellman equation, the solution of which we found in an explicit form. The results of the model analysis can be found in [4–6]. The solution of the Bellman equation estimates the company's value on the basis of discounting the future financial flows of the company as a function of the company's performance indicators and parameters of market conditions. In terms of the model, averaged performance indicators of production, comparable with the data of the official reporting of companies, are calculated. As a result of the research, a system of model algebraic equations is constructed. The set of input parameters and output variables are interpreted in terms of official reporting of companies. The solution of the system equations of the model allows us to promptly estimate the real value of the company on the basis of its performance in the current market conditions [1,3]. The proposed approach makes it possible to provide scenario calculations that allow one to assess the impact of the market indicators (for example, interest rates, the structure of prices for domestic and imported products), as well as management decisions regarding the company (for example, production reduction) on the company's value. The results are actual in the context of financial and economic instability typical for Russia and other countries with catching-up economies.

We developed a methodology for identifying and verifying the model according to the official reporting of companies (IFRS) [3]. The model is used to analyze the dynamics of capitalization of large companies in the automotive industry of Russia and Italy [1]. In this paper, in Sect. 3, we present the results of the analysis of the capitalization dynamics for the Kamaz company by means of the model.

Based on the results of the model research, we can propose a new methodology for estimating the value of enterprises in the high-tech sector of the catching-up country. The methodology is the development of a classical income approach to business valuation. The advantage of our approach is the consideration of the specifics of financial mechanisms for production sector regulating. The methodology, in contrast to the classical approach, allows one to take promptly into account changes in market conditions and their impact on the company's value, as well as conduct an immediate assessment of the impact of large management decisions on the value of the business.

In this paper, we also present the results of study of the modified model that takes into account the impact of the debt burden on the enterprise. In a crisis situation, the growth of debt burden can lead to the bankruptcy of the company. Therefore, the methodology for assessing the value of a company in a crisis situation should take into account the impact of the accumulated debt burden. The results of the study of the modified model are presented in Sect. 4.

2   Description of the Model

Consider the industrial enterprise that functions with the capacity of η and issues a uniform product. Let's define: y - product unit cost, p - product price, Y^* - the maximum amount of one-time product sale, Y_0 - current value of trade inventories in stock. Assume that

- the product sales moments form a Poisson stochastic process with parameter λ;
- the replenishment of current assets is possible by a credit line only;
- from the moment of a product sale the company functions at full capacity by a credit line $K(t)$ under percent r during the period τ chosen by the company's owner; if after the time period τ a sale didn't come, production stops until the buyer's request;
- at the time of the product sale the producer receives revenue, pays the cumulative loan debt and appeals to bank behind the credit line again;
- if the income from sales doesn't allow to pay the accumulated loan debt, then the debt is not fully extinguished and the company's debt load is growing;
- the profit after borrowing repayment is removed from turnover;
- the credit is used on the current assets replenishment only.

Then

$$K(t) = y\eta\theta(x), \text{where } \theta(x) = \begin{cases} 1, & x > 0, \\ 0, & x \le 0. \end{cases}$$

The output value at the time t is

$$\frac{dY(t)}{dt} = \frac{1}{y}K(t), \ Y(0) = Y_0$$

that is

$$Y(t) = Y_0 + \eta \min(t, \tau). \tag{1}$$

The dynamic of loan debt is defined as the solution of the following equation

$$\begin{cases} \frac{dL(t)}{dt} = K(t) + rL(t), \\ L(0) = L_0. \end{cases}$$

Consequently

$$L(t) = \frac{y\eta}{r}\left(e^{rt} - e^{r(t-\tau)+}\right) + L_0 e^{rt}. \tag{2}$$

The goal of the company's owner is maximization of discounted with $1 > \Delta > 0$ mean value of income $V(Y_0, L_0)$ by the choice of time period τ during which production costs are borrowed:

$$V(Y_0, L_0) = \sup_{\tau \geq 0} \int_0^{+\infty} \lambda e^{-(\lambda+\Delta)t} \left[(p \min (Y(t), Y^*) - L(t))_+ + \right.$$

$$\left. + V \left((Y(t) - Y^*)_+ , (L(t) - p \min (Y(t), Y^*))_+ \right) \right] dt. \quad (3)$$

The Bellman equation (3) describes the dynamic of company's capitalization $V(Y_0, L_0)$ with product inventory Y_0 and debt L_0 depending on production indicators of company's activities and parameters of economic environment (interest rate, price structure, etc). More precisely, the value $V(Y_0, L_0)$ corresponds to the fundamental capitalization and characterizes the company's real position in the market and the level of its creditworthiness excepting influence of the speculative component of market capitalization value. The case $L_0 = 0$ is investigated in [4–6], and the case $Y^* = +\infty$ is investigated in [7].

3   Methodology for Estimation of Company's Capitalization in a High-Technology Sector of Catching-Up Economy

In the case of $L(0) = 0$ we obtain $V(Y_0, L_0) = W(Y_0)$ and the model takes the form of the following Bellman equation

$$W(Y_0) = \sup_{\tau \geq 0} \int_0^{+\infty} \lambda e^{-(\lambda+\Delta)t} \left[p \min (Y_0 + \eta \min(t, \tau), Y^*) - \right.$$

$$\left. - \frac{y\eta}{r} \left(e^{rt} - e^{r(t-\tau)+} \right) + W \left((Y_0 + \eta \min(t, \tau) - Y^*)_+ \right) \right] dt. \quad (4)$$

The solution $W(Y_0)$ of Bellman equation (4) can be interpreted as estimate of company's value with the inventory Y_0. The value $\frac{\lambda W(0)}{y\eta}$ characterizes the ratio of the fundamental component of the company's capitalization to current assets of the company (not including the inventory value). The change of this indicator leads to company's market position and the level of its creditworthiness change. Values of $\frac{\lambda W(0)}{y\eta}$, close to 1, correspond to the limit of company's profitability. We proved in [5] that the Bellman equation (4) has a unique solution. As a result of model investigation, we obtain a complete system of three algebraic equations [1,3,8]. The sets of input parameters and output variables we interpreted in terms of official statistics [1,3]. The set of output model variables that define the base point for the selected year is: λ - the characteristic of market conditions, ζ_0 - the characteristic of an optimum inventory level (it equals $(\zeta_0 + 1)Y^*$), $\frac{\lambda Y^*}{\eta}$ - the ratio of demand and production capabilities. The set of input parameters of the system: r - interest rate, Δ - deflator of the company's income,

R - profitability value of the company ($R = \frac{p - y\frac{\lambda+\Delta}{\lambda+\Delta-r}}{p}$), u - average capacity utilization coefficient, $\frac{yQ}{yu\eta}$ - turnover of producer's trade inventories (yQ-average inventory level of producer). In the case $0 \leq \zeta_0 < 2$ the system of model equations we can write in an explicit analytical form (see [3]). The algorithm for the numerical solution of the system we give in [1,3,8]. The model calculations show that the analysis of solutions of the system over the specified range is sufficient for research purposes. Based on the model research results, we propose the following scheme for analyzing the impact of economic environment indicators on the producer's activity. As a result of the model identification process we obtain the set of input parameters of the model per year t. The solution of the model equations system determines the base point of the model $\lambda, \zeta_0, \frac{\lambda Y^*}{\eta}$ in a year t, which allows us to calculate a company's fundamental capitalization value in relation to current assets $\frac{\lambda W(0)}{y\eta}$ in a year t.

In this part, we apply the model to research of capitalization dynamic of large Russian car maker Kamaz from 2011 to 2017. We identified the model according to the Kamaz annual reports 2011–2017. By means of the model the fundamental component of the company's capitalization in relation to current assets $\frac{\lambda W(0)}{y\eta}$ according to conditions 2011–2017 is constructed (a continuous fat curve, a triangular point marker, Fig. 1a). The fundamental component of the capitalization differs from the market expectations which take into account a speculative factor. In Fig. 1a the dashed line (round point marker) corresponds to the market estimate of Kamaz's capitalization in relation to current assets funds. The value of market capitalization is based on the year averaged share value of Kamaz (Moscow exchange). Using the model we analyze fundamental and market capitalization discrepancy of line items of Soros reflexivity theory [9]. The theory contains the detailed humanitarian analysis of mechanisms of mutual influence between fundamental and speculative components of the company's capitalization. In terms of the model, the reflection of the market expectations (characteristic of a speculative component) is the deflator of the company income Δ. Calculations by comparative statics method confirm that Δ has a significant influence on the fundamental capitalization of the company [1].

In Fig. 1b the dynamics of an income deflator Δ of the company corresponding to the basic scenario (continuous fat curve, triangular point marker) and the real market capitalization calculated in the model (dotted curve, round point marker) are shown. The thin curve (Fig. 1a, transparent point marker) corresponds to the company's profitability limit, i.e. to the values $\frac{\lambda W(0)}{y\eta}$ close to 1. The same curve in Fig. 1b determines the corresponding values of the company's income deflator Δ calculated in the model. In Fig. 1c results of the model calculation of the company's product demand dynamics $\frac{\lambda Y^*}{\eta}$ are presented. The calculations show that in the conditions of sales recession expectations in 2012 because of WTO accession the market capitalization decreased while the fundamental capitalization component grew (Fig. 1a). The market expectations of 2012 correspond to the high growth rate of income deflator Δ (dotted curve, round point marker, Fig. 1b). However, the expectations were false - in 2012

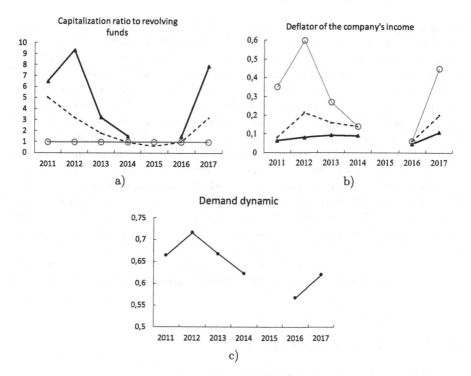

Fig. 1. Calculation results, Kamaz

the KAMAZ production sales grew (Fig. 1c). It led to some improvement of the
market expectations and Δ decreased in 2013 (dotted curve, Fig. 1b) against the
background of further fall of the market capitalization (dotted curve, Fig. 1a).
The fall in demand of 2013 (Fig. 1) and interest rates growth led to an essential decrease in the fundamental component of the company's capitalization in
2013 (continuous fat curve, triangular point marker, Fig. 1a). A further decline
in demand in the conditions of the crisis of 2014 led to a decrease in the fundamental and market capitalization. However previous high rates of fundamental
capitalization were expressed in a further fall of Δ which reflects market expectations (dotted curve, Fig. 1b). In 2014 KAMAZ was at the limit of profitability
in spite of state support. The market expectations of capitalization in 2014 were
lower than the fundamental capitalization which still corresponded to a prof-
itable value of Δ (triangular marker, Fig. 1b). In 2015, despite the large state
order, the fundamental company indicators were under the profitability limit
(model calculation of 2015 is impossible). Strong state support of KAMAZ in
2015–2016 (subsidies for credits interest payment and production components)
led to the fact that the company came to profitability limit in 2016 and the
fundamental indicator $\frac{\lambda W(0)}{y\eta}$ was slightly higher than the market expectations
(Fig. 1a) even in the conditions of a continuing demand decrease (Fig. 1c). At
the same time, the value of Δ which corresponds to fundamental capitalization

is practically at profitability limit unlike 2014 (Fig. 1b, triangular and transparent markers 2014, 2016). It characterizes a company's position as less stable in comparison with 2014. We'll notice that the state support of KAMAZ allowed it's entrance to the profitability area in 2016 with a small decrease in capacity utilization from $u = 0.54$ in 2014 to $u = 0.49$ in 2016. Upgrade of production lines in the 2016–2017th and an increase in demand for products of the company (Fig. 1c) led to significant improvement of indicators of activities of Kamaz in 2017. Capacity utilization of the company grew almost by 12% in 2017 in comparison with 2016. As a result of the fundamental component of capitalization sharply increased (continuous fat curve, triangular point marker, Fig. 1a). The unstable position of the company and proximity of fundamental indicators to limit of profitability in 2014–2016 led to the fact that growth of market capitalization in 2017 was significantly lower (dotted curve, Fig. 1a). Therefore the market expectations of capitalization in 2017 are worse than the corresponding fundamental capitalization characteristic Δ (dotted curve is higher than the fat curve in 2017, Fig. 1b). Technological upgrade and improvement of market conditions (demand growth) led in 2017 to the recovery of stability level of the company at the pre-crisis value (2011–2012) (transparent circle marker in 2017 is on the level of 2011–2012, Fig. 1b).

4   Crisis Management Model

4.1   Model Without Infrastructure Restriction

In crisis conditions many enterprises of the processing sector work under the order and their main goal is debt repayment. In terms of the model it means that $Y^* = +\infty$ and $\Delta = r$. The Bellman equation (3) is modified to the following form

$$
\begin{aligned}
W(Y_0, L_0) = \sup_{\tau \geq 0} \int_0^{+\infty} & \lambda e^{-(\lambda+r)t} \Big[(pY_0 + p\eta \min(t, \tau) \\
& - \frac{\eta y}{r} \left(e^{rt} - e^{r(t-\tau)+} \right) - L_0 e^{rt} \Big)_+ \\
& + W \left(0, \left(\frac{\eta y}{r} \left(e^{rt} - e^{r(t-\tau)+} \right) + L_0 e^{rt} - pY_0 - p\eta \min(t, \tau) \right)_+ \right) \Big] dt. \quad (5)
\end{aligned}
$$

Suppose that $W(Y_0, L_0)$ satisfies a priori conditions: continuity, increase and concavity with respect to the first argument, and also $\frac{\partial W}{\partial Y_0}(Y_0, L_0) \leq p$. We denote the class of such functions by $G[0, +\infty)$. In [5] the meaning of these assumptions is discussed. It follows from Eq. (5) that in order to find the function $W(Y_0, L_0)$, it is sufficient to investigate the equation for $V(L_0) = W(0, L_0)$:

$$
\begin{aligned}
V(L_0) = \sup_{\tau \geq 0} \int_0^{+\infty} & \lambda e^{-(\lambda+r)t} \Big[\left(p\eta \min(t, \tau) - \frac{\eta y}{r} \left(e^{rt} - e^{r(t-\tau)+} \right) - L_0 e^{rt} \right)_+ \\
& + V \left(\left(\frac{\eta y}{r} \left(e^{rt} - e^{r(t-\tau)+} \right) + L_0 e^{rt} - p\eta \min(t, \tau) \right)_+ \right) \Big] dt.
\end{aligned}
$$

In [7] it was proved that this equation has a unique solution in the class $F([0, L_0^*], h)$ – class of functions that are continuous, concave, decreasing, with bounded one-sided derivatives:

$$q = \max\left(-\frac{p\lambda}{y(\lambda + r)}, -\frac{\lambda + r}{\lambda}\right) \leq \frac{dV}{dL_0}(L_0) \leq -1 \quad \forall L_0 \geq 0,$$

and linear with coefficient $-1 \geq h \geq q$ when $L_0 \geq L_0^*$. The analytical form of the solution for $h = -1$ is also given there, and numerical calculations for other values of the parameter h, which characterizes the accounting of company debts during bankruptcy and is regulated by law, are carried out in [8].

4.2 Modification of the Model. Adding an Infrastructure Restriction

In this section analysis of more general model is presented. We introduce an infrastructural restriction in previous subsection model: now we assume that at one time the manufacturer can realize production in a volume of not more than Y^*. Then the Bellman equation (5) will change as follows:

$$W(Y_0, L_0) = \sup_{\tau \geq 0} \int_0^{+\infty} \lambda e^{-(\lambda + r)t} \left[(p\min(Y(t), Y^*) - L(t))_+ \right.$$

$$\left. + W\left((Y(t) - Y^*)_+, (L(t) - p\min(Y(t), Y^*))_+\right)\right] dt, \quad (6)$$

where

$$L(t) = \eta y/r\left(e^{rt} - e^{r(t-\tau)_+}\right) + L_0 e^{rt}, \quad Y(t) = Y_0 + \eta\min(t, \tau).$$

We obtain three properties of the solution of the Bellman equation (6), reflecting the influence of the infrastructural constraint. In the model that takes into account the debt load, but does not take into account the presence of a limit on the maximum volume of the product sold at one time, the non-stop production mode was optimal, i.e. $\tau = +\infty$. For the Bellman equation (6), the optimal period τ is planned to be searched numerically together with the solution of the Eq. (6), but we obtain the lower bound for τ (Theorem 1), which depends on the infrastructure constraint. It was also shown (Theorem 2) that the solution of the Bellman equation (6) does not exceed the solution of the Bellman equation (5) for the model without taking into account the infrastructure constraint, as evidenced by the economic interpretation of the company's cost estimate. The research of the Eq. (5) was reduced to the study of the Bellman equation for a function of one variable (debts). In the formulation discussed in this section, such a transition is impossible. Since it is natural to assume for $W(Y_0, L_0)$ increasing for the first argument, the proven boundedness of the solution of the Bellman equation (6) (Theorem 3) is an interesting fact from economic point of view, and will also be used in further studies of this formulation. We now prove the above properties.

As before, we assume that $W(Y_0, L_0)$ satisfies the a priori conditions: continuity, increase and concavity with respect to the first argument, and also $\frac{\partial W}{\partial Y_0}(Y_0, L_0) \leq p$. And for an arbitrary fixed $Y \in [0, +\infty)$, $W(Y, L_0) \in G[0, +\infty)$. We denote the class containing functions satisfying the described conditions for Q. We denote by $W_\infty(Y_0, L_0)$ the estimate of the manufacturing company value, which corresponds to the Bellman equation (5) with no infrastructure constraints. Next we prove several properties of the solution of the Bellman equation (6).

Theorem 1. *In the Eq.* (6), *for* $W(Y_0, L_0) \in Q$, *the optimal value of* τ_0 *is in the domain* $\tau_0 \geq \frac{(Y^* - Y_0)_+}{\eta}$.

Proof. Suppose the contrary, then we rewrite the original equation, taking into account $\tau_0 \leq \frac{(Y_* - Y_0)_+}{\eta}$:

$$W(Y_0, L_0) = \sup_{0 \leq \tau \leq \frac{(Y_* - Y_0)_+}{\eta}} \int_0^{+\infty} \lambda e^{-(\lambda + r)t} \left[(pY_0 + p\eta \min(t, \tau) \right.$$
$$- \frac{\eta y}{r} \left(e^{rt} - e^{r(t-\tau)_+} \right) - L_0 e^{rt} \right)_+$$
$$\left. + W \left(0, \left(\frac{\eta y}{r} \left(e^{rt} - e^{r(t-\tau)_+} \right) + L_0 e^{rt} - pY_0 - p\eta \min(t, \tau) \right)_+ \right) \right] dt.$$

Repeating the arguments from the proof of Theorem 1 in [7], we can show that the right-hand side of the equation is increasing by τ function, therefore, the optimal value τ_0 can not lie to the left of $\frac{(Y_* - Y_0)_+}{\eta}$.

Theorem 2. *For* $W(Y_0, L_0) \in Q$, *the following relation holds:*

$$W(Y_0, L_0) \leq W_\infty(Y_0, L_0), \quad \forall Y_0, L_0 \in [0, +\infty].$$

Proof. We analyze the dependence on Y^* for the function $I(Y_0, L_0)$, obtained from $W(Y_0, L_0)$ for fixed τ, obviously assuming that $\tau \geq \tau^* = \frac{Y^* - Y_0}{\eta}$ (follows from Theorem 1):

$$I(Y_0, L_0) = \int_0^{+\infty} \lambda e^{-(\lambda + r)t} \left[(p \min (Y_0 + \eta \min(t, \tau), Y^*) \right.$$
$$- \frac{\eta y}{r} \left(e^{rt} - e^{r(t-\tau)_+} \right) - L_0 e^{rt} \right)_+ + W \left((Y_0 + \eta \min(t, \tau) - Y^*)_+ \right),$$
$$\left. \left(\frac{\eta y}{r} \left(e^{rt} - e^{r(t-\tau)_+} \right) + L_0 e^{rt} - p \min (Y_0 + \eta \min(t, \tau), Y^*) \right)_+ \right) \right] dt.$$

Consider the equation for t:

$$pY_0 + p\eta \min(t, \tau) - \frac{\eta y}{r} \left(e^{rt} - e^{r(t-\tau)_+} \right) - L_0 e^{rt} = 0.$$

This equation has maximum two roots: $\theta_{min}, \theta_{max}$. Then we consider three possible cases of the location of the time moment $\tau^* = \frac{Y^* - Y_0}{\eta}$ relative to the roots of the equation. If for the given parameters this equation has one root or no roots, then the proof will be absolutely analogous to the case with two roots. We introduce the notation:

$$a = (Y_0 + \eta \min(t, \tau) - Y^*)_+,$$

$$b = \left(\frac{\eta y}{r} \left(e^{rt} - e^{r(t-\tau)+}\right) + L_0 e^{rt} - p \min\left(Y_0 + \eta \min(t, \tau), Y^*\right)\right)_+.$$

Let us consider the first case when $\tau^* \leq \theta_{min}$. In this case:

$$\frac{dI(Y_0, L_0)}{dY^*} = \int_{\tau^*}^{\theta_{min}} \lambda e^{-(\lambda+r)t} \left(-\frac{\partial W(a,b)}{\partial Y_0} - p\frac{\partial W(a,b)}{\partial L_0}\right) dt$$

$$+ \int_{\theta_{min}}^{\theta_{max}} \lambda e^{-(\lambda+r)t} p \, dt - \int_{\theta_{min}}^{\theta_{max}} \lambda e^{-(\lambda+r)t} \frac{\partial W(a,b)}{\partial Y_0}$$

$$+ \int_{\theta_{max}}^{+\infty} \lambda e^{-(\lambda+r)t} \left(-\frac{\partial W(a,b)}{\partial Y_0} - p\frac{\partial W(a,b)}{\partial L_0}\right) dt \geq 0.$$

This is true, since $\frac{\partial W(Y_0, L_0)}{\partial Y_0} \leq p$ and $\frac{\partial W(Y_0, L_0)}{\partial L_0} \leq -1$, $\forall Y_0, L_0 \in [0, +\infty]$.
Let us now consider the case when $\theta_{min} \leq \tau^* \leq \theta_{max}$, than

$$\frac{dI(Y_0, L_0)}{dY^*} = \int_{\tau^*}^{\theta_{max}} \lambda e^{-(\lambda+r)t} p \, dt - \int_{\tau^*}^{\theta_{max}} \lambda e^{-(\lambda+r)t} \frac{\partial W(a,b)}{\partial Y_0}$$

$$+ \int_{\theta_{max}}^{+\infty} \lambda e^{-(\lambda+r)t} \left(-\frac{\partial W(a,b)}{\partial Y_0} - p\frac{\partial W(a,b)}{\partial L_0}\right) dt \geq 0.$$

The last case remains, when $\theta_{max} \leq \tau^*$, in this case we have:

$$\frac{dI(Y_0, L_0)}{dY^*} = \int_{\tau^*}^{+\infty} \lambda e^{-(\lambda+r)t} \left(-\frac{\partial W(a,b)}{\partial Y_0} - p\frac{\partial W(a,b)}{\partial L_0}\right) dt \geq 0.$$

Therefore, in all cases $I(Y_0, L_0)$ will increase by Y^*.

Theorem 3. *When $W(Y_0, L_0) \in Q$ the following relation holds:*

$$W(Y_0, L_0) \leq \frac{\lambda}{r} p Y^*, \quad \forall Y_0, L_0 \in [0, +\infty].$$

Proof. We carry out the proof by mathematical induction. We show that for any n the following inequality holds:

$$W(Y_0, L_0) \leq \sum_{i=1}^{n+1} \left(\frac{\lambda}{\lambda+r}\right)^i pY^* + \frac{p\eta}{r} \frac{\lambda^{n+2}}{(\lambda+r)^{n+2}}$$

$$+ \frac{\lambda^{n+1}}{(\lambda+r)^{n+1}} pY_0 + \frac{(n+1)p\eta\lambda^{n+1}}{(\lambda+r)^{n+2}}. \quad (7)$$

We first prove the following for $n = 1$:

$$W(Y_0, L_0) = \sup_{\tau \geq 0} \int_0^{+\infty} \lambda e^{-(\lambda+r)t} \left[(p\min(Y_0 + \eta\min(t,\tau), Y^*) - \right.$$

$$\frac{\eta y}{r}\left(e^{rt} - e^{r(t-\tau)+}\right) - L_0 e^{rt}\right)_+ + W\left((Y_0 + \eta\min(t,\tau) - Y^*)_+,\right.$$

$$\left.\left(\frac{\eta y}{r}\left(e^{rt} - e^{r(t-\tau)+}\right) + L_0 e^{rt} - p\min(Y_0 + \eta\min(t,\tau), Y^*)\right)_+\right)\right] dt$$

$$\leq \sup_{\tau \geq 0} \int_0^{+\infty} \lambda e^{-(\lambda+r)t} \left[\left(p\min(Y_0 + \eta\min(t,\tau), Y^*) - \frac{\eta y}{r}\left(e^{rt} - e^{r(t-\tau)+}\right) - L_0 e^{rt}\right)_+\right.$$

$$\left. + W_\infty (Y_0 + \eta t, 0)\right] dt \leq \sup_{\tau \geq 0} \int_0^{+\infty} \lambda e^{-(\lambda+r)t} \left[pY^* + W_\infty (Y_0 + \eta t, 0)\right] dt.$$

Taking into account the properties of functions in the class Q, the following inequality holds:

$$W_\infty (Y_0, 0) = \sup_{\tau \geq 0} \int_0^{+\infty} \lambda e^{-(\lambda+r)t} \left[\left(pY_0 + p\eta\min(t,\tau) - \frac{\eta y}{r}\left(e^{rt} - e^{r(t-\tau)+}\right)\right)_+\right.$$

$$+ W_\infty \left(0, \left(\frac{\eta y}{r}\left(e^{rt} - e^{r(t-\tau)+}\right) - pY_0 - p\eta\min(t,\tau)\right)_+\right)\right] dt$$

$$\leq \int_0^{+\infty} \lambda e^{-(\lambda+r)t} \left[pY_0 + p\eta t + W_\infty (0,0)\right] dt. \quad (8)$$

We estimate separately $W_\infty(0,0)$:

$$W_\infty(0,0) \leq \int_0^{+\infty} \lambda e^{-(\lambda+r)t} \left[p\eta t + W_\infty (0,0)\right] dt,$$

$$W_\infty(0,0) \leq \frac{\lambda(\lambda+r)}{r} \int_0^{+\infty} e^{-(\lambda+r)t} p\eta t\, dt = \frac{\lambda}{r} p\eta \frac{1}{\lambda+r}.$$

Using the obtained inequality, we continue (8)

$$W_\infty(Y_0, 0) \le \int_0^{+\infty} \lambda e^{-(\lambda+r)t} \left[pY_0 + p\eta t\right] dt + \frac{\lambda}{\lambda+r} W_\infty(0,0) = \frac{p\eta}{r} \frac{\lambda^2}{(\lambda+r)^2}$$

$$+ \frac{\lambda}{\lambda+r} pY_0 + \frac{p\eta\lambda}{(\lambda+r)^2}.$$

$$W(Y_0, L_0) \le \sup_{\tau \ge 0} \int_0^{+\infty} \lambda e^{-(\lambda+r)t} \left[pY^* + \frac{p\eta}{r} \frac{\lambda^2}{(\lambda+r)^2} + \frac{\lambda}{\lambda+r} p(Y_0 + \eta t)\right.$$

$$\left. + \frac{p\eta\lambda}{(\lambda+r)^2}\right] dt = \frac{\lambda}{\lambda+r} pY^* + \frac{p\eta}{r} \frac{\lambda^3}{(\lambda+r)^3} + \frac{\lambda^2}{(\lambda+r)^2} pY_0 + \frac{2p\eta\lambda^2}{(\lambda+r)^3}.$$

Consequently, for $n = 1$, our assumption holds. Suppose now that for an arbitrary n, the following holds:

$$W(Y_0, L_0) \le \sum_{i=1}^{n} \left(\frac{\lambda}{\lambda+r}\right)^i pY^* + \frac{p\eta}{r} \frac{\lambda^{n+1}}{(\lambda+r)^{n+1}} + \frac{\lambda^n}{(\lambda+r)^n} pY_0 + \frac{np\eta\lambda^n}{(\lambda+r)^{n+1}}.$$

Let us prove that the relation remains valid for $n + 1$:

$$W(Y_0, L_0) \le \int_0^{+\infty} \lambda e^{-(\lambda+r)t} \left[pY^* + \frac{p\eta}{r} \frac{\lambda^{n+1}}{(\lambda+r)^{n+1}} + \frac{\lambda^n}{(\lambda+r)^n} p(Y_0 + \eta t)\right.$$

$$\left. + \frac{np\eta\lambda^n}{(\lambda+r)^{n+1}} + \sum_{i=1}^{n} \left(\frac{\lambda}{\lambda+r}\right)^i pY^*\right] dt$$

$$= \sum_{i=1}^{n+1} \left(\frac{\lambda}{\lambda+r}\right)^i pY^* + \frac{p\eta}{r} \frac{\lambda^{n+2}}{(\lambda+r)^{n+2}} + \frac{\lambda^{n+1}}{(\lambda+r)^{n+1}} pY_0 + \frac{(n+1)p\eta\lambda^{n+1}}{(\lambda+r)^{n+2}}.$$

Consequently, we have proved (7). In order to prove the original assertion, it is necessary to pass to the limit in the relation (7) for $n \to +\infty$.

In the absence of an infrastructure constraint in [7], the existence and uniqueness of the solution of the Bellman equation in subclass Q containing functions linear in L_0 for fixed reserves and sufficiently large debts were proved. For the model with the infrastructural restriction, the linear solution of the Bellman equation (6) was found in [5]. For the Eq. (6), its impossible to find other solutions linear in L_0 for all Y_0 because of the infrastructural constraint Y^*, which plays the role of a delay on the right-hand side of the equation in the second argument of the company's valuation function. The uniqueness of the found linear solution requires a separate investigation.

References

1. Obrosova, N.K., Shananin, A.A.: About an estimation of company's capitalization in the conditions of prices changing. In: Evtushenko, Y.G., Khachay, M.Y., Khamisov, O.V., Kochetov, Y.A., Malkova, V.U., Posypkin, M.A. (eds.) Proceedings of the VIII International Conference on Optimization and Applications (OPTIMA-2017), vol. 1987, pp. 420–427. CEUR Workshop Proceedings, Petrovac, Montenegro (2017)
2. Houthakker, H.S.: The Pareto distribution and the Cobb-Douglas production function in activity analysis. Rev. Econ. Stud. **23**(1), 27–31 (1955)
3. Obrosova, N.K., Shananin, A.A.: Models of production, taking into account the shortage of working capital. Impact of market expectations on the company's capitalization. In: Pospelov, I.G., Shatrov, A.V. (eds.) Materials of the Conference ECOMOD-2016, pp. 42–51, Kirov, Russia (2016)
4. Obrosova, N.K., Shananin, A.A.: A production model in the conditions of instable demand taking into account the influence of trading infrastructure. Dokl. Math. **87**(3), 372–375 (2013)
5. Obrosova, N.K., Shananin, A.A.: Study of the Bellman equation in a production model with unstable demand. Comput. Math. Math. Phys. **54**(9), 1411–1440 (2014)
6. Obrosova, N.K., Shananin, A.A.: Production model in the conditions of unstable demand taking into account the influence of trading infrastructure: ergodicity and its application. Comput. Math. Math. Phys. **55**(4), 699–723 (2015)
7. Alimov, D.A.: On the existence and uniqueness of a solution of the Bellman equation in a model of operation of a manufacturing company with regard to the debt load. Differ. Equ. **54**(3), 392–400 (2018)
8. Alimov, D.A., Obrosova, N.K., Shananin, A.A.: Modeling of the processing sector production with account taken of current assets deficit. Proc. MIPT **9**(3), 105–114 (2017)
9. Soros, G.: The Alchemy of Finance: Reading the Mind of the Market. Wiley, New York (1994)

Dynamic Marketing Model: Optimization of Retailer's Role

Igor Bykadorov[1,2,3](✉)

[1] Sobolev Institute of Mathematics, 4 Koptyug Avenue, 630090 Novosibirsk, Russia
bykadorov.igor@mail.ru
[2] Novosibirsk State University, 2 Pirogova St., 630090 Novosibirsk, Russia
[3] Novosibirsk State University of Economics and Management,
Kamenskaja street 56, 630099 Novosibirsk, Russia

Abstract. We study a vertical control distribution channel in which a manufacturer sells a single kind of good to a retailer. The state variables are the cumulative sales and the retailer's motivation. The manufacturer chooses wholesale price discount while retailer chooses pass-through. We assume that the wholesale price discount increases the retailer's sale motivation thus improving sales. In contrast to previous settings, we focus on the maximization of retailer's profit with respect to pass-through. The arising problem is linear with respect to both cumulative sales and the retailer's motivation, while it is quadratic with respect to wholesale price discount and pass-through. We obtain a complete description of optimal strategies and optimal trajectories. In particular, we demonstrate that the number of switches for change in the type of optimal policy is no more than one.

Keywords: Retailer · Pricing · Pass-through · Sale motivation

1 Introduction

To earn a reasonable profit the members of a distribution channel often adopt rather simple pricing techniques. For example, manufacturers may use cost-plus pricing, simply defining the price to be added to the desired profit margin to (variable) production costs; in a similar fashion, retailers very often use to determine shelf prices adding a fixed percentage markup to the wholesale price.

We study dynamic marketing model based on the ideas of [1]. The paper [2] should be recognize as the first unit work in this direction. Among many works on this subject, let us note [3–9].

In [1] we consider the concept of retailer's motivation, with the stimulation of the retailer (wholesale discount) as the manufacturer's control.

The presented article is devoted to the development of this approach.

© Springer Nature Switzerland AG 2019
Y. Evtushenko et al. (Eds.): OPTIMA 2018, CCIS 974, pp. 399–414, 2019.
https://doi.org/10.1007/978-3-030-10934-9_28

2 Problem

2.1 Profits and Motion

Consider a vertical distribution channel. There are a manufacturer, retailer and consumer on the market. The firm produces and sells a single product during the time period $[t_1, t_2]$. Let p be the unit price in a situation where the firm sells the product directly to the consumer, bypassing the retailer, $p > 0$. To increase its profits, the firm uses the services of a retailer. To encourage the retailer to sell the commodity, the firm provides it with wholesale discount $\alpha(t)$, $\alpha(t) \in [A_1, A_2] \subset [0, 1]$. Thus, the wholesale price of the goods is $p_w(t) = (1 - \alpha(t))p$. In turn, the retailer directs pass-through, i.e., a part $\beta(t)$ of the discount $\alpha(t)$ to reduce the market price of the commodity, $\beta(t) \in [B_1, B_2] \subset [0, 1]$. Therefore, the retail price of the commodity is equal to $(1 - \beta(t)\alpha(t))p$. Then the difference between retail price and wholesale price is the retailer's profit per unit from the sale and equals $\alpha(t)(1 - \beta(t))p$. Thus, the pricing process can be schematically represented as

$$p \longrightarrow p_w = (1 - \alpha)p = (1 - \beta\alpha)p - (1 - \beta)\alpha p. \tag{1}$$

Let

$x(t)$ be a state variable representing the amount of goods sold by the time t (that is, accumulated sales during the period $[t_1, t]$);

c_0 be a unit production cost.

At the end of the selling period, i.e., in time t_2, the total profit of the firm is

$$\int_{t_1}^{t_2} (p_w(t) - c_0)\, \dot{x}(t)dt,$$

while the total profit of the retailer is

$$p \int_{t_1}^{t_2} \dot{x}(t)\alpha(t)(1 - \beta(t))dt.$$

We assume that the motivation of the retailer increases when the consumer demand and the wholesale discount increase. More precisely, we assume that the motivation of the retailer is determined by the state variable $M(t)$, and her dynamics is given by the differential equation

$$\dot{M}(t) = \gamma\dot{x}(t) + \varepsilon\left(\alpha(t) - \overline{\alpha}\right),$$

where

γ is the sales productivity in terms of motivation, $\gamma > 0$,

ε is the discount productivity in terms of motivation, $\varepsilon > 0$.

Parameter $\overline{\alpha} \in [A_1, A_2]$ takes into account the fact that the retailer has some expectations about the wholesale discount. The motivation is reduced if the retailer is dissatisfied with the wholesale discount, i.e., if $\alpha(t) < \overline{\alpha}$; on the contrary, the motivation increases if the retailer is satisfied with the wholesale discount, i.e., if $\alpha(t) > \overline{\alpha}$.

The dynamics of the total amount of goods sold, $x(t)$, is determined by the differential equation

$$\dot{x}(t) = -\theta x(t) + \delta M(t) + \eta \alpha(t)\beta(t),$$

where

θ is the saturation parameter of the market, $\theta > 0$, i.e., this parameter reflects the effect that the demand decreases when the accumulated sales increase;

δ is the retailer's selling skill, $\delta > 0$,

η is the discount productivity in terms of sales (the market sensitivity to shelf price discounts), $\eta > 0$.

2.2 Maximization of Manufacturer's Profit Under Constant Pass-Through

In [1] we studied the case with constant $\beta(t) = \beta$ and maximized the manufacturer's profit w.r.t. $\alpha(t)$. Let us denote $\eta_\beta = \eta\beta$. This way the problem is
Manufacturer's problem:

$$\int_{t_1}^{t_2} (p_w(t) - c_0)\,\dot{x}(t)dt \longrightarrow \max$$
$$\dot{x}(t) = -\theta x(t) + \delta M(t) + \eta_\beta \alpha(t),$$
$$\dot{M}(t) = \gamma \dot{x}(t) + \varepsilon\left(\alpha(t) - \overline{\alpha}\right),$$
$$x(t_1) = 0,\ M(t_1) = \overline{M} > 0,$$
$$\alpha(t) \in [A_1, A_2] \subset [0, 1].$$

Under rather natural conditions (non negativity of profit and concavity of cumulative sales for constant wholesale price, see details in [1]), we assume

$$A_2 = \frac{p - c_0}{p}, \quad \theta \geq \left(\frac{\varepsilon(A_2 - \overline{\alpha})}{\delta \overline{M} + \eta_\beta A_2} + \gamma\right)\delta. \tag{2}$$

In particular, we obtain

Proposition 1 [1]. *Let be*

$$\delta_0 = \frac{\theta \eta_\beta}{\eta_\beta \gamma + \varepsilon}, \quad \lambda = \sqrt{\frac{(\theta - \gamma\delta)\delta\varepsilon}{\eta_\beta}}.$$

Case 1: The inefficient retailer.

If the retailer's efficiency is low, i.e. $\delta \leq \delta_0$, then the optimal trade discount policy is a continuous increasing function given by

$$\alpha^*(t) = \begin{cases} A_1 & t \in [t_1, \tau_1) \\ D_1 e^{\lambda t} + D_2 e^{-\lambda t} + \dfrac{\overline{\alpha} + A_2}{2} & t \in [\tau_1, t_2] \end{cases}$$

where $\tau_1 \in [t_1, t_2]$, $D_1 \geq 0$ and $D_2 < 0$ are constants.

Case 2: The efficient retailer.

If the retailer's efficiency is high, i.e. $\delta > \delta_0$, then the optimal trade discount policy is a continuous function

$$\alpha^*(t) = \begin{cases} A_1 & t \in [t_1, \tau_1) \\ D_1 e^{\lambda t} + D_2 e^{-\lambda t} + \dfrac{\overline{\alpha} + A_2}{2} & t \in [\tau_1, \tau_2] \\ A_1 & t \in (\tau_2, t_2] \end{cases}$$

where $t_1 \leq \tau_1 \leq \tau_2 \leq t_2$, $D_1 < 0$ and $D_2 < 0$ are constants. Function $\alpha^*(t)$ is strictly concave in $[\tau_1, \tau_2]$. If $\tau_1 = t_1$ then $\alpha^*(\tau_1) \geq \alpha^*(\tau_2)$.

Thus, the optimal wholesale discount is either constant or a linear combination of two exponents.

2.3 Maximization of Retailer's Profit Under Constant Wholesale Discount

Instead, now we study the case when $\alpha(t)$ is constant and maximize the retailer's profit with respect to $\beta(t)$.

Thus, let $\alpha(t) = \alpha \ \forall t$. Let us denote the difference between the granted and the minimum expected wholesale discounts by $\Omega = \alpha - \overline{\alpha}$. Moreover let $\eta_\alpha = \eta\alpha$.

Let us now formulate the problem of the retailer to maximize the profit.

Retailer's problem:

$$\alpha p \int_{t_1}^{t_2} (1 - \beta(t))\dot{x}(t)dt \longrightarrow \max$$
$$\dot{x}(t) = -\theta x(t) + \delta M(t) + \eta_\alpha \beta(t),$$
$$\dot{M}(t) = \gamma \dot{x}(t) + \varepsilon\Omega,$$
$$x(t_1) = 0, \ M(t_1) = \overline{M} > 0,$$
$$\beta(t) \in [B_1, B_2] \subset [0, 1].$$

The problem is linear w.r.t. the state variables $x(t), M(t)$ and quadratic with respect to the control $\beta(t)$.

2.4 Main Result

In this paper we consider the case when

$$\Omega > 0, \tag{3}$$

i.e., the retailer is satisfied with the sales process.

Under rather natural conditions (non negativity of profit and concavity of cumulative sales for constant pass-through, see details in Sect. A.1), we assume that (cf. (2))

$$B_2 = 1, \tag{4}$$

$$\theta \geq \left(\frac{\varepsilon \Omega}{\delta \overline{M} + \eta_\alpha B_1} + \gamma \right) \delta. \tag{5}$$

Let be

$$a = \theta - \gamma \delta, \tag{6}$$

$$b = a \eta_\alpha, \tag{7}$$

$$c = \delta \varepsilon \Omega > 0. \tag{8}$$

Note that $c > 0$ due to (3). Moreover, $a > 0$, see details in Sect. A.1 (in particular, (15)).

The main result of the paper is

Proposition 2. *Let the wholesale discount α be constant and moreover (3) holds, i.e., the retailer is satisfied with the sales process. Then under (4) and (5) the optimal pass-through is a continuous convex function given by*

$$\beta^*(t) = \begin{cases} \left(\dfrac{a^2 c}{4b} \cdot t + T_1(\rho_1) \right) \cdot (t - \rho_1) + B_1, & t \in [t_1, \rho_1] \\ B_1, & t \in [\rho_1, t_2] \end{cases}$$

or

$$\beta^*(t) = \begin{cases} B_1, & t \in [t_1, \rho_2] \\ \left(\dfrac{a^2 c}{4b} \cdot t + T_2(\rho_2) \right) (t - \rho_2) + B_1, & t \in [\rho_2, t_2] \end{cases}$$

where

$$T_1(\rho_1) = \frac{a}{4b} \cdot \left(\frac{c + 4(a\delta\overline{M} + bB_1 - c)}{q_1 + 1} - c \cdot (a t_1 - 1) \right),$$

$$T_2(\rho_2) = -\frac{a}{4b} \cdot \left(\frac{c + 4b(B_1 - 1)}{q_2} + c \cdot (a t_2 + 1) \right),$$

$q_1 = a(\rho_1 - t_1) \in [0; a(t_2 - t_1)]$ *is the root of the equation*

$$e^{q_1 - a(t_2 - t_1)} = \frac{\dfrac{c}{4} \cdot (q_1)^2 + c q_1 + a\delta\overline{M} + bB_1}{b(1 - B_1)(q_1 + 1)}, \tag{9}$$

while $q_2 = a(t_2 - \rho_2) \in [0; a(t_2 - t_1)]$ is the root of the equation

$$e^{q_2 - a(t_2 - t_1)} = -\frac{\dfrac{c}{4} \cdot (q_2)^2 + c q_2 + b(B_1 - 1) + c}{(q_2 + 1)(a\delta\overline{M} + bB_1 - c)}. \tag{10}$$

Let us note that it is possible to get the conditions of the solvability of Eqs. (9) and (10). For example, if $b(B_1 - 1) + c > 0$ (it means, for example, too big low boundary of pass-through, or too large retailer's skill,...) then neither (9) nor (10) have positive roots.

Example. Let $t_1 = 0, t_2 = 2, \delta = 0.8, \overline{M} = 0.5, \theta = 1.4, \eta_\alpha = 0.8, \gamma = 0.6, \Omega = 1.0, \varepsilon = 0.5, B_1 = 0.2$. Then the solution of (9) is $q_1 = 1.7834$ while the solution of (10) is $q_2 = 0.3528$. Hence $\rho_1 = 1.9385, \rho_2 = 1.6165$. Note that $\rho_1 > \rho_2$. This agrees with Proposition 2.

Discussion. Comparing Proposition 2 with Proposition 1, we can conclude the following. The form of optimal pass-through with a fixed trade discount is simpler than the form of optimal trade discount for a constant pass-through (a branch of a parabola vs a linear combination of two exponents). In our opinion, the reason for this phenomenon lies in the structure of the pricing process, see (1). Indeed, if trade discount is constant then, when pass-through varies, only retail price and retailer's profit per unit change. Instead, if pass-through is constant then, when trade discount varies, not only retail price and retailer's profit per unit change but also wholesale price changes.

3   Conclusion

In this paper, we consider the situation when the wholesale discount is constant. The control is pass-through, i.e., a part of the discount that the retailer directs to reduce the retail price. The accumulative sales level and motivation of the retailer are state variables. The profit of the retailer is maximized. The model is linear in state variables and quadratic in control. We show that the number of times for changing the type of optimal control does not exceed one. Thus, for this case we obtain results of the same type as that in [1] for the case when we maximize manufacturer's profit w.r.t. wholesale discount for the constant pass-through.

As for the topics of further research, let us note that it seems interesting to consider the problem of a "different level". Suppose that in the retailer's problem, for a fixed α, we find optimal control $\beta_\alpha^*(t)$ and optimal state variables $x_\alpha^*(t)$ and $M_\alpha^*(t)$. Then the problem of maximizing the profit of the manufacturer is

$$(p(1 - \alpha) - c_0)x_\alpha^*(t_2) \longrightarrow \max_{\alpha \in [A_1, A_2]},$$

where c_0 is unit production cost. Thus, we get the value of the optimal discount, at which the manufacturer will get the maximum profit, knowing the strategy of the retailer. The above problem is a one-dimensional optimization problem. However, the value $x_\alpha^*(t_2)$ is not explicitly known since in many cases it is impossible to compute explicitly the moment of change the optimal control type.

Further, we can consider the maximization the manufacturer's profit w.r.t. to the wholesale discount and maximization the retailer's profit w.r.t. pass-through jointly. Thereby a differential game appears.

The results can be generalized to another marketing models: the optimization of communication expenditure [10], the effectiveness of advertising [11], and to the monopolistic competition models: retailing [12], market distortion [13], investments in R&D [14–16], international trade [17].

Acknowledgments. The work was supported in part by the Russian Foundation for Basic Research, projects 16-01-00108, 16-06-00101 and 18-010-00728, by the program of fundamental scientific researches of the SB RAS No I.5.1, project 0314-2016-0018, and by the Russian Ministry of Science and Education under the 5-100 Excellence Programme.

A   Appendix

A.1   Assumptions About the Input Data of the Model

In this Section we show (4) and (5).

The Choice of Pass-Through Upper Boundary. The choice of B_2 is carried out from the following considerations. It is clear that the retailer can not receive a negative profit. We will assume that in the "most unfavorable" case for the retailer, i.e., when $\beta(t) = B_2 \ \forall t \in [t_1, t_2]$, the profit is zero, i.e.,

$$\alpha p \int_{t_1}^{t_2} (1 - B_2)\dot{x}(t)dt = \alpha p(1 - B_2)(x(t_2) - x(t_1)) = 0.$$

Hence, since $x(t_1) = 0$, we obtain $\alpha p(1 - B_2) = 0$. Therefore, $B_2 = 1$.

Thus, (4) holds.

The Concavity of $x(t)$ Under the Constant Pass-Through. Let the control $\beta(t)$ be constant, i.e., $\beta(t) = \overline{\beta} \ \forall t \in [t_1, t_2]$. This means that the wholesale price is constant throughout the sales period. Then it makes sense to assume that buyers "forget" about the discount provided by the firm (and retailer). We assume that in this case the function $x(t)$ is concave, i.e., $\ddot{x}(t) \leq 0 \ \forall t \in [t_1, t_2]$. This happens when the market is mature and saturation occurs. Moreover, since $x(t)$ is accumulated sales, we assume that the coefficients of the model are such that $\dot{x}(t) \geq 0$. The latter should be performed, in particular, for any constant control.

Consider the system

$$\begin{aligned}
\dot{x}(t) &= -\theta x(t) + \delta M(t) + \eta_\alpha \overline{\beta}, \\
\dot{M}(t) &= \gamma \dot{x}(t) + \varepsilon \Omega, \\
x(t_1) &= 0, \quad M(t_1) = \overline{M}.
\end{aligned} \tag{11}$$

We get
$$\ddot{x}(t) = -\theta \dot{x}(t) + \delta \dot{M}(t) = -a\dot{x}(t) + \delta \varepsilon \Omega, \tag{12}$$

If $a = 0$ then it is not possible to obtain concavity of $x(t)$ due to (12).

Let $a \neq 0$. Then we get from (11)
$$x(t) = \frac{C(\bar{\beta})}{a^2} \cdot (1 - e^{a(t_1 - t)}) + \frac{\delta \varepsilon \Omega}{a}(t - t_1), \tag{13}$$

where $C(\bar{\beta}) = a\delta\overline{M} + b\bar{\beta} - c$.

Due to (13), $\ddot{x}(t) = -C(\bar{\beta})e^{a(t_1-t)}$. So the cumulative sales function is concave for every $\bar{\beta} \in [B_1, 1]$ if and only if $C(\bar{\beta}) \geq 0 \ \forall \bar{\beta} \in [B_1, 1]$, i.e.,
$$a\delta\overline{M} + b\bar{\beta} \geq c \ \forall \bar{\beta} \in [B_1, 1]. \tag{14}$$

Note that (14) can be only if the condition
$$a > 0 \tag{15}$$

holds. Moreover if (15) holds then (14) is true if and only if
$$a\delta\overline{M} + bB_1 \geq c \tag{16}$$

holds. Now we get (5) by substituting (6) to (16). Remark that assumption (16) implies (15).

A.2   Solution of Retailer's Problem

Equivalent Problem. Let us introduce the new control
$$u(t) = \beta(t) - 1 \leq 0$$

and the new state variable
$$y(t) = -\theta x(t) + \delta M(t) + \eta_\alpha.$$

Then the Retailer's problem can be rewritten as
$$-\alpha p \int_{t_1}^{t_2} u(t)(y(t) + \eta_\alpha u(t))dt \longrightarrow max,$$
$$\dot{x}(t) = y(t) + \eta_\alpha u(t),$$
$$\dot{y}(t) = -ay(t) - bu(t) + c,$$
$$x(t_1) = 0, \quad y(t_1) = y_1,$$
$$u(t) \in [u_1, 0] \subseteq [-1, 0],$$

where $u_1 = B_1 - 1 < 0$, $y_1 = \delta\overline{M} + \eta_\alpha > 0$ while definition of a, b and c see in (6), (7) and (8). Note that
$$ay_1 + bu_1 - c = a\left(\delta\overline{M} + \eta_\alpha B_1\right) - c.$$

So due to (16) we get
$$ay_1 + bu_1 - c \geq 0. \tag{17}$$

Solution of Equivalent Problem. Consider the Hamiltonian function

$$H = -p\alpha(y(t) + \eta_\alpha u(t))u(t) + z_0(t)(y(t) + \eta_\alpha u(t)) + z(t)(-ay(t) - bu(t) + c)$$

and the Lagrangian function

$$L = H + \mu_1(t)(u(t) - u_1) - \mu_2(t)u(t).$$

Due to the Pontryagin Maximum Principle one has that if $u^*(t)$ is the optimal control and $x^*(t), y^*(t)$ are optimal state variables, then continuous and piece-wise continuously differentiable functions $z_0(t)$ and $z(t)$ and piece-wise continuous functions $\mu_1(t)$ and $\mu_2(t)$ must exist such that (cf. p. 546 of [1])

$$\dot{z}_0(t) = -\frac{\partial L^*}{\partial x}, \quad z_0(t_2) = 0, \quad \dot{z}(t) = -\frac{\partial L^*}{\partial y}, \quad z(t_2) = 0, \quad \frac{\partial L^*}{\partial u} = 0, \quad (18)$$

where, as usual, $L^* = L(y^*, x^*, u^*, z, z_0, \mu_1, \mu_2)$; moreover slackness complementary conditions

$$\mu_1(t) \geq 0, \quad \mu_1(t)(u^*(t) - u_1) = 0, \quad \mu_2(t) \geq 0, \quad \mu_2(t)u^*(t) = 0 \quad (19)$$

hold. Note that in our case $\dfrac{\partial L^*}{\partial x} \equiv 0$. Therefore from (18) we get $z_0(t) \equiv 0$. So equations $\dot{z}(t) = -\dfrac{\partial L^*}{\partial y}$ and $\dfrac{\partial L^*}{\partial u} = 0$ in (18) are

$$\dot{z}(t) = az(t) + p\alpha u^*(t), \quad (20)$$

$$\mu_1(t) - \mu_2(t) = pay^*(t) + bz(t) + 2p\alpha\eta_\alpha u^*(t). \quad (21)$$

Lemma 1. *Functions $\mu_1(t)$ and $\mu_2(t)$ are continuous.*

Proof (Cf. p. 546 of [1].). Function u^* is continuous since the Hamiltonian function H is strictly concave w.r.t. u (see e.g. p. 86 of [18]). So also $\mu_1(t) - \mu_2(t)$ are continuous due to (21). From (19) we get $\mu_1(t)(u^*(t) - u_1) = \mu_2(t)u^*(t)$ therefore $(\mu_1(t) - \mu_2(t))u^*(t) = \mu_1 u_1$. Thus, functions $\mu_1(t)$ and $\mu_2(t)$ are continuous.

Lemma 2. $\mu_2(t) = 0 \ \forall t \in [t_1, t_2]$.

Proof (Cf. pp. 546–547 of [1].). Suppose, conversely, that $\exists \tilde{t} \in (t_1, t_2) : \mu_2(\tilde{t}) > 0$. Due to Lemma 1, an interval exists where μ_2 is positive. Let $(\rho_1, \rho_2) \subset (t_1, t_2)$ be the interval of maximum length such that

$$\mu_2(t) > 0, \ t \in (\rho_1, \rho_2). \quad (22)$$

Due to (19), $u^*(t) = \mu_1(t) = 0 \ \forall t \in (\rho_1, \rho_2)$. Hence in (ρ_1, ρ_2) (21) becomes

$$\mu_2(t) = -pay^*(t) - bz(t) = -pa\left(C_y e^{-at} + \frac{c}{a}\right) - bC_z e^{at}, \quad (23)$$

where

$$C_y = \left(y(\rho_1) - \frac{c}{a}\right)e^{a\rho_1}, \quad C_z = z(\rho_2)e^{-a\rho_2}. \quad (24)$$

Let us consider the all possible cases.

– Let $\rho_1 = t_1, \rho_2 = t_2$. Since $y(t_1) = y_1 > 0, z(t_2) = 0$, we get from (23), (24) and (15)

$$\mu_2(t) = -p\alpha \left(y_1 e^{a(t_1-t)} + \frac{c}{a} \cdot \left(1 - e^{a(t_1-t)} \right) \right) < 0, \ t \in [t_1, t_2], \qquad (25)$$

But (25) contradicts (19).
– Let $\rho_1 = t_1, \rho_2 < t_2$. Then

$$\mu_2(t_1) > 0, \ \mu_2(\rho_2) = 0 \qquad (26)$$

and, moreover, for $t \in [t_1, \rho_2]$ we get

$$\mu_2(t) = -p\alpha \left(y_1 e^{a(t_1-t)} + \frac{c}{a} \cdot \left(1 - e^{a(t_1-t)} \right) \right) - a\eta_\alpha z(\rho_2) e^{a(t-\rho_2)}. \qquad (27)$$

Hence to keep (22) we need

$$z(\rho_2) < 0. \qquad (28)$$

We get from (27)

$$\ddot{\mu}_2(t) = a^2 \cdot \left(p\alpha \left(a y_1 - c \right) e^{a(t_1-t)} - \eta_\alpha z(\rho_2) e^{a(t-\rho_2)} \right), \ t \in [t_1, \rho_2]. \qquad (29)$$

From (17), (28) and (29) we get $\ddot{\mu}_2(t) > 0, \ t \in [t_1, \rho_2]$, which contradicts (26).
– Let $\rho_1 < t_1, \rho_2 = t_2$. Then

$$\mu_2(\rho_1) = 0, \ \mu_2(t_2) > 0 \qquad (30)$$

and

$$\mu_2(t) = -p\alpha \left(y(\rho_1) e^{a(\rho_1-t)} + \frac{c}{a} \cdot \left(1 - e^{a(\rho_1-t)} \right) \right), \ t \in [\rho_1, t_2]. \qquad (31)$$

Moreover, (31) becomes due to (30)

$$\mu_2(t) = -\frac{p\alpha c}{a} \cdot \left(1 - e^{a(\rho_1-t)} \right), \ t \in [\rho_1, t_2]. \qquad (32)$$

From (32) we get $\dot{\mu}_2(t) < 0, \ t \in [\rho_1, t_2]$, which contradicts (30).
– Let $\rho_1 < t_1, \rho_2 < t_2$. Then

$$\mu_2(\rho_1) = \mu_2(\rho_2) = 0. \qquad (33)$$

Moreover

$$\ddot{\mu}_2(t) = a^2 \left(-p\alpha C_y e^{-at} - a\eta_\alpha C_z e^{at} \right) = a^2 \left(\mu_2(t) + \frac{\delta\varepsilon\Omega}{a} \right), \ t \in (\rho_1, \rho_2).$$

Hence, due to (15) and (22), we get $\ddot{\mu}_2(t) > 0, \ t \in (\rho_1, \rho_2)$, i.e., function $\mu_2(t)$ is strictly convex in $[\rho_1, \rho_2]$ which contradicts (22) and (33).

Thus, (19) and (21) are

$$\mu_1(t) \geq 0, \quad \mu_1(t)(u^*(t) - u_1) = 0, \tag{34}$$

$$\mu_1(t) = pay^*(t) + bz(t) + 2pa\eta_\alpha u^*(t). \tag{35}$$

Lemma 3. *Let for some $\rho_0 < \rho_1 < \rho_2 < \rho_3$ be*

$$\mu_1(t) \begin{cases} = 0 \; t \in [\rho_0, \rho_1], \\ > 0 \; t \in (\rho_1, \rho_2), \\ = 0 \; t \in [\rho_2, \rho_3]. \end{cases} \tag{36}$$

Then

$$u^*(t) = \begin{cases} E_2 t^2 + D_1 t + D_0 \; t \in [\rho_0, \rho_1], \\ u_1 \hspace{3.2cm} t \in [\rho_1, \rho_2], \\ E_2 t^2 + E_1 t + E_0 \; t \in [\rho_2, \rho_3], \end{cases} \tag{37}$$

where $E_2 = \dfrac{ac}{4\eta_\alpha}$ while D_1, D_0, E_1, E_0 are some constants.

Proof. Due to (34) and (35), we get from (36)

$$u^*(t) = \begin{cases} -\dfrac{1}{2pa\eta_\alpha} \cdot (pay^*(t) + bz(t)) \; t \in [\rho_0, \rho_1], \\ u_1 \hspace{4.3cm} t \in [\rho_1, \rho_2], \\ -\dfrac{1}{2pa\eta_\alpha} \cdot (pay^*(t) + bz(t)) \; t \in [\rho_2, \rho_3]. \end{cases}$$

Hence $y^*(t), z(t)$ and $u^*(t)$ in $[\rho_0, \rho_1] \cup [\rho_2, \rho_3]$ satisfy

$$\begin{cases} \dot{y}^*(t) = -ay^*(t) - bu^*(t) + c, \\ \dot{z}(t) = az(t) + pau^*(t), \\ u^*(t) = -\dfrac{1}{2pa\eta_\alpha} \cdot (pay^*(t) + bz(t)). \end{cases} \tag{38}$$

In $[\rho_0, \rho_1]$ and in $[\rho_2, \rho_3]$, the solution of (38) has a form

$$\begin{cases} y^*(t) = -\dfrac{ac}{4} \cdot t^2 + C_1 \cdot t + C_2, \\ z(t) = \dfrac{pa}{b} \cdot \left(-\dfrac{ac}{4} \cdot t^2 + (C_1 - c) \cdot t + \dfrac{2C_1 - 2ca + aC_2}{a} \right), \\ u^*(t) = \dfrac{1}{b} \cdot \left(\dfrac{a^2 c}{4} \cdot t^2 + \dfrac{a(c - 2C_1)}{2} \cdot t - (C_1 - c + aC_2) \right). \end{cases} \tag{39}$$

Hence (37) holds.

Note that $y^*(t)$ and $z(t)$ in $[\rho_1, \rho_2]$ satisfy

$$\begin{cases} \dot{y}^*(t) = -ay^*(t) - bu_1 + c, \\ \dot{z}(t) = az(t) + pau_1, \end{cases} \tag{40}$$

i.e.,

$$\begin{cases} y^*(t) = C_y \cdot e^{-at} + \dfrac{c - bu_1}{a}, \\ z(t) = C_z e^{at} - \dfrac{pau_1}{a}, \end{cases} \tag{41}$$

where

$$C_y = \left(y(\rho_1) - \frac{c - bu_1}{a} \right) e^{a\rho_1}, \quad C_z = \left(z(\rho_2) + \frac{pau_1}{a} \right) e^{-a\rho_2}.$$

Lemma 4. *Either $\mu_1(t) = 0$ $\forall t \in [t_1, t_2]$ or only one interval $(\rho_1, \rho_2) \subset [t_1, t_2]$ exists such that $\mu_1(t) > 0$, $t \in (\rho_1, \rho_2)$.*

Proof. Due to Lemma 3, if for some $t_1 \le \rho_0 < \rho_1 < \rho_2 < \rho_3 \le t_2$ (37) holds, then (see (39)) $u^*(t) = u_1$ $\forall t \in [\rho_1, \rho_2]$ while function $u^*(t)$ is strictly larger than u_1 and strictly convex in $[\rho_0, \rho_1] \cup [\rho_2, \rho_3]$. Hence function $u^*(t)$ strictly decreases in $[\rho_0, \rho_1]$ and strictly increases in $[\rho_2, \rho_3]$. Hence function $u^*(t)$ strictly decreases in $[t_1, \rho_1]$ and strictly increases in $[\rho_2, t_2]$. Hence $\mu_1(t) = 0$ $\forall t \in [t_1, \rho_1] \cup [\rho_2, t_2]$.

From Lemmas 3 and 4 we get

Corollary 1. *The optimal control u^* has the form*

$$u^*(t) = \begin{cases} E_2 t^2 + D_1 t + D_0 & t \in [t_1, \rho_1], \\ u_1 & t \in [\rho_1, \rho_2], \\ E_2 t^2 + E_1 t + E_0 & t \in [\rho_2, t_2], \end{cases} \tag{42}$$

where $E_2 = \dfrac{a^2 c}{4b}$ while D_1, D_0, E_1, E_0 are some constants.

Lemma 5. *The optimal control can has no more than one switch.*

Proof. Let, by contradiction,

$$t_1 < \rho_1 < \rho_2 < t_2 \tag{43}$$

and

$$y^*(t) = \begin{cases} -\dfrac{ac}{4} \cdot t^2 + C_1 \cdot t + C_2 & t \in [t_1, \rho_1], \\ y^*(\rho_1) \cdot e^{a(\rho_1 - t)} + \dfrac{c - bu_1}{a} \cdot \left(1 - e^{a(\rho_1 - t)} \right) & t \in [\rho_1, \rho_2], \\ -\dfrac{ac}{4} \cdot t^2 + D_1 \cdot t + D_2 & t \in [\rho_2, t_2], \end{cases}$$

$$z(t) = \begin{cases} \dfrac{pa}{b} \cdot \left(-\dfrac{ca}{4} \cdot t^2 + (C_1 - c) \cdot t + (C_1 - c) \cdot \dfrac{2}{a} + C_2 \right) & t \in [t_1, \rho_1], \\ z(\rho_2) \cdot e^{a(t - \rho_2)} - \dfrac{pau_1}{a} \cdot \left(1 - e^{a(t - \rho_2)} \right) & t \in [\rho_1, \rho_2], \\ \dfrac{pa}{b} \cdot \left(-\dfrac{ca}{4} \cdot t^2 + (D_1 - c) \cdot t + (D_1 - c) \cdot \dfrac{2}{a} + D_2 \right) & t \in [\rho_2, t_2], \end{cases}$$

$$u^*(t) = \begin{cases} \frac{1}{b} \cdot \left(\frac{a^2c}{4} \cdot t^2 - \left(C_1 - \frac{c}{2} \right) a \cdot t - (C_1 - c + aC_2) \right) & t \in [t_1, \rho_1], \\ u_1 & t \in [\rho_1, \rho_2], \\ \frac{1}{b} \cdot \left(\frac{a^2c}{4} \cdot t^2 - \left(D_1 - \frac{c}{2} \right) a \cdot t - (D_1 - c + aD_2) \right) & t \in [\rho_2, t_2]. \end{cases}$$

Since $y^*(t_1) = y_1$ and $z(t_2) = 0$ we get

$$\begin{cases} C_2 = y_1 + \frac{ac}{4} \cdot (t_1)^2 - C_1 \cdot t_1, \\ D_2 = \frac{ca}{4} \cdot (t_2)^2 - (D_1 - c) \cdot t_2 - (D_1 - c) \cdot \frac{2}{a}. \end{cases} \tag{44}$$

Due to (44) and continuity of y^* and z we get

$$\begin{cases} y^*(\rho_1) = y_1 - \frac{ac}{4} \cdot \left((\rho_1)^2 - (t_1)^2 \right) + C_1 \cdot (\rho_1 - t_1), \\ z(\rho_2) = \frac{p\alpha}{b} \cdot \left(-\frac{ca}{4} \cdot (\rho_2 + t_2) + D_1 - c \right) \cdot (\rho_2 - t_2). \end{cases} \tag{45}$$

Due to (44), (45) and continuity of u^* we get

$$\begin{cases} C_1 = \frac{1}{z_1} \cdot \left(\frac{a^2c}{4} \cdot \left((\rho_1)^2 - (t_1)^2 \right) + \frac{ac}{2} \cdot \rho_1 - ay_1 - bu_1 + c \right), \\ D_1 = \frac{1}{z_2} \cdot \left(\frac{a^2c}{4} \cdot \left((t_2)^2 - (\rho_2)^2 \right) + \frac{ac}{2} \cdot (2t_2 - \rho_2) + bu_1 + c \right). \end{cases} \tag{46}$$

where

$$\begin{cases} z_1 = a(\rho_1 - t_1) + 1 > 1, \\ z_2 = a(t_2 - \rho_2) + 1 > 1. \end{cases} \tag{47}$$

Due to (44)–(47) and continuity of y^* and z we get

$$\begin{cases} e^{a(\rho_1 - \rho_2)} = -\dfrac{z_2 + 2 + \dfrac{L_2}{z_2}}{z_1 - 2 + \dfrac{L_1}{z_1}}, \\[4mm] e^{a(\rho_1 - \rho_2)} = -\dfrac{z_1 + 2 + \dfrac{L_1}{z_1}}{z_2 - 2 + \dfrac{L_2}{z_2}}, \end{cases} \tag{48}$$

where $L_1 := 1 + 4 \cdot \dfrac{ay_1 + bu_1 - c}{c}$ and $L_2 = 1 + \dfrac{4bu_1}{c}$. Note that

$$L_1 \geq 1 \tag{49}$$

due to (17). Let us rewrite (48) as

$$\begin{cases} e^{a(\rho_1-\rho_2)} = -\dfrac{z_2 + 2 + \dfrac{L_2}{z_2}}{z_1 - 2 + \dfrac{L_1}{z_1}}, \\ \left(z_1 + \dfrac{L_1}{z_1}\right) = \pm\left(z_2 + \dfrac{L_2}{z_2}\right). \end{cases} \tag{50}$$

Let $z_1 + \dfrac{L_1}{z_1} = -\left(z_2 + \dfrac{L_2}{z_2}\right)$. Then due to (50) we get $e^{a(\rho_1-\rho_2)} = 1$ in contradiction with (43). Let $z_1 + \dfrac{L_1}{z_1} = z_2 + \dfrac{L_2}{z_2}$. Then due to (50) we get

$$e^{a(\rho_1-\rho_2)} = -\frac{(z_1+1)^2 + L_1 - 1}{(z_1-1)^2 + L_1 - 1}.$$

Since $z_1 > 1$ and $a(\rho_1 - \rho_2) < 0$, we get $(z_1+1)^2 + L_1 - 1 > 0 > (z_1-1)^2 + L_1 - 1$ and moreover $(z_1+1)^2 + L_1 - 1 < -(z_1-1)^2 - L_1 + 1$, i.e., $L_1 < -(z_1)^2 < -1$, in contradiction with (49).

Due to Lemma 5, only two cases are possible, namely $\rho_2 = t_2$ and $\rho_1 = t_1$.

Lemma 6. *(i) If $\rho_2 = t_2$ then the optimal control is*

$$u^*(t) = \begin{cases} \left(\dfrac{a^2 c}{4b} \cdot t + T_1(\rho_1)\right) \cdot (t - \rho_1) + u_1, & t \in [t_1, \rho_1], \\ u_1, & t \in [\rho_1, t_2], \end{cases}$$

where

$$T_1(\rho_1) = \frac{a}{4b} \cdot \left(\frac{c + 4(ay_1 + bu_1 - c)}{q_1 + 1} - c \cdot (at_1 - 1)\right)$$

while $q_1 = a(\rho_1 - t_1) \in [0; a(t_2 - t_1)]$ is the root of the equation

$$e^{q_1 - a(t_2 - t_1)} = \frac{\dfrac{c}{4} \cdot (q_1)^2 + cq_1 + ay_1 + bu_1}{-bu_1(q_1 + 1)}.$$

(ii) If $\rho_1 = t_1$ then the optimal control is

$$u^*(t) = \begin{cases} u_1, & t \in [t_1, \rho_1], \\ \left(\dfrac{a^2 c}{4b} \cdot t + T_2(\rho_2)\right)(t - \rho_2) + B_1, & t \in [\rho_1, t_2], \end{cases}$$

where

$$T_2(\rho_2) = -\frac{a}{4b} \cdot \left(\frac{c + 4b(B_1 - 1)}{q_2} + c \cdot (at_2 + 1)\right)$$

while $q_2 = a(t_2 - \rho_2) \in [0; a(t_2 - t_1)]$ *is the root of the equation*

$$e^{q_2 - a(t_2 - t_1)} = -\frac{\dfrac{c}{4} \cdot (q_2)^2 + cq_2 + bu_1 + c}{(q_2 + 1)(ay_1 + bu_1 - c)}.$$

Proof. As in the proof of Lemma 5, we can write the form of $y^*(t), z(t)$ and $u^*(t)$. But now we have only one switch. To finish the proof we need only use the conditions $y^*(t_1) = y_1, z(t_2) = 0$, continuity of optimal trajectories and optimal control, and straightforward calculations.

References

1. Bykadorov, I., Ellero, A., Moretti, E., Vianello, S.: The role of retailer's performance in optimal wholesale price discount policies. Eur. J. Oper. Res. **194**(2), 538–550 (2009)
2. Nerlove, M., Arrow, K.J.: Optimal advertising policy under dynamic conditions. Economica **29**(144), 129–142 (1962)
3. Bala, P.K.: A data mining model for investigating the impact of promotion in retailing. In: 2009 IEEE International Advance Computing Conference, IACC 2009, pp. 670–674, 4809092 (2009)
4. Giri, B.C., Bardhan, S.: Coordinating a two-echelon supply chain with price and inventory level dependent demand, time dependent holding cost, and partial backlogging. Int. J. Math. Oper. Res. **8**(4), 406–423 (2016)
5. Giri, B.C., Bardhan, S., Maiti, T.: Coordinating a two-echelon supply chain through different contracts under price and promotional effort-dependent demand. J. Syst. Sci. Syst. Eng. **22**(3), 295–318 (2013)
6. Printezis, A., Burnetas, A.: The effect of discounts on optimal pricing under limited capacity. Int. J. Oper. Res. **10**(2), 160–179 (2011)
7. Routroy, S., Dixit, M., Sunil Kumar, C.V.: Achieving supply chain coordination through lot size based discount. Mater. Today: Proc. **2**(4–5), 2433–2442 (2015)
8. Ruteri, J.M., Xu, Q.: The new business model for SMEs food processors based on supply chain contracts. In: International Conference on Management and Service Science, MASS 2011, 5999355 (2011)
9. Sang, S.: Bargaining in a two echelon supply chain with price and retail service dependent demand. Eng. Lett. **26**(1), 181–186 (2018)
10. Bykadorov, I., Ellero, A., Moretti, E.: Minimization of communication expenditure for seasonal products. RAIRO Oper. Res. **36**(2), 109–127 (2002)
11. Bykadorov, I., Ellero, A., Funari, S., Moretti, E.: Dinkelbach approach to solving a class of fractional optimal control problems. J. Optim. Theory Appl. **142**(1), 55–66 (2009)
12. Bykadorov, I.A., Kokovin, S.G., Zhelobodko, E.V.: Product diversity in a vertical distribution channel under monopolistic competition. Autom. Remote Control **75**(8), 1503–1524 (2014)
13. Bykadorov, I., Ellero, A., Funari, S., Kokovin, S., Pudova, M.: Chain store against manufacturers: regulation can mitigate market distortion. In: Kochetov, Y., Khachay, M., Beresnev, V., Nurminski, E., Pardalos, P. (eds.) DOOR 2016. LNCS, vol. 9869, pp. 480–493. Springer, Cham (2016). https://doi.org/10.1007/978-3-319-44914-2_38

14. Antoshchenkova, I.V., Bykadorov, I.A.: Monopolistic competition model: the impact of technological innovation on equilibrium and social optimality. Autom. Remote Control **78**(3), 537–556 (2017)
15. Bykadorov, I.: Monopolistic competition model with different technological innovation and consumer utility levels. In: CEUR Workshop Proceeding 1987, pp. 108–114 (2017)
16. Bykadorov, I., Kokovin, S.: Can a larger market foster R&D under monopolistic competition with variable mark-ups? Res. Econ. **71**(4), 663–674 (2017)
17. Bykadorov, I., Gorn, A., Kokovin, S., Zhelobodko, E.: Why are losses from trade unlikely? Econ. Lett. **129**, 35–38 (2015)
18. Seierstad, A., Sydsæter, K.: Optimal Control Theory with Economic Applications. North-Holland, Amsterdam (1987)

The Berge Equilibrium in Cournot Oligopoly Model

Konstantin Kudryavtsev[1]([⊠]) [iD], Viktor Ukhobotov[2], and Vladislav Zhukovskiy[3]

[1] South Ural State University, Lenin prospekt 76, Chelyabinsk 454080, Russia
kudrkn@gmail.com
[2] Chelyabinsk State University, Bratiev Kashirinykh st. 129,
Chelyabinsk 454001, Russia
ukh@csu.ru
[3] M.V. Lomonosov Moscow State University, Leninskie Gory, Moscow 119991, Russia
zhkvlad@yandex.ru

Abstract. More than a hundred years ago, the first models of oligopolies were described. Modeling of oligopolies continues to this day in many modern papers. The main approach meets the concept of the Nash equilibrium and is actively used in modeling the behavior of players in a competitive market. The exact opposite of such "selfish" equilibrium is the "altruistic" concept of the Berge equilibrium. At the moment, many works are devoted to a Berge equilibrium. However, all of these items are limited to purely theoretical issues, or, in general, to psychological applications. Papers devoted to the study of Berge equilibrium in economic problems were not seen until now. In this paper, the Berge equilibrium is considered in the Cournot oligopoly, and its relationship to the Nash equilibrium is studied. Cases are revealed in which players gain more profit by following the concept of the Berge equilibrium, then by using strategies dictated by the Nash equilibrium.

Keywords: Nash equilibrium · Berge equilibrium · Cournot oligopoly

1 Introduction

In many major areas of the economy (such as oil production and refining, electronics and metallurgy), the main competition takes place amongst a few large companies that dominate the market. The first models of such markets, oligopolies, were described more than a hundred years ago in articles by Bertrand [2], Hotelling [7] and Cournot [4]. Modeling of oligopolies continues to this day in many modern works. Moreover, the 2014 Nobel Prize in Economics was awarded to Jean Tirole for his "analysis of market power and regulation in sectors with

The work was supported by Grant of the Foundation for perspective scientific researches of Chelyabinsk State University (2018) and by Act 211 Government of the Russian Federation, contract N 02.A03.21.0011.

© Springer Nature Switzerland AG 2019
Y. Evtushenko et al. (Eds.): OPTIMA 2018, CCIS 974, pp. 415–426, 2019.
https://doi.org/10.1007/978-3-030-10934-9_29

few large companies". He is the author of one of the best modern textbooks on the theory of imperfect competition called "The Theory of Industrial Organization" (see [15]).

The main concept of all these publications, studying the behavior of oligopolies, is that *every company is primarily concerned with its own profits.* This approach meets the concept of the Nash equilibrium (see Nash [12]) and is actively used in modeling the behavior of players in a competitive market. The exact opposite of such "selfish" equilibrium is the "altruistic" concept of the Berge equilibrium. In this approach, each player, without having to worry about himself, chooses his actions or strategies by trying to maximize the profits of all other market participants. The idea of altruistic equilibrium was first introduced in the Claude Berge monograph [1]. Its mathematical description, called Berge equilibrium, first appeared in 1994 in Russia. The first works on the concept of the Berge equilibrium were published by Zhukovskii and Vaisman (see [16–18]). Outside Russia, the concept of "Berge equilibrium" is only now slowly gaining popularity. Today, the number of publications related to this equilibrium is already measured in the tens (Nessah [13], Musy [11], Lung [10], Courtois [5] et al.). The review of publications on the Berge equilibrium is given in [9] However, all of these items are limited to purely theoretical issues, or, in general, to psychological applications (see Colman et al. [3]). The Berge equilibrium as a model of the Golden Rule is in [19] and the Berge equilibrium as a model of mutual support is in [6]. Works devoted to the study of Berge equilibrium in economic problems were not seen until now. This is probably a consequence of Martin Shubik's highly negative review [14] ("... no attention is paid to the application to the economy. ... the book is of little interest for economists"). As a consequence, it "frightened" economists for a considerable period. However, the reality is more complex. In this article, the Berge equilibrium is considered in the Cournot oligopoly, and its relationship to the Nash equilibrium is studied. Cases are revealed in which players gain more profit by following the concept of the Berge equilibrium than by using strategies dictated by the Nash equilibrium.

2   Basic Notations and Definitions

Consider a non-cooperative N-person game, which is identified as

$$\Gamma = \langle \{\mathbb{N}\}, \{\mathbf{X}_i\}_{i\in\mathbb{N}}, \{f_i(x)\}_{i\in\mathbb{N}} \rangle. \tag{1}$$

Here $\mathbb{N} = \{1, 2, \ldots, N\}$ is a set of players numbers, where $N > 1$; every one of N players choose his strategy (action) $x_i \in \mathbf{X}_i \subseteq \mathbb{R}^{n_i}$ where the symbol $\mathbb{R}^k$, $k \geq 1$, stands for k-dimensional Euclidean real arithmetic space, in which elements are ordered sets of k real numbers, presented in columns, with the standard inner product and Euclidean norm without cooperating with others. The result of this choice is a *strategy profile*

$$x = (x_1, \ldots, x_N) \in \mathbf{X} = \prod_{i\in\mathbb{N}} \mathbf{X}_i \subseteq \mathbb{R}^n \ \ (n = \sum_{i\in\mathbb{N}} n_i);$$

a payoff function $f_i(x)$ is defined on the set $\mathbf{X}$, which numerically evaluates the functioning quality of player number i ($i \in \mathbb{N}$); further more $(x\|z_i) = (x_1, \ldots, x_{i-1}, z_i, x_{i+1}, \ldots, x_N)$ and $f = (f_1, \ldots, f_N)$.

Definition 1. *The pair* $(x^e, f^e) = ((x_1^e, \ldots, x_N^e), (f_1(x^e), \ldots, f_N(x^e))) \in \mathbf{X} \times \mathbb{R}^N$ *is called the Nash equilibrium in the game* (1), *if*

$$\max_{x_i \in \mathbf{X}_i} f_i(x^e \| x_i) = f_i(x^e) \quad (i \in \mathbb{N}); \tag{2}$$

further more x^e *is a Nash equilibrium strategy profile in the game* (1).

Definition 2. *The pair* $(x^B, f^B) = ((x_1^B, \ldots, x_N^B), (f_1(x^B), \ldots, f_N(x^B))) \in \mathbf{X} \times \mathbb{R}^N$ *is called the Berge equilibrium in the game* (1), *if*

$$\max_{x \in \mathbf{X}} f_i(x \| x_i^B) = f_i(x^B) \quad (i \in \mathbb{N}); \tag{3}$$

further more x^B *is a Berge equilibrium strategy profile in the game* (1).

Let us proceed with Example 1 of a two-player matrix game in which the players have higher payoffs in the Berge equilibrium than in the Nash equilibrium (in the game sense, this is an analog of the Prisoner's Dilemma).

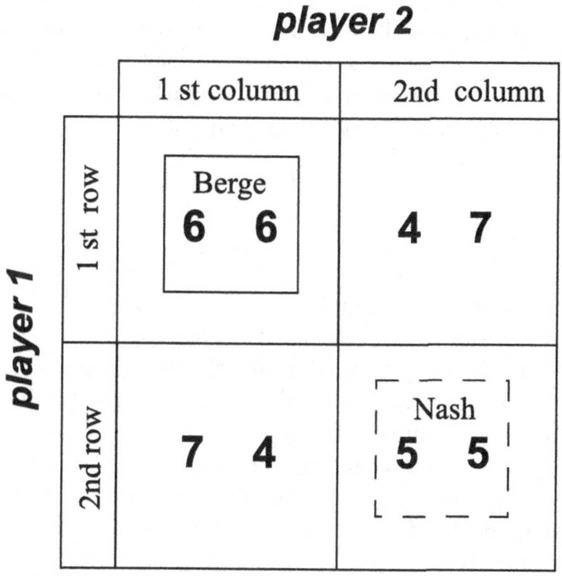

Fig. 1. Comparison of the Berge equilibrium profits with Nash equilibrium.

Example 1. In this bimatrix game, player 1 has two strategies, i.e., chooses between rows 1 and 2; accordingly, the strategies of player 2 are columns 1 and 2. For example, the choice of the strategy profile $(1, 2)$ means that the payoffs of players 1 and 2 are 4 and 7, respectively (Fig. 1).

According to the above definitions, in this bimatrix game the strategy profiles $(2, 2)$ and $(1, 1)$ are Nash and Berge equilibria, respectively. As $6 > 5$, the payoffs of both players in the Berge equilibrium are strictly greater than their counterparts in the Nash equilibrium. The same result occurs in the Prisoner's Dilemma, a well-known bimatrix game. Note that the paper [3] gave some examples of 2×2 bimatrix games where the payoffs in a Nash equilibrium are greater than in a Berge equilibrium, or they coincide.

3   Cournot Oligopoly and Equilibrium Strategies

In 1838 in his work [4] Cournout considered a market having several dominating large players, each producing the same product. In [4] players compete only with the amount of the product supplied to the market. The price is defined by the balance between demand and supply. Later this model was called *the Cournot oligopoly*.

So, according to this model of pricing we consider the Cournot oligopoly, namely, the market of a uniform product having N players i.e. manufacturers. We assume that all of them are assigned serial numbers from 1 to N. A set of players $\{1, 2, \ldots, N\}$ is denoted as $\mathbb{N}$. The volume of the product issued by the i manufacturer $(i \in \mathbb{N})$ during the defined time period is q_i. In addition, each of the players cannot put the goods onto the market for an amount less than $\alpha > 0$, and more than β, i.e. the following inequalities

$$\alpha \le q_i \le \beta \quad (i = 1, \ldots, N) \tag{4}$$

are correct.

The second inequality in (4) just states the fact that, the production capacity of each of the producers is physically limited. For simplicity, we assume that all of these facilities are the same. The essence of the left-hand inequality in (4) is that there is an arbiter in the model, for example, the electrical supply market which is controlled by the state. The arbiter allows into the market just enough large players. This guarantees a supply of goods having not less than a predetermined value α regardless of the current market price.

Production costs of the player number $i\,(i \in \mathbb{N})$ are assumed to be linearly dependent on the quantity of the product supplied q_i and can be represented as $cq_i + d$, where c and d respectively are the average variable and fixed costs.

In the market, the price of products is set depending upon the demand, which is also considered linearly dependent upon the total amount of

$$\bar{q} = q_1 + q_2 + \ldots + q_N$$

incoming goods for sale. We believe that the price of p linearly depends on the product supply, and so we present it in the form

$$p(\bar{q}) = a - b\bar{q}, \tag{5}$$

where $a = \text{const} > 0$—the initial price of the goods, and constant positive elasticity coefficient $b > 0$ shows how the price falls relative to the availability of the product unit.

Assume that the price is determined in such a way as to balance supply and demand. This means that each manufacturer sells everything he produces. Revenue of player number i $(i \in \mathbb{N})$ in this case is

$$p(\bar{q})q_i = (a - b\bar{q})q_i = \left[a - b\sum_{k \in \mathbb{N}} q_k\right]q_i,$$

and his *profit* (revenue minus costs) is

$$\pi_i(q_i, \ldots, q_N) = \left[a - b\sum_{k \in \mathbb{N}} q_k\right]q_i - (cq_i + d). \tag{6}$$

It is also presumed that the manufacturer's management when defining its scope of production, is focused on the rational behavior of its competitors.

A mathematical model of considered interaction is the N-person non-cooperative game

$$\langle \mathbb{N}, \{\mathbf{Q}_i = [\alpha; \beta]\}_{i \in \mathbb{N}}, \{\pi_i(q_1, \ldots, q_N) \div (6)\}_{i \in \mathbb{N}} \rangle. \tag{7}$$

There, as in (1), $\mathbb{N} = \{1, 2, \ldots, N\}$ is set of players' numbers, $\mathbf{Q}_i = [\alpha; \beta]$ is set of player's i strategies $(i \in \mathbb{N})$. Strategy profile

$$q = (q_1, \ldots, q_N) \in \mathbf{Q} = \mathbf{Q}_1 \times \mathbf{Q}_2 \times \ldots \times \mathbf{Q}_N,$$

and payoff function of player number i $\pi_i(q) = \pi_i(q_1, \ldots, q_N)$ is defined in (6).

Proposition 1. *If $a > c$, then the Berge equilibrium strategy profile in (7) is $q^B = (q_1^B, q_2^B, \ldots, q_N^B)$, where $q_i^B = \alpha$ $(i \in \mathbb{N})$, a player's profits are*

$$\pi_i^B = \pi_i(q^B) = [a - Nb\alpha]\alpha - (c\alpha + d) = [a - c]\alpha - bN\alpha^2 - d.$$

Proof. The Berge equilibrium strategy profile in the game (7) is defined by the system of N inequalities:

$$\pi_i(q\|q_i^B) \leq \pi_i(q^B) \quad \forall q \in \mathbf{Q} \quad (i \in \mathbb{N}), \tag{8}$$

where $(q\|q_i^B) = (q_1, q_2, \ldots, q_{i-1}, q_i^B, q_{i+1}, \ldots, q_N)$.

According to (6) inequalities (8) are presented as

$$\begin{cases} \left[a - b\left(q_1^B + q_2 + \ldots + q_N\right)\right]q_1^B - (cq_1^B + d) \leq \\ \qquad \leq \left[a - b\left(q_1^B + q_2^B + \ldots + q_N^B\right)\right]q_1^B - (cq_1^B + d), \\ \left[a - b\left(q_1 + q_2^B + \ldots + q_N\right)\right]q_2^B - (cq_2^B + d) \leq \\ \qquad \leq \left[a - b\left(q_1^B + q_2^B + \ldots + q_N^B\right)\right]q_2^B - (cq_2^B + d), \\ \cdot \quad \cdot \quad \cdot \quad \cdot \quad \cdot \quad \cdot \quad \cdot \quad \cdot \quad \cdot \quad \cdot \\ \left[a - b\left(q_1 + q_2 + \ldots + q_N^B\right)\right]q_N^B - (cq_N^B + d) \leq \\ \qquad \leq \left[a - b\left(q_1^B + q_2^B + \ldots + q_N^B\right)\right]q_N^B - (cq_N^B + d) \end{cases}$$

and hold for every $q_i \in \mathbf{Q}_i$ $(i \in \mathbb{N})$.

Note that the Berge equilibrium strategy profile in (7) is $q^B = (\alpha, \alpha, \ldots, \alpha)$, and the Berge equilibrium payoff π_i^B $(i \in \mathbb{N})$ are

$$\pi_i^B = \pi_i(q^B) = [a - Nb\alpha]\alpha - (c\alpha + d) = [a - c]\alpha - bN\alpha^2 - d.$$

Indeed, by reducing the supply of their own products to the market as much as possible, each player number i $(i \in \mathbb{N})$ thus increases the profits of all other participants in the game (7). □

We now turn to a Nash equilibrium in the game (7).

Proposition 2. *If $a > c$, then the Nash equilibrium strategy profile in (7) is*

$$q^e = (q_1^e, q_2^e, \ldots, q_N^e),$$

where for all $i \in \mathbb{N}$ equilibrium strategy

$$q_i^e = \begin{cases} \alpha, & if \ \dfrac{a - c}{b(N + 1)} \le \alpha, \\ \dfrac{a - c}{b(N + 1)}, & if \ \alpha < \dfrac{a - c}{b(N + 1)} < \beta, \\ \beta, & if \ \dfrac{a - c}{b(N + 1)} \ge \beta. \end{cases} \tag{9}$$

In this case player's profits are

$$\pi_i^e = \pi_i(q^e) = \begin{cases} (a - c)\alpha - bN\alpha^2 - d, & if \ \dfrac{a - c}{b(N + 1)} \le \alpha, \\ \dfrac{(a - c)^2}{(N + 1)^2 b} - d, & if \ \alpha < \dfrac{a - c}{b(N + 1)} < \beta, \\ (a - c)\beta - bN\beta^2 - d, & if \ \dfrac{a - c}{b(N + 1)} \ge \beta, \end{cases}$$

where $i \in \mathbb{N}$.

Proof. The Nash equilibrium in (7) is defined by the system of equations (see. Definition 1)

$$\begin{aligned} \pi_1(q^e) &= \max_{q_1 \in [\alpha; \beta]} \pi_1(q^e \| q_1) = \\ &= \max_{q_1 \in [\alpha; \beta]} \{[a - b(q_1 + q_2^e + \ldots + q_N^e)]q_1 - (cq_1 + d)\}, \\ \pi_2(q^e) &= \max_{q_2 \in [\alpha; \beta]} \pi_2(q^e \| q_2) = \\ &= \max_{q_2 \in [\alpha; \beta]} \{[a - b(q_1^e + q_2 + \ldots + q_N^e)]q_2 - (cq_2 + d)\}, \\ &\hspace{1cm} \cdot \quad \cdot \quad \cdot \quad \cdot \quad \cdot \quad \cdot \quad \cdot \quad \cdot \quad \cdot \\ \pi_N(q^e) &= \max_{q_N \in [\alpha; \beta]} \pi_N(q^e \| q_N) = \\ &= \max_{q_N \in [\alpha; \beta]} \{[a - b(q_1^e + q_2^e + \ldots + q_N)]q_N - (cq_N + d)\}. \end{aligned} \tag{10}$$

In (10), as well as the above, for all $i \in \mathbb{N}$ we denote $(q^e \| q_i)$ the strategy profile q^e, where the strategy of player number i q_i^e is replaced by q_i.

For each $i \in \mathbb{N}$ the maximum of $\pi_i(q^e \| q_i)$ with respect to q_i is attained if two requirements are met:

$$\frac{\partial \pi_i(q^e \| q_i)}{\partial q_i}\bigg|_{q_i=q_i^e} = [a - 2bq_i - b\,(q_1^e + \dots$$
$$\dots + q_{i-1}^e + q_{i+1}^e + \dots + q_N^e) - c]\big|_{q_i=q_i^e} = 0, \qquad (11)$$
$$\frac{\partial^2 \pi_i(q^e \| q_i)}{\partial q_i^2}\bigg|_{q_i=q_i^e} = -2b < 0.$$

The second condition (11) holds, since the coefficient of elasticity $b > 0$, and from the first equation for each $i \in \mathbb{N}$ we obtain a system of N of linear equations:

$$\begin{cases} 2q_1^e + q_2^e + q_3^e + \dots + q_N^e = \dfrac{a-c}{b}, \\ q_1^e + 2q_2^e + q_3^e + \dots + q_N^e = \dfrac{a-c}{b}, \\ \quad \cdot \quad \cdot \quad \cdot \quad \cdot \quad \cdot \quad \cdot \quad \cdot \\ q_1^e + q_2^e + q_3^e + \dots + 2q_N^e = \dfrac{a-c}{b}, \end{cases}$$

solution for which is the strategy profile

$$q^e = (q_1^e, q_2^e, \dots, q_N^e) = \left(\frac{a-c}{(N+1)b}, \frac{a-c}{(N+1)b}, \dots, \frac{a-c}{(N+1)b} \right).$$

When the condition $\alpha < \dfrac{a-c}{b(N+1)} < \beta$ is held, the found values $q_i^e (i \in \mathbb{N})$ deliver maximum value to the functions $\pi_i(q^e \| q_i)$ in the interval $[\alpha; \beta]$, and, consequently, these are the Nash equilibrium strategies in the game (7).

If $\dfrac{a-c}{b(N+1)} \le \alpha$, then, by the second inequality of (11) in the interval $[\alpha; \beta]$ each of the functions $\pi_i(q^e \| q_i)(i \in \mathbb{N})$ is monotonically decreasing. Consequently, the maximum in (10) is achieved for $q_i^e = \alpha(i \in \mathbb{N})$.

In the case when $\dfrac{a-c}{b(N+1)} \ge \beta$, functions $\pi_i(q^e \| q_i)$ for all $i \in \mathbb{N}$ will be monotonically increasing on $[\alpha; \beta]$. Accordingly, the Eq. (10) are satisfied for $q_i^e = \beta(i \in \mathbb{N})$.

Further, combining all three cited cases, we see that the Nash equilibrium strategies in (7) are defined by (9).

We proceed to the construction of the profits $\pi_i(q^e)$ in the found strategy profile (9) $(i \in \mathbb{N})$. Substituting (9) in the payoff function (6), for each player number $i(i \in \mathbb{N})$ we obtain the Nash equilibrium's profits in the game (7). Namely, if $\dfrac{a-c}{b(N+1)} \le \alpha$ we have

$$\pi_i^e = \pi_i(\alpha, \alpha, \dots, \alpha) = [a - bN\alpha]\,\alpha - (c\alpha + d) = (a-c)\alpha - bN\alpha^2 - d.$$

If $\dfrac{a-c}{b(N+1)} \geq \beta$, then the Nash equilibrium's profit of player i π_i^e is

$$\pi_i^e = \pi_i(\beta, \beta, \ldots, \beta) = [a - bN\beta]\,\beta - (c\beta + d) = (a - c)\beta - bN\beta^2 - d.$$

Finally, if $\alpha < \dfrac{a-c}{b(N+1)} < \beta$ it is

$$\pi_i^e = \pi_i\left(\frac{a-c}{(N+1)b}, \frac{a-c}{(N+1)b}, \ldots, \frac{a-c}{(N+1)b}\right) =$$

$$= \left[a - bN\frac{a-c}{(N+1)b}\right]\frac{a-c}{(N+1)b} - \left(c\frac{a-c}{(N+1)b} + d\right) =$$

$$= \frac{(a-c)^2}{(N+1)b} - \frac{(a-c)^2}{(N+1)b} \cdot \frac{N}{N+1} - d = \frac{(a-c)^2}{(N+1)^2 b} - d.$$

Thus, the profit of player number i in the Nash equilibrium is

$$\pi_i^e = \pi_i(q^e) = \begin{cases} (a-c)\alpha - bN\alpha^2 - d, & \text{if } \dfrac{a-c}{b(N+1)} \leq \alpha, \\[2mm] \dfrac{(a-c)^2}{(N+1)^2 b} - d, & \text{if } \alpha < \dfrac{a-c}{b(N+1)} < \beta, \\[2mm] (a-c)\beta - bN\beta^2 - d, & \text{if } \dfrac{a-c}{b(N+1)} \geq \beta. \end{cases}$$

$\square$

4   Comparison of the Berge Equilibrium Profits with the Nash Equilibrium

In this section, we shall compare the profits of the players, chosen according to the Berge equilibrium strategy profile, with the profits that await them in the Nash equilibrium. To do this, we again consider three cases.

Case I. At $\dfrac{a-c}{b(N+1)} \leq \alpha$ Nash equilibrium strategy x_i^e, defined in (9) coincides with the Berge equilibrium strategies $x_i^B = \alpha(i \in \mathbb{N})$. Therefore, players, adhering to the Nash equilibrium, will receive the same prizes as they would get for adhering to the Berge equilibrium, namely:

$$\pi_i^e = \pi_i^B \quad \text{for} \quad \frac{a-c}{b(N+1)} \leq \alpha. \tag{12}$$

Case II. If inequality $\alpha < \dfrac{a-c}{b(N+1)} < \beta$ is held, then the profit of player number i $(i \in \mathbb{N})$ in the Nash equilibrium strategy profile is

$$\pi_i^e = \pi_i(q^e) = \frac{(a-c)^2}{(N+1)^2 b} - d,$$

and the Berge equilibrium profit is

$$\pi_i^B = \pi_i(q^B) = \alpha[a - c - Nb\alpha] - d.$$

Their difference is

$$\pi_i^e - \pi_i^B = \left(\frac{(a-c)^2}{(N+1)^2 b} - d \right) - \left[(a-c)\alpha - bN\alpha^2 - d \right] =$$

$$= bN\alpha^2 - (a-c)\alpha + \frac{(a-c)^2}{(N+1)^2 b} = bN \left(\alpha - \frac{a-c}{(N+1)b} \right) \left(\alpha - \frac{a-c}{N(N+1)b} \right).$$

Since the number of players $N > 1$ and $a - c > 0$, then

$$\frac{a-c}{N(N+1)b} < \frac{a-c}{(N+1)b}.$$

And due to the fact that $\alpha > 0$ and the coefficient of elasticity $b > 0$, then difference $\pi_i^e - \pi_i^B$ is positive for

$$0 < \alpha < \frac{a-c}{N(N+1)b},$$

negative for

$$\frac{a-c}{N(N+1)b} < \alpha < \frac{a-c}{(N+1)b}$$

and zero if

$$\alpha = \frac{a-c}{N(N+1)b}.$$

Thus, for all $i \in \mathbb{N}$

$$\begin{cases} \pi_i^e > \pi_i^B, & \text{if } \alpha < \dfrac{a-c}{N(N+1)b} \text{ and } \dfrac{a-c}{(N+1)b} < \beta, \\ \pi_i^e = \pi_i^B, & \text{if } \alpha = \dfrac{a-c}{N(N+1)b} \text{ and } \dfrac{a-c}{(N+1)b} < \beta, \\ \pi_i^e < \pi_i^B, & \text{if } \dfrac{a-c}{N(N+1)b} < \alpha < \dfrac{a-c}{(N+1)b} < \beta. \end{cases} \tag{13}$$

Finally, we consider the last case.

Case III. When $\alpha < \beta \leq \dfrac{a-c}{(N+1)b}$ the Nash equilibrium strategy of the manufacturer i is to supply the largest quantity of goods possible to the market, i.e., $x_i^e = \beta \, (i \in \mathbb{N})$, and its profit in the Nash equilibrium will be

$$\pi_i^e = (a-c)\beta - bN\beta^2 - d.$$

In a Berge equilibrium strategy profile, the player has to reduce the supply of the product to the minimum allowed, namely, $x_i^B = \alpha \, (i \in \mathbb{N})$, and its profit in the Berge equilibrium will be

$$\pi_i^B = (a-c)\alpha - bN\alpha^2 - d.$$

Consider the difference in profits, which the participants of oligopoly (7) will have in the Nash equilibrium and Berge equilibrium. For example, for a player i ($i \in \mathbb{N}$) it will be

$$\pi_i^e - \pi_i^B = (a - c)\beta - bN\beta^2 - d - \left[(a - c)\alpha - bN\alpha^2 - d\right] =$$
$$= (a - c)(\beta - \alpha) - bN(\beta^2 - \alpha^2) = (\beta - \alpha) \cdot [a - c - bN(\beta + \alpha)].$$

Since $\beta > \alpha$, then the sign of the difference between the above coincides with that of a decreasing linear function

$$a - c - bN(\alpha + \beta).$$

This function changes the sign at $\alpha + \beta = \dfrac{a - c}{bN}$. Hence, the difference $\pi_i^e - \pi_i^B$ is zero for $\alpha + \beta = \dfrac{a - c}{bN}$ (section LM on Fig. 2). For $\alpha + \beta < \dfrac{a - c}{bN}$ difference $\pi_i^e - \pi_i^B$ is positive, and in the case $\alpha + \beta > \dfrac{a - c}{bN}$ it is negative.

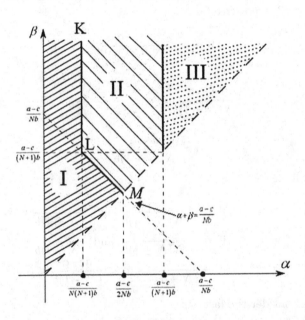

Fig. 2. Comparison of the Berge equilibrium profits with Nash equilibrium.

Therefore, when $\alpha < \beta \leq \dfrac{a - c}{(N + 1)b}$ for profit of player i in Berge equilibrium π_i^B and in Nash equilibrium π_i^e the following ratio takes place

$$\begin{cases} \pi_i^e > \pi_i^B, & \text{if } \alpha + \beta < \dfrac{a - c}{Nb} \text{ and } \alpha < \beta, \\ \pi_i^e = \pi_i^B, & \text{if } \alpha + \beta = \dfrac{a - c}{Nb} \text{ and } \alpha < \beta, \\ \pi_i^e < \pi_i^B, & \text{if } \alpha + \beta > \dfrac{a - c}{Nb} \text{ and } \alpha < \beta. \end{cases} \qquad (14)$$

Furthermore, by combining all three cases, i.e. the formulas (12), (13) and (14) we can obtain a full comparison of the Berge equilibrium and Nash equilibrium profits of player number i ($i \in \mathbb{N}$) in the game (7) (see Fig. 2).

In area I, within the Fig. 2, his profit π_i^e in the Nash equilibrium will be more than profit π_i^B, from the Berge equilibrium.

In area II, on the other hand, the Berge equilibrium gives the player i ($i \in \mathbb{N}$) more profit, than the Nash equilibrium.

In area III and on the line KLM players profits in the Berge equilibrium and the Nash equilibrium are equal.

Note 1. Propositions 1, 2 and Sect. 4 justify the following practices for the selection decision (set of strategies of the players) in a model of Cournot oligopoly.

I stage. Using constants a, b, c and N, find four numbers:

$$\frac{a-c}{N(N+1)b}, \quad \frac{a-c}{2Nb}, \quad \frac{a-c}{(N+1)b}, \quad \frac{a-c}{Nb}.$$

II stage. Using these numbers one may build Fig. 2, and highlight areas I, II, III.

III stage. Determine values α^* and β^*, forming a supply corridor q_i.

IV stage. Find dot (α^*, β^*) on Fig. 2. Then find out in which of the three areas is the point (α^*, β^*).

Finally, with the help of Propositions 1, 2 and Sect. 4 we can write out the explicit form of the equilibrium solution, i.e. an equilibrium strategy profile and also the profits of the players in it.

5 Conclusion

In conclusion, we note once again that this work disputes the claim by the Martin Shubik that the altruistic concept of the Berge equilibrium is not applicable to the problems of the economy. This work is the first to show that, on the contrary, in economic models the Berge equilibrium may be in fact more effective than the Nash equilibrium, and so it is our hope that the Berge equilibrium will take its rightful place in the economy, as it already has, both in the field of psychology as well as in sociology (see the review [3]). We also note that it is interesting to construct a Berge equilibrium in the fuzzy [8] model of Cournot's oligopoly.

References

1. Berge, C.: Théorie générale des jeux á n personnes games. Gauthier-Villars, Paris (1957)
2. Bertrand, J.: Book review of theorie mathematique de la richesse sociale and of recherches sur les principes mathematiques de la theorie des richesses. Journal des Savants **67**, 499–508 (1883)

3. Colman, A.M., Körner, T.W., Musy, O., Tazdaït, T.: Mutual support in games: some properties of Berge equilibria. J. Math. Psychol. **55**(2), 166–175 (2011). https://doi.org/10.1016/j.jmp.2011.02.001

4. Cournot, A.: Recherches sur les principes mathématiques de la théorie des richesses. Hachette, Paris (1838)

5. Courtois, P., Nessah, R., Tazdaït, T.: Existence and computation of Berge equilibrium and of two refinements. J. Math. Econ. **72**, 7–15 (2017). https://doi.org/10.1016/j.jmateco.2017.04.004

6. Crettez, B.: On Sugden's "mutually beneficial practice" and Berge equilibrium. Int. Rev. Econ. **64**(4), 357–366 (2017). https://doi.org/10.1007/s12232-017-0278-3

7. Hotelling, H.: Stability in competition. Econ. J. **39**, 41–57 (1929)

8. Kudryavtsev, K., Stabulit, I., Ukhobotov, V.: A bimatrix game with fuzzy payoffs and crisp game. In: CEUR Workshop Proceedings 1987, pp. 343–349 (2017)

9. Larbani, M., Zhukovskii, V.I.: Berge equilibrium in normal form static games: a literature review. Izv. IMI UdGU **49**, 80–110 (2017). https://doi.org/10.20537/2226-3594-2017-49-04

10. Lung, R.I., Suciu, M., Gaskó, N., Dumitrescu, D.: Characterization and detection of *epsilon*-Berge-Zhukovskii equilibria. PLoS ONE **10**(7), e0131983 (2015). https://doi.org/10.1371/journal.pone.0131983

11. Musy, O., Pottier, A., Tazdaït, T.: A new theorem to find Berge equilibria. Int. Game Theory Rev. **14**(01), 1250005 (2012). https://doi.org/10.1142/s0219198912500053

12. Nash, J.F.: Equilibrium points in N-person games. Proc. Natl. Acad. Sci. USA **36**, 48–49 (1950)

13. Nessah, R., Larbani, M., Tazdaït, T.: A note on Berge equilibrium. Appl. Math. Lett. **20**(8), 926–932 (2007). https://doi.org/10.1016/j.aml.2006.09.005

14. Shubik, M.: Review of C. Berge, General theory of n-person games. Econometrica **29**(4), 821 (1961)

15. Tirole, J.: The Theory of Industrial Organization. MIT press, Cambridge (1988)

16. Vaisman, K.S.: The Berge equilibrium. Abstract of Cand. Sci. (Phys.-Math.). Dissertation, St. Petersburg (1995). (in Russian)

17. Vaisman, K.S.: The Berge equilibrium for linear-quadratic differential game. In: Multiple Criteria Problems Under Uncertainty: Abstracts of the Third International Workshop, Orekhovo-Zuevo, Russia, p. 96 (1994)

18. Zhukovskii, V.I., Chikrii, A.A.: Linear-quadratic differential games. Naukova Dumka, Kiev (1994). (in Russian)

19. Zhukovskiy, V.I., Kudryavtsev, K.N.: Mathematical foundations of the Golden Rule. I. Static case. Autom. Remote. Control. **78**(10), 1920–1940 (2017). https://doi.org/10.1134/S0005117917100149

The Model of the Russian Banking System with Indicators Nominated in Rubles and in Foreign Currency

Nikolai Pilnik[1,3,5] , Stanislav Radionov[1,3,5(✉)] , and Artem Yazikov[1,2,4]

[1] National Research University Higher School of Economics,
20 Myasnitskaya Street, Moscow, Russia
u4d@yandex.ru, saradionov@edu.hse.ru, ayazikov@hse.ru
[2] Federal Research Center "Computer Science and Control" of the RAS,
40 Vavilova Street, Moscow, Russia
[3] Lebedev Physical Institute of the RAS, 53 Leninsky Avenue, Moscow, Russia
[4] Moscow Institute of Physics and Technology (Technical University),
9 Institutskiy lane, Dolgoprudny, Moscow Region, Russia
[5] Financial Research Institute of the Ministry of Finance of the Russian Federation,
3 Nastasyinsky Lane, Moscow, Russia

Abstract. We propose a model of the Russian banking system. It is based on the problem of a macroeconomic agent "bank" which is modeled according to the principles of aggregated description, optimality, and perfect foresight. To derive the equations of the model, we use the original method of relaxation of complementary slackness conditions. The model successfully reproduces main indicators of the banking system, such as total loans, deposits, settlement accounts, reserves and profits nominated both in rubles and in foreign currency.

Keywords: Banking system · Optimal control
Complementary slackness conditions

1 Introduction and Literature Review

After the financial crisis of 2007, it became clear that the financial sector is a much more significant part of the economy that it was thought before. It has also become obvious that macroeconomic theory has no instruments to perform a rigorous analysis of the state of the banking system and its impact of the rest of the economy. Pre-crisis works incorporating banking sector in macroeconomic models such as [4,5,7,15] treated it as a mere intermediary which thus could not influence the economy in a non-trivial way. More advanced post-crisis models were presented in [6,11–14]. In these models, the banking sector is subject to different kinds of distortions and the source of distortions for the rest of the economy itself. Other notable work which is probably closest to ours is [3]. It presents the DSGE model containing a non-trivial description of banks which seek to maximize their net worth by choosing the optimal lending contracts.

The research was supported by the Russian Scientific Fund (project 14-11-00432).

Y. Evtushenko et al. (Eds.): OPTIMA 2018, CCIS 974, pp. 427–438, 2019.
https://doi.org/10.1007/978-3-030-10934-9_30

Banks conform to government regulation—in the case of insufficient net worth they have to pay a penalty.

We also consider the optimization problem of the agent "bank", but with a wider set of available controls. Bank can not only lend to producers as in [3], but also to consumers and can also take savings from both. Savings and loans can be done both in national currency (Russian rubles) and in a foreign one (actually in a currency basket containing US dollars and euros). The bank is subject to liquidity constraint and has the obligation to keep a certain level of reserves. We do not present a full general equilibrium model as in [3] but our description of the bank is developed in such a way that it can be used as a block of that kind of model.

Identifying the parameters of models of this type is extremely difficult. Very often, the performance of the model is demonstrated only on the interval at which the parameters were evaluated. However, the question of the quality of forecasts outside this interval is much more important, because it allows answering the question whether it is possible to use the model for policy evaluations.

The procedure of parameter identification is also rather complex. For example, in the DSGE framework, the linearization of the model is required. Use of the second order approximation significantly complicates the computations. In the framework presented in this article the procedure for estimating the parameters of the model is carried out precisely for the same system of equations that was obtained when solving the bank's problem in continuous time and transforming it into a discrete variant for comparison with the data.

This work continues our research in the Russian banking sector presented in [1, 2, 17]. Our contribution in this article is threefold. First, we present a new methodology of relaxation of complementary slackness conditions and the interpretation behind it. Second, we consider a wider set of macroeconomic variables whose trajectories are successfully replicated by our model. Third, we provide a comparison of two methods of identification of the model described later and standard econometric methods. We find that the set of parameters derived via multi-step forecasting technique proposed in [17] is much more robust in a sense discussed later than the ones derived via the standard technique of identification of parameters of dynamic models. Econometric models such as AR, ARIMA, VAR, VARX perform much worse than above-mentioned methods and probably cannot be used for a policy analysis, ARIMAX with properly chosen exogenous variables performs about the same but contains significantly more parameters.

2   The Model: Statement

We consider the model of the whole banking system as a single agent. We will reduce it to the usual dynamic model that determines the demand of the banking system for deposits and the supply of loans depending on the current state,!! as well as on interest rates and other external factors on the market.

Consider the bank, which by the time t attracted deposits $S(t)$ and issued loans $L(t)$. The average terms for which deposits are attracted and loans are issued (durations), will be denoted, respectively, by $(\beta_S(t))^{-1}$ and $(\beta_L(t))^{-1}$. The variables $\beta_S(t)$ and $\beta_L(t)$ will be referred below as the inverse durations and interpreted as the average frequency of the return of deposits and loans, respectively. Then the dynamics of loans and deposits is described by equations

$$\frac{d}{dt}L(t) = K(t) - \beta_l(t)L(t),\ \frac{d}{dt}S(t) = V(t) - \beta_s(t)S(t), \tag{1}$$

where $0 \leq K(t), 0 \leq V(t)$ are the flows of newly issued loans and newly attracted deposits.

With respect to loans granted, the bank receives interest payments $r_L(t)L(t)$, where $r_L(t)$ is the effective interest rate on loans. The bank pays interest on borrowed funds $r_S(t)S(t)$, where $r_S(t)$ is the effective interest rate on deposits.

In addition to loans and deposits in national currency (rubles), the bank also attracts deposits $vS(t)$ and issues loans $vL(t)$ in dollars. The dynamics of foreign currency loans and deposits is described by equations

$$\frac{d}{dt}vL(t) = vK(t) - \beta_{vl}(t)vL(t),\ \frac{d}{dt}vS(t) = vV(t) - \beta_{vs}(t)vS(t). \tag{2}$$

Here $\beta_{vl}(t)$, $\beta_{vs}(t)$ are the reverse durations of currency loans and deposits, respectively. Newly issued foreign currency loans and attracted foreign currency deposits are denoted as $0 \leq vK(t)$, $0 \leq vV(t)$. The effective interest rate on foreign currency deposits is denoted by $r_{vs}(t)$, on foreign currency loans—by $r_{vl}(t)$.

In addition to deposits, the bank also attracts funds in the form of interest-free balances of settlement accounts $N(t)$. There is no regulating value for the size of these balances, and the bank should simply focus on the supply from the clients. Therefore, this value is considered to be bounded exogenously:

$$N(t) \leq N_n(t), \tag{3}$$

where $N_n(t)$ is the supply of the balances of settlement accounts known to the bank.

Other sources of raising funds for the bank are interest funds of the Central Bank $Sc(t)$, whose volume is also an exogenous variable determined by the regulator, denoted by $Sc_{sc}(t)$. Thus

$$Sc(t) \leq Sc_{sc}(t). \tag{4}$$

The interest rate on Central Bank deposits in the banking system is $r_{sc}(t)$.

For servicing operations related to loans and deposits, the bank uses two types of liquid funds nominated in ruble $A(t)$ and in foreign currency $vQ(t)$. It is assumed that their volumes in the balance sheet are proportional to ruble and foreign currency loans and deposits, respectively. These proportions are defined by the coefficients $\tau_l(t)$, $\tau_s(t)$, $\tau_{vl}(t)$, $\tau_{vs}(t)$:

$$\tau_{vs}vS(t) + \tau_{vl}vL(t) \leq vQ(t),\ \tau_s S(t) + \tau_l L(t) \leq A(t). \tag{5}$$

Part of the attracted deposits should be placed as mandatory reserves in the Central Bank $Rc(t)$. The mandatory reserve requirements for ruble and foreign currency deposits may differ and equal $n_s(t)$ and $n_{vs}(t)$.

$$n_s(t) S(t) + n_{vs}(t) w_w(t) vS(t) \le Rc(t), \tag{6}$$

where $w_w(t)$ is the exchange rate.

Other expenses of the bank are denoted by $OC_o(t)$ and are considered as an exogenous variable.

Thus, the financial balance of the bank can be written as

$$
\begin{aligned}
\frac{d}{dt} A(t) &= -K(t) + \beta_l(t) L(t) + r_l(t) L(t) - w_w(t)(vK(t) - \beta_{vl}(t) vL(t)) \\
&+ w_w(t) r_{vl}(t) vL(t) + V(t) - \beta_s(t) S(t) - r_s(t) S(t) \\
&+ w_w(t)(vV(t) - \beta_{vs}(t) vS(t)) - w_w(t) r_{vs}(t) vS(t) + \frac{d}{dt} Sc(t) \\
&- r_{sc}(t) Sc_{sc}(t) + \frac{d}{dt} N(t) - \left(\frac{d}{dt} vQ(t)\right) w_w(t) - \frac{d}{dt} Rc(t) - OC_o(t) - Z(t),
\end{aligned}
\tag{7}
$$

where $Z(t)$ is the amount of undistributed profit received by the bank.

The above relationships represent the limitations imposed within the model on the bank's ability to chose the values of its planned variables (controls):

$$
\begin{aligned}
&K(t), L(t), V(t), S(t), vK(t), vL(t), vV(t), vS(t), Sc(t), Z(t), N(t), \\
&Rc(t), A(t), vQ(t).
\end{aligned}
\tag{8}
$$

According to the principle of rational expectations underlying the models of intertemporal equilibrium, when planning its control variables, the bank can rely on an accurate forecast of information variables:

$$
\begin{aligned}
&N_n(t), OC_o(t), Sc_{sc}(t), \beta_l(t), \beta_s(t), \beta_{vl}(t), \beta_{vs}(t), n_s(t), n_{vs}(t), r_l(t), \\
&r_s(t), r_{sc}(t), r_{vl}(t), r_{vs}(t), w_w(t).
\end{aligned}
\tag{9}
$$

As a result, the choice of the planned variables by the bank actually determines the supply of loans to them, as well as its demand for attracted and liquid funds as a function of current and future values of information variables (primarily interest).

The goal of the bank in the model is to maximize the total discounted utility from the received undistributed profit taking into account the given deflator $\zeta_{liq}(t)$. Therefore, the functional of the bank can be written in the form

$$\int_{t0}^{T} \frac{e^{-\Delta t}}{1 - \eta} \left(\frac{Z(t)}{\zeta_{liq}(t)}\right)^{1-\eta} dt. \tag{10}$$

For the solvability of the constraint problem it is necessary to supplement with terminal conditions, which can be defined as a growth condition for some linear form of the phase variables:

$$
\begin{aligned}
&\gamma \Omega(t0) \le \Omega(T), \Omega(t) = A(t) + L(t) - N(t) + Rc(t) - S(t) - Sc(t) \\
&+ w_w(t) vL(t) + w_w(t) vQ(t) - w_w(t) vS(t).
\end{aligned}
\tag{11}
$$

This is an analog of no Ponzi condition written in terms of bank's own capital which is a difference between its assets and liabilities.

To derive the solution of the maximization problem of the function (10) under constraints (1)–(7) and the terminal condition (11) with respect to the variables (8), we compose a Lagrange functional and find its saddle point. The dual variables are denoted by $\Phi 2(t), \Phi 4(t), \Phi 6(t), \Phi 8(t)$. The obtained system, which is a set of sufficient conditions for optimality, is given below.

$$[\Phi 2(t)\,\xi(t)][K(t)], \ [\Phi 4(t)\,\xi(t)\,w_w(t)][vK(t)],$$

$$[\Phi 6(t)\,\xi(t)][V(t)], \ [\Phi 8(t)\,\xi(t)\,w_w(t)][vV(t)],$$

$$\frac{d}{dt}L(t) = K(t) - \beta_l(t)L(t), \ \frac{d}{dt}vL(t) = vK(t) - \beta_{vl}(t)vL(t),$$

$$\frac{d}{dt}S(t) = V(t) - \beta_s(t)S(t), \ \frac{d}{dt}vS(t) = vV(t) - \beta_{vs}(t)vS(t),$$

$$\frac{d}{dt}\Phi 2(t) = \beta_l(t)\Phi 2(t) - \tau_l\rho(t) + r_l(t) + \Phi 2(t)\rho(t) - \rho(t),$$

$$\frac{d}{dt}\Phi 4(t) = \beta_{vl}(t)\Phi 4(t) - \rho(t)\tau_{vl} + \frac{\left(\frac{d}{dt}w_w(t)\right)\tau_{vl}}{w_w(t)} + r_{vl}(t) + \rho(t)\Phi 4(t)$$

$$- \frac{\left(\frac{d}{dt}w_w(t)\right)\Phi 4(t)}{w_w(t)} - \rho(t) + \frac{\frac{d}{dt}w_w(t)}{w_w(t)},$$

$$\frac{d}{dt}\Phi 6(t) = \beta_s(t)\Phi 6(t) - \rho(t)n_s(t) - \tau_s\rho(t) - r_s(t) + \Phi 6(t)\rho(t) + \rho(t),$$

$$\frac{d}{dt}\Phi 8(t) = \beta_{vs}(t)\Phi 8(t) - \rho(t)\tau_{vs} - n_{vs}(t)\rho(t) + \frac{\left(\frac{d}{dt}w_w(t)\right)\tau_{vs}}{w_w(t)} - r_{vs}(t)$$

$$+ \Phi 8(t)\rho(t) - \frac{\left(\frac{d}{dt}w_w(t)\right)\Phi 8(t)}{w_w(t)} + \rho(t) - \frac{\frac{d}{dt}w_w(t)}{w_w(t)},$$

$$\frac{d}{dt}A(t) = -K(t) + \beta_l(t)L(t) + r_l(t)L(t) - w_{vl}(t)(vK(t) - \beta_{vl}(t)vL(t))$$

$$+ w_{vl}(t)r_{vl}(t)vL(t) + V(t) - \beta_s(t)S(t)$$

$$- r_s(t)S(t) + w_{vs}(t)(vV(t) - \beta_{vs}(t)vS(t)) - w_{vs}(t)r_{vs}(t)vS(t) + \frac{d}{dt}Sc(t)$$

$$- r_{sc}(t)Sc_{sc}(t) + \frac{d}{dt}N(t) - \left(\frac{d}{dt}vQ(t)\right)w_w(t) - \frac{d}{dt}Rc(t) - OC(t) - Z(t),$$

$$\frac{d}{dt}Z(t) = -\frac{Z(t)\Delta}{\eta} + \frac{Z(t)\rho(t)}{\eta} - \frac{Z(t)\frac{d}{dt}\zeta_{liq}(t)}{\zeta_{liq}(t)\eta} + \frac{Z(t)\frac{d}{dt}\zeta_{liq}(t)}{\zeta_{liq}(t)},$$

$$A(t) = \tau_s S(t) + \tau_l L(t), \ Rc(t) = n_s(t)S(t) + n_{vs}(t)w_w(t)vS(t),$$

$$vQ(t) = \tau_{vs}vS(t) + \tau_{vl}vL(t),$$

where $[a][b]$ means $a \geq 0, b \geq 0, ab = 0$.

These conditions fall into four groups:

1. initial differential equations, the initial conditions for which are assumed to be given;

2. conjugate differential equations for variables dual to the constraints of the problem;
3. transversality conditions that define terminal conditions for conjugate differential equations;
4. complementary slackness conditions for inequalities.

Briefly describe the main stages of analysis and transformation of the problem (it should be noted that they are described in a more detailed way in the accompanying paper). The main difficulty in using equations in continuous time when working with statistical data is the presence of direct and dual time variables in these equations. As part of our approach, we replace these derivatives by increments. Moreover, for direct variables, backward increments are used, and for dual variables, forward increments are used. This leads to considering some of the direct variables not in the current, but at the previous period. Next, we transform the differential equations for the dual variables using the so-called turnpike property. This transformation allows us to eliminate the impact of future values of dual variables on the present ones and also guarantee the fulfillment of the terminal condition.

A separate discussion is required for the complementary slackness conditions. In fact, they are degenerate functional dependencies, the plot of which delineates the boundaries of the first quarter, and, consequently, the optimal solutions of our system will "jump on the corners". From an economic point of view, they describe the infinitely elastic functions of the agent's demand or supply. But from the point of view of the subsequent calibration of the model, these relationships represent a significant difficulty, increasing the instability of estimates of coefficients by an order of magnitude and, as a consequence, constructed according to the forecast model. Nevertheless, with the help of natural assumptions about the alternation of regimes determined by the method of resolution by the condition of complementary non-rigidity, it turns out to be possible to proceed to more regular and convenient relations from the point of view of model calibration.

Preliminary analysis of the data showed that there is a strong seasonal component in some series, such as are ruble deposits and ruble liquidity, showing clear peaks at the end of each year. As the part of our approach, we do not make a preliminary adjustment to this seasonality, but instead, where necessary, we introduce a special dummy variable $Seas(t)$ that takes unit values every December and vanishes at all other points.

After all the transformations described, we proceed to the following system (a variable $B(t)$ was introduced for convenience)

$$L(t) - L(t-1) = K(t) - \beta_l(t)\ L(t-1),$$

$$vL(t) - vL(t-1) = vK(t) - \beta_{vl}(t)\ vL(t-1),$$

$$S(t) - S(t-1) = V(t) - \beta_s(t)\ S(t-1),$$

$$vS(t) - vS(t-1) = vV(t) - \beta_{vs}(t)\ vS(t-1),$$

$$Z(t) - Z(t-1) = -\frac{Z(t-1)\,\Delta}{\eta} + \frac{Z(t-1)\,\rho(t)}{\eta} - \frac{Z(t-1)\,(\zeta_{liq}(t) - \zeta_{liq}(t-1))}{\zeta_{liq}(t-1)\,\eta}$$

$$+ \frac{Z(t-1)\,(\zeta_{liq}(t) - \zeta_{liq}(t-1))}{\zeta_{liq}(t-1)},$$

$$B(t) = Sc(t) - Sc(t-1) - r_{sc}(t)\,Sc(t-1) + N_n(t) - N_n(t-1) - OC_o(t) - Z(t),$$

$$vV(t) = \left(-r_{vs}(t)\,a1 + b1\,\beta_{vs}(t) - \frac{(b1 - (\tau_{vs} - 1)\,a1)\,(w_w(t) - w_w(t-1))}{w_w(t-1)} \right) vS(t-1)$$

$$+ (((1 - n_{vs}(t) - \tau_{vs})\,a1 + b1)\,\rho(t))\,vS(t-1) - \frac{c1\,B(t)}{w_w(t)\,(1 - n_{vs}(t) - \tau_{vs})},$$

$$V(t) = (-r_s(t)\,a2 + b2\,\beta_s(t) + ((1 - n_s(t) - \tau_s)\,a2 + b2)\,\rho(t))\,S(t-1)$$

$$- \frac{c2\,B(t)}{1 - n_s(t) - \tau_s} + d2\,Seas(t) + d3\,Seas(t-1),$$

$$vK(t) = \left(r_{vl}(t)\,a3 + b3\,\beta_{vl}(t) + \frac{((\tau_{vl} + 1)\,a3 - b3)\,(w_w(t) - w_w(t-1))}{w_w(t-1)} \right) vL(t-1)$$

$$- ((\tau_{vl} + 1)\,a3 - b3)\,\rho(t)\,vL(t-1) + \frac{c3\,B(t)}{w_w(t)\,(\tau_{vl} + 1)},$$

$$K(t) = (r_l(t)\,a4 + b4\,\beta_l(t) + ((-\tau_l - 1)\,a4 + b4)\,\rho(t))\,L(t-1) + \frac{c4\,B(t)}{\tau_l + 1},$$

$$A(t) = \tau_s S(t) + \tau_l L(t) + d1\,Seas(t),$$

$$Rc(t) = n_s(t)\,S(t) + n_{vs}(t)\,w_w(t)\,vS(t),$$

$$vQ(t) = \tau_{vs} vS(t) + \tau_{vl} vL(t).$$

The variable $\rho(t)$ that appears in the final system is the rate of the falling of the dual variable to the financial balance and is treated as profitability. A rather cumbersome equation for it can be found by substituting the above ratios into the financial balance:

$$A(t) - A(t-1) = -K(t) + \beta_l(t)\ L(t-1) + r_l(t)\ L(t-1) - w_w(t)\ vK(t)$$

$$+ w_w(t)\ \beta_{vl}(t)\ vL(t-1) + w_w(t)\,r_{vl}\ (t)\ vL(t-1) + V(t) - \beta_s(t)\ S(t-1)$$

$$- r_s(t)\ S(t-1) + w_w(t)\ vV(t) - w_w(t)\ \beta_{vs}(t)\ vS(t-1)$$

$$- w_w(t)\ r_{vs}(t)\ vS(t-1) + Sc(t) - Sc(t-1) - r_{sc}(t)\ Sc(t-1) + N_n(t)$$

$$- N_n(t-1) - (vQ(t) - vQ(t-1))\,w_w(t) - Rc(t) + Rc(t-1) - OC_o(t) - Z(t).$$

3   The Model: Identification

3.1   Statistical Data

Our main source of statistical data is the Form 101 (turnover sheet) published by the Bank of Russia on a monthly basis starting from January 2004. It contains information on approximately 1400 accounts for about 1200 banks. From 2007, apart from account balance, the data on account turnover is also available both in rubles and in foreign currency. We use this data on the period from January 2010 to January 2018. We aggregate these approximately 1400 accounts into 14 basic model aggregates. In the presented model we do not distinguish between counterparties—for example, loans of households, firms, and all other economic agents are included in L.

To calculate the duration of loans and deposits, the information on their classification by the terms of the returns from Form 101 was used. Data on interest rates, the norm of mandatory reservation and the exchange rate are taken from the official site of the Bank of Russia.

The review of different aspects of the Russian banking system can be found in [8–10, 16, 18].

3.2   Econometric Approach

When analyzing the accuracy of data reproduction by the model, we use econometric models as a benchmark for comparison. We use standard econometric models (AR, ARIMA, VAR), as well as models containing some exogenous variables (ARIMAX, VARX). For accuracy analysis, the sample was divided into training and testing intervals. On the training interval, the coefficients of the models were estimated, then a forecast was made on the testing interval and its closeness to the actual data was checked. Such an experiment was conducted for different training intervals, each of which began in January 2010, and ended in one of the months from January 2016 to June 2017. Accordingly, the forecast was calculated for the next six months.

As a measure of accuracy, the percentage deviation of the predicted value of the indicator from the actual one was used. By averaging the deviations of all the projections by a certain number of months in advance, obtained in the evaluation at different intervals, we obtain for each indicator the mean absolute percent error (MAPE) with a prediction for a given depth.

Based on the results of the comparison of econometric models, it was found out that the models from the ARIMAX class demonstrate the greatest accuracy in forecasting. Models AR, ARIMA, VAR are worse because they do not take into account the significant information contained in exogenous variables. The VARX model contains too many coefficients, which leads to a strong instability of estimates and forecasts. Therefore, a further comparison of the accuracy of forecasts will be carried out only for ARIMAX models.

The determination of the model parameters for each indicator was carried out according to the following procedure. In the first step, using the KPSS test, the order of integration of the series was determined. In the second step, after taking the appropriate number of differences using the criteria AIC and BIC, the optimal structure of the autoregressive part was found. At the third step, exogenous variables were added to the model. It should be noted that, unlike the other econometric models, in this case, the exogenous variables for different variables differed. Their choice was carried out also on the basis of informative considerations. For credits and deposits, the appropriate duration and interest rates were used as exogenous variables. For currency variables, the exchange rate was also added. A seasonal component has been added for deposits and liquidity. For ruble and foreign currency deposits, the mandatory reserve ratio was taken into account. Finally, settlement accounts and central bank deposits were added to all equations as an exogenous variable.

3.3 Multi-step Forecasting

The standard approach to estimating the parameters of models of this type is to fix the initial values of the phase variables of the model, calculate, at given initial values of the parameters of the values of the variables at all other instants of time, according to the dynamic system written out and gradually select these parameters to minimize a certain quality criterion for fitting the model. However, due to the nonstationarity of the simulated variables and the instability of the equations, such an approach resulted in unstable values of parameters and poor quality of forecasts. For these reasons, we use another method for identifying model parameters.

We apply a methodology of multi-step forecasting, which, to the best of our knowledge, was used only in econometric models before. For every period $t \in t0, ..., T-1$ we calculate forecasts for a fixed number of periods. The parameters of the model minimize the total error of all forecasts:

$$\sum_{t=t0}^{T-1} \sum_{\gamma=1}^{\tau} \sum_{u \in \{U\}} \left(\frac{u(t, t+\gamma) - u_{st}(t+\gamma)}{u_{st}(t+\gamma)} \right)^2 \rightarrow \min_{P}.$$

Here $u(t, t+\tau)$ is a forecast value calculated at the period t for the period $t + \tau$, u_{st} is the statistical value to be approximated, τ is the desired length of forecast, $U = \{L, S, vL, vS, A, Rc, vQ\}$ is a list a variables we want to forecast, $P = \{a1, a2, a3, a4, b1, b2, b3, b4, c1, c2, c3, d1, d2, d3\}$ is a list of parameters we want to determine.

For the calibration of the model, the MATLAB computing environment was used. Since the parameters of the model can take any values, we face the problem of unconstrained optimization. It was solved using the lsqnonlin command. The search for model parameters was carried out by the Monte Carlo method—the components of the initial point were assumed to be distributed uniformly in segments taken from a priori considerations about their order. After a large number of iterations of the Monte Carlo method, a set of parameters which provides a good quality of fitting the data both in-sample and out-of-sample was found.

As for econometric models, different values of T were considered, from July 2016 to July 2017. Different lengths of forecast τ, from 1 to 5, were also tested. Surprisingly enough, the optimal values of parameters were about the same for any considered values of T and τ. This important stability property ensures us that the model can be used in policy analysis.

3.4   Comparison of Results

The following table shows the average percentage deviation of the forecast values from the actual forecast for each of the six months, as well as the mean values for the entire forecasting period (Table 1).

Table 1. Mean absolute percentage error for each of the six months

		1	2	3	4	5	6	Mean
L	Model	0.7	1.4	1.4	1.1	1.5	1.7	1.3
L	ARIMAX	0.5	0.7	1.0	1.3	1.8	2.3	1.3
S	Model	1.4	1.9	2.9	4.5	5.3	6.1	3.7
S	ARIMAX	0.9	1.2	1.1	1.3	1.4	1.5	1.2
vL	Model	0.6	1.7	2.9	3.9	5.2	6.8	3.5
vL	ARIMAX	1.2	2.0	2.8	3.8	4.9	6.0	3.5
vS	Model	2.7	4.6	6.9	9.9	12.4	15.8	8.7
vS	ARIMAX	1.1	2.0	2.7	3.8	4.4	5.0	3.2
A	Model	5.1	12.2	5.9	8.9	8.5	4.8	7.6
A	ARIMAX	8.6	9.7	9.3	8.7	7.6	7.8	8.6
Rc	Model	0.3	0.6	1.0	1.1	1.4	1.3	1.0
Rc	ARIMAX	2.7	2.8	3.7	4.6	4.9	5.4	4.0
vQ	Model	12.0	5.0	6.5	12.4	19.5	34.5	15.0
vQ	ARIMAX	3.8	6.1	8.3	8.4	9.6	11.0	7.9

As we can see, the quality of the forecast of the presented model is less than econometric for only three indicators: ruble and foreign currency deposits and currency liquidity. For all other indicators, the quality of forecasts is the same or even higher for the presented model. This result allows us to state that our model operates at the level of ARIMAX models and at the same time outperforms such models as AR, ARIMA, VAR, VARX. At the same time, the total number of parameters for all the estimated ARIMAX models was 43 (19 are used in ARIMA parts and 24 more for exogenous variables). In the presented model, 21 parameters were estimated, that is, half the size.

4   Conclusions

We describe the model of the banking system of the Russian Federation and the technology of identification of its parameters. The new technology of multi-step forecasting is described. It is shown that the quality of in-sample and out-of-sample data fitting of this model is about the same as for the ARIMAX model and significantly better than AR, ARIMA, VAR, VARX. It is also worth noting that it contains much fewer parameters than ARIMAX, which leaves a significant reserve for increasing the accuracy. We believe this model can be useful for modeling the effects of public policy.

References

1. Andreyev, M., Pilnik, N., Pospelov, I.: Modelirovanie deyatelnosti sovremennoy rossiyskoy bankoskoy sistemy. HSE Econ. J. **13**(2), 143–171 (2009). (in Russian)
2. Andreyev, M., Pilnik, N., Pospelov, I.: Silniy magistralny effekt v modeli ratsionalnykh ozhidaniy sovremennoy bankovskoy sistemy rossii. J. New Econ. Assoc. **1**(2), 70–84 (2009). (in Russian)
3. Benes, J., Kumhof, M.: Risky bank lending and optimal capital adequacy regulation. Technical report, IMF Working Papers 11/130, International Monetary Fund (2011)
4. Bernanke, B., Gertler, M.: Agency costs, net worth, and business fluctuations. Am. Econ. Rev. **79**(1), 14–31 (1989)
5. Bernanke, B., Gertler, M., Gilchrist, S.: The financial accelerator in a quantitative business cycle framework. In: Taylor, J.B. Woodford, M. (eds.) Handbook of Macroeconomics, 1 edn, vol. 1, Chap. 21, pp. 1341–1393 (1999)
6. Brunnermeier, M.K., Sannikov, Y.: A macroeconomic model with a financial sector. Am. Econ. Rev. **104**(2), 379–421 (2014)
7. Carlstrom, C., Fuerst, T.: Agency costs, net worth, and business fluctuations: a computable general equilibrium analysis. Am. Econ. Rev. **87**(5), 893–910 (1997)
8. Dubinin, S.: Rossiiskaia bankovskaia sistema - ispytanie finansovym krizisom. Money Credit **1**, 9–12 (2015). (in Russian)
9. Fungacova, Z., Solanko, L.: The Russian banking industry after the 2008–2009 financial crisis-what next? Technical report, 74, Russian analytical digest (2010)
10. Fungacova, Z., Solanko, L., Weill, L.: Market power in the Russian banking industry. Econ. Int. **4**, 127–145 (2011)
11. Gertler, M., Karadi, P.: A model of unconventional monetary policy. J. Monet. Econ. **58**(1), 17–34 (2011)
12. Gertler, M., Kiyotaki, N.: Financial intermediation and credit policy in business cycle analysis. In: Friedman, B.M., Woodford, M. (ed.) Handbook of Monetary Economics, 1 edn, vol. 3, Chap. 11, pp. 547–599 (2010)
13. Gertler, M., Kiyotaki, N., Queralto, A.: Financial crises, bank risk exposure and government financial policy. J. Monet. Econ. **59**(5), 17–34 (2012)
14. He, Z., Krishnamurthy, A.: A model of capital and crises. Rev. Econ. Stud. **79**(2), 735–777 (2012)
15. Kiyotaki, N., Moore, J.: Credit cycles. J. Polit. Econ. **105**(2), 211–248 (1997)
16. Mamonov, M.: Rynok kreditov naseleniyu: identifikatsiya sprosa i predlozheniya v ramkah vecm-analiza. HSE Econ. J. **21**(2), 251–282 (2017)

17. Pilnik, N., Radionov, S.: O novykh podkhodah k identifikacii blokov modeley obshego ravnovesiya. MIPT Proc. **9**(3), 151–160 (2017). (in Russian)
18. Zhuravleva, T., Leonov, M.: Bankovskaya sistema v rossii v poslednie gody: obshy i regionalny vzglyad. Financ. J. **28**(6), 47–58 (2015)

Research on the Model of Population Groups Human Capital Dynamics

Igor G. Pospelov[1] and Ivan G. Kamenev[1,2]

[1] Dorodnicyn Computing Centre, Federal Research Center
"Computer Science and Control" of Russian Academy of Sciences, Moscow, Russia
[2] FGAEI HE National Research University "Higher School of Economics",
Moscow, Russia
igekam@gmail.com

Abstract. This article describes the dynamic optimization model with human capital as a group educational characteristic (along with these groups population) and as the main factor of their production. The main feature of this model is inequality in qualification which leads towards the run for the middle as unlinear dynamics of educational effectiveness for different groups. The research of the simulation model in one specific regime allowed to describe two different scenarios. They include the development of the groups and run for the middle dynamics. These results allow stating conceptual usability of the model for real society dynamics description.

Keywords: Education · Human capital · Group human capital
Mathematical modeling · Optimization

1 Introduction

The modern post-industrial economy is the economy of knowledge and information. At the same time, it is the economy of the human factor. The key role of man in the modern economy can be attributed to him being a carrier of knowledge (in particular, the implicit knowledge) and ensures the service sector functioning. Thus, it is necessary to develop economic models, in which not physical capital, but human capital would play this key role.

The concept of "Human capital" means a combination of production factors, which are inalienable from man. This concept has been evolved for a long time, so various researchers use this term in a completely different sense. As a result, the term is used along with the concepts of "labor resources", "human resources", "human potential", "social capital", "intellectual capital" etc., often as synonyms. Among the various human capital conceptualizations, it is essential to distinguish three types of this term interpretations depending on the modeled subject (agent):

This work was supported by the Russian Science Foundation, project 14-11-00432. The authors are grateful to G.K. Kamenev for mathematical calculation tools applied in this paper.

© Springer Nature Switzerland AG 2019
Y. Evtushenko et al. (Eds.): OPTIMA 2018, CCIS 974, pp. 439–452, 2019.
https://doi.org/10.1007/978-3-030-10934-9_31

- the individual human capital (see, for example, publications [1,3]);
- the group, community, strata human capital (see, for example, publications [4, 6]);
- the society, national economy human capital (see, for example, publications [1–4]).

It should be noted that the first interpretation is purely microeconomic, and the third one combines human capital with human potential. In this study, the second interpretation will be applied. Consequently, human capital is interpreted as the competence of the group, its ability of useful products creation. This interpretation tends to be applied to the firm's human capital (both in microeconomic and macroeconomic models). In this study, however, we propose to apply this interpretation to the simplest macroeconomic model in which households are the only economic agents. More specifically, these are large groups called clans. Each subject has exactly 2 characteristics: quantity and qualification.

The dynamic model is to describe their changes. Each clan at any given time distributes its members between production, education, and service in accordance with the chosen strategy. On the one hand, the description of society as the collection of large clans (families, communities) can be applied to human civilization in the early stages of development (starting from the primitive community and finishing with the Renaissance workshop). The completely different interpretation of this approach (like strata whose members collective behavior is regulated by general social norms) can be applied to modern society.

Let us briefly formulate this model basic premises:

1. physical capital (created goods) is a "fast" factor, fully used for consumption and training at the time of production;
2. human resources (the component of human capital associated with health) is a "slow" factor of production, which transfer in the next period occurs through the demographic function;
3. qualification (not health-related human capital component) is a "slow" factor of production which transfer in the next period occurs through the educational function;
4. economic entities are considered as autarky groups (clans) that produce and consume a homogeneous, non-stored product;
5. clans are not interested in trade because of material goods homogeneous and non-stored nature;
6. each clan's rational interest is to minimize the probability of extinction;
7. natural resources are not limited quantitatively;
8. clans do not come into conflict with each other for natural or labor force resources.

The last premises are not fundamental and can be eliminated with further model development. The model with the clans autarkic resources corresponds to the extensive social development stages, with the development of new territories, resources etc. The product autarky corresponds to communities with different cultural and economic habits. This difference makes useless one groups products for the other groups. Also, groups might be large enough for their internal labor division making external labor division benefit unimportant. Thus, with the further model development, it will be reasonable to introduce the struggle for resources concept to extend model applicability. But it is unnecessary in the case of the trade since the model is knowingly oriented toward the description of large groups (including aggregated ones). However, the clans in the model are not autarky in the general cultural level of the scientific development. This dependency will be described in the qualification section of this article.

2   Variables

The model includes 7 types of variables:

1. $N[i](t)$ the clan i population in year t (variable type: stock; dimension: man-hour above 0)
2. $q[i](t)$ the clan i qualification in year t (stock; qualification units above 0)
3. $L[i](t)$ the clan i labor force in year t (flow; man-hour above 0)
4. $Y[i](t)$ the clan i production in year t (flow; money above 0)
5. $C[i](t)$ the clan i consumption in year t (flow; money above 0)
6. $E[i](t)$ the clan i education spending in year t (flow; money above 0)
7. $q[a](t)$ society average qualification in year t (flow; qualification units above 0)

Neither of variables are negative nor discrete. To ensure non-discretion, the size of the clan is measured not in man, but in man-hours (non-integer values are considered as the partial capacity for some members of the clan). The stock variables are linked by dynamic equations, while flow variables by balance equations. The article also uses a large number of calibration constants. Some of them simultaneously carry the equation dimension. For the convenience of in-text orientation, we briefly list them, while a detailed description will be given directly together with the equations.

1. b and c0 are constants that reflect the positive and negative changes in the clan size (Eq. 1)
2. h and a are constants that carry the dimension and scale, and the efficiency of labor quantity substitution by its qualification in the production function (Eq. 2)
3. k1 and k2 are constants that reflect the time consumed by created goods consumption in the time balance (Eq. 3)
4. R - the discount factor in the minimized functional (Eq. 13) Used in the derivation of equations or re-designation for the simulation model, the other constants are introduced with appropriate constructions and calculations.

The [i] index reflects the variable belonging to a particular clan, and (t) index corresponds to a specific moment of time. Since the only variable that connects different clans $q[a](t)$ does not have a clan index, the clan's membership index is omitted in all further formulas and calculations.

3 Functions

3.1 Demographic Function

The population dynamics of clans in the model is determined by consumption:

$$\frac{d}{dt}N(t) = N(t) * b * \frac{(c(t) - \frac{c0}{b})}{c(t)} \tag{1}$$

Here, $c(t) = C(t)/N(t)$, i.e. consumption per capita. The constant c0 being strictly above 0 (money per capita) represents the required minimum level of consumption, without which the clan will extinct (so it can be interpreted as mortality and simultaneously as a level of consumption that makes it possible to compensate mortality). The calibration constant b being strictly above 0 (dimensionless) represents the consumption influence at the clan population (can be associated with natural birth rate). Consumption $c(t)$ is considered as an investment in clan members health, including children. If $b * c(t)$ is strictly below c0, the clan's population decreases, otherwise it increases. If b is big, the growth is rapid, and the extinction is slow. If b is small, the growth is slow, and the level of consumption which is necessary for any increment is much bigger. The nature of the clan sizes dependence from consumption is determined by the probability of death or withdrawal from the clan, as well as the birth of new clan members as a kinetic equation for a set of independent events. His detailed conclusion goes beyond this articles limits, and it does not have a fundamental scientific novelty. Essential limitations of the model are: the absence of any population pyramid (which makes it impossible to include external shocks like a drought in model); and the assumption that there is no possible mobility between the clans (caste system), or such movement occurs in a balanced manner (the same number of people leave the clan as join).

3.2 Production Function

The volume of each clans production in the model depends on two factors: the number of labor forces L and their skill level q. Since the product is considered homogeneous, we can use the simplest version (linearly homogeneous) of the classical Cobb-Douglas production function with constant returns to scale, whose properties are common knowledge [7]). The nonlinearity of the dependence reflects a negative effect: with the increase in the quantity, it is increasingly difficult to coordinate the work of people.

$$Y(t) = h * q(t)^a * L(t)^{1-a} \tag{2}$$

According to this, the (dimensionless) constant a (0 strictly below a strictly below 1) represents the efficiency of resources substitution (elasticity coefficient). It is intended for model calibration. Its value is considered externally specified. Finally, the constant h is used to restore the dimension (from exponent multiplication of man-hours and qualification units to money hence, reflects the estimation of these production factors value).

3.3   Qualification Function

The qualification dynamics is the most difficult part of the model, for the reason that the qualification is not only transmitted between periods but also connecting clans with each other through an average qualification:

$$q[a](t) = \frac{\sum_{i=1}^{x}(q[i](t) * N[i](t))}{\sum_{i=1}^{x}(N[i](t))} \tag{3}$$

The ratio between clan qualification q(t) and average qualification q[a](t) is a key tool for setting up the model. Therefore, it requires additional analysis. In general, this problem was called "race for the middle": how the strategy chosen by each subject (clan) affects the other clan's position. The similar problem was considered earlier by some researchers (e.g., Shananin [5]), but applied to technology and physical capital, and usually in models with the trade.

The qualification dynamics is determined by education investments per capita E(t)/N(t) multiplied by some efficiency coefficient. However, in the model, this coefficient is considered to be not constant, but depending on the ratio between clan qualification q(t) and the average qualification q[a](t).

$$\frac{d}{dt}q(t) = q1(\frac{q(t)}{q[a](t)}) * \frac{E(t)}{N(t)} \tag{4}$$

The choice of this dependence nature determines the race for the middle dynamics, consequently, it should be specially researched. Let us consider extreme cases. The q(t)/q[a](t) ratio is strictly above zero since both variables in it are strictly above zero. For q(t)/q[a](t) tending to 0, the clans development stays heavily behind from the general level of society culture. For q(t)/q[a](t) tending to infinity, the clan is heavily ahead of the general level of society culture. Depending on whether the development gap and advancement being considered factors that increase or decrease the effectiveness of education, the race for the middle takes different trajectories.

First of all, we should normalize q1:

$$q1(\frac{q(t)}{q[a](t)}) = y * q2(\frac{q(t)}{q[a](t)}) \tag{5}$$

Here y being strictly above 0 is the calibration constant of investments in qualification efficiency (with the dimension: reverse investment per capita). Then q2 is a dimensionless function 0 strictly below q2 strictly below 1 that determines the race after the middle trajectory, i.e. relative effectiveness of education.

3.4   Race for the Middle

In this study, we simulate two effects which affect it:

1. The observation and repetition effect. The stronger the clan falls behind in development from the average level, the easier it is for him to improve his skills, adopting the most widely known practices.

2. The competencies concentration effect. Clans that form the intellectual elite are the easiest to come up with fundamentally new technologies, scientific discoveries, etc., which makes them innovation drivers.

These oppositely acting effects can be described by a single function as two exponentials sum. One of them reaches a maximum which is equal to 1, while it approaches 0, and the other while it tends to infinity. On the contrary, clans with "mediocre" qualification indicators growth with the greatest difficulty. Obviously, such qualification model, which takes into account the heterogeneity of the society qualification distribution, works correctly only with the number of clans x being above 2.

$$q2\left(\frac{q(t)}{q[a](t)}\right) = g3^{g2*\frac{q(t)}{q[a](t)}} + g1 * \left(1 - g3^{\frac{q(t)}{q[a](t)}}\right) \tag{6}$$

Here g1 being strictly above 0, g2 being strictly above 0 and g3 being strictly above 0 are calibration constants, with g3 = g3 (g1, g2) in such a way that min (q2) = q2(1): g3 = (g1/g2)exponent(1/(g2-1)); min(q2) = q2(1) = g3exponent(g2) + g1exponent(1-g3); lim(q2, q(t)/q[a](t) tends to infinity) = g1. This function is continuous and smooth. It is integrable and differentiable on any segment, including the q2(1) point. The race after the middle can be visually represented (Fig. 1) on the graph q2(q(t)/q[a](t)) because any clan can be uniquely identified one of the points on the function trajectory.

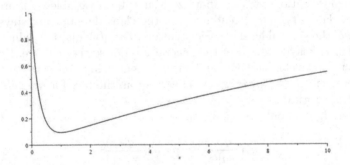

Fig. 1. Dependence between q2 (y-axis) and q(t)/q[a](t) (x-axis)

Substituting q2 in the initial qualification function, we get:

$$\frac{d}{dt}q(t) = y * g3^{g2*\frac{q(t)}{q[a](t)}} + g1 * \left(1 - g3^{\frac{q(t)}{q[a](t)}}\right) * \frac{E(t)}{N(t)} \tag{7}$$

This dependence creates the problem of the clans expectations: changing one of the clans policies leads to unexpected changes in the average qualification. The clan may lose the advantage because of the rapprochement with the average qualification, which is possible both due to changes in the qualifications of other clans and due to their quantity changes. The task of analyzing expectations can be examined in detail with Game theory, but we confine ourselves to the simplest assumption: each clan extrapolates the other clans existing policies for the future (which is as natural for people as primitive). This approach is based on the assumption on the existence of the turnpike effect, which is ordinary for economic systems [8]. This assumption is performed for some model regimes provided by relatively realistic initial choices of expectations in the model. As will be shown below in the simulation model study, the turnpike effect exists. However, unrealistic (for example, random) initial expectations choice can introduce a noticeable imbalance, causing high mortality in the first few periods. But even after that, the model reaches the main path.

4   The Balance of Time and Products

The total clan population N(t) is distributed between the labor force L(t) and the services: the consumption services (C) and the educational services (E).

$$L(t) + k1 * C(t) + k2 * E(t) \leq N(t) \tag{8}$$

The calibration constants k1strictly being above 0 (with the dimension man-hour/money) and k2 being strictly above 0 (man-hour/qualification unit) represent the number of services required to use the created unit of consumer goods (C) and educational goods (E). Created by production factors product is distributed between consumption and training.

$$0 = Y(t) - C(t) - E(t) \tag{9}$$

The balance of the time is in the form of inequality since formal logic does not prevent the clan from leaving part of the people who are unemployed, not engaged in any useful labor. However, it is not difficult to show that under existing model design it is always more advantageous for the clan to redistribute unemployed between creating educational goods and providing educational services.

The qualification growth cannot be disadvantageous to the clan since qualification is a non-decreasing stock (in contrast to the demographic function that allows the decrease, if there are not enough produced goods).

The bime balance (in equality form) and the products balance are technically reducible to one equation through the production function. However, a separate entry is more convenient for an understanding of models basic dependencies.

5   Functional and Optimization

For the case of optimization, we should add the control to the model.

$$L(t) = u(t) * N(t) \tag{10}$$

Also we need to renormalize E, y, and $k2$, transferring the coefficient to the products balance equation (here and below the coefficient h is taken as 1 and omitted, since any other of its values can be compensated by remaining coefficients renormalization):

$$N(t) = L(t) + k1 * C(t) + E(t) \tag{11}$$

$$Y(t) = C(t) + p * E(t) \tag{12}$$

The given control u value (0 below u(t) below 1) uniquely determines remaining variables dynamics, taking into account the above equations that give dq(t)/dt and dN(t)/dt, but solutions with nonnegative variables N, E and C exist not for all controls. Therefore, there is an optimal solution problem.

$$J(t) = c0 * \int_t^\infty \frac{e^{-Rs}}{c(s)} ds, t > 0 \tag{13}$$

The solution optimality principle is determined through the functional minimized by the clan, which is the discounted inverse relative consumption per capita c(t)/c0. The calibration factor R (between 0 and 1) reflects the consumption significance for the clans future generations. In real calculations, the infinite upper limit is replaced by some maximum time T. This functional optimization corresponds to a decrease in the clans death probability. More precisely, for some calibration constants values, minimizing this functional, the clan maximizes the duration of its life (until the clan members quantity reaches 0), while in others it makes a safety margin that can be needed when the external conditions worsen. At any time t, each clan selects the control u(t), minimizing the value of J(t) provided that for any s above t the u(s) other clans remain values are constant and equal to u(t), i.e. other clans continue the current moment economic policy. Therefore it can be shown that:

$$E(t) = \frac{(1 - u(t)) * N(t) - k1 * Y(t)}{1 - k1 * p} \tag{14}$$

$$C(t) = \frac{Y(t) - p * (1 - u(t) * N(t))}{1 - k1 * p} \tag{15}$$

Since for u, (0 below u(t) below 1) is valid (u(t) below u(t)exponent(a)), so the requirement $p * E(t)$ below Y(t) leads to the control lower limit. And the upper control limit follows from the requirement E(t) above 0 (or C(t) above 0).

$$u(t) > u_{\min}(t) = (1 + \frac{(\frac{q(t)}{N(t)})^{1-a}}{p})^{-\frac{1}{a}} \tag{16}$$

$$u(t) \leq u_{\max}(t) = (1 + k1 * (\frac{q(t)}{N(t)})^{1-a})^{-1} \tag{17}$$

6   Simulation Model

6.1   Simulation Model Construction

The model uses many standard functions, but the qualification function requires non-standard solutions. So we constructed a dynamic simulation model for main model primary research and actual processes describing usage verification. It is discrete (time becomes a discrete variable, so integration and differentiation are replaced by the corresponding summation operations).

Since J(t) is a complex nonlinear function, the minimization is performed by searching for u(t), umin(t), umax(t), over a uniform net with a given nodes number.

The model dynamics was researched for various constants values and starting conditions. This research's details are omitted in this publication. At this research stage, it is essential the very existence of a model regime that allows clans population and qualification realistic dynamics. So for further analysis, we chose a particular model regime with the following coefficients values:

a = 0,4; h = 1; b = 0.01; c0 = 0,01; g1 = 1; g2 = 100; g3 = 0,95455; k1 = 0,1; k2 = 1 means p = 1; y = 100; R = 0,01;

In this regime, we considered two typical scenarios. They describe the race for the average with an uneven society qualifications distribution. Both scenarios assume the presence of two clans: Clan 1 includes an elite with high qualification and small population, and Clan 2 includes a common people with a low qualification and a big population. The difference between the scenarios is in this unevenness scale.

6.2   Simulation Model First Scenario

The race for the middle dynamics In the first scenario, the clans characteristics are:

q[1](0) = 50; q[2](0) = 1000; n[1](0) = 99; n[2](0) = 1;

This combination of factors allows Clan 1 to gradually build up its qualification (Fig. 3) while losing its population (Fig. 2). Meanwhile, the second clan rapidly increase both its population and qualification.

If we describe the race for the middle on the graph, where the X-axis is the q[i](t)/q[a](t) ratio, and q2(q[i]/q[a]) is located along the Y-axis, we will get the following dynamics (Fig. 4):

Obviously, Clan 1 is initially very close to the Y-axis minimum, i.e. its qualification investment is ineffective (which leads to its slow population reduction), while Clan 2 is in a moderately effective investment zone. However, no matter how much Clan 2 invests in qualification, the effectiveness of its investments gradually decreases, as its increase in population also increases the average qualification q[a]. On the contrary, Clan 1 is gradually staying behind the average and therefore shifting in the high investment efficiency range direction. This allows him after a while to begin its population growth.

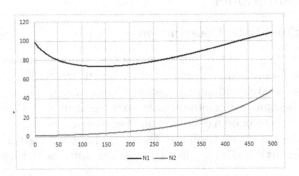

Fig. 2. Clan 1 and Clan 2 population dynamics in the first (normal) scenario

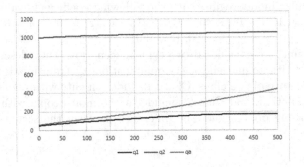

Fig. 3. Clan 1 and Clan 2 qualification and average qualification dynamics in the first (normal) scenario

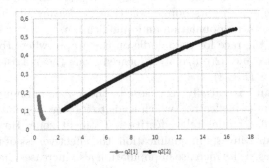

Fig. 4. Clan 1 and Clan 2 relative education effectiveness in the first (normal) scenario

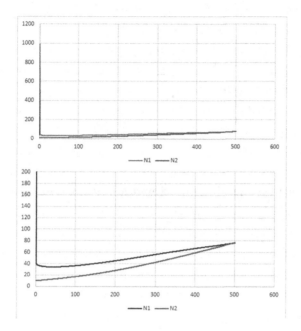

Fig. 5. Clan 1 and Clan 2 population dynamics in the second (catastrophic) scenario in different scales

6.3 Simulation Model Second Scenario

In the second scenario, the clans characteristics are:

$q[1](0) = 5$; $q[2](0) = 100$; $n[1](0) = 990$; $n[2](0) = 10$;

Clan 1 here is in a state where the population requires such a quantity of consumer goods, which can't be provided by the entire clan size due to its low qualification. At this point, the optimal choice is a large part of the population rapid death (Fig. 5) (which can be interpreted, for example, as a civil war for the few remaining consumer goods), while all possible resources are used to improve clans qualification (Fig. 6). The population reduction rate is determined by the b and c0 ratio.

The race for the middle nature is also different (Fig. 7): instead of moving uniformly along the curve, we observe a sudden increase in the average qualification, which immediately places Clan 2 into a low education efficiency zone, and Clan 1 - into a relatively high-efficiency zone (from where it then gradually returns to the low-efficiency zone).

6.4 Scenarios Interrelation

The considered scenarios are interrelated and can transform from one into another. Obviously, the model provides an equalization of clans unrealistic initial population values, and under realistic values allows you to simulate a

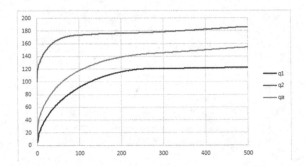

Fig. 6. Clan 1 and Clan 2 qualification and average qualification dynamics in the second (catastrophic) scenario

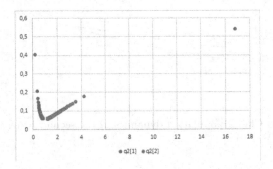

Fig. 7. Clan 1 and Clan 2 relative education efficiency in the second (catastrophic) scenario

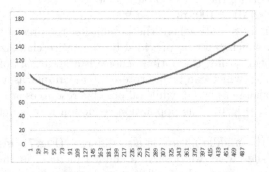

Fig. 8. The first (normal) scenario total population

demographic transition as a result of the gradual population and qualification accumulation in more qualified clans (Fig. 8).

The race for the middle common feature in both scenarios is the qualification gap $(q[2]-q[1])/q[a]$ reduction. In the catastrophic scenario, it happens much faster (Fig. 9).

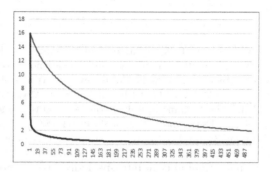

Fig. 9. Qualification gap between Clan1 and Clan 2 in normal (top line) and catastrophic (bottom line) scenarios

Considering other model regimes, cyclic jumps in the clans population are possible as they approach the critical numbers.

7 Conclusion

The article describes a dynamic model of the development of social groups and communities, taking into account their qualification and population. The different group's dynamics are linked through the average qualification indicator. It is shown that there is a model's mode (the combination of the calibration coefficients values) which provides realistic groups dynamic.

On the basis of the series of simulation experiments, we choose two typical scenarios of the groups mutual influence (race for the middle): "normal" and "catastrophic". It is shown that in the normal scenario, the educational effectiveness of the highly qualified group is gradually slowing down, and for the low-qualified group is increasing, while the qualification gap between them is slowly decreasing. In the catastrophic scenario, when a low-qualified group suddenly reduces its population, the effectiveness of its education initially increases, after which it gradually decreases, while the highly qualified clan dynamics remains the same.

The following questions are of the greatest interest in the further model research and development:

1. The coefficients multidimensional metric analysis, their identification and the establishment of their restrictions in accordance with the available social data.
2. The calibration conditions, ensuring realistic population and race for the middle dynamics.
3. The calibration conditions that determine the cyclic amplitude of the model.
4. The conditions of initial clans characteristics, which determine their development in "normal" and "catastrophic" dynamics.
5. The nature of the dynamics of interaction between 3 or more clans.
6. The changes of models dynamics on the assumption of including several clans with the same characteristics.

References

1. Kamenev, I.G.: Problems and tasks of human capital modeling. In: Modeling the Nature and Society Co-evolution: Problems and Experience. To the 100th N.N. Moiseev Academician Birth Anniversary (Moiseyev-100) Works of the All-Russian Scientific Conference, pp. 227–237. GBI FRC CC of RSA, Moscow (2017)
2. Kapelyushnikov, R.I.: The human capital of Russia: evolution and structural features. Bull. Public Opin. Data Anal. Discuss. **4**, 46–54 (2005)
3. Koritsky, A.V.: Human Capital as a Russian Regions Economic Growth Factor. Siberian University of Consumer Cooperatives, Novosibirsk (2010)
4. Korchagin, Y.A.: Human Capital is an Intensive Individual, Economy, Society, and Country Social and Economic Development Factor. NRU HSE, Moscow (2011)
5. Shananin, A.A.: Technological changes on the macroeconomic level – mathematical modeling. Discret. Dyn. Nat. Soc. **6**, 291–304 (2001)
6. Tuguskina, G.N.: Evaluation of enterprises investments effectiveness in the human capital. Pers. Manag. **3**, 73–77 (2009)
7. Cobb, C.W., Douglas, P.H.: A theory of production. Am. Econ. Rev. **18**, 139–165 (1928)
8. McKenzie, L.: Turnpike theory. Econometrica **44**, 841–865 (1976)

Maximization of the Accumulated Extraction in a Gas Fields Model

Alexander K. Skiba[✉]

Federal Research Center "Computer Science and Control"
of Russian Academy of Sciences, Vavilov st. 40, 119333 Moscow, Russia
a.k.skiba@mail.ru

Abstract. A continuous dynamic long-term model of the gas fields group is considered. Two problems are set and solved: the problem of maximizing accumulated production for a gas fields group over a fixed period and the problem of maximizing the length of the general "shelf" for fields group. The problems proposed for the study belong to the class of optimal control problems with mixed constraints. The basic mathematical apparatus is Pontryagin's maximum principle in Arrow's form, in which Lagrange's multipliers are applied. The obtained results are analyzed.

Keywords: Optimal control · Maximum principle
K. Arrow's proposition · Mixed constraint
Maximization of accumulated extraction

1 Introduction

Russia is one of the richest countries in the world in terms of its natural resources and minerals. Among their extensive list, natural gas plays a special role. However, there are characteristic features peculiar to Russia. Long distances divide the extraction sites from the places of consumption of this mineral. A natural gas pipeline is used to transport natural gas. At the same time, the flow of gas is limited by the capacity of the pipeline.

There are two problems. The first problem is to maximize the accumulated gas production of a group of gas fields over a fixed period with a limit on the capacity of the pipeline. The second problem is to maximize the length of the "shelf" for a group of gas fields. Both tasks are interrelated. The latter problem means the following. It is necessary to ensure the uniform flow of natural gas to the sales markets as long as possible.

A dynamic model is used in the mathematical formulation of problems. Modeling of the field development processes is given a significant attention in the scientific literature. Many dynamic processes occurring in the depths of the deposit are well studied [1].

© Springer Nature Switzerland AG 2019
Y. Evtushenko et al. (Eds.): OPTIMA 2018, CCIS 974, pp. 453–469, 2019.
https://doi.org/10.1007/978-3-030-10934-9_32

However, a large number of input parameters that vary with time are required for their calculations. Therefore, dynamic processes are, at best, predictable for a short-term period. Moreover, some of the input information may not be available for fields not currently being developed Hence it is necessary to use an approximate simplified continuous dynamic model with a small number of previously known stationary input parameters. Such a model, based on balance ratios, is described in [2,3].

In 1968, an article by Nobel Laureate Arrow [4] was published, which played a significant role in the further promotion of the use of optimal control in the theory of economic growth. The article caused great scientific resonance.

To solve problems of optimal control with mixed constraints, K. Arrow formulates propositions that are a modification of the Pontryagin maximum principle. Some elements of nonlinear programming are additionally included in the propositions: Lagrangian, the Lagrange multipliers and the complementary slackness condition.

The paper pays great attention to the finite and infinite horizon, existence theorems, necessary and sufficient optimality conditions.

Based on the principle of maximum Pontryagin in K. Arrow's form, the author of this article in 1978 published in the journal "Econometrica" [5], which is included in the course of lectures by the Massachusetts Institute of technology on macroeconomics. Concepts such as "Skiba points" and "Skiba sets" have been introduced in the economy.

Among other works on optimal control, one should pay attention to the book by Ter-Krikorov [6]. It deals with the problem of optimal control with mixed constraints. The method of research of problems is based on the theory of linear programming in the space conjugate Banach space. Problems with continuous and discrete time are investigated.

The monograph by Aseev and by Kryazhemskii [7] is interesting. This work is devoted to the theory of Pontryagin's maximum principle applied to the problems of optimal economic growth. In the center of attention is the characterization of the behavior of the conjugate variable and the Hamiltonian at infinity.

Among foreign authors, attention should be paid to the work of Bolder [8], in which the existence theorem for optimal control problems with an infinite horizon is proved.

As the basic mathematical apparatus, we will use the Pontryagin maximum principle in the Arrow form.

2   Arrow's Proposition

Proposition 1. *Let $v^*(t)$ be a choice of instruments $(0 \leq t \leq T)$ which maximizes $\int_0^T U[x(t), v(t), t]\, dt$ subject to the conditions,*

(a) $\dot{x} = S[x(t), v(t), t], \quad 0 \leq t \leq T,$

a set of constraints,

(b) $F[x(t), v(t), t] \geq 0, \quad 0 \leq t \leq T,$

on the instruments possibly involving the state variables, initial conditions on the state variables, and the terminal conditions $x(T) \geq 0$. If the Constraint Qualification holds, then there exist auxiliary variables $p(t)$, such that, for each t,

(c) $v^*(t)$ *maximizes* $H[x(t), v(t), p(t), t]$ *subject to the constaints (b), where* $H(x, v, p, t) = U(x, v, t) + pS(x, v, t);$

(d) $\dot{p}_i = -\partial L/\partial x_i,$ *evaluated at* $x = x(t), v = v^*(t), p = p(t),$ *where*

(e) $L(x, v, p, t) = H(x, v, p, t) + qF(x, v, t),$ *and the Lagrange multipliers q are such that*

(f) $\partial L/\partial v_k = 0,$ *for* $x = x(r), v = v^*(t), p = p(t), q(t) \geq 0,$

$q(t)F[x(t), v^*(t), t] = 0,$ *and*

(g) $p(T) \geq 0, p(T)x(T) = 0.$

3 Model Description and Problem Posing

In 2009, article [9] was published in the journal Vestnik of the Peoples' Friendship University, in which described the model considered in this paper. Despite this, we will describe it again, pursuing two goals. First, there are still some differences in the descriptions of the models. Second, it is desirable to keep the integrity of the conducted researches description.

Let us consider a model for the functioning of a gas fields group with interacting wells [2,3]. The group consists of n deposits. We introduce the following notation:

N_i : Fund of producing wells at the i-th field
$\bar{N}_i$: Upper limit on the Fund of producing wells by the i-th field
q_i : Average flow rate of producing wells at the i-th field
V_i : Recoverable gas reserve at the i-th field.

A differential relationship is established between the variables. We describe it in the form of differential equations system

$$\dot{V}_i = -q_i(t)N_i(t), \quad \dot{q}_i = -\frac{q_i^0}{V_i^0}q_i(t)N_i(t), \qquad i = 1, 2, \ldots, n$$

under the initial conditions $V_i^0 > 0, q_i^0 > 0$. The producing wells fund have the following restrictions $0 \leq N_i \leq \bar{N}_i$. We assume in this case $\bar{N}_i > 0$. We also suppose that the following values are different $\frac{q_i^0}{V_i^0}\bar{N}_i, \quad i = 1, 2, \ldots, n$.

There is a general limitation to a common "shelf" of deposits

$$\sum_{i=1}^{n} q_i(t)N_i(t) \leq \bar{Q}.$$

The accumulated gas production is calculated by the formula

$$\sum_{i=1}^{n} \int_{0}^{T} q_i(t) N_i(t) \, dt.$$

The task of accumulated extraction maximizing is actual as with practical, and from a theoretical point of view. This fact has been repeatedly stressed in many works, including in paper [2]. We turn to the mathematical description of the optimal control problem with mixed constraints.

On the Maximization of Accumulated Production for a Group of Gas Fields

Problem 1. It is required to maximize the functional

$$\sum_{i=1}^{n} \int_{0}^{T} q_i(t) N_i(t) \, dt \tag{1}$$

on the fixed interval $[0, T]$
under the differential connections

$$\dot{q}_i = -\frac{q_i^0}{V_i^0} q_i(t) N_i(t), \qquad 0 \le t \le T, \qquad i = 1, 2, \dots, n \tag{2}$$

with the initial conditions

$$q_i^0 > 0, \qquad i = 1, 2, \dots, n \tag{3}$$

and with the constraints on the controls

$$0 \le N_i(t) \le \bar{N}_i, \qquad 0 \le t \le T, \qquad i = 1, 2, \dots, n, \tag{4}$$

$$\sum_{i=1}^{n} q_i(t) N_i(t) \le \bar{Q}, \qquad 0 \le t \le T. \tag{5}$$

We also assume that

$$\sum_{i=1}^{n} q_i^0 \bar{N}_i > \bar{Q}. \tag{6}$$

Here $q_i(t)$ for $i = 1, 2, \dots, n$ are phase variables. $\bar{Q}$ is a constant value.

The controls $N_i(t)$ for $i = 1, 2, \dots, n$ belong to the set of measurable functions. The right end of the phase trajectory is free. The quantities $\frac{q_i^0}{V_i^0} \bar{N}_i$ differ between themselves.

The existence of an optimal control follows, for example, from the theorem given in scientific book [10, Sect. 4.2].

Note that the number of wells $N_i(t)$ in the model (2)–(6) is a control, while in original model [2], it is a phase variable. This fact simplifies the original model without losing the efficiency and adequacy. The simplification used allows us to reduce the number of input parameters, which as a result leads to the possibility of an analytical solution of the optimal control problem with mixed constraints [4].

We proceed to describe the solution of the above-formulated problem.

4   Investigation of Problem 1

First let us introduce some notation

$$\alpha_i^0 = \frac{q_i^0}{V_i^0}, \qquad i = 1, 2, \ldots, n. \tag{7}$$

Further, we order the phase variables in the order of increase of the quantities $\alpha_i^0 \bar{N}_i$. This order is unique because of the conditions in the statement of the problem.

According to Arrow's proposition, we write down the Hamiltonian and the Lagrangian

$$H(q, N, \psi) = \sum_{i=1}^{n} [q_i N_i - \psi_i \alpha_i^0 q_i N_i], \tag{8}$$

$$L(q, N, \psi, \gamma, \delta, \beta) = \sum_{i=1}^{n} [q_i N_i - \psi_i \alpha_i^0 q_i N_i + \gamma_i(\bar{N}_i - N_i) + \delta_i N_i] + \beta[\bar{Q} - \sum_{i=1}^{n}(q_i N_i)].$$

Next, we describe the set $G(q)$ of admissible controls on which the Hamiltonian (8) is maximized

$$G(q) = \{N \in \mathbb{R}^n \mid 0 \le N \le \bar{N}, \sum_{i=1}^{n} q_i N_i \le \bar{Q}\}. \tag{9}$$

The main objective of the study is to find the vector function $\psi(t)$ and the control $\tilde{N}(t)$ that satisfy the system of adjoint differential equations (12) given below, the transversality conditions

$$\psi(T)q(T) = 0 \tag{10}$$

and for each $t \in [0, T]$ the control $\tilde{N}(t)$ maximize the Hamiltonian

$$H(\tilde{q}(t), \tilde{N}(t), \psi(t)) = \max_{N \in G(\tilde{q}(t))} H(\tilde{q}(t), N, \psi(t))$$

$$= \max_{N \in G(\tilde{q}(t))} \sum_{i=1}^{n} [\tilde{q}_i(t)N_i - \alpha_i^0 \psi_i(t)\tilde{q}_i(t)N_i]. \tag{11}$$

The system of adjoint equations is represented in the form

$$\dot{\psi}_i(t) = -\partial L/\partial q_i = -\tilde{N}_i(t) + \alpha_i^0 \psi_i(t)\tilde{N}_i(t) + \beta(t)\tilde{N}_i(t)$$
$$= [\alpha_i^0 \psi_i(t) + \beta(t) - 1]\tilde{N}_i(t), \quad i = 1, 2, \ldots, n. \tag{12}$$

Let us introduce the following notation

$$a_i(t) = \alpha_i^0 \tilde{N}_i(t), \quad i = 1, 2, \ldots, n, \tag{13}$$

$$b_i(t) = [\beta(t) - 1]\tilde{N}_i(t), \quad i = 1, 2, \ldots, n, \tag{14}$$

then the system of adjoint differential equations (12) is rewritten in as follows

$$\dot{\psi}_i(t) = a_i(t)\psi_i(t) + b_i(t), \quad i = 1, 2, \ldots, n. \tag{15}$$

According to Arrow's proposition the Lagrange multipliers $\beta(t), \gamma(t), \delta(t)$ and the controls N_i must satisfy the equality:

$$\partial L/\partial N_i = \tilde{q}_i(t) - \alpha_i^0 \psi_i(t)\tilde{q}_i(t) - \beta(t)\tilde{q}_i(t) - \gamma_i(t) + \delta_i(t) = 0, \quad i = 1, 2, \ldots, n,$$

or

$$\tilde{q}_i(t)[1 - \alpha_i^0 \psi_i(t) - \beta(t)] = \gamma_i(t) - \delta_i(t), \quad i = 1, 2, \ldots, n; \tag{16}$$

$$\beta(t)[\bar{Q} - \sum_{i=1}^{n} \tilde{q}_i(t)\tilde{N}_i(t)] = 0, \quad \beta(t) \geq 0; \tag{17}$$

$$\gamma_i(t)[\bar{N}_i - \tilde{N}_i(t)] = 0, \quad \gamma_i(t) \geq 0, \quad i = 1, 2, \ldots, n; \tag{18}$$

$$\delta_i(t)\tilde{N}_i(t) = 0, \quad \delta_i(t) \geq 0, \quad i = 1, 2, \ldots, n. \tag{19}$$

In nonlinear programming, the above relations (17), (18), (19) are defined as conditions for complementary non-rigidity.

As can be seen from the system of differential equations (2), the phase variables $q(T)$ are positive for any finite values of T. Therefore, the transversality conditions (10) are represented in the following form:

$$\psi_i(T) = 0, \quad i = 1, 2, \ldots, n. \tag{20}$$

Next, we solve adjoint differential equations (15) given the transversality conditions (20). As a result, we get

$$\psi_i(t) = -\exp\left[-\int_t^T a_i(\theta)\,d\theta\right]\left\{\int_t^T b_i(\vartheta)\exp\left[\int_\vartheta^T a_i(\theta)\,d\theta\right]d\vartheta\right\},$$
$$i = 1, 2, \ldots, n, \tag{21}$$

where $a_i(t)$ and $b_i(t)$ are defined by the formulas (13) and (14), respectively.

Next, we define two sets:

$$M = \{t \mid \sum_{i=1}^{n} \tilde{q}_i(t)\tilde{N}_i(t) = \bar{Q},\ 0 \leq t \leq T\}; \tag{22}$$

$$H = \{t \mid \sum_{i=1}^{n} \tilde{q}_i(t)\tilde{N}_i(t) < \bar{Q},\ 0 \leq t \leq T\}. \tag{23}$$

The following propositions are true.

Proposition 2. *Let the inequality*

$$\bar{Q} - \sum_{i=1}^{n} \tilde{q}_i(t)\tilde{N}_i(t) > 0, \tag{24}$$

be satisfied on the optimal trajectory with some value $t = t^* \in [0,T]$ *that is, the set (23) is not empty. Then the value of the vector optimal controls are determined by* $\tilde{N}(t^*) = \bar{N}$.

Proof. Suppose that the strict inequality holds for the k-th component of the control vector $\tilde{N}(t)$: $\tilde{N}_k(t^*) < \bar{N}_k$. From (17) and (24) for $t = t^*$ follows $\beta(t^*) = 0$. With allowance for (16), (18) and (19) we get $\delta_k(t^*) \geq 0, \gamma_k(t^*) = 0$ and $\alpha_k^0 \psi_k(t^*) \geq 1$.

We rewrite k-th adjoint differential equation (12)

$$\alpha_k^0 \dot{\psi}_k(t) = \alpha_k^0 \tilde{N}_k(t)([\alpha_k^0 \psi_k(t) - 1] + \alpha_k^0 \tilde{N}_k(t)\beta(t). \tag{25}$$

Let us introduce the following notation: $x(t) = \alpha_k^0 \psi_k(t) - 1$; $c(t) = \alpha_k^0 \tilde{N}_k(t)$; $h(t) = \alpha_k^0 \tilde{N}_k(t)\beta(t)$. In the new notation, adjoint differential equation (25) is rewritten in the following form: $\dot{x} = c(t)x + h(t)$. We solve this adjoint equation with initial condition $x(t^*) \geq 0$. As a result, we get

$$x(t) = \left\{ x(t^*) + \int_{t^*}^{t} h(\vartheta) \exp\left[-\int_{t^*}^{\vartheta} c(\theta)\,d\theta\right] d\vartheta \right\} \exp\left[\int_{t^*}^{t} c(\theta)\,d\theta\right]. \tag{26}$$

From (26) taking into account the non-negativity of the coefficients $c(t)$ and $h(t)$, we obtain $x(t) = \alpha_k^0 \psi_k(t) - 1 \geq x(t^*) \geq 0$. Hence the transversality condition (20) for k of the component of $\psi(t)$ is not satisfied. This completes the proof.

Corollary 1. *The value of the optimal controls vector on the set (23) is uniquely determined by the equality* $\tilde{N}(t) = \bar{N}$.

Proposition 3. *Let* $t^* \in [0,T]$ *and* $t^* \in H$, *then* $[t^*, T] \subset H$.

Proof. Suppose that for some value $t' \in (t^*, T]$ the following equality

$$\bar{Q} - \sum_{i=1}^{n} \tilde{q}_i(t)\tilde{N}_i(t) = 0, \tag{27}$$

holds, that is, $t' \in M$, then there exists at least one k-th component of the control vector $\tilde{N}(t)$ such that $\tilde{N}_k(t^*) < \bar{N}_k$. This is due to a drop in flow rate of producing wells over time. From Proposition 2 follows $\tilde{N}_k(t^*) = \bar{N}_k$. We arrive at a contradiction.

Proposition 4. *Suppose at some value $t = t^* \in [0, T]$ the following inequality*

$$\bar{Q} - \sum_{i=1}^{n} \tilde{q}_i(t)\bar{N}_i \geq 0 \tag{28}$$

is satisfied, then $\tilde{N}(t) = \bar{N}$ for all $t \in [t^*, T]$ and $t^* \in \bar{H}$, where $\bar{H}$ is a closure of the set H.

Proof. Due to the decreasing of the phase variables $q_i(t)$ and taking into account the inequality (28), the binding constraint (5) is inactive on any admissible controls at $t > t^*$. In this case, the Problem 1 is divided into n independent optimization problems, the solution of which is trivial and it is defined by the values of the optimal control vector $\tilde{N}(t) = \bar{N}$. The belonging of t^* to the closure of the set (23) is obvious.

Proposition 5. *Let $t^* \in [0, T]$ and $t^* \in M$, then $[0, t^*] \subset M$.*

The proof of Proposition 5 follows from Proposition 3.

Proposition 6. *There exists a value of T such that the set (23) constructed on the optimal trajectory is not empty.*

Proof. As the value of T, it suffices to take $T' = (\sum_{i=1}^{n} V_i^0)/\bar{Q}$. It is easy to show that there exists a value of $t \in (0, T')$ such that the inequality (24) holds. Hence, the set (23) is not empty.

Proposition 7. *Suppose that for at least one k-th component of the optimal control vector $\tilde{N}(t)$, the strict inequality $\tilde{N}_k(t^*) < \bar{N}_k$ holds at time $t^* \in (0, T)$. Then there exists $t' \in (t^*, T]$ such that $[0, t'] \subset M$.*

The proof of Proposition 7 is easily obtained from Propositions 2 and 3.

Proposition 8. *Let the sets (22) and (23) are not empty, then there exists $T_{max} \in M$ such that $\tilde{N}(T_{max}) = \bar{N}$. The set M precedes the set H.*

Proof. We take $T_{max} = \inf H$. In this case $T_{max} = \max M$. In view of the decrease and continuity of the phase variables $q(t)$, we obtain $\tilde{N}(T_{max}) = \bar{N}$. For this reason, the set of M is preceded by the set H.

Corollary 2. *Let the sets (23) and (22) is not empty, then the optimal trajectory passes through the plane described by equation*

$$\sum_{i=1}^{n} q_i \bar{N}_i = \bar{Q}, \tag{29}$$

where the vector q belongs to the positive orthant of the space $\mathbb{R}^n$.

Theorem 1. *There exist only one of three possible options of optimal trajectory behavior.*

(1) If the inequality $\sum_{i=1}^{n} q_i^0 \bar{N}_i \leq \bar{Q}$, is fulfilled at initial time, then the set (23) is not empty, and the set (22) is empty. The vector of optimal controls $\tilde{N}(t) = \bar{N}$ for all values of $t \in [0,T]$. The sum $\sum_{i=1}^{n} \tilde{q}_i(t)\bar{N}_i < \bar{Q}$ for $t \in (0,T]$, which decreases over time.

(2) If the strict inequality $\sum_{i=1}^{n} q_i^0 \bar{N}_i > \bar{Q}$ is fulfilled at initial time and there exists an admissible trajectory, on which equality

$$\bar{Q} - \sum_{i=1}^{n} q_i(t)N_i(t) = 0 \tag{30}$$

is satisfied for all values of $t \in [0,T]$. Then this trajectory is optimal and the maximum value of the functional (1) is equal to $\bar{Q}T$.

(3) If the inequality $\sum_{i=1}^{n} q_i^0 \bar{N}_i > \bar{Q}$ is fulfilled at initial time and the set (23) is not empty, then there exists T_{max} such the equality

$$\bar{Q} - \sum_{i=1}^{n} \tilde{q}_i(t)\tilde{N}_i(t) = 0$$

holds for all values of $t \in [0, T_{max}]$. The sum $\sum_{i=1}^{n} \tilde{q}_i(t)\bar{N}_i < \bar{Q}$ for all values of $t \in (T_{max}, T]$, which decreases over time.

The proof of Theorem 1 follows from the above propositions and corollaries.

Theorem 2. *Suppose the set (23) is not empty and the phase variables q_i are ordered in ascending order of magnitudes $\alpha_i^0 \bar{N}_i$. Then the optimal trajectory is calculated at each time $t \in [0,T]$ by formulas*

$$\tilde{q}_i(t) = q_i^0 \exp[-\alpha_i^0 \int_0^t \tilde{N}_i(\theta)\, d\theta], \quad I = 1, 2, \dots, n, \tag{31}$$

and the optimal control vector $\tilde{N}(t)$ is uniquely determined at the time t according to the following rules:

(1) if the condition $\sum_{i=1}^{n} \tilde{q}_i(t)\bar{N}_i \leq \bar{Q}$ is satisfied, then $\tilde{N}(t) = \bar{N}$;

(2) otherwise $(\bar{Q} < \sum_{i=1}^{n} \tilde{q}_i(t)\bar{N}_i)$, there exists an integer $k \in \{1,\ldots,n\}$ such that the double inequality

$$\sum_{i=1}^{k-1} \tilde{q}_i(t)\bar{N}_i \leq \bar{Q} < \sum_{i=1}^{k} \tilde{q}_i(t)\bar{N}_i$$

is performed, and

$$\tilde{N}_i(t) = \bar{N}_i \quad i = 1,2,\ldots,k-1, \tag{32}$$

$$\tilde{N}_i(t) = 0 \quad i = k+1,\ldots,n, \tag{33}$$

$$\tilde{N}_k(t) = [\bar{Q} - \sum_{i=1}^{k-1} \tilde{q}_i(t)\bar{N}_i]/\tilde{q}_k(t). \tag{34}$$

Proof. First let us introduce some notation $\varphi_i(t) = \alpha_i^0 \psi_i(t)$ for $i = 1,2,\ldots,n$. Adjoint equations (12) and their solutions (21) are rewritten in the following form:

$$\dot{\varphi}_i(t) = [\varphi_i(t) + \beta(t) - 1]\alpha_i^0 \bar{N}_i(t), \quad i = 1,2,\ldots,n. \tag{35}$$

$$\varphi_i(t) = -\alpha_i^0 \exp\left[-\int_t^T a_i(\theta)\,d\theta\right]\left\{\int_t^T b_i(\vartheta)\exp\left[\int_\vartheta^T a_i(\theta)\,d\theta\right]d\vartheta\right\},$$

$$i = 1,2,\ldots,n. \tag{36}$$

Further, speaking of differential equations (35), we will use the word "adjoint" in sentences by taking it in quotation marks. This will also apply to the variable $\varphi_i(t)$.

Let $t \in H$, then optimal controls $\tilde{N}(t) = \bar{N}$ and Lagrange's multiplier $\beta(t) = 0$. These equalities hold on (23) and follow from Proposition 2 and from the condition (17). We substitute these values in formulas (13) and (14). From solutions (21) of the system of adjoint differential equations (15), taking into account formulas (13) and (14), we get

$$\varphi_i(t) = \alpha_i^0 \psi_i(t) = 1 - \exp[-\alpha_i^0 \bar{N}_i(T-t)], \quad i = 1,2,\ldots,n. \tag{37}$$

Further, taking into account the ordering of the phase variables in the order ascending values of $\alpha_i^0 \bar{N}_i$ we compare the values of functions $\varphi_i(t)$ at a fixed value of t. As a result, we come to the following conclusion. The values of functions $\varphi_i(t)$ strictly increase with increasing ordinal number of the component.

Assume that $\bar{Q} < \sum_{i=1}^{n} q_i^0 \bar{N}_i$. The opposite inequality is not interesting because of solving Problem 1 simplicity.

We set $\varphi_1(0) < \varphi_2(0) < \ldots < \varphi_n(0)$. In this case $t = 0 \in M$. We show below that the same order relation of the functions $\varphi_i(t)$ is conserved on the whole set M, that is, inequalities $\varphi_1(t) < \varphi_2(t) < \ldots < \varphi_n(t)$ hold for any value of $t \in M$.

Under the condition $t \in M$, we introduce the set G', on which is to maximize Hamiltonian (11):

$$G'(q) = \{N \in \mathbb{R}^n \mid 0 \le N \le \bar{N}, \sum_{I=1}^{n} q_i N_i = \bar{Q}\}. \tag{38}$$

We transform Hamiltonian (11)

$$H(\cdot) = \max_{N \in G'(\tilde{q}(t))} \sum_{I=1}^{n} [\tilde{q}_i(t)N_i - \varphi_i(t)\tilde{q}_i(t)N_i]$$

$$= \bar{Q} + \max_{N \in G'(\tilde{q}(t))} \sum_{I=1}^{n} [-\varphi_i(t)\tilde{q}_i(t)N_i]. \tag{39}$$

Maximization of Hamiltonian (39) is reduced to function minimization

$$\min_{N \in G'(\tilde{q}(t))} \sum_{I=1}^{n} [\varphi_i(t)\tilde{q}_i(t)N_i]. \tag{40}$$

Now we introduce the following notation: $u = \tilde{q}N$; $\bar{u} = \tilde{q}\bar{N}$. As a result, the Hamiltonian maximization is (39) reduced to the following linear programming problem:

$$\sum_{i=1}^{n} \varphi_i u_i \to \min \tag{41}$$

under conditions

$$\sum_{i=1}^{n} u_i = \bar{Q}, \tag{42}$$

$$0 \le u \le \bar{u}. \tag{43}$$

Moreover, the coefficients φ_i are ordered in the order of strict ascending. Taking into account (42), we transform the linear function (41). As a result, we obtain the following linear programming problem for a maximum

$$\sum_{i=1}^{n-1} (\varphi_n - \varphi_i)u_i \to \max. \tag{44}$$

Note that the coefficients $\varphi_i' = \varphi_n - \varphi_i$ in the description of the linear function (44) are positive and are decreased with increasing order number of vector u component.

The following procedure allows us to find such optimal vector $\tilde{u}$, which delivers the maximum of linear function (44) and, correspondingly, the minimum of linear function (41). First on the first step we choose $\tilde{u}_1$. If $\bar{u}_1 \ge \bar{Q}$, then $\tilde{u}_1 = \bar{Q}$

and $\tilde{u}_i = 0, i = 2, \ldots, n$. In this case, we determine all the optimal values of vector $\tilde{u}$. If $\bar{u}_1 < \bar{Q}$, then $\tilde{u}_1 = \bar{u}_1$ and make substitution $\bar{Q} = \bar{Q} - \bar{u}_1$. Next, we proceed to the second component u_2 and repeat the procedure of the first step, but with the changed value $\bar{Q}$. In the second step, the changed value of $\bar{Q}$ is again subject to check.

If we stopped at the k-th step, then $\tilde{u}_i = \bar{u}_i$, $i = 1, 2, \ldots, k-1, \tilde{u}_k = \bar{Q}, \tilde{u}_i = 0$, $i = k+1, \ldots, n$. If we have gone all the steps from one to $n-1$, then $\tilde{u}_i = \bar{u}_i$, $i = 1, 2, \ldots, n-1, \tilde{u}_n = \bar{Q}$.

From the proofs given above follows validity of three assertions:

(1) if $\tilde{u}_k = \bar{u}_k$, then $\tilde{N}_k(t) = \bar{N}_k$;

(2) if $\tilde{u}_k = 0$, then $\hat{N}_k(t) = 0$;

(3) if $\tilde{u}_k \in (0, \bar{u}_k)$, then $\tilde{N}_k(t) \in (0, \bar{N})$.

Let us consider in dynamics the behavior of the optimal trajectory on the whole set M, i.e. from 0 to T_{max}.

We will carry out further proof only if the restriction $q_1^0 \bar{N}_1 > \bar{Q}$ is satisfied. The use of the contradictory inequality in the analysis does not substantially change the essence of the proof, but only complicates its presentation.

Let $q_1^0 \bar{N}_1 > \bar{Q}$ at the initial instant of time, then $\tilde{N}_1(0) \in (0, \bar{N}_1)$, $\tilde{N}_i(0) = 0$ for $i = 2, 3, \ldots, n$.

Due to the decreasing of the phase variable $\tilde{q}_1(t)$, there exists a time $\tau_1 > 0$ such that,

(1) on the half-open interval $t \in [0, \tau_1)$ the following relations

$$\tilde{N}_1(t) \in (0, \bar{N}_1) \text{ and } \tilde{N}_i(t) = 0 \text{ for } i = 2, 3, \ldots, n$$

hold;

(2) $\tilde{N}_1(\tau_1) = \bar{N}_1$ and $\tilde{N}_i(\tau_1) = 0$ for $i = 2, 3, \ldots, n$.

From the system of "adjoint" differential equations (35) with allowance for (16), (18), (19) for $t \in [0, \tau_1)$ we obtain $\gamma_1(t) = \delta_1(t) = 0$; $\beta(t) = 1 - \varphi_1(t) = \beta_1 = const$; $\varphi_i(t) = const$ for $i = 1, 2, \ldots, n$. In this case, the order relation established among the components vector-valued functions $\varphi(t)$ by their values at time $t = 0$, is not changes on the segment $[0, \tau_1]$.

Further, there exists a time τ_2 such that,

(1) on the interval (τ_1, τ_2) the following relations

$$\tilde{N}_1(t) = \bar{N}_1, \ \tilde{N}_2(t) \in (0, \bar{N}_2) \text{ and } \tilde{N}_i(t) = 0 \text{ for } i = 3, \ldots, n,$$

hold;

(2) $\tilde{N}_1(\tau_2) = \bar{N}_1$, $\tilde{N}_2(\tau_2) = \bar{N}_2$ and $\tilde{N}_i(\tau_2) = 0$ for $i = 3, \ldots, n$.

From the system of "adjoint" equations (35) with allowance for (16), (18), (19) for $t \in (\tau_1, \tau_2)$ we obtain $\gamma_2(t) = \delta_2(t) = 0$; $\beta(t) = 1 - \varphi_2(t) = \beta_2 = const$; $\varphi_i(t) = const$ for $i = 2, 3, \ldots, n$; the function $\varphi_1(t)$ decreases on the interval (τ_1, τ_2). The latter follows from the inequality $\beta_1 = 1 - \varphi_1(\tau_1) > \beta_2 = 1 - \varphi_2(\tau_1)$. This means that the order relations established among the components of the vector function $\varphi(t)$ by their values at time $t = 0$, are stored on the interval $[0, \tau_2]$, and so on.

Thus, the period $[0, T_{max}]$ splits into n segments $[0, \tau_1], [\tau_1, \tau_2], ..., [\tau_{n-1}, \tau_n]$. Here $\tau_n = T_{max}$. Equality

$$\sum_{i=1}^{k} q_k(t)\bar{N}_k = \bar{Q}, \quad k = 1, \ldots, n.$$

is performed at the end of each k-th segment.

The following proposition holds.

Proposition 9. *Let* $t \in (0, \tau_1)$, *then*

$$\beta(t) = 1 - \varphi_1(t) = 1 - \varphi_1(0) = \beta_1;$$

$$\tilde{N}_1(t) = \frac{\bar{Q}}{\tilde{q}_1(t)} \text{ and } \tilde{N}_i(t) = 0 \text{ for } i = 2, 3, \ldots, n;$$

$$\varphi_i(t) = \varphi_i(0) \text{ for } i = 1, 2, \ldots, n.$$

Let $t \in (\tau_1, \tau_2)$, *then*

$$\beta(t) = 1 - \varphi_2(t) = 1 - \varphi_2(0) = \beta_2 < \beta_1;$$

$$\tilde{N}_1(t) = \bar{N}_1, \quad \tilde{N}_2(t) = \frac{\bar{Q} - \tilde{q}_1(t)\bar{N}_1}{\tilde{q}_2(t)} \text{ and } \tilde{N}_i(t) = 0 \text{ for } i = 3, \ldots, n;$$

$$\varphi_1(t) = \varphi_2(\tau_1) - [\varphi_2(\tau_1) - \varphi_1(\tau_1)] \exp[\alpha_1^0 \bar{N}_1(t - \tau_1)] = \varphi_2(0) - [\varphi_2(0) - \varphi_1(0)] \exp[\alpha_1^0 \bar{N}_1(t - \tau_1)],$$

$$\varphi_i(t) = \varphi_i(0) \text{ for } i = 2, \ldots, n.$$

Let $t \in (\tau_{k-1}, \tau_k)$, $k = 2, ..., n$, *then*

$$\beta(t) = 1 - \varphi_k(t) = 1 - \varphi_k(0) = \beta_k < \beta_{k-1};$$

$$\tilde{N}_i(t) = \bar{N}_i \text{ for } i = 1, \ldots, k-1,$$

$$\tilde{N}_k(t) = \frac{\bar{Q} - \sum_{i=1}^{i=k-1} \tilde{q}_i(t)\bar{N}_i}{\tilde{q}_k(t)} \text{ and } \tilde{N}_i(t) = 0 \text{ for } i = k+1, \ldots, n;$$

$$\varphi_i(t) = \varphi_k(\tau_{k-1}) - [\varphi_k(\tau_{k-1}) - \varphi_i(\tau_{k-1})] \exp[\alpha_i^0 \bar{N}_i(t - \tau_{k-1})]$$

$$= \varphi_k(0) - [\varphi_k(0) - \varphi_i(\tau_{k-1})] \exp[\alpha_i^0 \bar{N}_i(t - \tau_{k-1})] \text{ for } i = 1, \ldots, k-1$$

$$\text{and } \varphi_i(t) = \varphi_i(0) \text{ for } i = k, \ldots, n.$$

Proof. Using relations (35), (16), (17), (18), (19), (31) and taking into account the continuity of vector function $\varphi(t)$, we write out the formulas for calculating variables $\beta(t), N(t), \varphi(t)$ at each of the n intervals. *End of proof.*

At time T_{max}, according to Proposition 8, $\tilde{N}(T_{max}) = \bar{N}$. Analyzing the dynamics of changes in Lagrange's multiplier $\beta(t)$, vector-valued functions $\varphi(t)$ and the vector of optimal controls $\tilde{N}(t)$ on set (22), we arrive at the following conclusions:

1. Lagrange's multiplier $\beta(t)$ is a piecewise constant decreasing function.
2. The strict order relations established among the components values of vector-valued function $\varphi(t)$ at initial time $t = 0$, remain unchanged on interval $[0, T_{max}]$.
3. At the interval $(0, T_{max})$, at each time t can be introduced into the development of no more than one deposit.

It is important that the established strict order relations among the components values of the vector function $\varphi(t)$ at time $t = 0$ coincide at time $t = T_{max}$ with strict order relations among components values (37) of the same vector function, but obtained with allowance for the transversality conditions for $t > T_{max}$.

Therefore, we can synthesize $\varphi(t)$ vector-valued functions at time T_{max}. If there were differences in the relations of strict order, then the synthesis of the vector functions $\varphi(t)$ at time T_{max} would be impossible.

We note that the "adjoint" functions $\varphi_i(t)$ $(i = 1, 2, \ldots, n)$ are initially constant, and then decrease. At time T, they reach zero.

Behavior dynamics of optimal controls $\tilde{N}_1(t), \tilde{N}_2(t), \tilde{N}_3(t)$, of functions $\varphi_1(t)$, $\varphi_2(t), \varphi_3(t)$ and of Lagrange's multiplier $\beta(t)$, obtained in solving Problem 1 of dimension $n = 3$, are schematically shown in Fig. 1.

Next, we describe an algorithm for computing continuous vector-valued function $\varphi(t)$. First, using Theorem 2, we find $\tau_1, \tau_2, \ldots, \tau_n = T_{max}$. On the half-open interval $(T_{max}, T]$ Lagrange's multiplier $\beta(t)$ is equal to 0. Substituting $t = T_{max}$ in (37), we define the value of vector function $\varphi(t)$ at T_{max}. Then it gives an opportunity to find vector function $\varphi(t)$ on segment $[t_{max}, T]$.

The further search algorithm is based on the results of Proposition 9. We know function $\varphi_n(t) = \varphi_n(T_{max})$ on segment $[0, T_{max}]$, and Lagrange's multiplier $\beta(t) = \beta_n = \varphi_n(T_{max})$ on half-open interval $(\tau_{n-1}, T_{max}]$. All the remaining components of vector function $\varphi(t)$ on the same half-interval are calculated by formulas

$$\varphi_i(t) = \varphi_n(T_{max}) - [\varphi_n(T_{max}) - \varphi_i(T_{max})] \exp[-\alpha_i^0 \bar{N}_i(T_{max} - t)],$$
$$i = 1, 2, \ldots, n - 1. \tag{45}$$

We know function $\varphi_{n-1}(t) = \varphi_{n-1}(\tau_{n-1})$ on segment $[0, \tau_{n-1}]$, and Lagrange's multiplier $\beta(t) = \beta_{n-1} = \varphi_{n-1}(\tau_{n-1})$ on half-open interval $(\tau_{n-2}, \tau_{n-1}]$. All the remaining $n - 2$ components of vector-valued function $\varphi(t)$ on the same half-interval are computed by formulas

$$\varphi_i(t) = \varphi_{n-1}(\tau_{n-1}) - [\varphi_{n-1}(\tau_{n-1}) - \varphi_i(\tau_{n-1})] \exp[-\alpha_i^0 \bar{N}_i(\tau_{n-1} - t)],$$
$$i = 1, 2, \ldots, n - 2, \tag{46}$$

and so on.

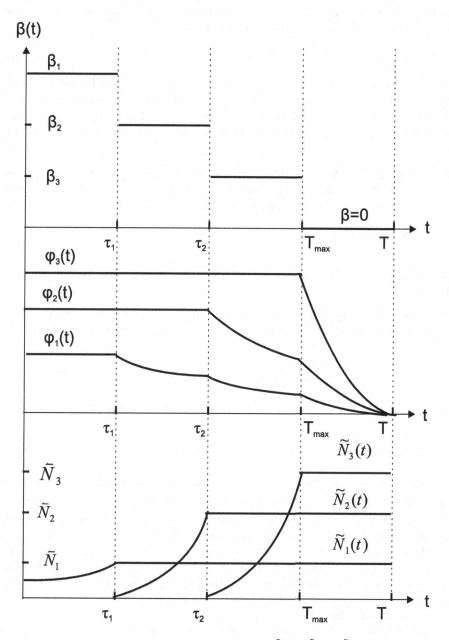

Fig. 1. Behavior dynamics of optimal controls $\tilde{N}_1(t), \tilde{N}_2(t), \tilde{N}_3(t)$, of functions $\varphi_1(t), \varphi_2(t), \varphi_3(t)$ and of Lagrange's multiplier $\beta(t)$ in solving problem 1 of dimension $n = 3$

We know function $\varphi_2(t) = \varphi_2(\tau_2)$ on segment $[0, \tau_2]$, and Lagrange's multiplier $\beta(t) = \beta_2 = \varphi_2(\tau_2)$ on half-open interval $(\tau_1, \tau_2]$. Function $\varphi_1(t)$ on the same half-interval are computed by formula

$$\varphi_1(t) = \varphi_2(\tau_2) - [\varphi_2(\tau_2) - \varphi_1(\tau_2)] \exp[-\alpha_1^0 \bar{N}_1(\tau_2 - t)], \tag{47}$$

We know function $\varphi_1(t) = \varphi_1(\tau_1)$ on segment $[0, \tau_1]$, and Lagrange's multiplier $\beta(t) = \beta_1 = \varphi_1(\tau_1)$ on half-open interval $(0, \tau_1]$. Thus, we completely define all the components of continuous vector-valued function $\varphi(t)$ on segment $[0, T]$ and, respectively, adjoint variables $\psi(t)$ on the same interval. Theorem 2 is proved.

Corollary 3. *Let the set (23) for Problem 1 be nonempty, then the optimal trajectories obtained in solving Problem 1 and Problem 2 given below coincide on segment $[0, T_{max}]$ and the maximum functional value of Problem 2 is T_{max}.*

The proof is trivial.

On Maximizing the Length of a Common "Shelf" for Group of Gas Fields

Problem 2. It is required to maximize the functional $T \longrightarrow max$ under differential connections (2), initial conditions (3) and constraints on control (4),

$$\sum_{i=1}^{n} q_i(t) N_i(t) = \bar{Q}, \quad t \in [0, T].$$

We assume that the initial conditions satisfy strict inequality (6).

5   Conclusion

The optimal policy for the development of gas fields is as follows. First, the first field with the lowest gas extraction rate $\frac{q_1^0}{v_1^0} \bar{N}_1$ is introduced into the development. If the capacity of the first field is not enough to produce gas at the "shelf" level, i.e. $q_1^0 \bar{N}_1 \leq \bar{Q}$, then the second field is introduced into the development.

If the capacity of the first field is enough to carry out gas production at the "shelf" level of the fields, i.e. $q_1^0 \bar{N}_1 > \bar{Q}$, then the development of the first field is carried out with production at the level of $\bar{Q}$. At the same time, part of wells is disconnected. However, with the extraction of gas, wells flow rate is reduced. Gas production is falling.

To maintain production at the level $\bar{Q}$, new unused wells are included in the development. There comes time point τ_1, when all the wells in the first field are involved in production, i.e. $\tilde{q}_1(\tau_1)\bar{N}_1 = \bar{Q}$. To compensate for the decline in production at the first deposit at time τ_1, a second field is being developed. The drop in production rates leads to the commissioning of new unused wells at the second deposit.

There comes a point in time τ_2, when all available wells in both the first and the second field are involved in production, i.e. $\sum_{i=1}^{2} \tilde{q}_i(\tau_2)\bar{N}_i = \bar{Q}$. This procedure continues as long as it is possible to maintain the total gas production at the level of $\bar{Q}$ by entering new wells. At the time T_{max}, all wells in all fields are included in the development, i.e. $\sum_{I=1}^{n} \tilde{q}_i(T_{max})\bar{N}_i = \bar{Q}$. Further, the total production at all fields falls and $\sum_{i=1}^{n} \tilde{q}_i(t)\bar{N}_i < \bar{Q}$ with $t \in (T_{max}, T]$.

Thus, the described algorithm allows finding the maximum accumulated gas production for a period of $[0, T]$. At the same time, the maximum length of the total "shelf" of deposits is T_{max}.

Solutions of the second problem are tested on numerous numerical experiments. The theoretical and numerical results coincided with each other.

References

1. Vyakhirev, R., Korotaev, Yu., Kabanov, N.: Theory and Experience of Gas Recovery. Nedra, Moscow (1998)
2. Margulov, R., Khachaturov, V., Fedoseev, A.: System Analysis in Long-Term Planning of Gas Production. Nedra, Moscow (1992)
3. Khachaturov, V., Solomatin, A., Zlotov, A.: Planning and Design of Development of Oil and Gas Producing Regions and Deposits: Mathematical Models, Methods, Application. LENAND, Moscow (2015)
4. Arrow, K.: Applications of control theory to economic growth. Math. Decis. Sci. **2**, 85–119 (1968)
5. Skiba, A.: Optimal growth with a convex-concave production function. Econometrica **3**(46), 527–539 (1978)
6. Ter-Krikorov, A.: Optimal Control and Mathematical Economics. Nauka, Moscow (1977)
7. Aseev, S., Kryazhimskii, A.: The Pontryagin maximum principle and optimal economic growth problems. Proc. Steklov Inst. Math. **257**, 1–255 (2007). MAIK Nauka/Interperiodica, Moscow
8. Balder, E.: An existence result for optimal economic growth problems. J. Math. Anal. Appl. **95**, 195–213 (1983)
9. Skiba, A.: The maximum principle in the problem of maximization of income for the gas field model. Bull. People's Friendsh. Univ. Russia Ser. "Math. Comput. Sci. Phys." **1**, 14–22 (2009)
10. Lee, E., Markus, L.: Foundations of Optimal Control Theory. Wiley, New York (1967)

Estimation of Multiproduct Models in Economics on the Example of Production Sector of Russian Economy

Ivan Stankevich[1,2], Alexei Ujegov[1(✉)], and Sergey Vasilyev[2,3]

[1] National Research University – Higher School of Economics, Moscow, Russia
vpvstankevich@yandex.ru, ujegov@gmail.com
[2] The Lebedev Physical Institute of the Russian Academy of Sciences,
Moscow, Russia
indianjoe92@gmail.com
[3] Federal Research Center "Computer Science and Control"
of the Russian Academy of Sciences, Moscow, Russia

Abstract. The model of the real sector of the Russian economy is presented. It allows for the separate description of GDP and its components by expenditure both in constant and in current prices. Unlike standard macroeconomic models, the model proposed considers a set of Trader agents in addition to Producer agent. Traders are based on a set of CES-functions and allow to decompose the statistics available into a set of unobserved components. The Producer is based on a specific production function that performs well for Russian data and works with financial variables, such as credits and bank accounts. In contrary to the standard approach, the model is not linearized to get estimates of model parameters but is estimated directly using a set of nonlinear equations. The optimization is performed numerically and allows to get both series of unobserved model products and their prices and model parameters. The stability of the solution found is checked on simulated data.

Keywords: Macroeconomic modeling · Nonlinear models
Gross Domestic Product (GDP) · Mathematical programming

1 Introduction

A very common approach in modern macroeconomics is to ignore the multiproduct structure of economics and to describe the whole economy as an entity producing one single product (usually it is Gross Domestic Product, GDP). However, such description often lacks both economic sense and forecasting accuracy of resulting models. The first is due to the structure of modern economic activity, where a country's production scopes far beyond goods that can be easily

The reported study was supported by the Russian Scientific Fund, research project No. 14-11-00432.

accounted for and ranges from physical assets (that are easy to measure) to services and other intangible assets.

The problems with forecasting accuracy are mostly the result of the GDP structure and the way prices are calculated. One of the most common ways to describe the GDP is what is called GDP by expenditure: the decomposition of GDP into consumption expenditures, investment, government consumption, exports, and imports. All these GDP components are initially calculated in current (observable) prices and then deflated to get numbers, that can be compared at different moments of time. This procedure is performed independently for each GDP component, thus deflators of components may differ significantly. Hence, a model with one product has forecasting accuracy that is naturally limited by the difference in deflators of GDP components.

The standard approach in conventional macroeconomics [6, 14] is to describe the economy as a set of macroeconomic agents, each of which has its own goals (maximization of expected discounted utility for the consumer, maximization of expected discounted profits for the producer) that interact via demand, supply and prices formation. The economy here produces a single product that is distributed between consumption, investment and other uses.

The interest to multiproduct nature of the economy is growing recently, with a growing number of researchers studying this issue. Speaking about macroeconomics modeling, especially DSGE (Dynamic Stochastic General Equilibrium) modeling, it is a range of models started by [5] and further developed mostly in terms of models studying product turnover (entry and exit), such as [3, 10, 11]. Producers here are described as entities working in the multiproduct economy, with an infinite, of finite but very large, set of goods produced and used.

The main problem arising here is that multiple goods produced by firms are consequently aggregated by a set of CES (constant elasticity of substitution) functions to some aggregate. CES function, widely used in economics since [7], is a homogeneous of degree 1 function $X = (\sum_{i=1}^{N} X_i^{\rho})^{1/\rho}$, it is widely used in economics as it contains as cases many widely used function, such as linear function, Cobb-Douglas function and Leontief function.

The standard practice of using the same parameters of these aggregating functions for different agents and sets of goods leads to the restriction of attention to de facto one single good again. Prices of all goods and their sets become equal and the analysis performed is restricted by one good with one price in the same fashion as it was described above.

Another common problem with conventional macroeconomic models is the way they are estimated. Due to the high complexity of problems and the resulting equations (rationality conditions of economic agents - solutions of their optimization problems) the standard practice is to log-linearize the model equations around some steady state (equilibrium condition of economics) and estimate the resulting linear equations using econometric techniques. This approach works well for stable periods of economic growth, but it normally cannot explain deviations from the equilibrium state of economics, such as economic shocks, crises, sharp changes in some indicators etc.

The latter problem is addressed in [2,13]. The models here are solved "as is" without linearization and are capable of forecasting during periods different than balanced growth path.

The current paper aims to solve the problems described above. We consider a multiproduct economy in some sense close to [11], but with no restrictions on coefficients of aggregating functions and we solve it directly in a fashion of [2]. Our approach introduces multiple products in a bit different fashion then it is done usually. We consider an economy with a finite number of intermediate products (three in this case, as calculations on real data show that this number is sufficient to get sufficiently high quality of model), where main economic agents (those that represent production – GDP itself and GDP components, such as consumption) work with one aggregated product (each agent, though, works with its own product).

The aggregation of intermediate products to final products and disaggregation of GDP produced into three intermediate products is performed by Traders - specialized agents, that purchase intermediate products, aggregate them to final product using a CES aggregation function and sell the final product directly to agents. The separation of traders and other agents allows us to solve their problems independently, making the solution and estimation process much simpler. The solution of traders' problems yields what is called in [12] multiproduct model decomposition – a methodology of obtaining unobserved intermediate products series from the available GDP statistics, based on a set of equations obtained as a solution of traders' problems.

The agents in this scheme may solve quite standard problems, for example, of maximization of discounted dividends and maximization of discounted utility for Producer and Consumer respectively. The simplicity of their problems allows us to estimate the nonlinear relationships, such as production function or agents' optimality conditions, directly, without linearization. In the same time, we continue to work in the multiproduct environment, thus, we still can model different GDP components as having different prices and model the dynamics of series both in constant and in current prices.

. The simple model economy presented in this paper consists of a set of Traders, one per each GDP component (consumption, investment, government expenditure, imports and exports), plus the Trader of final good, and two agents: the Producer, that produces GDP using labor and capital as inputs, and Aggregate Consumer that maximizes specific multiproduct utility function with the possibility to have deposits as financial instrument. The usual practice is to define labor as an exogenous variable, e.g. standard Ramsey-Kass-Koopmans model. In this paper, the consumer model has endogenous consumption and labor, which makes more sense regarding consumer behavior modeling. One of the most common utility function types is CRRA function which is widely used in macroeconomic models, e.g. DSGE models [1]. Consumption component usually follows standard CRRA form, while labor is additive in the utility function. Moreover, labor function form may vary from linear too, for example, CRRA [8,9]. In this paper, labor is included multiplicatively as CRRA function.

2 Traders and Multiproduct Decomposition

2.1 Description of a Typical Trader Agent

The standard Trader agent solves a problem of the following type. It minimizes its expenditures on the purchase of intermediate goods X_a, X_b, X_c at prices p_a, p_b, p_c

$$p_x(t)X(t) = p_a(t)X_a(t) + p_b(t)X_b(t) + p_c(t)X_c(t) \to \min_{X_a(t), X_b(t), X_c(t)} \quad (1)$$

With the restriction that the total amount of final good $X(t)$ to be sold is fixed and is determined as a CES function of intermediate goods:

$$X(t) = \left(\alpha_a \left(\frac{X_a(t)}{X_a(0)} \right)^\rho + \alpha_b \left(\frac{X_b(t)}{X_b(0)} \right)^\rho + (1 - \alpha_a - \alpha_b) \left(\frac{X_c(t)}{X_c(0)} \right)^\rho \right)^{1/\rho} \quad (2)$$

By $X(t)$ we denote here one of the GDP components. So, in the case of the trader of consumer good $X(t)$ is consumption expenditure, for the trader of investment good it is investment expenditure, etc.

Equivalently, the problem can be stated as a maximization of the amount of final good $X(t)$ sold with the restriction that the total expenditure is fixed.

Both problems give the same solution. Denoting

$$\Omega^{aX}(t) = \left(\frac{(1 - \alpha_a - \alpha_b)p_a(t)X_a(0)}{\alpha_a p_c(t)X_c(0)} \right)^{\frac{1}{\rho - 1}} \quad (3)$$

and

$$\Omega^{bX}(t) = \left(\frac{(1 - \alpha_a - \alpha_b)p_b(t)X_b(0)}{\alpha_b p_c(t)X_c(0)} \right)^{\frac{1}{\rho - 1}} \quad (4)$$

the solution of a typical trader problem looks as following:

$$X_a(t) = X_a(0) \frac{p_X(t)X(t)\Omega^{aX}(t)}{p_a(t)X_a(0)\Omega^{aX}(t) + p_b(t)X_b(0)\Omega^{bX}(t) + p_c(t)X_c(0)} \quad (5)$$

$$X_b(t) = X_b(0) \frac{p_X(t)X(t)\Omega^{bX}(t)}{p_a(t)X_a(0)\Omega^{aX}(t) + p_b(t)X_b(0)\Omega^{bX}(t) + p_c(t)X_c(0)} \quad (6)$$

$$X_c(t) = X_c(0) \frac{p_X(t)X(t)}{p_a(t)X_a(0)\Omega^{aX}(t) + p_b(t)X_b(0)\Omega^{bX}(t) + p_c(t)X_c(0)} \quad (7)$$

These equations can be used to decompose GDP components into unobserved components if we know the prices $p_a(t), p_b(t), p_c(t)$.

The model estimate for price deflator of each GDP component is

$$\hat{p}_X(t) =$$

$$\left(\alpha_a \left(\frac{p_a(t)}{p_a(0)} \right)^{\frac{\rho}{\rho - 1}} + \alpha_b \left(\frac{p_b(t)}{p_b(0)} \right)^{\frac{\rho}{\rho - 1}} + (1 - \alpha_a - \alpha_b) \left(\frac{p_c(t)}{p_c(0)} \right)^{\frac{\rho}{\rho - 1}} \right)^{\frac{\rho - 1}{\rho}} \quad (8)$$

2.2   The Decomposition Scheme

Given $p_a(t), p_b(t), p_c(t)$, we can calculate model series for price deflators of GDP components using (8) and actual decomposition of GDP components into unobservable intermediate products using (5)–(7). For each moment of time we have 5 equations of type (8) (one for each GDP component) and 3 unknown values of prices $p_a(t), p_b(t), p_c(t)$. Thus, we have more equations then unknown values and we can estimate prices of intermediate goods using model price conditions.

The estimation procedure is organized as follows: we select some initial values for $p_a(t), p_b(t), p_c(t)$, calculate model estimates $\hat{p}_X(t)$ for GDP components' prices using (8) and compute the sum of relative errors, that can be minimized to get estimates of intermediate goods prices.

$$\sum_{t=1}^{T} \sum_{X \in \{C, I, G, Im, Ex\}} \frac{p_X(t) - \hat{p}_X(t)}{p_X(t)} \to min \qquad (9)$$

The optimization over more than 200 parameters is performed numerically in R, using the SPG method, originally presented in [4], with R adaptation by [15]. Initial values for parameters are selected randomly, several different sets of initial parameters were tested.

To get more stable estimates, one of the intermediate goods (good c) was fixed to be used only in government spending. Thus, we get a decomposition into three goods, where one of them can be interpreted as a government good.

The results of estimation are presented on Figs. 1, 2 and 3. The accuracy of model estimates is shown for the series with the lowest quality (consumption), the accuracy for other GDP components is even higher. MAE (mean absolute error) for consumption is 0.17 trillion roubles, for consumption deflator is around 3%.

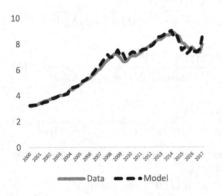

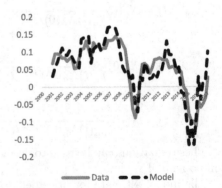

Fig. 1. Accuracy of the model for consumption deflator

Fig. 2. Accuracy of the model for consumption deflator growth rate

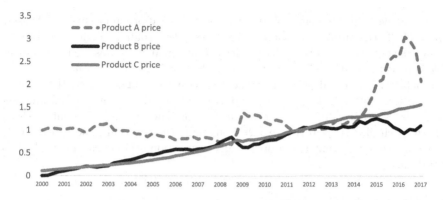

Fig. 3. Model prices of intermediate products

2.3   The Stability of Solution

To check the stability of the solution, we conducted an experiment on simulated data. Prices of intermediate products were generated as random walk processes, GDP components and their price deflators were calculated using (3)–(6) and used to calculate prices of intermediate goods using the decomposition scheme described above. For simplicity of calculations, the stability was checked for the case of two intermediate goods. Initial parameters for optimization were randomly generated, new set for each new iteration.

The stability of the solution found on simulated data is demonstrated in Fig. 4. Real prices are plotted in black (solid and dashed lines), model estimates obtained with different sets of initial parameters are plotted in gray.

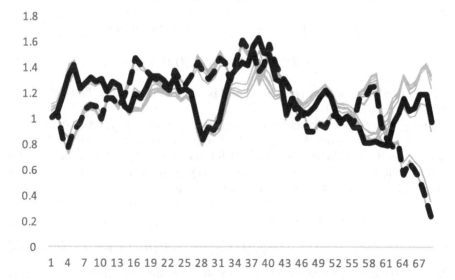

Fig. 4. Stability of decomposition problem solution of simulated data

The convergence to some solution was present in approximately 80% of cases, with MAPE (mean absolute percentage errors) less than 1%. All the solutions found converge to one of the prices (prices of intermediate goods are interchangeable here as there are no links between any of the prices and any other indicator). The behavior of actual prices is replicated with very high quality, the level of series differs a bit, but it is mostly a matter of correct identification of parameters for $t = 0$: different values at the base moment of time can give different levels to otherwise similar trajectories. Apart from that, the solution looks quite stable so we can conclude that the proposed decomposition scheme can be applied to real data.

3   Aggregate Producer

3.1   The Model and Solution

The Producer produces one final good $Y(t)$, using capital $M(t)$ and labor $R(t)$ as inputs. Total investment is divided into capital investment $J_m(t)$ that is used in capital formation, and current investment $J_u(t)$ that is to determine the capital utilization rate. Total investment $J(t) = J_m(t) + J_u(t)$, both investment components are purchased by the producer at price $p_J(t)$. Labor is purchased at price $W_w(t)$.

Production function is a kind of Cobb-Douglas production function with exogenous technological progress:

$$Y_p(t) = Ae^{\gamma t} \left(J_u(t) - u_0 M(t)\right)^{b\alpha} \left(M(t)\right)^{\alpha(1-b)} \left(R(t)\right)^{1-\alpha} \tag{10}$$

where A, γ are some constants that determine the base production level and technological progress respectively, $(J_u(t) - u_0 M(t))$ is the capital utilization rate.

Capital formation is determined by capital investment and amortization:

$$\frac{d}{dt} M(t) = J_m(t) - \delta_{am}(t)M(t) \tag{11}$$

Final product is sold at price $p_Y(t)$ and is taxed at rate $\tau_Y(t)$. Thus, Producer earns $(1 - \tau_Y(t))p_Y(t)Y(t)$.

The producer can take credits, the current amount of credit is denoted as $L(t)$, the credit amount rises when the Producer takes new credits $K(t)$ and lowers when it pays the credit $HL(t)$ with interest $r_l(t)L(t)$. The Producer also has a current account $N(t)$. The producer also pays dividends to its shareholders $Div(t)$. The change in Producer's current account is

$$\frac{d}{dt} N(t) = K(t) - HL(t) - r_l(t)L(t) - OC(t) + p_Y(t)Y(t)$$
$$-p_J(t)J_u(t) - p_J(t)J_m(t) - W_w(t)R(t) + pS(t) - Div(t) \tag{12}$$

where $OC(t)$ are other costs, $pS(t)$ are new shares sold by the producer at current prices.

The aim of the Producer is to maximize the utility of the flow of future dividends payed to shareholders, deflated at final good price:

$$\int_0^T U\left(\frac{Div(t)}{p_Y(t)}\right) e^{\Delta t} \tag{13}$$

With respect to restrictions (3)–(5).
The analytical solution of the problem yields the following results:

$$R(t) = \frac{p_Y(t)(\alpha - 1)(\tau_Y(t) - 1)Y(t)}{W_w(t)} \tag{14}$$

$$\frac{\frac{d}{dt}p_J(t)}{p_J(t)} - r_l(t) =$$

$$\delta_{am}(t) - \frac{\alpha p_Y(t)(J_u(t)b + u_0 M(t) - J_u(t))(\tau_Y(t) - 1)Y(t)}{(J_u(t) - u_0 M(t))M(t)p_J(t)} \tag{15}$$

$$J_u(t) - u_0 M(T) = \frac{\alpha p_Y(t)b(\tau_Y(t) - 1)Y(t)}{p_J(t)} \tag{16}$$

Available statistics allows to get series for all prices, $Y(t)$, $J_u(t)$, $J_m(t)$, $M(t)$, $R(t)$, $\tau_Y(t)$. So, we need to estimate only constant parameters of production function.

3.2   Calibration of Model

The model is estimated numerically, with the sum of squared errors in the GDP prediction using production function (10) as a target function that is minimized over production function parameters. The accuracy of resulting model is demonstrated on Figs. 5 and 6.

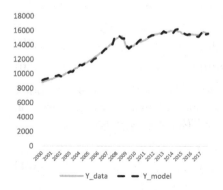

Fig. 5. Accuracy of the model for production function

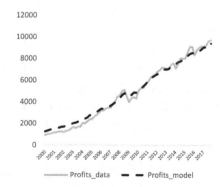

Fig. 6. Accuracy of the model for profits in the economy

Table 1. Aggregate producer model accuracy based on MAE and MAPE coefficients

Variable	MAE	MAPE
GDP Fig. 5	162.66	0.013
Profits Fig. 6	275.24	0.108

The model demonstrates incredibly high accuracy for GDP, with mean absolute percentage error a bit more than 1% and a quite high accuracy for profits predictions. As we can see from the graph, the main problems with profits are due to cyclical fluctuations that start around 2010 and some overestimation of the profits by the model before 2006. The source of the first problem can be in some omitted seasonality (and the problem can be solved by a more thorough work with original data), the second problem is not so serious as we are mostly interested in the more recent data (Table 1).

Other indicators, such as the demand for labor (14), can be calculated using the production function coefficients' estimates and available data.

4 Aggregate Consumer

4.1 The Model and Solution

Aggregate consumer maximizes utility function

$$\int_0^T \frac{C(t)^{1-\beta}}{1-\beta} \frac{R(t)^{1-\alpha}}{1-\alpha} e^{-\delta t} dt \tag{17}$$

choosing both aggregate consumption and labor. In addition, consumer decides the amount of cash $M(t)$ and deposits $S(t)$ taking into account financial balance:

$$\frac{d}{dt}M(t) = \omega(t)R(t) - p(t)C(t) + r_S(t)S(t) - \frac{d}{dt}S(t) - OC(t) \tag{18}$$

considering

$$M(t) \geq 0 \tag{19}$$

$$S(T) \geq \gamma S(0) \tag{20}$$

Wage rate $\omega(t)$, consumption deflator $p(t)$, deposit rate $r_S(t)$ and other incomes $OC(t)$ are known for the whole period $[0, T]$. Solving given optimization problem the following trajectories for $C(t)$ and $R(t)$ can be obtained:

$$C(t) = C(0) \left[\left(\frac{p(t)}{p(0)} \right)^\alpha \left(\frac{\omega(t)}{\omega(0)} \right)^{1-\alpha} e^{-\int_0^t r_S(u)du} e^{\delta t} \right]^{\frac{1}{1-\alpha-\beta}}, \tag{21}$$

$$R(t) = R(0) \left[\left(\frac{p(t)}{p(0)} \right)^{1-\beta} \left(\frac{\omega(t)}{\omega(0)} \right)^\beta e^{-\int_0^t r_S(u)du} e^{\delta t} \right]^{\frac{1}{1-\alpha-\beta}} \tag{22}$$

Logarithmizing and then taking derivatives the following equations can be written:

$$C_t = C_{t-1} \frac{1}{1 - \frac{\alpha}{1-\alpha-\beta}\frac{p_t - p_{t-1}}{p_t} - \frac{1-\alpha}{1-\alpha-\beta}\frac{w_t - w_{t-1}}{w_{t-1}} + \frac{1}{1-\alpha-\beta}r_S(t) - \frac{1}{1-\alpha-\beta}\delta} \tag{23}$$

$$R_t = R_{t-1} \frac{1}{1 - \frac{1-\beta}{1-\alpha-\beta}\frac{p_t - p_{t-1}}{p_t} - \frac{\beta}{1-\alpha-\beta}\frac{w_t - w_{t-1}}{w_{t-1}} + \frac{1}{1-\alpha-\beta}r_S(t) - \frac{1}{1-\alpha-\beta}\delta} \tag{24}$$

Assuming (18) it is easy to rewrite discrete equation for $S(t)$:

$$S_t = r_{S_t}S_{t-1} + S_{t-1} + w_t R_t - p_t C_t - OC_t \tag{25}$$

4.2   Calibration of Model

Model defines optimal trajectories for consumption $C(t)$ (23), labor $R(t)$ (24) and deposits $S(t)$ (25) with the knowledge of exogenous variables consumption deflator $p(t)$, wage rate $w(t)$, deposit rate $r_S(t)$. α, β, δ, $C(0)$, $R(0)$ are calibration parameters that are calculated using minimization of sum of squared errors procedure:

$$\sum_{i=1}^{3}\sum_{t=1}^{T} \frac{(X_i(t) - \hat{X}_i(t))^2}{X_i(T)^2} \tag{26}$$

where $X_i(t) = \{C(t), R(t), S(t)\}$, $\hat{X}_i(t) = \{\hat{C}(t), \hat{R}(t), \hat{S}(t)\}$. $X_i(t)$ stands for statistics and $\hat{X}_i(t)$ stands for model estimated variable. Consumption and labor data is provided by Russian Federation Federal State Statistics Service and deposits data is provided by Central Bank of Russian Federation. Model results are presented by Figs. 7, 8 and 9.

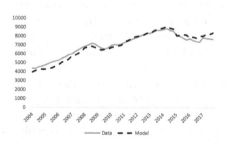

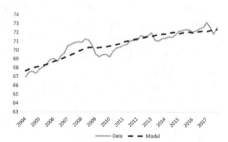

Fig. 7. Accuracy of the model for aggregate consumption

Fig. 8. Accuracy of the model for labor

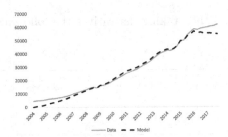

Fig. 9. Accuracy of the model for consumer deposits

Table 2. Aggregate consumer model accuracy based on MAE and MAPE coefficients

Variable	MAE	MAPE
Aggregate consumption Fig. 7	281.36	0.044
Labor Fig. 8	0.496	0.007
Consumer deposits Fig. 9	1867.978	0.153

5   Conclusion

We have presented a methodology for the estimation of multiproduct models
in economics. It is based on a set of Trader agents that decompose the original
single good produced in the economy into several intermediate products that
are unobserved in reality but do constitute in different proportions the GDP
and its components. This methodology allows getting the estimates of series of
unobserved intermediate products as well as estimates of coefficients of models.

The main advantage of the suggested approach is that it allows taking into
account the multiproduct nature of the economy without the necessity to introduce the multiproductivity to the models of main economic agents directly. Using
the additional layer of specialized Trader agents that convert multiple intermediate products to the single final product (different for each of the macroeconomic agents) we are able to model multiple products with different prices while
still working with conveniently simple single product models of macroeconomic
agents (Table 2).

The resulting model is estimated numerically using modern optimization
techniques and yields results that are shown to be quite stable and reliable.
The accuracy of resulting models is high, and it allows to use them both to
explain macroeconomic processes and evaluate policy measures and to make
high-quality forecasts of macroeconomic indicators.

References

1. An, S., Schorfheide, F.: Bayesian analysis of DSGE models. Econ. Rev. **26**(2–4), 113–172 (2007)
2. Andreev, M., Pilnik, N., Pospelov, I.G., Vrzheshch, V.P., Masyutin, A.: Intertemporal three-product general equilibrium model of Russian economy. Int. J. Arts Sci. **6**(1), 125–145 (2013)
3. Bilbiie, F.O., Ghironi, F., Melitz, M.J.: Endogenous entry, product variety, and business cycles. J. Polit. Econ. **120**(2), 304–345 (2012)
4. Birgin, E.G., Martinez, J.M., Raydan, M.: Algorithm 813: SPG-software for convex-constrained optimization. ACM Trans. Math. Softw. (TOMS) **27**(3), 340–349 (2001)
5. Chatterjee, S., Cooper, R.: Entry and exit, product variety and the business cycle (No. w4562). National Bureau of Economic Research (1993)
6. Christiano, L., Eichenbaum, M., Rebelo, S.: When Is the government spending multiplier large? J. Polit. Econ. **119**, 78–121 (2011)
7. Dixit, A.K., Stiglitz, J.E.: Monopolistic competition and optimum product diversity. Am. Econ. Rev. **67**(3), 297–308 (1977)
8. Etro, F., Rossi, L.: New-Keynesian Phillips curve with Bertrand competition and endogenous entry. J. Econ. Dyn. Control **51**, 318–340 (2015)
9. Guo, J.T., Krause, A.: Changing social preferences and optimal redistributive taxation. Oxford Econ. Pap. **70**(1), 73–92 (2017)
10. Hamano, M., Zanetti, F.: Endogenous product turnover and macroeconomic dynamics. Rev. Econ. Dyn. **26**, 263–279 (2017)
11. Minniti, A., Turino, F.: Multi-product firms and business cycle dynamics. Eur. Econ. Rev. **57**, 75–97 (2013)
12. Pilnik, N., Pospelov, I., Stankevich, I.: Multiproduct Model Decomposition of Components of Russian GDP. NRU Higher School of Economics. Series WP BRP "Economics/EC". No. WP BRP 111/EC/2015 (2015)
13. Pilnik, N., Radionov, S.: On new approaches to the identification of blocks of general equilibrium models. Proc. MIPT **3**(35), 151–161 (2017)
14. Smets, F., Wouters, R.: Shocks and frictions in US business cycles: a Bayesian DSGE approach. Am. Econ. Rev. **97**(3), 586–606 (2007)
15. Varadhan, R., Gilbert, P.: BB: an R package for solving a large system of nonlinear equations and for optimizing a high-dimensional nonlinear objective function. J. Stat. softw. **32**(4), 1–26 (2009)

Optimization of Transmission Systems for Chain-Type Markets

Alexander Vasin[(✉)] and Nikita Tsyganov

Faculty of Computational Mathematics and Cybernetics,
Lomonosov Moscow State University, GSP-1, Leninskie Gory, 119991 Moscow, Russia
foravas@yandex.ru, nikita-93@mail.ru

Abstract. We consider a market for a homogeneous good with a chain-type transmission network. Every node corresponds to a local market with a perfect competition characterized by supply and demand functions. The initial transmission capacity, the cost of the capacity expansion and the unit transmission cost are given for every transmission line. The cost of the capacity expansion includes fixed and variable components. We examine the social welfare optimization problem for such a market. The welfare corresponds to the difference between the total consumption utility and the costs of production, transportation, and expansion of the transmission lines. Due to fixed costs of lines expansion, the problem is in general NP-hard with respect to the number of the nodes. We generalize the concept of supermodularity of the welfare function on the set of expanded lines and propose an algorithm for the solution of the problem. Results of computer simulation confirm the statistical efficiency of the algorithm.

Keywords: Mixed optimization problem · Social welfare
Transmission network

1 Introduction

Natural gas, oil and electricity markets play an important role in the economies of many countries. Any such market includes sets of consumers and producers located at different nodes as well as a transmission system connecting such nodes. The transmission cost component in the final resource price is usually high, so the problem of transmission system optimization is of great practical interest.

In this paper we consider the problem of the social welfare maximization, taking into account production costs, consumption utility and costs of transmission capacity increment. The main difficulty of this problem is that an expansion of any line requires considerable fixed costs [1]. The problem is in general NP-hard with respect to the number of the nodes [2]. Under a fixed set of expanded lines, the problem becomes convex, so it is possible to apply the known numeric

This research is supported by RFBR project 16-01-00353/16.

methods [3]. In several cases, the welfare function meets supermodularity or sub-modularity properties on the set of expanded lines [4]. The paper [1] describes an algorithm for the transmission system optimization in the case of a chain-type market with unidirectional flows. In that case, the welfare function is super-modular. The present paper generalizes this algorithm for a case of bidirectional flows and also presents results of calculational experiments that confirm its efficiency. Bidirectional flows on an underlying path appear in the Nordic Pool, some electricity and gas transmitting chains crossing Siberia.

2   Setting of the Problem

Consider a market consisting of $m+1$ consequently connected local markets and a network transmission system. Let $N = \{0, 1, 2, \ldots, m\}$ be the set of nodes corresponding to local perfectly competitive markets and $L = \{\{0, 1\}, \{1, 2\}, \ldots, \{m-1, m\}\}$ be the set of edges in this network. For every node let the demand function $D_i(p_i)$ depend on the price value p_i at node i and characterize consumers, and let the cost function $c_i(v_i)$ depend on the production volume v_i and characterize producers. The demand function is non-increasing, and its value equals to 0 for sufficiently large price values p_i. It relates to the consumption utility function of consumers at node i: $U_i(v_i^d) = \int_0^{v_i^d} D_i^{-1}(u)du$, where v_i^d is the quantity of the good consumed. So the inverse demand function $D_i^{-1}(u)$ shows the marginal utility of consumption. The cost function $c_i(v_i)$ is non-decreasing and convex and shows the minimal production cost of the volume v_i at node i. It relates to the supply function $S_i(p)$ that determines the optimal (profit-maximizing) production volume at node i, i.e. $S_i(p) = Arg \max_{v \geq 0}(pv - c_i(v))$.

Each transmission line $l \in L$ is characterized by the initial transmission capacity Q_l^0, the unit transmission cost e_l^t, and the function of the transmission capacity increment costs $e_l(\Delta Q_l)$ that includes fixed cost e_l^f and the variable cost $e_l^v(\Delta Q_l)$, where $e_l^v(\Delta Q_l)$ is a monotonous convex function, depending on the increase $\Delta Q_l = Q_l - Q_l^0$, Q_l is the new transmission capacity, $e_l^v(0) = 0$. The cost of the line expansion equals to the overnight construction cost OC_l amortized over the life-time T_l of the line with discount rate r: $e_l = \frac{rOC_l}{1-e^{-rT_l}}$ [5]. Note that considered model can be easily generalized to the case of the unit transmission costs e_l^t that depend on the flow directions.

For every line $l = \{i, i+1\}$ let $q_{i,i+1}$ be the flow (the amount of transmitted goods) from market i to market $i+1$, and $q_{i+1,i} = -q_{i,i+1}$ be the flow from market $i+1$ to market i. Under any fixed flows $q = (q_{i,i+1}, i = 0, \ldots, m-1)$ and production volumes $v = (v_i, i \in N)$ the social welfare for this market is determined as follows:

$$W(q, v) = \sum_{i \in N}[U_i(v_i - \sum_{j|\{i,j\}\in L} q_{i,j}) - c_i(v_i)] - \sum_{i=0}^{m-1} E_{\{i,i+1\}}(q_{i,i+1}),$$

where the total transportation costs for line $\{i, i+1\}$ include the commodity transportation costs and the costs of the line expansion:

$$E_{\{i,i+1\}}(q_{i,i+1}) = \begin{cases} e^f_{\{i,i+1\}} + e^v_{\{i,i+1\}}(|q_{i,i+1}| - Q^0_{\{i,i+1\}}) + e^t_{\{i,i+1\}}|q_{i,i+1}|, \\ \qquad\qquad\qquad\qquad\qquad\qquad\qquad |q_{i,i+1}| > Q^0_{\{i,i+1\}}, \\ e^t_{\{i,i+1\}}|q_{i,i+1}|, \qquad\qquad\qquad\quad |q_{i,i+1}| \le Q^0_{\{i,i+1\}}, \end{cases}$$

Thus, the social welfare is the total utility of consumption less the total costs of production and total transportation costs. The aim of the present paper is to develop efficient algorithms for resolving the social welfare maximization problem:

$$\max_{q,v} W(q,v). \tag{1}$$

Consider an auxiliary problem with a fixed set of expanded lines $R \subseteq L$:

$$\max_{q,v} W(q,v,R), \tag{2}$$

where, unlike (1), $|q_{i,i+1}| \le Q^0_{\{i,i+1\}}$ for $\{i, i+1\} \in L \setminus R$, and the fixed costs are always included in $E_{\{i,i+1\}}(q_{i,i+1})$ for $\{i, i+1\} \in R$. Let $\widetilde{W}(R)$ denote the maximal welfare value of problem (2). Then the initial problem (1) reduces to searching of the optimal set of expanded lines:

$$L^* \in Arg \max_{R \subseteq L} \widetilde{W}(R). \tag{3}$$

Consider the notion of the competitive equilibrium for a market with a fixed set R of expanded lines. Let $\Delta S_i(p) = S_i(p) - D_i(p)$ denote the supply-demand difference at node i. The combination of price vector $p = (p_i, i \in N)$, production volume vector $v = (v_i, i \in N)$ and flow vector $q = (q_{i,i+1}, \{i, i+1\} \in L)$, is called a competitive equilibrium of the market if, for any node $i \in N$ the production volume v_i maximizes the profit of producers under a given price p_i, the price p_i balances the supply and the demand at this node taking into account incoming and outcoming flows:

$$\begin{cases} v_i = S_i(p_i), \\ \Delta S_i(p_i) = \sum_{j | \{i,j\} \in L} q_{i,j}, \end{cases} \tag{4}$$

and for every line $\{i, i+1\} \in L$ either the flow is zero and the unit transmission cost exceed the price difference between the incident nodes, making reselling unprofitable, or marginal total transportation costs have come up to the price difference, or the transmission capacity limit is reached (if $\{i, i+1\} \in L \setminus R$):

$$\begin{cases} q_{i,i+1} = 0, & p_{i+1} - p_i \in [0, e^t_{\{i,i+1\}}), \\ q_{i,i+1} \in [0, Q^0_{\{i,i+1\}}], & p_{i+1} - p_i = e^t_{\{i,i+1\}}, \\ q_{i,i+1} = Q^0_{\{i,i+1\}}, & p_{i+1} - p_i > e^t_{\{i,i+1\}}, \quad \{i, i+1\} \in L \setminus R, \\ p_{i+1} - p_i = e^t_{\{i,i+1\}} + e^v_{\{i,i+1\}}{}'(|q_{i,i+1}| - Q^0_{\{i,i+1\}}), \\ \qquad\qquad\qquad q_{i,i+1} > Q^0_{\{i,i+1\}}, \quad \{i, i+1\} \in R, \end{cases} \tag{5}$$

$$\begin{cases} q_{i,i+1} = 0, & p_{i+1} - p_i \in (-e^t_{\{i,i+1\}}, 0], \\ q_{i,i+1} \in [-Q^0_{\{i,i+1\}}, 0], & p_{i+1} - p_i = -e^t_{\{i,i+1\}}, \\ q_{i,i+1} = -Q^0_{\{i,i+1\}}, & p_{i+1} - p_i < -e^t_{\{i,i+1\}}, \quad \{i, i+1\} \in L \setminus R, \\ p_{i+1} - p_i = -e^t_{\{i,i+1\}} + e^v_{\{i,i+1\}}{}'(|q_{i,i+1}| - Q^0_{\{i,i+1\}}), \\ \qquad\qquad\qquad q_{i,i+1} < -Q^0_{\{i,i+1\}}, \quad \{i, i+1\} \in R. \end{cases} \quad (6)$$

System (5) corresponds to the case of the good's flow from node i to node $i + 1$, and system (6) corresponds to the reverse. According to [1], problem (2) with a fixed set of expanded lines R is a convex programming problem, and its solution corresponds to the competitive equilibrium of the market, i.e. conditions (4)–(6) are met. Therefore, for solving problem (2) known numerical methods can be applied [3]. Below we solve problem (3) of searching for the optimal set of expanded lines L^*. For this reason we define the flow structure invariance condition (FSIC) [4], which is important for our study. For any vector $Q = (Q_{i,i+1}, \{i, i+1\} \in L)$ of transmission capacities, consider the market without construction costs and with restrictions $|q_{i,i+1}| \leq Q_{i,i+1}, \{i, i+1\} \in L$. Let $p_i(Q)$, $i \in N$, denote the equilibrium prices corresponding to this market.

Definition 1. *The considered model meets the FSIC if, for any $Q \geq Q^0$,*

$$sign(p_{i+1}(Q) - p_i(Q)) = sign(p_{i+1}(Q^0) - p_i(Q^0)),$$

$\{i, i+1\} \in L.$

In other words, a market meets the FSIC if for any expansion the equilibrium flow directions for every line stays unchanged. In [4] it is proved that if initial equilibrium prices $p_i(Q^0)$ are monotonous in i, i.e. if for initial transmission capacities all the flows in the market are unidirectional, then this market meets the FSIC, and the welfare function $\widetilde{W}(R)$ is supermodular on L (i.e. for any sets $L_1, L_2 \in L$ $\widetilde{W}(L_1) + \widetilde{W}(L_2) \leq \widetilde{W}(L_1 \cup L_2) + \widetilde{W}(L_1 \cap L_2)$). Paper [1] proposes an algorithm, which, using the supermodularity property, calculates the exact solution to the problem (3) for a case of unidirectional flows. A numerical experiment there demonstrates that for a set of randomly generated supermodular functions the average number of solved subproblems (2) is well approximated by a cubic polynomial depending on the number of nodes in the market.

The present paper considers a general case in which flows can take the both directions. Consider equilibrium flow directions of the initial market under initial transmission capacities ($Q = Q^0$). Divide the set L into set L_1 with left-to-right flow directions (lines of the first type) and set L_2 with right-to-left flow directions (lines of the second type): $L_1 = \{\{i, i+1\} \subseteq L| \ p_i(Q^0) \leq p_{i+1}(Q^0)\}$, $L_2 = \{\{i, i+1\} \subseteq L| \ p_i(Q^0) > p_{i+1}(Q^0)\}$. An example of such a market is presented in Fig. 1. In [1] the necessary and sufficient condition of the FSIC is obtained in the form of the following theorem.

Theorem 1. *A chain-type market meets the FSIC if and only if for any $l = \{i, i+1\} \in L_1$ $p_i(Q^0_{L_1}, Q^\infty_{L_2}) \leq p_{i+1}(Q^0_{L_1}, Q^\infty_{L_2})$, and for any $l = \{i, i+1\} \in L_2$ $p_i(Q^\infty_{L_1}, Q^0_{L_2}) \geq p_{i+1}(Q^\infty_{L_1}, Q^0_{L_2})$, where Q_{L_1} and Q_{L_2} are the projections of the vector Q on the corresponding subspace, and Q^∞ is the vector for which all the components are equal to plus infinity.*

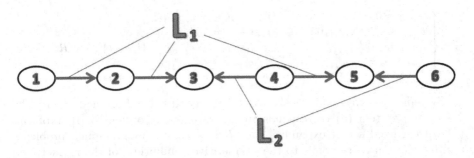

Fig. 1. An example of a network market with bidirectional flows

Assume that the model meets the FSIC. Now we describe an algorithm that finds an optimal set L^* of expanded lines. For any set $\tilde{L} \subseteq L$ let $L_1(\tilde{L}) = L_1 \cap \tilde{L}$, $L_2(\tilde{L}) = L_2 \cap \tilde{L}$. Let $L_1^* = L_1(L^*)$, $L_2^* = L_2(L^*)$. The algorithm works with a lower bound L_1^{min} and with an upper bound L_1^{max} on the set L_1^*, and also with a lower bound L_2^{min} and an upper bound L_2^{max} on the set L_2^* and is based on the consequently performed steps of two types. The first type aims to expand the current set L_1^{min} and to narrow down the current set L_2^{max}. The second type, vice versa, aims to narrow down the current set L_1^{max} and to expand the current set L_2^{min}. The line is called *included* if it belongs to the current lower bound, and *excluded* if it doesn't belong to the current upper bound. The line is called *determined* if it is included or excluded. The step of the first type is executed if the number of determined lines for it is less than for the step of the second type. Otherwise the step of the second type is executed. Finally, the optimal set L^* is found in the form of $L_1^* \cup L_2^*$, and L_1^* and L_2^* meet conditions $L_1^{min} \subseteq L_1^* \subseteq L_1^{max}$ and $L_2^{min} \subseteq L_2^* \subseteq L_2^{max}$.

The function $\widetilde{W}(R)$ for a chain-type market with bidirectional flows meets the following property, which is an analog of the property of supermodularity for a market with unidirectional flows.

Theorem 2. *Let* $R \subseteq L$, $R_1 = L_1(R)$, $R_2 = L_2(R)$, $S_1^+ \subseteq L_1 \backslash R_1$, $S_2^+ \subseteq L_2 \backslash R_2$, $S_1^- \subseteq R_1$, $S_2^- \subseteq R_2$, $S_1 = S_1^+ \cup S_1^-$, $S_2 = S_2^+ \cup S_2^-$, $R_1^+ \subseteq (L_1 \backslash R_1) \backslash S_1$, $R_2^+ \subseteq (L_2 \backslash R_2) \backslash S_2$, $R_1^- \subseteq R_1 \backslash S_1^-$, $R_2^- \subseteq R_2 \backslash S_2^-$. *Then the following implications hold:*

$$\text{if } \widetilde{W}((R_1 \cup S_1^+) \cup (R_2 \backslash S_2^-)) \geq \widetilde{W}(R_1 \cup R_2), \text{ then}$$
$$\widetilde{W}((R_1 \cup R_1^+ \cup S_1^+) \cup ((R_2 \backslash R_2^-) \backslash S_2^-)) \geq \widetilde{W}((R_1 \cup R_1^+) \cup (R_2 \backslash R_2^-)); \tag{7}$$

$$\text{if } \widetilde{W}((R_1 \backslash S_1^-) \cup (R_2 \cup S_2^+)) \geq \widetilde{W}(R_1 \cup R_2), \text{ then}$$
$$\widetilde{W}(((R_1 \backslash R_1^-) \backslash S_1^-) \cup (R_2 \cup R_2^+ \cup S_2^+)) \geq \widetilde{W}((R_1 \backslash R_1^-) \cup (R_2 \cup R_2^+)). \tag{8}$$

Property (7) means that if adding a set S_1^+ to the initial set R_1 of the lines of the first type and excluding a set S_2^- from the initial set R_2 of lines of the second type leads to social welfare increase, then adding the set S_1^+ to the expanded initial set $R_1 \cup R_1^+$ of lines of the first type and excluding the set S_2^- from the narrowed initial set $R_2 \setminus R_2^-$ of the lines of the second type also leads to social welfare increase. Property (8) is similar.

3   Algorithm

The algorithm steps are based on the following theorem which is a corollary of Theorem 2.

Theorem 3. *Let L_1^{min} and L_1^{max} be the current lower and the current upper bounds of the set L_1^* respectively, and let L_2^{min} and L_2^{max} be the current lower and the current upper bounds of the set L_2^* respectively. Let $S_1 \subseteq L_1^{max} \setminus L_1^{min}$, $S_2 \subseteq L_2^{max} \setminus L_2^{min}$, $S = S_1 \cup S_2$.*
  I. If the following inequalities hold:

$$\widetilde{W}((L_1^{min} \cup S_1) \cup (L_2^{max} \setminus S_2)) \geq \widetilde{W}(L_1^{min} \cup L_2^{max}),$$
$$\widetilde{W}((L_1^{min} \cup L_1(R)) \cup (L_2^{max} \setminus L_2(R))) < \widetilde{W}(L_1^{min} \cup L_2^{max})$$

for any nonempty set $R \subset S$, then the refined lower bound L_{1r}^{min} and upper bound L_{2r}^{max} are determined as follows: $L_{1r}^{min} = L_1^{min} \cup S_1$, $L_{2r}^{max} = L_2^{max} \setminus S_2$.
  II. If the following inequalities hold:

$$\widetilde{W}((L_1^{max} \setminus S_1) \cup (L_2^{min} \cup S_2)) \geq \widetilde{W}(L_1^{max} \cup L_2^{min}),$$
$$\widetilde{W}((L_1^{max} \setminus L_1(R)1) \cup (L_2^{min} \cup L_2(R))) < \widetilde{W}(L_1^{max} \cup L_2^{min})$$

for any nonempty set $R \subset S$, then the refined upper bound L_{1r}^{max} and lower bound L_{2r}^{min} are determined as follows: $L_{1r}^{max} = L_1^{max} \setminus S_1$, $L_{2r}^{min} = L_2^{min} \cup S_2$.

Proof. We present evidence for item I. Item II is proved in a similar way. From the theorem assumptions it follows that, for any $R \subseteq S$ (among them $R = \emptyset$), the following inequality holds:

$$\widetilde{W}((L_1^{min} \cup S_1) \cup (L_2^{max} \setminus S_2)) \geq \widetilde{W}((L_1^{min} \cup L_1(R)) \cup (L_2^{max} \setminus L_2(R))). \quad (9)$$

It's sufficient to check that for any subsets $\hat{S}_1 \subseteq L_1^{max} \setminus L_1^{min}$ and $\hat{S}_2 \subseteq L_2^{max} \setminus L_2^{min}$ the following inequality is true:

$$\widetilde{W}((L_1^{min} \cup \hat{S}_1 \cup S_1) \cup ((L_2^{max} \setminus \hat{S}_2) \setminus S_2)) \geq \widetilde{W}((L_1^{min} \cup \hat{S}_1) \cup (L_2^{max} \setminus \hat{S}_2)). \quad (10)$$

From (9) it follows that

$$\widetilde{W}((L_1^{min} \cup S_1) \cup (L_2^{max} \setminus S_2)) \geq \widetilde{W}((L_1^{min} \cup (S_1 \cap \hat{S}_1)) \cup (L_2^{max} \setminus (S_2 \cap \hat{S}_2))),$$

that is equivalent to the inequality

$$\widetilde{W}((L_1^{min} \cup (S_1 \cap \hat{S}_1) \cup (S_1 \setminus \hat{S}_1)) \cup ((L_2^{max} \setminus (S_2 \cap \hat{S}_2)) \setminus (S_2 \setminus \hat{S}_2))) \geq$$
$$\widetilde{W}((L_1^{min} \cup (S_1 \cap \hat{S}_1)) \cup (L_2^{max} \setminus (S_2 \cap \hat{S}_2))).$$

Then, using (7) with S_1^+, S_2^- and R_1^+, R_2^- replaced by $S_1 \setminus \hat{S}_1$, $S_2 \setminus \hat{S}_2$ and $\hat{S}_1 \setminus S_1$, $\hat{S}_2 \setminus S_2$ respectively, we obtain that

$$\widetilde{W}((L_1^{min} \cup (S_1 \cap \hat{S}_1) \cup (\hat{S}_1 \setminus S_1) \cup (S_1 \setminus \hat{S}_1))$$
$$\cup(((L_2^{max} \setminus (S_2 \cap \hat{S}_2)) \setminus (\hat{S}_2 \setminus S_2)) \setminus (S_2 \setminus \hat{S}_2)))$$
$$\geq \widetilde{W}((L_1^{min} \cup (S_1 \cap \hat{S}_1) \cup (\hat{S}_1 \setminus S_1)) \cup ((L_2^{max} \setminus (S_2 \cap \hat{S}_2)) \setminus (\hat{S}_2 \setminus S_2))),$$

that is equivalent to (10). □

Consider the algorithm that determines the optimal set of expanded lines. We use the following variables: L_1^{min}, L_1^{max}, L_2^{min}, L_2^{max} are the current lower and upper bounds of the sets L_1^* and L_2^* respectively; k_1 and k_2 are the numbers of determined lines for the step of the first type and for the step of the second type respectively; Q is the set of not yet determined lines; $T \in \{1, 2\}$ is the type of the executing step, s is the step number.

1. Assign $Q := L$, $L_1^{min} := L_2^{min} := \emptyset$, $L_1^{max} := L_1$, $L_2^{max} := L_2$, $k_1 := k_2 := 1$, $T := 1$, $s := 1$.
2. If $s = m$, then $L^* := \underset{R \in \{L_1^{min} \cup L_2^{max}, L_1^{max} \cup L_2^{min}\}}{\text{Arg max}} \widetilde{W}(R)$. Finish.
3. If $T = 2$, then go to item 6.
4. Consequently consider every set S for which $S \subseteq Q$, $|S| = k_1$. If for some considered set S $\widetilde{W}((L_1^{min} \cup L_1(S)) \cup (L_2^{max} \setminus L_2(S))) \geq \widetilde{W}(L_1^{min} \cup L_2^{max})$, then assign $L_1^{min} := L_1^{min} \cup L_1(S)$, $L_2^{max} := L_2^{max} \setminus L_2(S)$, $Q := Q \setminus S$, $k_1 := 1$, $s := s + 1$, go to item 2.
5. Assign $k_1 := k_1 + 1$, $s := s + 1$. If $k_1 > k_2$, then assign $T := 2$. Go to item 2.
6. Consequently consider every set S for which $S \subseteq Q$, $|S| = k_2$. If for some considered set S $\widetilde{W}(L_1^{max} \setminus L_1(S)) \cup (L_2^{min} \cup L_2(S))) \geq \widetilde{W}(L_1^{max} \cup L_2^{min})$, then assign $L_1^{max} := L_1^{max} \setminus L_1(S)$, $L_2^{min} := L_2^{min} \cup L_2(S)$, $Q := Q \setminus S$, $k_2 := 1$, $s := s + 1$, go to item 2.
7. Assign $k_2 := k_2 + 1$, $s := s + 1$. If $k_1 < k_2$, then assign $T := 1$. Go to item 2.

Note that, if the set S under consideration meets inequality $\widetilde{W}((L_1^{min} \cup L_1(S)) \cup (L_2^{max} \setminus L_2(S))) \geq \widetilde{W}(L_1^{min} \cup L_2^{max})$, then, according to Theorem 3, the set $L_1(S)$ can be included into the lower bound L_1^{min}, and the set $L_2(S)$ can be excluded from the upper bound L_2^{max}. Indeed, every unempty subset $R \subset S$ is already considered at previous stages and meets inequality $\widetilde{W}((L_1^{min} \cup L_1(R)) \cup (L_2^{max} \setminus L_2(R))) < \widetilde{W}(L_1^{min} \cup L_2^{max})$ (otherwise the set $L_1(R)$ would be included in the set L_1^{min}, and the set $L_2(R)$ would be excluded from the set L_2^{max}). Excluding the set $L_1(S)$ and including the set $L_2(S)$ are similar.

4   Evaluation of the Algorithm's Efficiency

The presented algorithm enables us to considerably reduce the number of solved subproblems (2) and reduce the time of solving the problem (1). In order to evaluate the number of solved subproblems depending on the number of nodes, we performed a numerical experiment for markets with piecewise linear demand functions and supply functions of the following types:

$$D_i(p_i) = \begin{cases} d_i^f - \frac{1}{2}c_i p_i, \ p_i \le 2\frac{d_i^f}{c_i}, \\ 0, \qquad\quad p_i > 2\frac{d_i^f}{c_i}, \end{cases} \qquad S_i(p_i) = \begin{cases} \frac{1}{2}c_i p_i, \qquad\quad p_i \le 2\frac{d_i^f}{c_i}, \\ -d_i^f + c_i p_i, \ p_i > 2\frac{d_i^f}{c_i}. \end{cases}$$

In this case for every node $i \in N$ the supply-demand difference is a linear function $\Delta S_i(p) = -d_i^f + c_i p_i$. For every line $\{i, i+1\} \in L$ the cost function of the transmission capacity increment is quadratic:

$$e_l^v(\Delta Q_l) = e_l^q \Delta Q_l^2.$$

Let Δp_i denote the difference between prices for nodes i and $i+1$ under zero initial transmission capacities: $\Delta p_i = p_{i+1}(0) - p_i(0)$, $\{i, i+1\} \in L$. Let $p_{min} = \min\limits_{i \in N} p_i(0)$. The initial market is completely determined by the following parameters: m, p_{min}, d_i^f ($i \in N$), Δp_i ($i = 1, \ldots, m$), Q_l^0, e_l^t, e_l^q, e_l^f ($l \in L$).

Parameters p_{min}, d_i^f, $|\Delta p_i|$, e_l^t, e_l^q, e_l^f are selected randomly under uniform distributions. Their minimum and maximum values are presented in Table 1. Q_l^0 is equal to 0, $l \in L$. For any $i = 1, \ldots, m$, the sign of Δp_i is equal to the sign of Δp_{i-1} with a probability of 0.9.

Table 1. Parameters of the distributions

Model parameter	Minimum value	Maximum value
p_{min}	0	10
d_i^f	10	20
$|\Delta p_i|$	0	10
e_l^t	0	4
e_l^q	0	4
e_l^f	0	4

In order for the market to meet the FSIC, we use the modified unit transmission costs $e_{i,i+1}^{tm}$ that depend on the flow directions:

$$e_{i,i+1}^{tm}(q_{i,i+1}) = \begin{cases} e_{i,i+1}^t, \ sign(q_{i,i+1}) = sign(\Delta p_i), \\ +\infty, \quad sign(q_{i,i+1}) \ne sign(\Delta p_i), \end{cases}$$

$\{i, i+1\} \in L$. We present the results of the performed experiment for various numbers of nodes (from 1 to 65). For every number, 1000 problems are generated. Figure 2 demonstrates the observed dependency between the number of nodes and the number of solved subproblems. Table 2 shows the sample observed average and maximum numbers of solved subproblems.

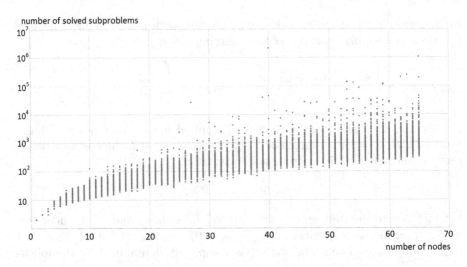

Fig. 2. Results of the numerical experiment

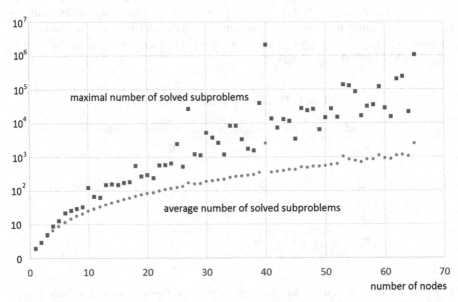

Fig. 3. Average (lower observations) and maximum (upper observations) numbers of solved subproblems

Denote the number of nodes by x and the average number of solved subproblems by y_{av}. To study the dependency of y_{av} on x (Fig. 3), we construct the regression using the weighted least-squares method [6] and obtain the following approximation: $y_{av} = 0.25142x^2 + 1.77108$ (Fig. 4). The corresponding determination coefficient R^2 is equal to 0.6357.

Table 2. Average and maximum numbers of solved subproblems

Number of nodes	Average number of solved subproblems	Maximum number of solved subproblems
1	2	2
3	4.599	5
5	8.8	13
7	14.334	26
9	20.468	33
11	28.281	67
13	37.992	148
15	48.44	150
17	60.34	180
19	73.675	264
21	86.587	237
23	105.937	567
25	123.306	2396
27	166.944	26368
29	161.458	1125
31	192.45	3663
33	214.08	1183
35	256.25	8466
37	274.971	1718
39	338.875	38695
41	348.4	13441
43	380.349	12834
45	405.324	3386
47	472.59	23590
49	505.125	6480
51	570.868	26248
53	985.463	134614
55	749.519	83523
57	819.707	31032
59	1074.316	117688
61	838.352	15151
63	1107.681	237354
65	2380.221	1049639

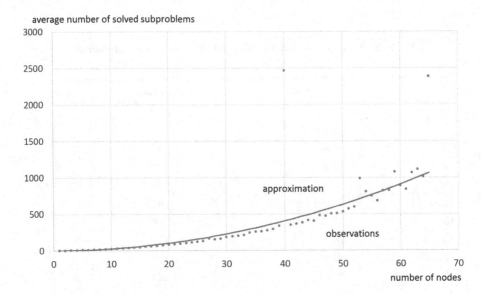

Fig. 4. Approximation of the average number of solved subproblems

Thus, the average number of solved subproblems is well approximated by the square function, that enables finding the solution to the initial problem (1) in a reasonable time.

5   Conclusion

In this paper, we proposed the algorithm solving the social welfare maximization problem for a transmission system of a chain-type competitive market. The performed statistical analysis has shown that, for a randomly generated set of such problems, the average number of convex optimization subproblems we had to solve in order to find the solution quadratically depends on the number of nodes in the chain. So the algorithm is statistically efficient. The challenging task is to generalize the method for markets with tree-type transmission networks.

References

1. Vasin, A.A., Grigoryeva, O.M., Tsyganov, N.I.: Optimization of an energy market transportation system. Doklady Math. **96**(1), 411–414 (2017)
2. Guisewite, G.M., Pardalos, P.M.: Minimum concave-cost network flow problems: applications, complexity, and algorithms. Ann. Oper. Res. **25**(1), 75–99 (1990)
3. Gasnikov, A.V.: A note on effective computability of competitive balances in transportation economic models. Math. Model. **27**(12), 121–136 (2015)
4. Vasin, A.A., Dolmatova, M.: Optimization of transmission capacities for multinodal markets. Procedia Comput. Sci. **91**, 238–244 (2016)

5. Stoft, S.: Power System Economics: Designing Markets for Electricity. IEEE/Wiley, New York (2002)
6. Magnus, Y.R., Katyshev, P.K., Peresetsky, A.A.: Econometrics. Beginner's Course. Tutorial. Delo, Moscow (2004)

Applications

Hypoelastic Stabilization of Variational Algorithm for Construction of Moving Deforming Meshes

Vladimir Garanzha[1,2(✉)] and Liudmila Kudryavtseva[1,2,3]

[1] Dorodnicyn Computing Center, FRC CSC RAS, Moscow 119333, Russia
garan@ccas.ru
[2] Moscow Institute of Physics and Technology, Moscow, Russia
[3] Keldysh Institute of Applied Mathematics RAS, Moscow, Russia
liukudr@yandex.ru
http://www.ccas.ru/gridgen/lab

Abstract. We suggest an algorithm for time-dependent mesh deformation based on the minimization of hyperelastic quasi-isometric functional. The source of deformation is time-dependent metric tensor in Eulerian coordinates. In order to attain the time continuity of deformation we suggest stress relaxation procedure similar to hypoelasticity where at each time step special choice of metric in Lagrangian coordinates eliminates internal stresses. Continuation procedure gradually introduces internal stresses back while forcing deformation to follow prescribed Eulerian metric tensor. At each step of continuation procedure functional is approximately minimized using few steps of preconditioned gradient search algorithm. Stress relaxation and continuation procedure are implemented as a special choice of factorized representation of Lagrangian metric tensor and nonlinear interpolation procedure for factors of this metric tensor. Thus we avoid solving time-dependent PDE for mesh deformation and under certain assumption guarantee that deformation from mesh on a one-time level to the next one converges to isometry when time step tending to zero.

Keywords: Mesh deformation · Quasi-isometric mapping
Metric interpolation · Hypoelasticity

1 Introduction

The equidistribution principle [7] is a basic component of moving adapting mesh algorithms [1]. The idea of equidistribution is to compute the distribution of the mesh size using information from certain PDE or from an estimate of interpolation error. In 1d equidistribution principle results in the solution of the linear elliptic equation of the second order with a scalar control function. Generalization

Supported by Russian Science Foundation 16-11-10350 grant.

Y. Evtushenko et al. (Eds.): OPTIMA 2018, CCIS 974, pp. 497–511, 2019.
https://doi.org/10.1007/978-3-030-10934-9_35

to multiple dimensions still remains controversial. Widely used generalization is based on the solution of Laplacian-like equations for mesh coordinates using scalar control function. In the formulation of S.K. Godunov, the equidistribution principle in multiple dimensions is equivalent to the construction of quasi-isometric mapping between metric spaces. Efficient approximate variational principle and solver for construction quasi-isometric mappings was suggested in the series of papers by V.A. Garanzha. With this solver, one can avoid limitations on the dimension and shape of the domain which is a curse for standard linear equidistribution mesh solvers. When trying to apply this variational principle to the problem of constructing moving adaptive meshes we have encountered serious drawbacks. The main one is related to the fact that nonlinear variational problem is never solved exactly. As a result, for small changes of metric per time step one still can get large deformations due to gradual minimization for a problem with relatively large residual. We abstain from the use of artificial time derivative in the mesh deformation solver since no mesh quality guarantees are available in this case. Moreover one generally has to use adaptive filters to suppress moving mesh instabilities which were described in the seminal paper by Coyle, Flaherty, Ludwig, 1986 [8]. Note recent paper [2] where variational mesh deformation is considered and stability is demonstrated in the case when functional does not increase along deformation trajectory.

As an alternative, we use special interpolation formulae for factorized metric tensors assuming that current mesh is absolute minimizer of variational principle using a special choice of the metric tensor. One can show that deformation from mesh on a one-time level to the next one tends to the isometric one when time-step is diminished resulting in the stable and continuous mesh deformation solver. While rigorous proof is not available, we were not able to detect instabilities predicted by Flaherty et al.

2   Multidimensional Equidistribution Principle and Quasi-Isometric Mappings

At the beginning of the 1990s, S.K. Godunov formulated the following basic principles of optimal variational meshing:

- mesh should be quasi-isometric deformation by itself and should converge to certain quasi-isometric mapping upon refinement;
- quasi-isometric mapping should be unique and stable solution of the variational problem;
- variational problem should use distortion measures based on principal invariants of metric tensor of deformation:
- discretized variational problem should also have a unique and stable solution;

Let us remind that by definition the ratio of the length of any simple rectifiable curve to the length of its image under quasi-isometric mapping is bounded above by a constant K and bounded below by $1/K$.

Implementation of these ideas was first presented in [4] using the conformal mapping technique for the problem of constructing parametrization of the curvilinear quadrangle. However, in the case of a general domain, such a problem has still not been solved even in the 2d case. These ideas inspired a series of papers including [5,6] where variational principle for the construction of multidimensional quasi-isometric mappings was suggested and justified theoretically and numerically.

3 Weighted Functional for Approximate Construction of Quasi-Isometric Mappings

We start with a brief introduction explaining how hyperelasticity principle can be used to construct quasi-isometric mappings.

Let $\xi_1, \xi_2, \ldots, \xi_d$ denote the Lagrangian coordinates associated with elastic material, and $x_1, x_2, \ldots, x_d$ denote the Eulerian coordinates of a material point. Spatial mapping $x(\xi) : \mathbb{R}^d \to \mathbb{R}^d$ defines a stationary elastic deformation. The Jacobian matrix of the mapping $x(\xi)$ is denoted by C, where $c_{ij} = \partial x_i / \partial \xi_j$.

We look for the elastic deformation $x(\xi)$ that minimizes the following weighted stored energy functional [6]

$$F(x) = \int_{\Omega_\xi} w(\xi) W(C) \, d\xi, \tag{1}$$

where $W(C)$ is polyconvex elastic potential (internal energy) which is a weighted sum of shape distortion measure and volume distortion measure [5]:

$$W(C) = (1 - \theta) \frac{\left(\frac{1}{d} \operatorname{tr}(C^T C)\right)}{\det C^{2/d}} + \frac{1}{2} \theta \left(\frac{1}{\det C} + \det C \right). \tag{2}$$

In most cases we set $\theta = 4/5$.

Since distortion measure (2) is minimized on the average, locally it can be quite large. In theory, it can be infinite on the set of zero measure. In practice, it means that with mesh refinement quality of mesh cell can locally deteriorate.

A simple transformation can be applied to the elastic potential to make the singular mappings inadmissible. Let $W(C)$ denote the elastic potential. For $1 < \alpha < +\infty$ consider the following transformation:

$$W_\alpha(C) = \begin{cases} \dfrac{1}{\alpha} \dfrac{(\alpha - 1)^2 W(I) W(C)}{\alpha W(I) - W(C)} & \text{when } W(C) < \alpha W(I) \\ +\infty & \text{when } W(C) \geq \alpha W(I) \end{cases} \tag{3}$$

Potential $W_\alpha(C)$ is finite only if

$$W(C) < \alpha W(I). \tag{4}$$

The Lame coefficients λ and μ of the potentials $W(C)$ and $W_\alpha(C)$ coincide meaning that in the limit of small deformations these potentials describe the same elastic material. Parameter α is used to control the upper bound of deformation (quasi-isometry constant). One can use special prolongation procedure which reduces α as suggested in [5]. It seems that this technique produces quasi-isometric mapping with lowest quasi-isometry constants, however, it is quite expensive.

In engineering practice, one can control the spatial distribution of distortion measure without actual contraction of the set of feasible mappings. Experience suggests that large values of distortion appear near boundaries and surfaces of material discontinuity Hence it is possible to introduce a weight function $w(\cdot)$ in the Lagrangian or Eulerian coordinates which takes large values in critical regions and is close to unity elsewhere.

In the process of minimization, elements with a larger weight tend to have a smaller value of distortion function $W(C)$. Hence, their shapes and sizes are very close to the target ones. This simple approach proved to be very efficient for mesh orthogonalization near the boundary [5]. A proper choice of the weight allows us to satisfy the Dirichlet boundary conditions and to approximate boundary orthogonality conditions and prescribed mesh element size in the normal direction very accurately.

Theoretical arguments suggest that in order to eliminate the local singularities of the distortion function the weight distribution should be singular. However, this singularity is only reached in the limit of mesh refinement and for any given finite mesh weight distribution is bounded. One cannot prove that resulting deformation is quasi-isometric but numerical evidence suggest that global mesh distortion bounds do not depend on the mesh size.

Let $G_\xi(\xi)$ and $G_x(x)$ denote the metric tensors defining linear elements and length of curves in Lagrangian and Eulerian coordinates in the domains Ω_ξ and Ω_x, respectively. Then, $x(\xi)$ is the mapping between metric manifolds M_ξ and M_x. The distortion functional (1) for this mapping can be written as

$$F(x) = \int_{\Omega_\xi} w(\xi) W(Q \nabla_\xi x H^{-1}) \det H d\xi, \tag{5}$$

where

$$H^T H = G_\xi, \det H > 0, \quad Q^T Q = G_x, \det Q > 0$$

are arbitrary matrix factorizations of metric tensors G_ξ and G_x. We assume that singular values of the matrices Q and H are uniformly bounded from below and from above.

Suppose that domain Ω_ξ can be partitioned into convex polyhedra U_k. Then stored energy functional (1) can be approximated by the following semi-discrete functional:

$$F(x_h(\xi)) = \sum_k \int_{U_k} w(\xi) W(\nabla x_h(\xi)) d\xi, \tag{6}$$

where $x_h(\xi)$ is continuous piecewise-smooth deformation.

In order to approximate integral over convex cell U_k one should use certain quadrature rules. As a result semi-discrete functional (6) is approximated by the discrete functional:

$$F(x_h(\xi)) \approx \sum_k \text{vol}(U_k) \sum_{q=1}^{N_k} w_q \beta_q W(Q_q C_q H_q^{-1}) det H_q = F^h(x_h(\xi)).$$

Here N_k is the number of quadrature nodes per cell U_k, C_q denotes the Jacobian matrix in q-th quadrature node of U_k, while β_q are the quadrature weights and w_q are values of weight function in the quadrature nodes.

In the case when matrices Q and H are constant, the following majorization property holds

$$F(x_h(\xi)) \leq F^h(x_h(\xi)). \tag{7}$$

This property can be used to prove that all intermediate deformations $x_h(\xi)$ providing finite values of discrete functional are homeomorphisms [6].

In the opposite case, exact majorization inequality can be violated and one should be careful with quadrature rules in order to guarantee certain relaxed formulation for majorization, say in the form

$$F(x_h(\xi)) \leq C F^h(x_h(\xi)), \tag{8}$$

where C is a constant. This inequality should guarantee that every intermediate iteration of the mesh generation method has the finite energy for mapping as a whole and not just for the finite set of quadrature nodes.

3.1   Preconditioned Minimization Algorithm

Minimization of the discrete functional can be formulated as a problem of minimization of function $F(Z)$ where argument is the vector Z such that $Z^T = (z_1^T \, z_2^T \, \ldots \, z_{n_v}^T)$ where $z_k \in \mathbb{R}^d$, $k = 1, \ldots, n_v$ are positions of mesh vertices. Hessian matrix $\tilde{H}$ of the function F is built of $d \times d$ blocks $\tilde{H}_{ij} = \frac{\partial^2 F}{\partial z_i \partial z_j^T}$. Here matrix $\tilde{H}_{ij}$ is placed on the intersection of i-th block row and j-th block column.

The Newton method for finding stationary point of the function can be written as follows

$$\sum_{j=1}^{n_v} \tilde{H}_{ij}(Z^l)\delta z_j + r_i(Z^l) = 0 \tag{9}$$

$$z_k^{l+1} = z_k^l + \tau_l \delta z_k, \quad k = 1, \ldots, n_v. \tag{10}$$

Here parameter τ_l is found as approximate solution of the following 1d problem

$$\tau_l = \arg\min_\tau F(Z^l + \tau \delta Z).$$

We use simple binary subdivision to find approximate minimum.

Let

$$B_i^T B_i = \tilde{H}_{ii} \tag{11}$$

denote the factorization of $d \times d$ matrix $\tilde{H}_{ii}$. We apply to the Hessian matrix $\tilde{H}$ the following block scaling

$$\tilde{H}_{ij}^B = B_i^{-T} \tilde{H}_{ij} B_j^{-1}. \tag{12}$$

This equality can be rewritten as $\tilde{H}^B = B^{-T} \tilde{H} B^{-1}$.

The next step is to apply permutation which allows to represent matrix $\tilde{H}^B$ as $d \times d$ block matrix with $n_v \times n_v$ blocks, which is illustrated below for $d = 3$:

$$\bar{H} = P \tilde{H}^B P^T, \bar{H} = \begin{pmatrix} \bar{H}_{11} & \bar{H}_{12} & \bar{H}_{13} \\ \bar{H}_{21} & \bar{H}_{22} & \bar{H}_{23} \\ \bar{H}_{31} & \bar{H}_{32} & \bar{H}_{33} \end{pmatrix}. \tag{13}$$

Setting off-diagonal blocks in $\bar{H}$ to zero, one obtains d independent linear systems with $n_v \times n_v$ matrices $\bar{H}_{ii}$.

Preconditioned conjugate gradient technique is used for approximate solution with second-order Cholesski factorization as a preconditioner

$$L_i L_i^T \approx \bar{H}_{ii}$$

It is well known that approximate PCG solution to a linear system

$$Ax = f$$

starting from zero initial guess which was obtained with relative error ϵ can be formally written as

$$x_\epsilon = R_\epsilon f,$$

where R_ϵ is a certain symmetric positive definite matrix which in some sense approximates matrix A^{-1}.

Hence we eventually replaced inverse of the full Hessian matrix $\tilde{H}$ by the following matrix

$$H' = B^{-1} P^T R_\epsilon P B^{-T}.$$

Thus increment vector which is used in (10) is defined by equality

$$\delta Z = H' R(Z^l), \quad R^T = (r_1^T r_2^T, \ldots, r_{n_v}^T).$$

For symmetric positive definite matrices the double scaling transformation is safe, while for the matrices which are not positive definite it is not. In order to illustrate this statement, consider simple example where 2×2 block model matrix $\tilde{H}$ is defined as follows

$$\tilde{H}_{11} = \begin{pmatrix} 1 & 1 \\ 1 & 2 \end{pmatrix}, \quad \tilde{H}_{22} = \begin{pmatrix} 2 & 0 \\ 0 & 2 \end{pmatrix}, \quad \tilde{H}_{12} = \tilde{H}_{21} = \begin{pmatrix} 1 & 2 \\ 2 & 1 \end{pmatrix}.$$

After 2×2 block-diagonal scaling (12) with Cholesky factors, we get

$$\tilde{H}_{ii}^B = I$$

while

$$\tilde{H}_{12}^B = \sqrt{2} \begin{pmatrix} \frac{1}{2} & 1 \\ -\frac{1}{2} & -1 \end{pmatrix}.$$

It means that after block permutation (13) one obtains diagonal 2×2 block which is not positive definite.

In general case in order to prove that diagonal blocks of the matrix $\bar{H}$ are positive definite one has constructed a mesh deformation in such a way that after a change of variables $d \times d$ diagonal blocks of the matrix $\tilde{H}$ are diagonal matrices. Search for such a deformation is not a trivial problem. One should also note that such a proof does not allow general factorization (11).

In order to avoid such complications we use special matrix assembly procedure, which guarantees that only positive semidefinite terms are added to global matrix from each quadrature node. Consider polyconvex function $W(C)$, where matrix C with columns $c_i \in \mathbb{R}^d$ is defined by $C = Q\nabla_\xi x H^{-1}$. It is assumed that $J = \det C > 0$. It is convenient to define dual basis c^i, $i = 1, \ldots, d$ defined by equality

$$c_i^T c^j = \delta_{ij} J$$

Then

$$J = c_i^T c^i, \ i = 1, \ldots, d \qquad (14)$$

When $d = 2$ dual basis vectors are defined by

$$c^1 = \Omega c_2, \ c^2 = -\Omega c_1, \ \Omega = \begin{pmatrix} 0 & 1 \\ -1 & 0 \end{pmatrix}.$$

For three-dimensional case

$$c^k = c_i \times c_j,$$

where i, j, k is cyclic permutation from $1, 2, 3$. Each quadrature node is associated with simplex T with vertices $p_0, p_1, \ldots, p_d$. On each simplex matrix C is linear function of vectors p_i, hence positive definiteness of contribution to global matrix $\tilde{H}$ from simplex T is defined by the properties of the $d \times d$ block matrix D consisting of 3×3 blocks

$$D_{ij} = \frac{\partial^2 F}{\partial c_i \partial c_j}.$$

In general this matrix is undefinite hence one has to eliminate certain terms when assembling D_{ij}. Namely, when computing second derivatives one needs to neglect the derivatives of c^k with respect to c_p. It means that we interpret determinant J as linear function of c_i as defined by equality (14). Since function $W(C)$ is polyconvex it can be written as

$$W(C) = W_e(C, J), \quad J = \det C$$

where W_e is convex function. The Hessian matrix of W_e is non-negative definite. Reduced matrix $\tilde{D}$ is defined by equalities

$$\tilde{D}_{ij} = \frac{\partial^2 W_e}{\partial c_i \partial c_j} + \frac{\partial^2 W_e}{\partial J^2} \frac{\partial J}{\partial c_i} \otimes \frac{\partial J}{\partial c_j} + \frac{\partial^2 W_e}{\partial c_i \partial J} \otimes \frac{\partial J}{\partial c_j} + \frac{\partial J}{\partial c_i} \otimes \frac{\partial^2 W_e}{\partial J \partial c_j}.$$

Non-negative definiteness of matrix $\tilde{D}$ immediately follows from convexity of W_e.

We have found that the presence of terms related to the second derivatives of J may essentially influence the behavior of the iterative scheme in the presence of very thin and elongated elements. Using full matrix in some cases may sharply improve the convergence, however, the risk to get undefinite diagonal blocks in the matrix led us to the choice of the filtered matrix. In this case, the matrix $\tilde{H}$ is positive definite and double scaling procedure is always stable. Moreover, there is a certain freedom in the choice of factorization (11). To this end, we use the simplest Cholesky factorization without reordering.

4   Metric Interpolation for Stable Time-Dependent Mesh Deformation

Time-dependent mesh deformation is introduced via time-dependent metric tensor $G_x(x, t) = Q^T(x, t)Q(x, t)$.

Formally, mesh deformation $x(\xi, t)$ can be found as the solution of the following variational problem

$$F(x(\xi, t)) = \int_{\Omega_\xi} wW(Q(x, t)\nabla_\xi x(\xi, t)H(\xi)^{-1}) \det H \, d\xi. \tag{15}$$

In functional (15) time t is just a parameter. Exact minimization of the discrete counterpart of variational problem (15) on each time level allows to obtain stable and accurate mesh deformation method.

Of course, in practice, exact minimization is not acceptable. We have no other choice but to assume that the approximate solution of the variational problem is far from exact minimizer. It means that approximate solutions $x(\xi, t - \delta t)$ and $x(\xi, t)$ can deviate sharply even for very small values of δt since norm of the gradient of the functional is not small and minimization process continues even for very small values of δt which means that the ratio

$$\frac{1}{\delta t}\|x(\xi, t) - x(\xi, t - \delta t)\|$$

can be very large or even tend to infinity with $\delta t \to 0$.

In order to resolve this problem we use special stress relaxation procedure which resembles hypoelastic behaviour of material. Suppose that we have l-th guess to the deformation at time t denoted by $x^l(\xi, t)$. We generously assume that $x^l(\xi, t)$ is quasi-isometric diffeomorphism and weight function

$$w_l = W^\alpha(Q(x^l, t)\nabla_\xi x^l(\xi, t)H^{-1}), \alpha \geq 1$$

is bounded everywhere.

Next guess to deformation is found as the solution of the following variational problem

$$F_\beta(x(\xi, t), x^l(\xi, t)) = \int_{\Omega_\xi} w_l W(Q(x, t)\nabla_\xi x(\xi, t)H^l_\beta(\xi)^{-1}) \det H^l_\beta \, d\xi \tag{16}$$

where $0 \geq \beta \geq 1$ is the stress relaxation parameter and

$$H_\beta^l = (1 - \beta)Q(x^l, t)\nabla_\xi x^l + \beta U_l H \tag{17}$$

is interpolated shape matrix. Here U_l is rotation matrix.

Shape matrix H_β^l now depends on parameter β. When $\beta = 1$ we recover variational problem (15) while in the case $\beta = 0$ exact solution of variational problem (16) is precisely $x^l(\xi, t)$. It is obvious since in this case distortion measure attains its minimum

$$W(Q(x^l, t)\nabla_\xi x^l(\xi, t)H_0^l(\xi)^{-1}) = 1.$$

In this case the stress tensor is zero everywhere.

When β is small parameter, then

$$W(Q(x^l, t)\nabla_\xi x^l(\xi, t)H_\beta^l(\xi)^{-1}) = 1 + O(\beta)$$

and stress tensor as well as the norm of the residual of the Euler-Lagrange equations for functional (16) has the order $O(\beta)$. For completeness sake let us explain the origin of this estimate. Consider distortion function (2) defined by equality

$$W(C) = (1 - \theta)\frac{(\frac{1}{d}\operatorname{tr}(C^T C))}{\det C^{2/d}} + \frac{1}{2}\theta(\frac{1}{\det C} + \det C).$$

Then

$$\frac{\partial W}{\partial c_i} = (1 - \theta)\frac{2}{d}\frac{c_i}{\det C^{2/d}} - (1 - \theta)\frac{2}{d}\frac{c^i \frac{1}{d}\operatorname{tr}(C^T C)}{\det C^{1 + \frac{2}{d}}} + \frac{1}{2}\theta((1 - \frac{1}{\det C^2})c^i).$$

When $C = I$, then vector $c_i = c^i$ is just i-th Cartesian basis vector e_i and

$$\frac{\partial W}{\partial c_i} = 0$$

In the case $C = I + \beta E$, where norm of the matrix E behaves as $O(1)$, we get

$$\det C = 1 + \beta \operatorname{tr} E + O(\beta^2), \quad \operatorname{tr} C^T C = d + 2\beta \operatorname{tr} E + O(\beta^2), \quad c^i = e_i + \beta\delta_i + O(\beta^2),$$

where $|\delta_i|^2 < \operatorname{tr}(E^T E)$. Using above estimates and assumption that $Q(x)$ is smooth function we get

$$W(C) = 1 + O(\beta), \quad \frac{\partial W}{\partial c_i} = O(\beta).$$

The Cauchy stress matrix is defined by equality

$$\frac{1}{\det C}\frac{\partial W}{\partial C}C^T.$$

Obviously norm of this matrix behaves as $O(\beta)$.

Similar estimates for derivatives with respect to columns of the matrix $\nabla_\xi x^l(\xi, t)$ follow from the above estimates using the chain rule and the fact that matrix factors Q and H has uniform lower and upper bounds on their singular values.

One can try to reformulate hypoelastic procedure (16), (17) as evolution equation for matrix H_β^l. Denote approximate solution of variational problem (16) by $x^{l+1}(\xi, t)$. Assuming that this deformation is quasi-isometric diffeomorphism, one obtains the following equality

$$Q(x^{l+1})\nabla_\xi x^{l+1} H_\beta^{l^{-1}} = I + \beta \Delta_l,$$

where Δ_l is the error matrix with norm bounded from above by a certain constant K. Using (17) one obtains the following discrete evolution equation

$$H_\beta^{l+1} = (1 - \beta)H_\beta^l(I + \beta \Delta_l) + \beta U_l H. \tag{18}$$

This iterative process converges when, say, all singular values of matrix Δ_l are less then unity. Discrete process (18) can be rewritten as

$$\frac{H_\beta^{l+1} - H_\beta^l}{\beta} = -H_\beta^l(I + \beta \Delta_l) + U_l H. \tag{19}$$

Since matrix H is responsible for internal stresses in our artificial elastic material, Eq. (19) can be considered as evolutionary equation for stress relaxation.

5   Numerical Experiments

Our main problem of interest is the modeling of moving deforming bodies on background mesh. It is required that the connectivity of the mesh is fixed and the mesh is compressed near the boundary of these bodies. It is assumed that such a mesh can be used by an immersed boundary flow solver in order to simulate time-dependent viscous flows. System of single or multiple bodies or just a domain $B \in \mathbb{R}^d$ is defined by implicit function $u(x, t)$ in such a way that the function u is negative inside body, positive outside it, and isosurface $u(x, t) = 0$ defines the boundary ∂B at the time t. We assume that $u(x, t)$ resembles signed distance function for the instant domain boundary. In theory one can assume existence of the quasi-isometric mapping $x(y) : \mathbb{R}^d \to \mathbb{R}^d$ such that the function $d_s(y) = u(x(y), t)$ is precisely the signed distance function. The norm of the vector $\nabla_x u$ in the vicinity of the boundary is bounded from below and from above.

Metric tensor $G(x, t)$ is defined as function of $u(x, t)$ in such a way that it attains largest value on the domain boundary and decrease using different laws inside and outside body. Mesh inside body is in general quite coarse since immersed boundary solver solution inside the body does not have physical meaning. One can use isotropic version of the metric tensor, when

$$G(x, t) = I(1 + C\gamma^2(x, t))$$

where $\gamma(x,t)$ is prescribed scale function defined via $u(x,t)$:

$$\gamma(x,t) = \phi(u(x,t)).$$

Here $\phi(\cdot) : \mathbb{R}^1 \to \mathbb{R}^1$ is prescribed function which defines mesh compression law near boundary of the body. Anisotropic version of the metric tensor can be constructed as follows

$$G(x,t) = I(1 + C\gamma^2(x,t)) + \gamma_1(x,t)\nabla_x u \nabla_x u^T \frac{1}{|\nabla_x u|^2},$$

where function $\gamma_1(x,t)$ should be equal to zero in the regions of the computational domain where norm of vector $\nabla_x u$ is not guaranteed to be bounded from below.

We present results for a planar test case where the ellipse is uniformly rotated inside the square domain initially covered by uniform Cartesian mesh. In order to define function $u(x,t)$ we use time-dependent affine transformation and distance function for a circle. Mesh adaptation to the boundary of the domain is controlled by an isotropic metric tensor. We use five iterations of stress relaxation algorithm per time step.

Initial mesh is constructed by solving variational problem (15) for the moment $t = 0$. This mesh is shown in Fig. 1.

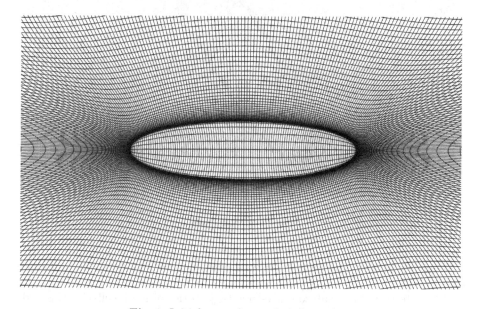

Fig. 1. Initial guess for mesh deformation.

Figure 2 shows enlarged fragments of initial mesh. One can see that compression law outside domain defines gradual cell size change while inside the body mesh size is changed rather sharply.

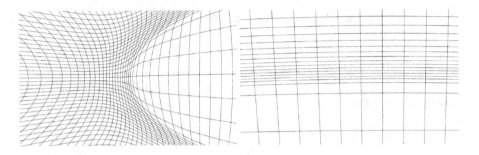

Fig. 2. Enlarged fragments of initial mesh for mesh deformation.

Figures 3 and 5 show adapted mesh at certain consecutive time moments with their enlarged fragments shown in Figs. 4 and 6, respectively.

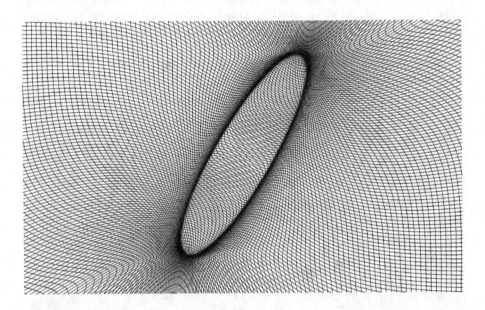

Fig. 3. Instant mesh.

One can see that the mesh compression zone precisely follows the boundary of the ellipse.

It is interesting to analyze space-time trajectories of mesh cells. To this end, we create 3d cells which are constructed by creating hexahedra with lower and upper faces defined by the same mesh cell on adjacent time levels. We use artificial time scaling to make figures more clear.

Figure 7 shows trajectories of two cells. In the process of ellipse rotation, these cells travel far across the domain to the boundary layer, move along the

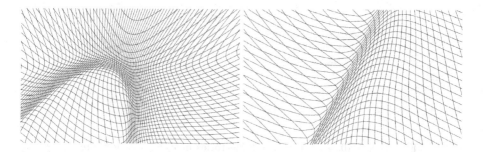

Fig. 4. Enlarged fragments of instant mesh.

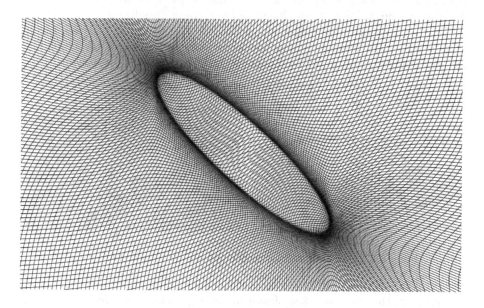

Fig. 5. Instant mesh.

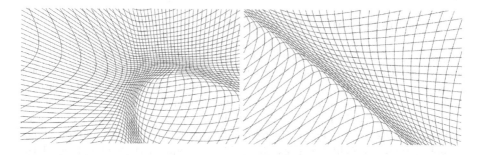

Fig. 6. Enlarged fragments of instant mesh.

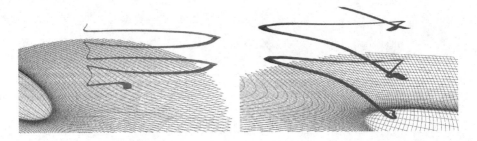

Fig. 7. Two distinct mesh cell trajectories in x_1, x_2, t coordinates.

layer and travel back. One can easily see that trajectories are periodic. Visible deviations from periodicity are due to perspective in visualization.

Figure 8 illustrates the trajectory of the cell which most of the time travels near or across the boundary layer and crosses the boundary of the body. One can see fragments which can be considered as instabilities or oscillations.

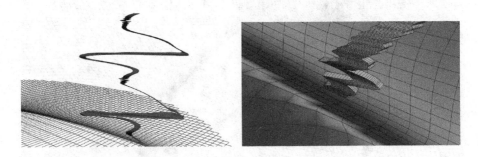

Fig. 8. Time-dependent mesh trajectory and its enlarged fragment.

However, the enlarged view shown in the right explains that this is just a complicated part of the trajectory related to the movement of deformed cells along the mesh boundary layer.

6   Conclusions and Discussions

We describe the algorithm which allows obtaining high quality stable time-dependent mesh deformations using the concept of stress relaxation for hyper-elastic material. We present elementary analysis and demonstrate the viability of the algorithm. It is clear that the presented algorithm is considerably slower compared to mesh solver based on linear elliptic equations [9]. Algorithms based on linear elliptic solver can attain reasonable mesh quality via a careful choice of metric tensor/weight functions [3] so it may happen that advantage of the presented method in terms of mesh quality does not overweight computational overhead.

However, unlike linear mesh solvers, the presented algorithm does not have any limitation on the shape of the domain and the type of mesh elements. It can be applied in the case of multiple dimensions and for high-order elements.

Reduction of computational cost for the presented algorithm and development of parallel version is the topic of ongoing research.

References

1. Huang, W., Ren, Y., Russell, R.D.: Moving mesh partial differential equations (MMPDES) based on the equidistribution principle. SIAM J. Numer. Anal. **31**(3), 709–730 (1994)
2. Huang, W., Kamenski, L.: On the mesh nonsingularity of the moving mesh PDE method. Math. Comput. **87**(312), 1887–1911 (2018)
3. Van Dam, A., Zegeling, P.A.: Balanced monitoring of flow phenomena in moving mesh methods. Commun. Comput. Phys. **7**(1), 138–170 (2010)
4. Godunov, S.K., Gordienko, V.M., Chumakov, G.A.: Quasi-isometric parametrization of a curvilinear quadrangle and a metric of constant curvature. Sib. Adv. Math. **5**(2), 1–20 (1995)
5. Garanzha, V.A.: The barrier method for constructing quasi-isometric grids. Comput. Math. Math. Phys. **40**(11), 1617–1637 (2000)
6. Garanzha, V.A., Kudryavtseva, L.N., Utyzhnikov, S.V.: Untangling and optimization of spatial meshes. J. Comput. Appl. Math. **269**, 24–41 (2014)
7. de Boor, C.: Good approximation by splines with variable knots II. In: Watson, G.A. (ed.) Conference on the Numerical Solution of Differential Equations. LNM, vol. 363, pp. 12–20. Springer, Heidelberg (1974). https://doi.org/10.1007/BFb0069121
8. Coyle, J.M., Flaherty, J.E., Ludwig, R.: On the stability of mesh equidistribution strategies for time-dependent partial differential equations. J. Comput. Phys. **62**(1), 26–39 (1986)
9. Tang, H.Z., Tang, T.: Adaptive mesh methods for one- and two-dimensional hyperbolic conservation laws. SIAM J. Numer. Anal. **41**(2), 487–515 (2003)

Approximate Coalitional Equilibria in the Bipolar World

Andrei Golman[1] and Daniil Musatov[1,2,3]($\boxtimes$)

[1] Moscow Institute of Physics and Technology, Dolgoprudny, Russia
andrewsgolman@gmail.com, musatych@gmail.com
[2] Russian Presidential Academy of National Economy and Public Administration,
Moscow, Russia
[3] Caucasus Mathematical Center at Adyghe State University, Maykop, Russia

Abstract. We study a discrete model of jurisdiction formation in the spirit of Alesina and Spolaore [1]. A finite number of agents live along a line. They can be divided into several groups. If a group is formed, then some facility is located at its median and every member x of a group S with a median m pays $\frac{1}{|S|} + |x - m|$.

We consider the notion of coalitional stability: a partition is stable if no coalition wishes to form a new group decreasing the cost of all members. It was shown by Savvateev et al. [4] that no stable partition may exist even for 5 agents living at 2 points. We now study approximately stable partitions: no coalition wishes to form a new group decreasing all costs by at least ϵ.

In this work, we define a relative measure of partition instability and consider bipolar worlds where all agents live in just 2 points. We prove that the maximum possible value of this measure is approximately 6.2%.

Keywords: Facility location · Group partition · Coalitional stability
Approximate equilibrium

1 Introduction

Partitioning of society into groups affects many economic, political and social processes. Everyone is a member of many groups but sometimes the groups must be disjoint. For instance, most people have only one citizenship and only one employer. Almost everywhere a person may be a member of at most one political party. It is very hard to be a fan of two football clubs. An airline may be a member of only one alliance. A bitcoin miner may be a member of only one mining pool.

Why do people and other economic agents unite themselves into groups (clubs, coalitions, communities, jurisdictions etc.)? The general answer is that a group provides some type of club good that is either infeasible or too expensive

Supported with RFBR 16-01-00362 and RaCAF ANR-15-CE40-0016-01 grants.

for an individual. In this paper, we study the case when this good is *horizontally differentiated*. It means that the good is described by a set of characteristics and different agents have different tastes. In this case, two opposite forces affect the size of a group. On the one hand, the larger the group the less fraction of the fixed cost is beared by an individual. On the other hand, a smaller group is typically more homogeneous, hence its members could be better satisfied by the club good characteristics.

The main theoretical question is whether these two forces always balance each other and yield a stable partition. The answer crucially depends on the underlying notion of stability. In some frameworks very general existence theorems are proven. In other cases, some tricky examples without stable partitioning are constructed. To the best of our knowledge, this paper is the first one where an *approximate* equilibrium is studied in this context. It is more or less obvious that an approximate equilibrium should always exist for coarse approximation factors and may not exist for fine approximation factors if no exact equilibrium exists. Our main contribution is to derive concrete bounds on these factors.

Our study is narrowed to a specific framework where the main features of the model are highlighted. Firstly, we analyze a *bipolar* world with only two types of public good. Secondly, we analyze coalitional stability: a group partition is stable if no new coalition wishes to break out and thus decrease the cost of all its members. The approximate analog is the following: no new coalition wishes to break out and thus decrease the cost of all its members *by a considerable amount*.

We illustrate our model by the following story: several sportspersons decide which game to play: football or handball. Some people prefer football, the others prefer handball, but everyone prefers to play than not to play. We ignore all limitations on the number of players: any group may rent a hall and play there. The hall is universal, so they may also mix the games: play football for some time and then switch to handball. The rent price is distributed equally among the players. If a player has to participate in the less preferred game, then he gets some additional disutility. The society may divide themselves into several groups. Each group rents a hall shares the payment equally and choose the game by majority voting. If an equal number of agents vote for both games then the playing time may be divided in any proportion.

1.1   Related Literature

Here we present a broad perspective of the studies in local public goods, facility location and group partition.

The study of local public goods was started in the 1950s by a discussion between Samuelson [15] and Tiebout [19]. Samuelson stated that a public good can never be financed by a free choice of all society members. The main idea is that if an agent chooses how much public good to purchase then she does not account for positive externalities and thus underfinances the good. Tiebout responded that the situation is different in the case of local public goods. In this case, the choice may be internalized by "foot voting": if an agent does not like the tax level and the amount of public good provided in his jurisdiction then he can migrate to another one. Tiebout's hypothesis was that in this case a social optimum is restored.

Unfortunately, Tiebout did not present a formal mathematical model. A subsequent line of research formulated several approaches that dealt with *vertical differentiation*: all agents value the local good but have different willingness to pay for it. The society becomes stratified by the level of public good provision and, consequently, taxes. This framework was analyzed by Westhoff [20], Bewley [3], Greenberg and Weber [8], among others. The latter paper was the first one to distinguish two notions of stability: migrational and coalitional. Under the former type of stability, no single agent wishes to change her jurisdiction. Under the latter one no group of agents would like to form a new jurisdiction.[1] Both notions are based on quotes from the Tiebout paper.

Another approach deals with *horizontal differentiation* of local goods and heterogeneous tastes of agents. Mas-Colell [13] introduced a model of determining a type of the public good within a community. In the seminal paper [1] Alesina and Spolaore elaborated a model of stable partitioning into several jurisdictions (nations). Their model is continuous and one-dimensional with a uniform distribution of agents. Each jurisdiction determines the location of its public good (the capital) by the median rule.[2] Every agent has two types of cost: monetary cost for providing the good and transportational cost for reaching its location. The paper also employs migrational and coalitional stability.

A bunch of subsequent papers analyzes the stability and efficiency issues of the model. Haimanko et al. [9] prove that if all coalitions may redistribute the cost in any manner then a coalitionary stable partition does always exist. Bogomolnaia et al. [4,5] study different notions of stability for models with finite number of agents and provide examples where no coalitionary stable structure exists. Savvateev [16] provides a counterexample for a more general notion of stability. Musatov et al. [14] explore a very general model with a continuum of agents and migrational stability. It is proven that if a benefit from the local good is bounded and an agent may refuse to join any group then no stable structure may exist even for a very narrow class of models. On the other hand, if participation is mandatory, then there always exists a partition satisfying the border indifference property: no agent wants to migrate to the neighboring community. Under some mild single-crossing conditions such a partition is also a migrational equilibrium: no agent wants to migrate to a distant community as well. The existence result was expanded by Marakulin [12] to the case of a continuous population density with atoms.

Finally we mention some notable works on multi-dimensional models which are due to Drèze et al. [7], Marakulin [11] and Savvateev et al. [18]. The first paper also employed some kind of approximate equilibrium notion.

---

[1] This is the type of stability developed by Aumann and Drèze [2].
[2] This rule was also employed in the migrational models due to Bolton and Roland [6] and Jehiel and Scotchmer [10].

2   The Model

2.1   General Setting

Here we formulate a specific model of splitting a bipolar world with a finite number of agents. This model is similar to the model in [17]. A more general model may be found, for instance, in [14]. In our model, there is a set X consisting of $N = L + R$ agents: L agents live at the point 0 (representing football) and R agents live at the point d (representing handball). Denote by $p(x)$ the point where agent x lives. A community S may consist of l agents from the left and r agents from the right. Every community should locate a facility somewhere between 0 and d (endpoints included). This facility represents the distribution of time between the games. We adopt the *median rule*: the facility is located at a median $m \in \text{med}(S)$, i.e., both $\{x \in S \mid p(x)x \leq m\}$ and $\{x \in S \mid p(x)x \geq m\}$ must contain at least half of members of S. It means that if $l > r$ then $m = 0$, if $l < r$ then $m = d$, and if $l = r$ then m may be any point in $[0, d]$. (In this case, we say that S is *indeterminate*). The median rule has two advantages. Firstly, a median is a minimizer for the total disutility. Secondly, it beats any other option by majority voting.

If a community is organized then all its members may use its facility. The facility is uncapacitated and the cost of its maintenance (payment for hall rent) does not depend on the number of users. Let this cost be equal to g. Apart from the maintenance cost, every member pays for transportation to the facility (disutility from a non-preferred game). If an agent is located at p and a facility is located at m then the agent pays $t \cdot |p - m|$. Thus, the total cost of agent x within coalition S with the facility at m equals

$$C(x, S, m) = \frac{g}{|S|} + t \cdot |p(x) - m|.$$

Suppose that a society is split into jurisdictions $S_1, \ldots, S_k$ with fixed medians $m_1, \ldots, m_k$. The configuration is *coalitionary stable* if no new community can emerge and decrease the cost of all its members. That is, there is no community S with its median m such that for all i and for all $x \in S \cap S_i$ it holds that

$$C(x, S, m) < C(x, S_i, m_i).$$

2.2   Fixing the Parameters

In our model, we have three arbitrary parameters: d (the distance between the two options), g (the cost of facility maintenance) and t (the unit cost of transportation). It is clear that d and t do not matter by themselves. Only the product $t \cdot d$ does influence the outcome. This value represents the disutility of football players from playing handball and vice versa. So, let us fix $t = 1$ and keep d variating. Now let us notice that g and d do not matter by themselves either, only g/d does matter. This value is the "exchange rate" between the hall rent and the disutility. So, let us fix $g = 1$ and keep (L, R, d) as the parameters of the model.

One could think that the model is also invariant with respect to multiplying L, R, and g by the same factor. Indeed, all costs remain the same after this multiplication. But the notion of stability may differ because new coalitional threats emerge. For instance, for $L = 30$ and $R = 20$ a community of size 27 may be a potential threat that is absent for $L = 3$ and $R = 2$. Nevertheless, we could suppose w.l.o.g. that $L \geq R$. (Football is not less popular than handball).

2.3   Approximate Coalitional Stability

Now we define the central notions for our analysis.

Definition 1. *Consider a configuration $\mathcal{P}$, i.e. a partition $X = S_1 \sqcup \cdots \sqcup S_k$ with fixed medians $m_1, \ldots, m_k$. Denote by $S(x)$ the group that contains x and by $m(x)$ the respective median. Denote by $C(x, \mathcal{P})$ the cost of x in $\mathcal{P}$, i.e. $C(x, S(x), m(x))$. Then define the absolute instability of the configuration as the value*

$$\Delta_{abs}(\mathcal{P}) = \max_{T \subset X, T \neq \emptyset, m \in \mathrm{med}(T)} \min_{x \in T} \big(C(x, \mathcal{P}) - C(x, T, m) \big). \tag{1}$$

The intuitive meaning is the following: if x agrees to join T with median m then she bears cost $C(x, T, m)$. The difference $C(x, \mathcal{P}) - C(x, T, m)$ is her advantage from joining T, or willingness to join T. The minimum over $x \in T$ is the minimal willingness to join T. We may also call it the willingness to secede of coalition T. If it is positive, then all agents in T would like to break out and establish this community. If it is true for at least one T, then the configuration is unstable. Thus, we have justified the following fact:

Proposition 1. *For any configuration $\mathcal{P}$ it holds that $\Delta_{abs}(\mathcal{P}) \geq 0$. Configuration $\mathcal{P}$ is stable if and only if $\Delta_{abs}(\mathcal{P}) = 0$.*

Proof. If we take $T = S_i$ then the willingness to join T is zero for all its members, hence the maximum is non-negative. The second part was already proved.

Definition 2. *The relative instability of configuration $\mathcal{P}$ is the value*

$$\Delta_{rel}(\mathcal{P}) = \max_{T \subset X, T \neq \emptyset, m \in \mathrm{med}(T)} \min_{x \in T} \frac{C(x, \mathcal{P}) - C(x, T, m)}{C(x, \mathcal{P})}. \tag{2}$$

The intuitive meaning is the same as before, but now the willingness to join is measured in terms of initial costs. Now we expand our notions from a configuration to the whole world.

Definition 3. *The (absolute, relative) instability of a bipolar world (L, R, d) is the minimal (absolute, relative) instability of all configurations $\mathcal{P}$ of this world. We use the same notation as before: $\Delta_{abs}(L, R, d)$ and $\Delta_{rel}(L, R, d)$.*

Our main question is the following:

Problem 1. Find the least upper bound on $\Delta_{abs}(L, R, d)$ and on $\Delta_{rel}(L, R, d)$. Which worlds are the least stable?

Partial solutions to this problem include computing the values of $\Delta_{abs}(L, R, d)$ and $\Delta_{rel}(L, R, d)$ for particular L, R and d, finding parameters with large instability and establishing some upper bounds.

3   Analysis of Approximate Equilibria: General Considerations

It was proven in [17, Theorem 2] that any stable configuration must belong to one of three types:

- "Union". The partition consists of one grand coalition. (One hall is rented and all play football).
- "Federation". The partition consists of two coalitions: all agents at 0 and all agents at d. (Two halls are rented and everyone plays their preferred game).
- "Mixed structure". The partition consists of two coalitions: the first one contains R agents from 0 and R agents from d, the second one contains $L - R$ agents from 0. The median of the first coalition lies somewhere between 0 and d. (Two halls are rented, the first one is used for football, the second one is used partially for football and partially for handball, no person who prefers handball plays in the first hall).

In the case of approximate stability, things become more complex. But the following theorem still holds (both for absolute and relative instability):

Theorem 1. *There exists a configuration $\mathcal{P}$ minimizing the instability of a bipolar world (L, R, d), such that:*

- *There are at most 3 coalitions in $\mathcal{P}$. Among them, at most one coalition has the median at 0, at most one has the median at d and at most one has the median strictly in between.*
- *Among the coalitions with medians at 0 and d at most one contains some agents from the other point.*

The proof employs several lemmas. We start with the following one:

Lemma 1. *Among all configurations that minimize the instability of a bipolar world there must exist a strictly Pareto optimal one, i.e., there could not exist another configuration $\mathcal{P}'$ such that for all x it holds that $C(x, \mathcal{P}') \leq C(x, \mathcal{P})$ and for some x it holds that $C(x, \mathcal{P}') < C(x, \mathcal{P})$. Also, every minimizing configuration must be weakly Pareto optimal, i.e., there could not exist another configuration $\mathcal{P}'$ such that for all x it holds that $C(x, \mathcal{P}') < C(x, \mathcal{P})$.*

Proof. Suppose that $\mathcal{P}$ is a configuration with minimal absolute instability and that $\mathcal{P}'$ weakly Pareto improves it. It means that for all x, T and m it holds that $C(x, \mathcal{P}') - C(x, T, m) \leq C(x, \mathcal{P}) - C(x, T, m)$. The inequality remains valid after minimizing both parts in $x \in T$ and then after maximizing in (T, m). Thus we obtain $\Delta_{abs}(\mathcal{P}') \leq \Delta_{abs}(\mathcal{P})$. If $\mathcal{P}$ is a minimizer then so must be $\mathcal{P}'$. If $\mathcal{P}'$ cannot be further Pareto improved, then it must be a strictly Pareto optimal minimizer. It can be easily shown that the set of possible tuples of utilities that Pareto improves the initial one is compact, and thus the mentioned $\mathcal{P}'$ exists. If the initial inequality were strict, then we would obtain $\Delta_{abs}(\mathcal{P}') < \Delta_{abs}(\mathcal{P})$ that contradicts to the choice of $\mathcal{P}$. Thus an optimal configuration cannot be strictly improved and thus must be weakly Pareto optimal. The proof for relative instability is similar.

This lemma immediately implies the following weak version of Theorem 1:

Lemma 2. *There exists a configuration $\mathcal{P}$ that minimizes the instability of a bipolar world (L, R, d), such that for any m there exists at most one coalition with median m, and among the coalitions with medians at 0 or d at most one contains agents from the opposite point.*

Proof. From Lemma 1 we may suppose that $\mathcal{P}$ is strictly Pareto efficient. If there are two coalitions with the same median then they may unite and thus reduce their costs without affecting any other agent. If there is a coalition S_1 with median 0 that contains agents from d and a coalition S_2 with median d that contains agents from 0 then the extra agents may switch their coalitions. The costs of these agents will fall and the costs of all others remain the same. If no such improvements could occur then we get the claimed configuration.

In order to get the full statement of Theorem 1, we must prove that there could not be two jurisdictions with medians strictly between 0 and d. The argument from [17, Theorem 1] is no longer valid since not all threats considered there lead to Pareto improvement. Instead, we employ the fact that the question is non-trivial only if no stable partition exists.

Lemma 3. *If a world (L, R, d) does not admit a stable configuration, then*

$$\frac{L}{R(L+R)} < d < \frac{1}{L}. \tag{3}$$

Proof. If no stable configuration exists, then, in particular, configurations "Union" and "Federation" are unstable. Consider the configuration "Union". Agents from 0 get least possible monetary cost and zero transportation cost. Therefore, a separating group may include only agents from the right.

The greater is the separating group the less monetary cost each of them bears. If some group would like to secede then the group of all right agents would like to secede all the more. This implies that if "Union" is unstable then

$$\frac{1}{R} < \frac{1}{L+R} + d. \tag{4}$$

By transposing the terms we get

$$d > \frac{1}{R} - \frac{1}{L+R} = \frac{L}{R(L+R)}$$

and thus establish the first inequality. In the sequel we suppose that (4) holds.

Now consider the configuration "Federation". Let S be the coalition having maximum willingness to secede. Let S contain l agents from the left and r agents from the right. Consider several cases:

1. S is indeterminate (i.e., $l = r$) and chooses median $m \in [0, d]$. If S wants to break out then it must hold that

$$\begin{cases} \frac{1}{L} > \frac{1}{l+r} + m & = \frac{1}{2r} + m; \\ \frac{1}{R} > \frac{1}{l+r} + d - m & = \frac{1}{2r} + d - m. \end{cases}$$

By summing up the two inequalities, we get

$$\frac{1}{L} + \frac{1}{R} > \frac{1}{r} + d.$$

Since $r \leq R$ we get

$$d < \frac{1}{L}, \tag{5}$$

as stated.

2. The median of S is at 0 and $l > r$. Because of (4), agents from d cannot win from joining S. But a group of left agents cannot win from breaking out since after seceding they get $\frac{1}{l}$ instead of $\frac{1}{L}$.

3. The median of S is at d and $l < r$. If $l = 0$, then S cannot win from seceding, like in the previous case. If $l > 0$, then consider the group S' consisting of S and $r - l$ agents from the left with the median still at d. Transportation cost in S' is the same as in S and monetary cost is lower. Thus, if S wins from seceding, then S' does win all the more. But S' is indeterminate and thus it implies $d < \frac{1}{L}$, as before.

To complete the proof of Theorem 1, we need only one more ingredient. Specifically, the case of two indeterminate coalitions must be excluded.

Lemma 4. *There exists a configuration $\mathcal{P}$ minimizing the instability of a bipolar world (L, R, d) that satisfies the properties from Lemma 2 and, moreover, contains at most one indeterminate coalition.*

Proof. Suppose that $\mathcal{P}$ contains at least two indeterminate coalitions with total population $2a$. We prove that after merging into one group they can locate the median such that all their members are better off. Suppose that, on the contrary, some member x is worse off for any location of the median. Suppose that she belongs to coalition B with population $2b$.

Firstly we prove that B must be a large coalition. Specifically, b is greater than $\frac{a}{2}$. Indeed, the initial cost of x must be at least $\frac{1}{2b}$. The cost beared by x in the merged group can be made $\frac{1}{2a} + \frac{d}{2}$. Since x must be worse off, we obtain

$$\frac{1}{2a} + \frac{d}{2} > \frac{1}{2b}. \tag{6}$$

But from (5) we have $\frac{d}{2} < \frac{1}{2L}$. Since $a \leq L$, we have $\frac{d}{2} < \frac{1}{2a}$. Plugging this into (6) we get $\frac{1}{2b} < \frac{1}{2a} + \frac{1}{2a} = \frac{1}{a}$, thus $b > \frac{a}{2}$, as stated.

Secondly, we prove that there could not be small coalitions. Specifically, any coalition C with population $c \leq \frac{a}{2}$ is better off after joining B. Indeed, the new cost is at most $\frac{1}{2b} + d$. Since $b > \frac{a}{2}$ and $d < \frac{1}{L}$, it is at most $\frac{1}{a} + \frac{1}{L}$. Since $a < L$, this is at most $\frac{2}{a} \leq \frac{1}{c}$. Since the old cost is at least $\frac{1}{c}$, all members of C prefer to join B. All members of B do also win if the median does not change: the transportation cost stays the same and the monetary cost decreases.

Thus there could be only two indeterminate coalitions. Finally we show that they are better off if they merge and locate the median at some point m. Suppose that their populations are $2a\gamma$ and $2a(1-\gamma)$, and the distance between their initial medians is q. Suppose that they merge and locate the new median between the old medians within the distances q_1 and q_2 to them, respectively. If the median moves towards an agent then she must be definitely better off. Consider the other agents. The members of the first coalition win $\frac{1}{2a\gamma} - \frac{1}{2a} - q_1$. The members of the second coalition win $\frac{1}{2a(1-\gamma)} - \frac{1}{2a} - q_2$. Both these value can be made non-negative iff their sum is non-negative. This is equivalent to the following:

$$\frac{1}{2a\gamma} + \frac{1}{2a(1-\gamma)} - \frac{1}{a} \geq q. \tag{7}$$

Note that $\frac{1}{\gamma} + \frac{1}{1-\gamma} \geq 4$ if $\gamma \in (0,1)$. Thus the left part of (7) is at least $\frac{2}{a} - \frac{1}{a} = \frac{1}{a}$. But $\frac{1}{a} \geq \frac{1}{L} > d$ and of course $d \geq q$. Thus (7) is established, the merger can be profitably done and Lemma 4 and Theorem 1 are proven.

4 Approximation Algorithm for Computing Instabilities

The results of the previous section crucially decrease the number of configurations that could potentially minimize the instability. The remaining possibilities can be looked through in polynomial time. Firstly show how to calculate the instability of a fixed configuration. Note that after the secession all members of the seceding coalition residing at the same point get the same cost. Thus the minima in (1) and (2) are achieved for x with minimal initial cost. Note also that we may consider only the case when the seceding coalition contains only the agents with maximal cost from the two poles. This justifies the following algorithm for computing $\Delta_{abs}(\mathcal{P})$:

1. Calculate $C(x,\mathcal{P})$ for all x. This step is done straightforwardly by definition.
2. Sort all agents from the left by the costs in nonincreasing order. Do the same with the agents from the right.
3. Assign value 0 to the variable Δ. For all $l \in [0, L]$ and $r \in [0, R]$ consider the lth agent from the left and the rth agent from the right in the obtained orders. Denote them by x and y respectively. Let T be the coalition of l agents from the left and r agents from the right. Do the following:
   - If $l > r > 0$ then $\Delta := \max\{\Delta, \min\{C(x,\mathcal{P}) - C(x,T,0), C(y,\mathcal{P}) - C(y,T,0)\}\}$;
   - If $r = 0$ then $\Delta := \max\{\Delta, C(x,\mathcal{P}) - C(x,T,0)\}$;
   - If $0 < l < r$ then $\Delta := \max\{\Delta, \min\{C(x,\mathcal{P}) - C(x,T,d), C(y,\mathcal{P}) - C(y,T,d)\}\}$;
   - If $l = 0$ then $\Delta := \max\{\Delta, C(y,\mathcal{P}) - C(y,T,d)\}$;
   - If $l = r$ then find m that maximizes $\min\{C(x,\mathcal{P}) - C(x,T,m), C(y,\mathcal{P}) - C(y,T,m)\}$. Usually it is just the root of linear equation $C(x,\mathcal{P}) - C(x,T,m) = C(y,\mathcal{P}) - C(y,T,m)$. Then assign $\Delta := \max\{\Delta, \min\{C(x,\mathcal{P}) - C(x,T,m), C(y,\mathcal{P}) - C(y,T,m)\}\}$.
4. Return Δ.

The algorithm for computing $\Delta_{rel}(\mathcal{P})$ is similar, but all terms of type $C(z,\mathcal{P}) - C(z,T,m)$ are replaced by $\frac{C(z,\mathcal{P})-C(z,T,m)}{C(z,\mathcal{P})}$. The following bound on the time complexity is straightforward:

Proposition 2. *The algorithm for computing $\Delta_{abs}(\mathcal{P})$ works for $O(N^2)$ steps for worlds with N agents.*

Proof. The first stage is performed in $O(NM)$ steps where M is the number of groups. Clearly $M = O(N)$, and for minimizing configurations we have even $M = O(1)$ by Theorem 1. The sorting on the second stage takes $O(N \log N)$ steps. On the third stage the algorithms looks through all pairs (l, r) and makes a small calculation. This takes $O(LR) = O(N^2)$ steps. The total running time is $O(N^2)$, as stated.

Now we describe an approximate algorithm for computing $\Delta_{abs}(L, R, d)$ (or $\Delta_{rel}(L, R, d)$) in a bipolar world with parameters (L, R, d). The algorithm also gets the parameter M—the maximum number of considered medians of the indeterminate coalition. The idea is to search through all configurations satisfying the condition of Theorem 1 and take the minimal instability. We denote by (l, r) the group consisting of l agents from the left and r agents from the right. If $l = r$, we attach the median m as the third parameter. The procedure is the following:

1. Check the validity of condition (3). If it does not hold, return 0.
2. Assign $\Delta := \infty$. For all $k \in [0, R]$ and for all $c \in [0, M]$:
   (a) For all $l \in (L - R, L - k]$:
       - $\mathcal{P} :=$ the configuration consisting of groups $(l, 0)$, $(L - k - l, R - k)$, $(k, k, \frac{c}{M}d)$;
       - $\Delta := \min\{\Delta, \Delta_{abs}(\mathcal{P})\}$.
   (b) For all $r \in [0, R - k]$:
       - $\mathcal{P} :=$ the configuration consisting of groups $(L - k, R - k - r)$, $(0, r)$, $(k, k, \frac{c}{M}d)$;
       - $\Delta := \min\{\Delta, \Delta_{abs}(\mathcal{P})\}$.
3. Return Δ.

Proposition 3. *The algorithm for computing $\Delta_{abs}(L, R, d)$ works for $O(N^4 M)$ steps for worlds with N agents.*

Proof. The algorithm considers all possible k, c and l (or r). There are less than $R \cdot M \cdot L = O(N^2 M)$ variants totally. The computation of $\Delta_{abs}(\mathcal{P})$ takes $O(N^2)$ steps, thus the total time is $O(N^4 M)$.

Now we estimate the precision of the described algorithm.

Proposition 4. *Let θ be $\frac{d}{M}$. If the algorithm for computing the absolute instability returns Δ, then $\Delta_{abs}(L, R, d) \geq \Delta - \frac{\theta}{2}$.*

Proof. Suppose that $\Delta_{abs}(L, R, d) = \Delta_{abs}(\mathcal{P})$, where $\mathcal{P}$ satisfies the condition of Theorem 1. If $\mathcal{P}$ does not contain an indeterminate coalition then it will be considered during the algorithm and $\Delta = \Delta_{abs}(L, R, d)$ exactly. Suppose that $\mathcal{P}$ contains an indeterminate coalition Q with median q and one or two other coalitions. During the algorithm we will consider the partition $\mathcal{P}'$ with the same coalitions and median q' of Q such that $|q - q'| \leq \frac{\theta}{2}$. Note that for all x it holds that $C(x, Q, q) \leq C(x, Q, q') + \frac{\theta}{2}$. Hence

$$C(x, \mathcal{P}') \leq C(x, \mathcal{P}) + \frac{\theta}{2} \tag{8}$$

for all x. This inequality still holds after subtracting $C(x, T, m)$, taking minimum and maximum. Thus $\Delta_{abs}(\mathcal{P}') \leq \Delta_{abs}(\mathcal{P}) + \frac{\theta}{2}$. Since $\mathcal{P}'$ is considered by the algorithm, we obtain $\Delta \leq \Delta_{abs}(\mathcal{P}')$. By our assumption, $\Delta_{abs}(\mathcal{P}) = \Delta_{abs}(L, R, d)$. Thus the inequality $\Delta_{abs}(L, R, d) \geq \Delta - \frac{\theta}{2}$ is established.

Proposition 5. *If the algorithm for computing the relative instability returns* Δ, *then* $\Delta_{rel}(L, R, d) \geq \Delta - \theta(L + R)$.

Proof. The argument proceeds along the same line as the previous one. The difference starts when we make implications about Δ_{rel} instead of Δ_{abs}. The inequality (8) still holds and thus

$$\frac{C(x, \mathcal{P}') - C(x, T, m)}{C(x, \mathcal{P}')} \leq \frac{C(x, \mathcal{P}) - C(x, T, m) + \frac{\theta}{2}}{C(x, \mathcal{P}')}$$

$$= \frac{C(x, \mathcal{P}) - C(x, T, m)}{C(x, \mathcal{P})} \cdot \frac{C(x, \mathcal{P})}{C(x, \mathcal{P}')} + \frac{\theta}{2C(x, \mathcal{P}')}. \tag{9}$$

The inequality (8) does also hold after switching $\mathcal{P}$ and $\mathcal{P}'$, i.e., $C(x, \mathcal{P}) \leq C(x, \mathcal{P}') + \frac{\theta}{2}$. Hence $\frac{C(x, \mathcal{P})}{C(x, \mathcal{P}')} \leq 1 + \frac{\theta}{2C(x, \mathcal{P}')}$. Thus the right part of (9) is at most $\frac{C(x, \mathcal{P}) - C(x, T, m)}{C(x, \mathcal{P})} + \frac{\theta}{2C(x, \mathcal{P}')} \left(1 + \frac{C(x, \mathcal{P}) - C(x, T, m)}{C(x, \mathcal{P})} \right) \leq \frac{C(x, \mathcal{P}) - C(x, T, m)}{C(x, \mathcal{P})} + \frac{\theta}{C(x, \mathcal{P}')} \leq \frac{C(x, \mathcal{P}) - C(x, T, m)}{C(x, \mathcal{P})} + \theta(L + R)$. The first inequality holds because $\frac{C(x, \mathcal{P}) - C(x, T, m)}{C(x, \mathcal{P})} \leq 1$ and the second one because $C(x, \mathcal{P}') \geq \frac{1}{L+R}$. Putting all together, we get

$$\frac{C(x, \mathcal{P}') - C(x, T, m)}{C(x, \mathcal{P}')} \leq \frac{C(x, \mathcal{P}) - C(x, T, m)}{C(x, \mathcal{P})} + \theta(L + R).$$

After sequential minimizing and maximizing we get $\Delta_{rel}(\mathcal{P}') \leq \Delta_{rel}(\mathcal{P}) + \theta(L + R)$, that implies the statement of the theorem, as before.

5   Analysis of Approximate Equilibria: Absolute Instability

Now we proceed in the following way: we narrow the set of possible worlds where the maximum instability can be achieved and search through the remaining set exhaustively.

Theorem 2. *The maximal possible absolute instability is achieved for some* (L, R, d) *with* $L < 100$, $R < 100$ *and* $d \leq 1$.

Proof. If $d > 1$ then "Federation" is stable. Indeed, every agent has the cost at most 1, but in any coalition with agents from different poles some agents has the cost greater than 1.

If $L \geq 100$ or $R \geq 100$ then in "Federation" some agents have cost at most 0.01. Thus the absolute instability is at most 0.01. But the configuration $L = 3$, $R = 2$, $d = \frac{14}{45}$ yields absolute instability $\frac{1}{90} > 0.01$. (The proof of this fact is straightforward but tedious and thus is omitted).

The remaining possibilities were analyzed in a brute-force manner. We found only three pairs (L, R) with $\Delta_{abs} > 0.01$ for some d. In all cases optimal d and Δ can be shown to be rational numbers. In Table 1 we summarize our findings for optimal distances:

Table 1. The worlds with maximal absolute instabilities.

L	R	d	Δ
3	2	$\frac{14}{45} \approx 0.311$	$\frac{1}{90} \approx 0.0111$
4	3	$\frac{5}{24} \approx 0.208$	$\frac{1}{48} \approx 0.0208$
5	4	$\frac{7}{40} = 0.175$	$\frac{1}{80} = 0.0125$

These findings justify the following theorem:

Theorem 3. *The maximal possible absolute instability in a bipolar world is* $\frac{1}{48}$.

6 Analysis of Approximate Equilibria: Relative Instability

The case of relative instability is much more complex. The numerical experiments show that sometimes the optimal configuration is neither "Federation", nor "Union", nor "Mixed structure". Specifically, the optimal configuration could be a "Pseudofederation": there are two communities, one of which contains $L - k$ agents from the left and the other contains k agents from the left and all agents from the right. Thus, an analog of Theorem 2 from [17] does not hold in our setting. On Fig. 1 we show which configuration is the most stable one for which parameters.

Theorem 4. *There exists a bipolar world with* $\Delta_{rel}(L, R, d) > 0.0615$.

Proof. By applying our algorithm for computing Δ_{rel} to various parameters we found the following example: $L = 73$, $R = 56$, $d = 0.0114$. The relative instability returned by the algorithm 0.0622. By subtracting the discrepancy we get the claimed lower bound. In Table 2 the relative instabilities for various configurations are shown.

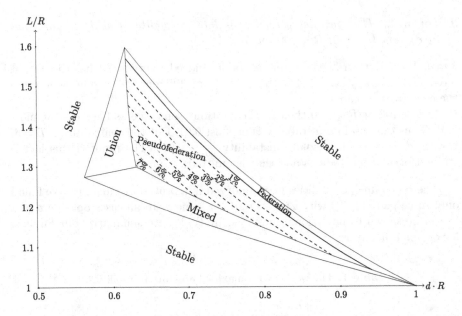

Fig. 1. The figure describes which configurations are the most stable ones for which parameters. Pseudofederation with $a\%$ means that the sizes of the groups are approximately $(L - 0.01aR, 0)$ and $(0.01aR, R)$.

Finally, we present an upper bound on Δ_{rel}.

Theorem 5. *For any bipolar world $\Delta_{rel}(L, R, d) \leq 0.063$.*

Proof (Sketch). The complete proof is rather technical and tedious. As before, it combines analytical and numerical methods. Here we present only the general idea. We consider the four main classes of configurations: "Union", "Federation", "Mixed" and "Pseudofederation". We prove that our bound hold even for these configurations. All the more it does hold for an arbitrary one.

For example, consider the case of "Federation". Here $C(x, \mathcal{P}) = \frac{1}{L}$ for the left agents and $C(x, \mathcal{P}) = \frac{1}{R}$ for the right agents. Let the seceding coalition T be indeterminate. If T is willing to secede, then the group (R, R) is willing stronger,

Table 2. Relative instabilities in the world $L = 73$, $R = 56$, $d = 0.0114$.

Configuration	Groups	Δ
Union	$(73, 56)$	0.066
Federation	$(73, 0) + (0, 56)$	0.074
Mixed	$(56, 56) + (9, 0)$	0.062
Optimal	$(69, 0) + (4, 56)$	0.062

so we may think that $T = (R, R)$. Let t be the median of T. Thus the new costs are $\frac{1}{2R} + t$ for the left agents and $\frac{1}{2R} + d - t$ for the right agents. If the relative instability equals δ then

$$
\begin{cases}
\delta \geq 1 - \dfrac{\frac{1}{2R} + t}{\frac{1}{R}} = \dfrac{1}{2} + Lt; \\[3mm]
\delta \geq 1 - \dfrac{\frac{1}{2R} + d - t}{\frac{1}{L}} = 1 - \dfrac{L}{R} \cdot \left(\dfrac{1}{2} + Ld - Lt \right).
\end{cases}
\tag{10}
$$

Note that the right parts of (10) depend not on L, R, d, t themselves, but on the composite values $\rho = \frac{L}{R}$, $\gamma = Ld$, $\eta = Lt$. Thus δ must be not less than

$$
\min \left\{ \dfrac{1}{2} + \eta, 1 - \rho \left(\dfrac{1}{2} + \gamma - \eta \right) \right\}.
\tag{11}
$$

This lower bound is maximal for $\eta = \frac{\frac{1}{2} - \rho(\frac{1}{2} + \gamma)}{1 + \rho}$. Plugging this into (11) we get a lower bound on δ.

Then we consider other seceding coalitions and obtain other lower bounds on δ. The maximum of these bounds is a lower bound on the relative instability of the configuration. Then we consider other configurations and take the minimal bound. We search for this minimum numerically and find the following values:

$$
\dfrac{L}{R} \approx 1.3065; \quad \dfrac{d}{R} \approx 0.6328; \quad \delta \approx 0.0623.
$$

Adding the discrepancy, we obtain $\Delta_{rel}(L, R, d) \leq 0.063$ for all (L, R, d), as claimed.

7 Conclusion

This work is the first contribution to the literature on jurisdiction partitions that analyzes approximate coalitional equilibria. We have analyzed the bipolar world and found out how far such a world could be from an equilibrium. We considered the cases of absolute and relative instability metrics and established the bounds on them using numerical methods. In the first case, the bound is exact. In the second case, the upper and the lower bounds almost coincide. We believe that the true value lies somewhere between 0.0622 and 0.0623. The future work should give the precise value in the analytic form and study the worlds other than bipolar ones. Our conjecture is that the bipolar worlds are the least stable.

Acknowledgments. We want to thank Alexei Savvateev for his support and advice during the work on this paper.

References

1. Alesina, A., Spolaore, E.: On the number and size of nations. Q. J. Econ. **112**(4), 1027–1056 (1997)
2. Aumann, R.J., Drèze, J.H.: Cooperative games with coalition structures. Int. J. Game Theory **3**(4), 217–237 (1974)
3. Bewley, T.F.: A critique of Tiebout's theory of local public expenditures. Econom. J. Econom. Soc. **49**, 713–740 (1981)
4. Bogomolnaia, A., Le Breton, M., Savvateev, A., Weber, S.: Stability under unanimous consent, free mobility and core. Int. J. Game Theory **35**(2), 185–204 (2007)
5. Bogomolnaia, A., Le Breton, M., Savvateev, A., Weber, S.: Stability of jurisdiction structures under the equal share and median rules. Econ. Theory **34**(3), 525–543 (2008)
6. Bolton, P., Roland, G.: The breakup of nations: a political economy analysis. Q. J. Econ. **112**(4), 1057–1090 (1997)
7. Drèze, J., Le Breton, M., Savvateev, A., Weber, S.: "Almost" subsidy-free spatial pricing in a multi-dimensional setting. J. Econ. Theory **143**(1), 275–291 (2008)
8. Greenberg, J., Weber, S.: Strong Tiebout equilibrium under restricted preferences domain. J. Econ. Theory **38**(1), 101–117 (1986)
9. Haimanko, O., Le Breton, M., Weber, S.: Voluntary formation of communities for the provision of public projects. J. Econ. Theory **115**(1), 1–34 (2004)
10. Jehiel, P., Scotchmer, S.: Constitutional rules of exclusion in jurisdiction formation. Rev. Econ. Stud. **68**(2), 393–413 (2001)
11. Marakulin, V.M.: On the existence of immigration proof partition into countries in multidimensional space. In: Kochetov, Y., Khachay, M., Beresnev, V., Nurminski, E., Pardalos, P. (eds.) DOOR 2016. LNCS, vol. 9869, pp. 494–508. Springer, Cham (2016). https://doi.org/10.1007/978-3-319-44914-2_39
12. Marakulin, V.M.: A theory of spatial equilibrium: the existence of migration proof country partition in an uni-dimensional world. Sib. J. Pure Appl. Math. **17**(4), 64–78 (2017). (in Russian)
13. Mas-Colell, A.: Efficiency and decentralization in the pure theory of public goods. Q. J. Econ. **94**(4), 625–641 (1980)
14. Musatov, D.V., Savvateev, A.V., Weber, S.: Gale-Nikaido-Debreu and Milgrom-Shannon: communal interactions with endogenous community structures. J. Econ. Theory **166**, 282–303 (2016)
15. Samuelson, P.A.: The pure theory of public expenditure. In: The Review of Economics and Statistics, pp. 387–389 (1954)
16. Savvateev, A.: Uni-dimensional models of coalition formation: non-existence of stable partitions. Mosc. J. Comb. Number Theory **2**(4), 49–62 (2012)
17. Savvateev, A.: An analysis of coalitional stability in a bipolar world. J. New Econ. Assoc. **17**, 10–44 (2013). (in Russian)
18. Savvateev, A., Sorokin, C., Weber, S.: Multidimensional free-mobility equilibrium: Tiebout revisited (2018). https://arxiv.org/abs/1805.11871
19. Tiebout, C.M.: A pure theory of local expenditures. J. Polit. Econ. **64**(5), 416–424 (1956)
20. Westhoff, F.: Existence of equilibria in economies with a local public good. J. Econ. Theory **14**(1), 84–112 (1977)

Author Index

Printed in the United States
By Bookmasters